Solving PDEs in C++

COMPUTATIONAL SCIENCE & ENGINEERING

Computational Science and Engineering (CS&E) is widely accepted, along with theory and experiment, as a crucial third mode of scientific investigation and engineering design. This series publishes research monographs, advanced undergraduate- and graduate-level textbooks, and other volumes of interest to a wide segment of the community of computational scientists and engineers. The series also includes volumes addressed to users of CS&E methods by targeting specific groups of professionals whose work relies extensively on computational science and engineering.

Series Volumes

Shapira, Yair, *Solving PDEs in C++: Numerical Methods in a Unified Object-Oriented Approach*

Solving PDEs in C++

Numerical Methods in a Unified Object-Oriented Approach

Yair Shapira
Technion–Israel Institute of Technology
Haifa, Israel

Society for Industrial and Applied Mathematics
Philadelphia

10 9 8 7 6 5 4 3 2 1

Library of Congress Cataloging-in-Publication Data

Shapira, Yair, 1960-

Solving PDEs in C++ : numerical methods in a unified object-oriented approach / Yair Shapira.

p. cm. — (Computational science and engineering)

Includes bibliographical references and index.

ISBN 0-89871-601-2 (pbk. : alk. paper)

1. Differential equations, Partial. 2. C++ (Computer program language) 3. Object-oriented programming (Computer science) I. Title: Solving partial differential equations in C++. II. Title. III. Series.

QA377.S466 2006

518'.64'02855133—dc22 2005054086

Partial royalties from the sale of this book are placed in a fund to help students attend SIAM meetings and other SIAM-related activities. This fund is administered by SIAM, and qualified individuals are encouraged to write directly to SIAM for guidelines.

Contents

List of Figures xiii

List of Tables xxi

Preface xxiii

I Programming 1

1 Introduction to C 7

1.1 Variables and Types . . . 7
1.2 Assignment . . . 8
1.3 Initialization . . . 8
1.4 Conversion . . . 9
1.5 Arithmetic Operations . . . 9
1.6 Functions . . . 11
1.7 Printing Output . . . 13
1.8 Conditions . . . 14
1.9 Scope of Variables . . . 15
1.10 Loops . . . 17
1.11 Examples with Loops . . . 19
1.12 Example: Reversed Integer . . . 20
1.13 Pointers . . . 22
1.14 Arrays . . . 23
1.15 Passing Arguments to Functions . . . 24
1.16 I/O . . . 25
1.17 Recursion . . . 26
1.18 Example: Binary Representation . . . 27
1.19 Example: Pascal's Triangle . . . 29
1.20 Example: Local Maximum . . . 30
1.21 Example: Arithmetic Expression . . . 36
1.22 Example: The Exponent Function . . . 40
1.23 Exercises . . . 43

2 Introduction to C++ 47
2.1 Objects . . . 47
2.2 Classes . . . 48
2.3 Constructors . . . 51
2.4 Explicit Conversion . . . 53
2.5 Implicit Conversion . . . 53
2.6 The Default Copy Constructor . . . 53
2.7 Destructor . . . 55
2.8 Member and Friend Functions . . . 55
2.9 References . . . 57
2.10 Copy Constructor . . . 59
2.11 Assignment Operators . . . 60
2.12 Operators . . . 62
2.13 Inverse Conversion . . . 63
2.14 Unary Operators . . . 63
2.15 Binary Operators . . . 65
2.16 Example: Complex Numbers . . . 68
2.17 Templates . . . 72
2.18 Example: The Vector Object . . . 75
2.19 Inheritance . . . 79
2.20 Example: The Matrix Object . . . 86
2.21 Determinant and Inverse of a Square Matrix . . . 88
2.22 Exponent of a Square Matrix . . . 90
2.23 Exercises . . . 90

3 Data Structures 93
3.1 Data Structures . . . 93
3.2 Templates in Data Structures . . . 94
3.3 Dynamic Vectors . . . 94
3.4 Lists . . . 99
3.5 Connected Lists . . . 102
3.6 The Merging Problem . . . 108
3.7 The Ordering Problem . . . 110
3.8 Vectors vs. Lists . . . 111
3.9 Two-Sided Connected Lists . . . 112
3.10 Trees . . . 112
3.11 Graphs . . . 114
3.12 Exercises . . . 115

II The Object-Oriented Approach 117

4 Object-Oriented Programming 121
4.1 Object-Oriented Language . . . 121
4.2 Example: The Graph-Coloring Problem . . . 122
4.3 Downward Implementation . . . 124

4.4 The C++ Implementation 126
4.5 Triangulation 128
4.6 Example: The Triangle-Coloring Problem 129
4.7 Downward Implementation 130
4.8 Separation of Information 131
4.9 Application in Numerical Schemes 133
4.10 Exercises 134

5 Algorithms and Their Object-Oriented Implementation 135
5.1 Ideas and Their Implementation 135
5.2 Multilevel Programming 136
5.3 Information and Storage 137
5.4 Example: The Polynomial Object 138
5.5 Multiplication of Polynomials 141
5.6 Calculation of a Polynomial 143
5.7 Algorithms and Their Implementation 143
5.8 Horner's Algorithm 144
5.9 Calculation of a Power 145
5.10 Calculation of Derivatives 147
5.11 The Taylor Expansion 148
5.12 Derivatives of a Product 151
5.13 Polynomial of Two Variables 152
5.14 Integration of a Polynomial 154
5.15 Exercises 156

6 Object-Oriented Analysis 161
6.1 ODEs 161
6.2 Stability in the ODE 162
6.3 System of ODEs 162
6.4 Stability in a System of ODEs 163
6.5 Stable Invariant Subspace 164
6.6 The Inhomogeneous Case 165
6.7 Numerical Solution 165
6.8 Difference Schemes 166
6.9 The Taylor Difference Scheme 167
6.10 Computational Error Estimates 168
6.11 Nonlinear ODEs 170
6.12 Object-Oriented Analysis 170
6.13 Application 171
6.14 Taylor Scheme with Error Estimates 172
6.15 Asymptotic Solution 174
6.16 Exercises 176

III Partial Differential Equations and Their Discretization 179

7 The Convection-Diffusion Equation 187
7.1 Initial-Boundary-Value Problems . . . 187
7.2 Finite Differences . . . 188
7.3 The Upwind Scheme . . . 189
7.4 Discrete Boundary Conditions . . . 190
7.5 The Explicit Scheme . . . 191
7.6 The Implicit Scheme . . . 193
7.7 The Semi-Implicit Scheme . . . 193
7.8 The Implementation . . . 194
7.9 Hierarchy of Objects . . . 198
7.10 List of Vectors . . . 198
7.11 The Time-Space Grid . . . 199
7.12 Difference Operators . . . 201
7.13 Two Spatial Dimensions . . . 205
7.14 Exercises . . . 207

8 Stability Analysis 209
8.1 Preliminaries . . . 209
8.2 Algebraic Representation . . . 211
8.3 Stability in Time Marching . . . 212
8.4 Accuracy and Adequacy . . . 214
8.5 Exercises . . . 216

9 Nonlinear Equations 219
9.1 Nonlinear PDEs . . . 219
9.2 The Riemann Problem . . . 219
9.3 Conflicting Characteristic Lines . . . 220
9.4 The Godunov Scheme . . . 222
9.5 The Random-Choice Scheme . . . 225
9.6 The N-Wave . . . 226
9.7 Singular Perturbation . . . 226
9.8 Linearization . . . 228
9.9 Adequacy in the Linearized Problem . . . 230
9.10 The Inhomogeneous Case . . . 232
9.11 System of Nonlinear PDEs . . . 233
9.12 Exercises . . . 235

10 Application in Image Processing 237
10.1 Digital Images . . . 237
10.2 The Denoising Problem . . . 237
10.3 The Nonlinear Diffusion Problem . . . 238
10.4 The Discretization . . . 239
10.5 Linearization . . . 240
10.6 Color Images . . . 240

10.7 Denoising Color Images . . . 241
10.8 Numerical Examples . . . 242
10.9 Exercises . . . 243

IV The Finite-Element Discretization Method **245**

11 The Weak Formulation **249**
11.1 The Diffusion Problem . . . 249
11.2 The Weak Formulation . . . 250
11.3 The Minimization Problem . . . 252
11.4 The Coercivity Property . . . 253
11.5 Existence Theorem . . . 254
11.6 Uniqueness Theorem . . . 255
11.7 Exercises . . . 256

12 Linear Finite Elements **259**
12.1 The Finite-Element Triangulation . . . 259
12.2 The Discrete Weak Formulation . . . 260
12.3 The Stiffness System . . . 260
12.4 Properties of the Stiffness Matrix . . . 261
12.5 Calculating the Stiffness Matrix . . . 263
12.6 Example: Rectangular Domain and Uniform Mesh . . . 266
12.7 M-Matrix Property in the Isotropic Case . . . 269
12.8 Highly Anisotropic Equations . . . 271
12.9 Example: Circular Domain . . . 274
12.10 Exercises . . . 276
12.11 Advanced Exercises . . . 277

13 Unstructured Finite-Element Meshes **281**
13.1 Concrete and Abstract Objects . . . 281
13.2 The Node Object . . . 284
13.3 The Finite-Element Object . . . 286
13.4 The Mesh Object . . . 291
13.5 Assembling the Stiffness Matrix . . . 293
13.6 Exercises . . . 296

14 Adaptive Mesh Refinement **297**
14.1 Local Refinement . . . 297
14.2 Adaptive Refinement . . . 299
14.3 The Adaptive-Refinement Algorithm . . . 299
14.4 Preserving Conformity . . . 301
14.5 Preserving Conformity in Practice . . . 302
14.6 Mesh Refinement in Practice . . . 304
14.7 Automatic Boundary Refinement . . . 307
14.8 Implementation of Automatic Boundary Refinement . . . 307

14.9	Nonconvex Domains	309
14.10	Exercises	312

15 High-Order Finite Elements 313

15.1	High-Order vs. Linear Finite Elements	313
15.2	Quadratic Finite Elements	314
15.3	Cubic Finite Elements	316
15.4	Pros and Cons	317
15.5	Exercises	318

V The Numerical Solution of Large Sparse Linear Systems of Equations 321

16 Sparse Matrices and Their Implementation 325

16.1	Sparse vs. Dense Matrices	325
16.2	The Matrix-Element Object	326
16.3	The Row Object	329
16.4	The Sparse-Matrix Object	334
16.5	Assembling the Sparse Stiffness Matrix	337
16.6	Exercises	338

17 Iterative Methods for Large Sparse Linear Systems 341

17.1	Iterative vs. Direct Methods	341
17.2	Iterative Methods	342
17.3	Gauss–Seidel Relaxation	344
17.4	Jacobi Relaxation	344
17.5	Symmetric Gauss–Seidel	346
17.6	The Normal Equation	346
17.7	Incomplete Factorization	347
17.8	The Multigrid Method	351
17.9	Algebraic Multigrid (AMG)	354
17.10	Implementation of Multigrid	355
17.11	Preconditioned Conjugate Gradients (PCGs)	359
17.12	Multigrid for Nonsymmetric Linear Systems	362
17.13	Domain Decomposition and Multigrid	362
17.14	Exercises	366

18 Parallelism 369

18.1	Parallel vs. Sequential Computers	369
18.2	Parallelizable Algorithms	370
18.3	Inherently Sequential Algorithms	370
18.4	Architecture	370
18.5	Cache-Oriented Relaxation	371
18.6	Schwarz Block Relaxation	373
18.7	Parallel Architectures	373
18.8	Shared Memory	373

18.9 Distributed Memory . . . 374
18.10 Communication Network . . . 375
18.11 Hypercube . . . 376
18.12 Example: Multiplication of Sparse Matrices . . . 378
18.13 Low-Level C++ Implementation . . . 379
18.14 Comparing Architectures . . . 380
18.15 Jacobi Relaxation . . . 382
18.16 Convergence Analysis . . . 383
18.17 Examples from Harwell–Boeing . . . 385
18.18 Block Jacobi Relaxation . . . 388
18.19 Exercises . . . 389

VI Applications **391**

19 Diffusion Equations **395**
19.1 The Boundary-Value Problem . . . 395
19.2 The Finite-Element Mesh . . . 396
19.3 The Linear-System Solver . . . 398
19.4 Implicit Time Marching . . . 399
19.5 Exercises . . . 401

20 The Linear Elasticity Equations **403**
20.1 System of PDEs . . . 403
20.2 The Strong Formulation . . . 403
20.3 The Weak Formulation . . . 404
20.4 The Minimization Problem . . . 405
20.5 Coercivity of the Quadratic Form . . . 405
20.6 The Finite-Element Discretization . . . 406
20.7 The Stiffness Matrix . . . 407
20.8 The Adaptive-Refinement Criterion . . . 408
20.9 The Modified Multigrid Algorithm . . . 408
20.10 Numerical Examples . . . 409
20.11 Exercises . . . 412

21 The Stokes Equations **413**
21.1 The Nabla Operator . . . 413
21.2 General Linear Elasticity . . . 416
21.3 Reduction to the Linear Elasticity Equations . . . 417
21.4 The Stokes Equations . . . 418
21.5 Continuation Process . . . 418
21.6 Adequacy Consideration . . . 419
21.7 Preconditioner for the Stokes Equations . . . 419
21.8 First-Order System Least Squares . . . 422
21.9 The Navier–Stokes Equations . . . 423
21.10 Exercises . . . 424

22 Electromagnetic Waves 425

22.1 The Wave Equation . . . 425
22.2 The Helmholtz Equation . . . 426
22.3 Finite-Difference Discretization . . . 427
22.4 Adequacy in Finite Differences . . . 427
22.5 Bilinear Finite Elements . . . 428
22.6 Adequacy in Bilinear Finite Elements . . . 429
22.7 The Measuring Problem . . . 430
22.8 The Nearly Rectangular Model . . . 431
22.9 The Nearly Circular Model . . . 432
22.10 Nonlinear Finite Elements . . . 433
22.11 The Maxwell Equations . . . 434
22.12 Reduction to Three-Dimensional Helmholtz Equations . . . 436
22.13 Exercises . . . 437

Appendix 439

A.1 Operations with Vectors . . . 439
A.2 Operations with Matrices . . . 440
A.3 Operations with Dynamic Vectors . . . 442
A.4 Two-Dimensional Convection-Diffusion Equation . . . 444
A.5 Stability in the Explicit Scheme . . . 456
A.6 Jordan Form of a Tridiagonal Matrix . . . 456
A.7 Denoising Digital Images . . . 458
A.8 Members of the Mesh Class . . . 460
A.9 Operations with Sparse Matrices . . . 462
A.10 Kacmarz Iteration . . . 465
A.11 ILU Iteration . . . 465
A.12 Multigrid Iteration . . . 466
A.13 Acceleration Techniques . . . 470
A.14 Parallel Implementation . . . 475
A.15 The Adaptive-Refinement Code . . . 476
A.16 The Diffusion Solver . . . 478
A.17 The Linear Elasticity Solver . . . 481

Bibliography 485

Index 489

List of Figures

I.1 The three basic elements of the digital computer. 4

1.1 An if-else structure. First, the condition at the top is checked. If it holds, then the commands on the right are executed. If not, then the commands on the left are executed, including another if-else question. 15
1.2 A loop: the same instruction is repeated for $i = 0, 1, 2, 3, 4, 5, 6$ (SIMD). 18
1.3 Nested loops: the outer loop goes over $i = 0, 1, 2$; for each particular i, the inner loop goes over $j = 0, 1, 2$. 21
1.4 Pascal's triangle. 29
1.5 An iteration of the bisection algorithm for choosing a subinterval in which the local maximum of f lies. f is calculated at three intermediate points. If the maximum at these points is obtained at the midpoint, then the middle subinterval is chosen (a). If it is obtained at the left point, then the left subinterval is chosen (b). If it is obtained at the right point, then the right subinterval is chosen (c). 32
1.6 An iteration of the golden-ratio algorithm for choosing a subinterval in which the local maximum of f lies. f is calculated at the intermediate points l and r. If $f(l) > f(r)$, then the left subinterval is chosen (a); otherwise, the right one is chosen (b). 34
1.7 The "fix()" function calculates $3 \cdot 7 + 12/3$ by scanning it backward until the '+' is found and splitting it into two subexpressions, $3 \cdot 7$ and $12/3$, which are calculated recursively and added. 37

2.1 Schematic representation of inheritance. 80
2.2 Schematic representation of inheritance from the base class "point" to the derived class "complex". 80
2.3 The three kinds of members of a class (public, protected, and private) and their access pattern. 82
2.4 Schematic representation of inheritance from the base class "vector<vector>" to the derived class "matrix". 87

3.1 Schematic representation of a list of objects. The arrows represent pointers, and the bullets represent constructed objects. i stands for the index in the array of pointers. 99

3.2 Schematic representation of a connected list: each item (denoted by a bullet) contains a pointer (denoted by an arrow) to the next item (except the last item, which contains the null (or zero) pointer). 102

3.3 Merging two connected lists while preserving order. The items in the top connected list (the current object) are scanned by the pointer "runner" in the outer loop. The items in the bottom connected list 'L' (the argument) are scanned by the pointer "Lrunner" in the inner loop and inserted in the right place. 109

3.4 The "order()" function that orders a disordered connected list: the original list is split into two sublists, which are ordered recursively and merged (while preserving order). 111

3.5 Schematic representation of a binary tree with three levels. The arrows represent pointers, and the bullets represent constructed objects. The circles at the lowest level stand for the null (zero) pointer. 113

3.6 Schematic representation of an oriented graph. 114

3.7 Schematic representation of a nonoriented graph. 114

4.1 Upward implementation of a graph. Nodes are implemented as objects and stored in an array. Edges are not implemented but only indicated by the function $A_{i,j}$, which takes a nonzero value if and only if i and j are connected by an edge. 124

4.2 Downward implementation of a graph. Edges are implemented and stored in an array. Each edge points to its two endpoints, which provides edge-to-node (but not node-to-edge) data access. 125

4.3 Downward implementation with complete data access. Edges have access to their endpoints, and nodes have access to edges that use them as endpoints. 126

4.4 Implementation of a graph in C++. Each node points to the nodes that are connected to it. The edges are implemented implicitly through these pointers. 127

4.5 A triangulation, or conformal mesh of triangles. 129

4.6 Naive upward implementation of triangulation. Nodes are implemented and stored in an array. Triangles are not implemented but only indicated in the function $A_{i,j,k}$, which takes a nonzero value if and only if i, j, and k form a triangle (cubic storage and time). 130

4.7 Downward implementation of triangulation. Each triangle is implemented as an object that points to its three vertices, providing triangle-to-node (but not node-to-triangle) data access (linear storage, quadratic time). 131

5.1 Schematic representation of inheritance from the base class "list" to the derived class "polynomial". 139

5.2 Short-memory process: for fixed x, each derivative can be calculated from data about the previous one only. 150

5.3 Horner's algorithm is preferable in a long-memory process, where the derivatives must be calculated and stored in advance anyway, but not in a short-memory process, where they are better used and dropped, as can be done only in the standard algorithm. 151

5.4 Long-memory process: for fixed x, the calculation of each derivative requires data about all the previous ones. 151

5.5 The triangle in which the polynomial $p(x, y)$ is integrated. 155

6.1 Numerical solution of the Kuramoto–Sivashinsky equation, projected onto the (u, v) plane. Truncated power series are used to start the numerical marching. Before diverging to $(-\infty, -\infty)$, the solution gets sufficiently close to the fixed point, which proves the existence of an asymptotic solution. 175

III.1 The entire process of producing the numerical approximation to the original physical phenomenon. 184

7.1 The uniform $M \times N$ time-space grid for the discretization of the convection-diffusion equation in the time-space domain $0 < x < L, 0 < t < T$. . . . 191

7.2 Hierarchy of objects for the convection-diffusion equation: the "domain" object uses an "xtGrid" object, which uses "dynamicVector" and "difference" objects. 199

7.3 Schematic representation of inheritance from the base class "list" (list of dynamic vectors) to the derived classes "xtGrid" and "difference". 202

7.4 The uniform $N_x \times N_y$ spatial grid for the discretization of the convection-diffusion terms in the unit square $0 \le x, y \le 1$. 206

9.1 The rarefaction wave that travels to the left ($a < b < 0$). 221

9.2 The rarefaction wave whose right part travels to the right and left part travels to the left ($a < 0 < b \le -a$). 221

9.3 The shock wave that travels to the right with speed $(a + b)/2 > 0$. 222

9.4 The square s on which the conservation law is integrated to produce the Godunov scheme. 223

9.5 The local Riemann problem in the two adjacent cells that contain the points j and $j + 1$ in the ith time step. This local problem produces the half time step used in the Godunov scheme. 225

9.6 The N-wave produced by the Godunov scheme at the 200th time step, with $\Delta t = h/2$ and a 200×200 x-t grid. 227

9.7 The N-wave produced by the random-choice scheme at the 200th time step, with $\Delta t = h/2$ and a 200×200 x-t grid. 227

9.8 The singularly perturbed N-wave produced by the explicit Godunov scheme at the 200th time step, with $\Delta t = h/2$, $\varepsilon = h/4$, and a 200×200 x-t grid. 228

9.9 The characteristic line (along which F is integrated to contribute to $u_{i+1/2,j+1/2}$ at the half time step) has slope $dx/dt = u_{i,j} > 0$ if the state $u_{i,j}$ on the left travels to the right (a), $dx/dt = u_{i,j+1} < 0$ if the state $u_{i,j+1}$ on the right travels to the left (b), or $dx/dt = 0$ if $u_{i,j} < 0 < u_{i,j+1}$ (c). . . . 233

12.1 The original finite element e in the mesh. . . . 263
12.2 The reference element r that is mapped to e by M_e. . . . 263
12.3 The uniform finite-element mesh. . . . 266
12.4 The original finite element e in the uniform mesh. The reference element r is mapped to e by M_e. . . . 267
12.5 The finite element t onto which the reference element r is mapped by M_t. . . . 268
12.6 The rotated finite element. . . . 270
12.7 The diffusion coefficients P and Q for the anisotropic diffusion equation. 273
12.8 The stretched finite-element mesh for the anisotropic diffusion equation. . 274
12.9 The initial coarse mesh that provides a poor approximation to the circular boundary. . . . 275
12.10 The next, finer, mesh that better approximates the circular boundary. . . . 275

13.1 The hierarchy of objects used to implement an unstructured mesh: the "mesh" object is a connected list of "finiteElement" objects, each of which is a list of (pointers to) "node" objects, each of which contains a "point" object to indicate its location in the Cartesian plane. . . . 283
13.2 Schematic representation of inheritance from the base class "connectedList" to the derived class "mesh". . . . 291

14.1 The initial coarse mesh that approximates the square poorly. . . . 298
14.2 The second, finer, mesh resulting from one step of local refinement. It is assumed here that extra accuracy is needed only at the origin. . . . 298
14.3 The third, yet finer, mesh resulting from the second local-refinement step. It is assumed here that extra accuracy is needed only at the origin. . . . 299
14.4 The adaptive-refinement algorithm: the numerical solution obtained by the multigrid preconditioner at a particular mesh is used to refine it further and produce the next level of refinement. . . . 300
14.5 The original coarse finite-element mesh. . . . 301
14.6 The adaptively refined finite-element mesh. . . . 301
14.7 The coarse triangle with vertices A, nI, and nJ (a) is divided into two smaller triangles by the new line leading from A to nIJ (b). In order to preserve conformity, its neighbor on the upper right is also divided by a new line leading from nIJ to B (c) in the "refineNeighbor()" function. . . 303
14.8 The coarse mesh that serves as input for the adaptive-refinement algorithm with automatic boundary refinement. . . . 308
14.9 The finer mesh, in which the upper-right triangle is refined and two extra triangles are also added to better approximate the upper-right part of the circular boundary. . . . 308

14.10 The nonconvex domain in which the PDE is defined. 309

14.11 The original coarse mesh that gives a poor approximation to the nonconvex domain. 310

14.12 The refined mesh that gives a better approximation to the nonconvex domain. 310

15.1 The reference triangle in the quadratic finite-element method. The nodes are numbered $1, 2, 3, 4, 5, 6$. A typical quadratic nodal function assumes the value 1 at one of these nodes and 0 at the others. 315

15.2 The reference triangle in the cubic finite-element method. The nodes are numbered $1, 2, 3, 4, 5, 6, 7, 8, 9, 10$. A typical cubic nodal function assumes the value 1 at one of these nodes and 0 at the others. 316

16.1 Schematic representation of inheritance from the base class "connectedList<rowElement>" to the derived class "row". 329

16.2 The hierarchy of objects used to implement the sparse matrix: the "sparseMatrix" object is a list of "row" objects, each of which is a connected list of "rowElement" objects, which use the template 'T' to refer to the type of value of the matrix elements. 334

16.3 Schematic representation of inheritance from the base class "list<row>" to the derived class "sparseMatrix". 335

17.1 The multigrid iteration has the shape of the letter V: first, relaxation is used at the fine grid; then, the residual is transferred to the next, coarser, grid, where a smaller V-cycle is used recursively to produce a correction term, which is transferred back to the fine grid; finally, relaxation is used again at the fine grid. 354

17.2 The "multigrid" object contains the square matrices A, U, and L and the rectangular matrices R and P, which are used to transfer information to and from the coarse grid. 356

17.3 The domain decomposition. The bullets denote nodes in the coarse grid c. 363

17.4 The first prolongation step, in which the known values at the bullets are prolonged to the line connecting them by solving a homogeneous subproblem in the strip that contains this line. 363

17.5 The second prolongation step, in which the known values at the bullets and edges are prolonged to the interior of the subdomain by solving a homogeneous Dirichlet subproblem. 364

17.6 Prolonging to node i by solving a homogeneous Dirichlet–Neumann subproblem in the "molecule" of finite elements that surround it. 365

18.1 The cache-oriented relaxation for a tridiagonal system: first part, in which the unknowns in each chunk are relaxed over and over again using data from the previous relaxation. 372

18.2 The cache-oriented relaxation for a tridiagonal system: second part, in which the unknowns in each intermediate (interface) chunk are relaxed over and over again using data from the previous relaxation and the first part described above. 372

18.3 Shared-memory architecture: each processor has direct access to the entire memory. 374

18.4 Distributed-memory architecture: each processor has access only to its own local memory but not to the memories of other processors. If a processor needs data from the memory of another processor, then it must send a request through the communication wires, indicated by straight lines. 375

18.5 A connection machine with four processors: every processor is connected to every other processor by a wire. 376

18.6 Four processors connected in a line: each processor is connected physically only to its immediate neighbors. 376

18.7 A three-dimensional hypercube (cube) with eight processors in its nodes, numbered by the binary integers 000 to 111. Each node is connected only to the three nodes that differ from it by one binary digit only. 377

18.8 Different kinds of digital computers and architectures. 378

18.9 The implementation of the "sparseMatrix" and "dynamicVector" objects in the low-level code is completely hidden from the algorithms that use them, so it can be parallelized with no need to change the high-level code, provided that the low-level interface functions still take and return the same types of arguments as before. 380

18.10 Fair comparison between parallel and sequential computers for a particular problem: the performance of each computer is estimated with the algorithm that is most suitable for it in terms of efficient implementation. 382

18.11 Symmetric Gauss–Seidel relaxation: schematic description of the triangular preconditioners L and U as windows through which elements of A are seen. 384

19.1 The domain, diffusion coefficients, and boundary conditions for the present diffusion example. 396

19.2 The initial (coarse) mesh that contains only 11 nodes and 9 triangles and is refined further in the adaptive-refinement algorithm both in the interior of the domain and at the circular boundary. 397

19.3 Distribution of nodes in the first five levels of refinement in the diffusion example. The adaptive refinement uses a threshold of 0.01 and automatic boundary refinement. The strong diffusion in the lower-left quarter of the domain prevents large variation, so no extra refinement is needed. 398

20.1 The staircase in Ω that leads from Γ_D to $(x, y) \in \Omega$ in the coercivity proof. 406

20.2 The adaptive-refinement algorithm for the linear elasticity equations: the numerical approximations to u and v obtained by the modified multigrid preconditioner at a particular mesh are used to refine it further and produce the next level of refinement. 409

20.3 The boundary segments Γ_D (where Dirichlet boundary conditions are given) and Γ_N (where Neumann boundary conditions are given) in the linear elasticity example. 410

20.4 Distribution of nodes at the first five levels of refinement (with automatic boundary refinement). The Poisson ratio is $\nu = 1/3$. 411

21.1 Both the Stokes and the linear elasticity equations can be viewed as special cases of general linear elasticity. 419

21.2 Distribution of nodes at the first five levels of refinement (with automatic boundary refinement). The Poisson ratio increases gradually from 0.9 at the first level to 0.99 at the seventh level (11000 nodes). 420

22.1 The bilinear finite-element mesh. 429

22.2 The reference element for the bilinear finite-element scheme. 429

22.3 The measuring problem: find the depth of the slit in the object on the left from the wave reflected from it. 430

22.4 The bilinear finite-element mesh for the measuring problem. 431

22.5 The nearly circular model for the measuring problem. 432

22.6 The nonlinear finite-element mesh in the half-circular domain. 433

22.7 The wave and Maxwell equations can be reduced to the Helmholtz equation by a Fourier transform in the time variable t and the assumption $\nabla \cdot V = -\frac{1}{c}\Phi_t$, respectively. 437

A.1 Hierarchy of objects for the convection-diffusion equation in two spatial dimensions: the "domain2" object uses an "xytGrid" object, which uses "dynamicVector2" and "difference2" objects. 444

A.2 Schematic representation of inheritance from the base class "dynamicVector" to the derived class "dynamicVector2". 445

A.3 Schematic representation of inheritance from the base class "list" (list of "dynamicVector2" objects) to the derived classes "xytGrid" and "difference2". 452

A.4 The coarse mesh that contains only 11 nodes and 9 triangles and serves as input for the adaptive-refinement algorithm to solve the diffusion problem. 479

List of Tables

3.1 Different kinds of vectors and lists and their pros and cons. 112

15.1 Pros and cons of linear and high-order finite elements. It is assumed that the same number of degrees of freedom (unknowns) is used in both methods. 318

18.1 Number of symmetric Gauss–Seidel (or Jacobi) relaxations used within PCG applied to four SPD stiffness matrices (examples 2 to 5 in the file "bcsstruc2.data" in the Harwell–Boeing collection), where "order" is the matrix order and "nonzeroes" is the number of nonzero elements in the entire matrix. 387

19.1 Number of PCG iterations used in each refinement level in the adaptive-refinement algorithm (with threshold of 0.01 and automatic boundary refinement) applied to the diffusion example. The computation time of a multigrid iteration is the same as that of five symmetric Gauss–Seidel iterations. (The setup time is negligible.) 399

19.2 Number of PCG iterations used in each refinement level in the adaptive-refinement algorithm (with threshold of 0.01 and automatic boundary refinement) applied to an implicit time-marching step (with $\Delta t = 1$) in the time-dependent diffusion example. 400

20.1 The adaptive-refinement algorithm (with automatic boundary refinement) applied to the linear elasticity equations in the circle (with Poisson ratio $\nu = 1/3$). The modified multigrid preconditioner costs the same as five symmetric Gauss–Seidel iterations. (The setup time is negligible.) 412

Preface

This book teaches C++ and uses it to solve partial differential equations (PDEs). Basic and advanced numerical methods are introduced and implemented easily and efficiently in a unified object-oriented approach.

The powerful features and advanced tools available in C++ are particularly useful for implementing the complex mathematical objects that are often used in numerical modeling. The code segments and their detailed explanations show how easy it is to implement advanced algorithms such as finite elements and multigrid.

The book contains such as six parts. The first two parts introduce C, C++, and the object-oriented approach. The third and fourth parts describe and implement finite-difference and finite-element schemes. The fifth part deals with numerical linear algebra and parallelism. The sixth and final part uses the present approach in advanced applications. Each chapter ends with exercises and solutions.

The book contains two theoretical chapters, which can be skipped by readers who are interested in only the practical aspects (Chapters 8 and 11 and also Sections 9.9, 12.7, 18.16, and 20.4–20.5). More advanced readers may find them most useful and relevant to the present methods.

Because the book already contains the required background in programming, PDEs, and numerical methods, the only prerequisites are calculus and linear algebra. It can thus be used as a textbook in courses on C++ for scientific computing, C++ for engineers, numerical analysis, and numerical PDEs in the advanced undergraduate and graduate levels. Because it is self-contained, it is also suitable for self-study by researchers and students in applied and computational sciences and engineering.

Acknowledgments. I wish to thank Dr. Dan Quinlan for inspiring me to use C++ at the Los Alamos National Laboratory in 1995. I wish to thank the referees for their valuable comments. I wish to thank Prof. Daniel Michelson for his advice in [37] (used in Chapter 6, Section 15). I wish to thank Prof. Matania Ben-Artzi for his advice in a course project at the Hebrew University in 1988 (Figures 9.6–9.7). I wish to thank Prof. Moshe Israeli of the Technion for introducing to me the models in Chapter 22, Sections 8 and 9. I wish to thank Mable Amador of the Los Alamos National Laboratory for improving my writing style in previous papers. Last, but not least, I wish to thank my wife Marina and my sons Roy and Amir for their valuable support.

Yair Shapira

Part I

Programming

One of the major inventions of the 20th century was no doubt the digital computer. This invention has enabled us not only to store huge amounts of information in easily accessible memory devices but also to complete huge computational tasks that had never been tackled before. This turning point in technology opened up new horizons in science, industry, and commercial services.

The invention of the digital computer completely transformed scientific and technological development in the modern world. The human mind no longer needs to be occupied with technical calculations that can be left to the computer and can enjoy much more freedom and time to develop ideas and theories.

Unlike the calculation machine built by Pascal in the 17th century, which is limited to arithmetic calculations, the digital computer is a general-purpose computer: it can in principle solve every computable problem, provided that the required data and solution method (algorithm) are available to it. Therefore, in order to benefit from the full capacity of the computer, there is a need to describe the problem and its solution method in its own language. This is the art of programming.

The algorithm to solve a computational problem is actually a finite sequence of instructions. At first, these instructions can be written in natural (human) language, which the computer cannot understand. The original algorithm is then rewritten in the computer's own language to form a program or code. This is done by the programmer; we also say that the code is the implementation of the algorithm on the computer.

Some problems, however, can't be solved by any algorithm in a reasonable amount of time. It was Allen Turing who first investigated the properties of computable problems. Turing defined a theoretical computer in the 1940s known as the Turing machine. It can be shown that not every problem is computable in the sense that it can be solved by some algorithm on some Turing machine. Moreover, even problems that are computable in the above sense may be not computable in a reasonable amount of time.

Some problems, although computable, require too much time to solve. For example, a problem that is given in terms of N data elements and requires $\exp(N)$ time units (exponential time) to solve is practically unsolvable, or uncomputable. On the other hand, if a problem can be solved in N^k (for some fixed k) time units (polynomial time), then the problem is computable not only theoretically but also practically.

One of the earliest digital computers was built by Von Neumann in the 1950s. This computer consists of three basic elements: memory to store data and programs, processor to execute these programs, and input/output (I/O) devices to obtain information and questions and return answers and solutions. Common I/O devices are keyboard, screen, printer, mouse, etc. (see Figure I.1).

The main advantage (or disadvantage) of the computer is that, unlike us humans, it lacks any common sense or sense of humor. Therefore, it does exactly what it is told to do. For this purpose, it must have a clear and well-ordered list of instructions. This list is stored in the memory as a program and executed by the processor.

The instructions that can be executed by the processor are of two main kinds: simple arithmetic operations on integer or real numbers and data fetching from the memory.

The memory is usually divided into two parts: primary memory, where only the data required for the currently executed program are stored, and secondary memory, where the rest of the data are stored. Primary memory is more easily accessed, which saves valuable

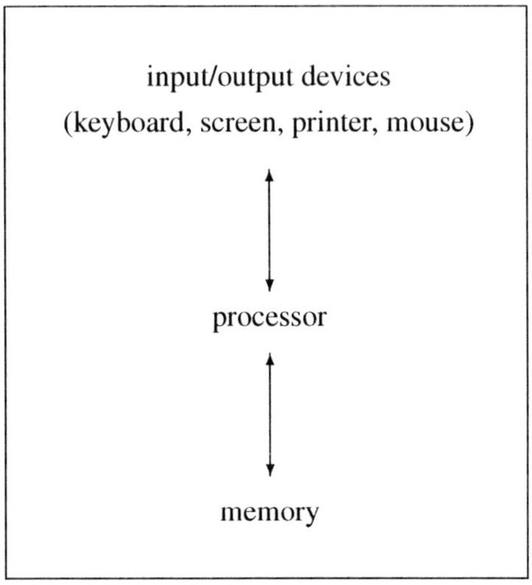

Figure I.1. *The three basic elements of the digital computer.*

time in the actual execution (run) of programs. Secondary memory, on the other hand, is more difficult to access and should thus be avoided in actual runs.

Even the most powerful computer is useless until it has received a program to execute through an input device. Furthermore, even after the computer has solved the problem, the solution is useless until it has been communicated back to the user through the output device. Thus, the I/O devices are as important as the memory and processor.

Modern computers use the same principles and basic elements as the original digital computer. In particular, they require a program written in their own language before they can run it.

In this part, we deal with programming: the art of writing algorithms in the computer's own language. The present examples are in the field of numerical algorithms for the solution of partial differential equations (PDEs). The problems in this field, although computable in polynomial time, require rather sophisticated algorithms and implementation, which are suitable for illustrating advanced programming techniques.

Programming is carried out in a so-called high-level language (like C or C++) that uses reserved words in English to form the logical structure of the program. The program is then translated in a compilation phase (compilation time) into a more elementary (machine or low-level) language that can be actually executed by the computer in "run time." In this machine language, the details of data fetching and update and elementary arithmetic operations are specified. Fortunately, the programmer may write in high-level language and avoid these details completely.

It is the responsibility of the programmer to write code that implements the original algorithm not only correctly and efficiently but also transparently and elegantly, so it can be read, understood, and modified later if necessary. For this purpose, it is particularly desirable for the program to be modular in the sense that different tasks are programmed in separate blocks of instructions (functions), which can then be invoked (called) by a single instruction from the main code segment that actually executes the algorithm. Furthermore, functions can use (call) yet other functions, which produces a useful hierarchy of functions. The main block of instruction that actually executes the algorithm is at the highest level. Each instruction in this block may invoke another block in the lower level, which in turn invokes blocks in the yet lower level, and so on.

Modular programs are easier to debug. Bugs are mistakes made by the programmer, resulting in a different algorithm, producing the wrong output. In large programs, bugs are particularly difficult to locate. The hierarchy of the functions could be useful here: first, the bug is located in the highest level in the hierarchy (the code segment that actually executes the algorithm) by running the program over and over again and comparing the data produced before and after each instruction in this segment. The bugged instruction in this segment usually calls another function; the block of this function is then debugged by repeating the same procedure recursively, and so on, until the bug is eventually found.

Some programming languages (like Fortran) are somewhat overstructured in the sense that memory must be allocated in advance in compilation time. This requirement is undesirable because in many applications the amount of memory needed is not known until run time. More flexible programming languages (like C and C++) also allow dynamic memory allocation in run time. This property is most important because it allows recursion: a function can call itself recursively, with data specified and memory allocated only in run time.

The C++ programming language is built on top of C; that is, C++ can use all the commands, types, and structures available in C. In addition, C++ can define new objects and associated functions. The objects can then be used much like standard types, which is most useful in the implementation of complex algorithms.

Objects are abstract concepts that one can operate on. In C, only numerical objects such as integer and real numbers are used; they can be passed as input arguments to functions and returned as output. In C++, on the other hand, programmers can construct much more complex objects (along with their own functions that express their properties) and use them in further code.

Thus, C++ is viewed as an object-oriented language that focuses on objects and their complete implementation rather than the algorithms and functions that use them. In C++, the programmer implements not only algorithms but also his or her own objects that can be used in many algorithms and applications. This approach is called object-oriented programming.

This part contains three chapters. The first two describe the C and C++ languages, assuming no previous knowledge whatsoever. The third chapter implements data structures that are often used in numerical algorithms.

Code Segments in this Book

The code segments in this book are fully debugged and tested. They are compiled with the standard GNU compiler. On the UNIX operating system, this compiler is invoked by the commands

```
>> g++ program.cxx
>> a.out
```

The output of the program named "program.cxx" is then printed onto the screen. When the program produces a lot of output, it can also be printed into a file named "Output" by the commands

```
>> g++ program.cxx
>> a.out > Output
```

The output can then be read from the file by using, e.g., the "vi" editor:

```
>> vi Output
```

One can also use the Windows® operating system to compile and run C++ programs, but this requires some extra linking commands.

The GNU compiler used here is one of the most widely used C++ compilers. Other compilers may require slight modifications due to some other restrictions, requirements, and properties. In principle, the code is suitable for other compilers as well.

Our convention is that words quoted from code are placed in quotation marks. Double quotation marks are used for strings (e.g., "const"), and single quotation marks are used for single characters (e.g., 'c'). When the word quoted from the code is a function name, it is often followed by "()", e.g., "min()".

Each command in the code ends with the symbol ';'. Commands that are too long are broken into several consecutive code lines. These code lines are interpreted as one continuous code line that lasts until the symbol ';' at the end of the command.

The code segments are presented in nested-block style; that is, an inner block is shifted farther to the right than the outer block that contains it. A code line that belongs to a particular inner block is also shifted to the right in the same way even when it is on its own to indicate that it is not just an isolated code line but actually belongs to this block.

Chapter 1

Introduction to C

C++ is built on top of C, so all the features of C are available in it as well. This is why we start with a concise description of C. In particular, we show how to write logical "if" conditions, loops, and functions. We conclude this chapter with several examples that use recursion in numerical applications, a feature that is used often throughout the book.

1.1 Variables and Types

A program in C is a sequence of commands or instructions to the computer. Usually, a command is written in a single code line; however, long commands can extend to two or even more code lines. Every command must end with the character ';'.

The most elementary command is the definition of a variable. This command contains two strings of characters to specify the type and name of the variable. The computer executes this command by allocating sufficient memory for the variable. This portion of memory can then be accessed through the name of the variable.

The variable is, thus, just a location in the memory of the computer with sufficient room to store the information that specifies its value, which is usually of type integer or real number. For example, if one writes in a C program

```
int i;
float a;
double x;
char c;
```

then locations in the memory become available to store an integer number, a real number, a double-precision real number, and a character, respectively.

The integer variable may take every integer value, whether negative, positive, or zero. Both "float" and "double" variables may take every real value. The character variable may take only nonnegative integer values between 0 and 255. Each of these potential values represents a character on the keyboard, such as letters in English (either lowercase or capital letters), digits, arithmetic symbols, and other special characters.

1.2 Assignment

Variables are referred to by their names ('i', 'a', 'x', and 'c' in the above example). Initially, they are assigned random, meaningless values. Meaningful values can be assigned later on, e.g.,

```
i = 0;
a = 0.;
x = 0.;
c = 'A';
```

Note that '0' stands for the integer number zero, whereas "0." stands for the real number zero.

A command in C is actually a function that not only carries out an instruction but also returns a value. In particular, the assignment operator '=' used above not only assigns a value to a particular variable but also returns the assigned value as output. This property can be used as follows:

```
x = a = i = 0;
```

This command is executed from right to left. First, the integer value 0 is assigned to the variable i. This assignment also returns the assigned value 0, which in turn is converted implicitly to the real number 0. and assigned to the variable a. This assignment also returns the (single-precision) real number 0., which in turn is converted implicitly to the (double-precision) real number 0. and assigned to the variable x. Thus, the above command is equivalent to three assignment commands.

1.3 Initialization

Actually, one could avoid the above assignment by initializing the variables with their required values upon definition:

```
int i = 0;
float a = 0.;
double x = 0.;
char c = 'A';
```

Here, the '=' symbol means not assignment to an existing variable as before but rather initialization of the new variable immediately with the correct value.

Unlike assignment, initialization returns no value, so it is impossible to write

```
double x = double y = 0.;   /* error! */
```

Here, the characters "/*" indicate the start of a comment, which ends with the characters "*/". Such comments are ignored by the C compiler. (C++ has another form of comment: the characters "//" indicate the start of a comment line ignored by the C++ compiler.)

Usually, comments are used to explain briefly what the code does. Here, however, the comment is used to warn us that the code is wrong. Indeed, because the initialization "double y = 0." on the right returns no value, it cannot be used to initialize 'x' on the left.

Initialization can also be used to define "read-only" types. Such types are obtained by writing the reserved word "const" before the type name:

```
const int i=1;
```

Here, 'i' is of type constant integer. Therefore, it must be initialized in its definition, and its value can never change throughout the block in which it is defined.

1.4 Conversion

We have seen above that the value returned from a function can be converted implicitly to the type required for assignment to some variable. Here, we see that variables can also be converted explicitly to another type. Conversion actually never changes the variable; it only returns its value, with the required type. Although "conversion" is an inexact term, it is commonly used to indicate change of type in a returned value.

Here is how explicit conversion is used:

```
i = 5;
x = (double)i;   /* or: double(i)  */
i = (int)3.4;    /* or: int(3.4)   */
```

Here, 'i' is an integer variable and remains so, and 'x' is a "double" variable and remains so. The prefix "(double)" before 'i' indicates a function that returns the value in 'i', converted to type "double". In the above example, the integer value 5 in 'i' is converted to 5. before being assigned to the "double" variable 'x'. Then, the prefix "(int)" before the real number 3.4 truncates and converts it into the integer number 3, which is then assigned to the integer variable 'i'.

The explicit conversion used in the above code is actually unnecessary. Exactly the same results would have been produced even without the prefixes "(double)" and "(int)". This is because when the C compiler encounters an assignment, it implicitly converts the assigned value to the right type.

1.5 Arithmetic Operations

The C compiler also supports standard binary arithmetic operations: multiplication (denoted by '*'), division ('/'), addition ('+'), and subtraction ('−'). It also supports the unary positive ('+') and negative ('−') operators. These arithmetic operators are actually functions of two (or one) variables that return a result of the same type as the arguments. For example, the C compiler invokes the integer-times-integer version of the multiplication operator to multiply the integers i and j and produce the integer result ij, and the "float"-times-"float" version to multiply the float variables a and b and produce the "float" result ab.

If variables of different types are multiplied by each other, then the variable of lower type is converted implicitly to the higher type before multiplication takes place. For example, to calculate the product "j*b" (where 'j' is integer and 'b' is "float"), the value in 'j' is first converted implicitly to "float" and then multiplied by 'b'.

The arithmetic operations are executed with the standard order of priority: multiplication and division have the same priority, and are prior to the modulus operator (%), which returns the residual in integer division. The modulus operator in turn is prior to addition and subtraction, which have the same priority.

Operations of the same priority are executed in the order in which they are written (left to right). For example, $1 + 8/4*2$ is calculated as follows. First, the division operator is invoked to calculate $8/4 = 2$. Then, the multiplication operator is invoked to calculate $2*2 = 4$. Finally, the addition operator is invoked to calculate $1 + 4 = 5$. (Parentheses can be used to change this order, if required.)

Division of integers is done with residual, and this residual can be obtained by the modulus operator, denoted by '%'. For example, $10/3$ gives the result 3, and 10%3 gives the residual 1.

For the sake of convenience, arithmetic symbols are often separated from the arguments by blank spaces. For example, "a + b" is the same as "a+b", but is slightly easier to read. When multiplication is used, however, one must be careful to use the blank spaces symmetrically. For example, both "a * b" and "a * b" mean 'a' multiplied by 'b', but "a * b" and "a * b" mean completely different things that have nothing to do with multiplication.

The result of an arithmetic operation, as well as the value returned by any other function, is stored in a temporary variable that has no name until the present command terminates. Thus, the returned value can be used only in the present command, before it vanishes with no trace. If the returned value is required for longer than that, then it must be stored in a properly defined variable through assignment or initialization.

The C compiler also supports some special arithmetic operations:

```
x += 1.;
x -= 1.;
x *= 1.;
x /= 1.;
++i;
--i;
i++;
i--;
```

These operations are the same as

```
x = x + 1.;
x = x - 1.;
x = x * 1.;
x = x / 1.;
i = i + 1;
i = i - 1;
i = i + 1;
i = i - 1;
```

(respectively). In the above, the arithmetic operation on the right-hand side is executed first, using the old values of 'x' and 'i', and stored in a temporary, unnamed variable. Then, this value is reassigned into 'x' or 'i' to update its value, and the temporary variables are destroyed.

The "+=", "−=", "*=", and "/=" operators are actually functions that not only change their first argument (the variable on the left) but also return the new (updated) value as output. This property is used in the following commands:

```
a = x += 1.;
j = ++i;
j = --i;
```

In each command, first the operation on the right-hand side is executed to update the value of 'x' or 'i' and return it as output. This output is then assigned to 'a' or 'j' on the left-hand side.

It is only with the unary operators "++" and "−−" that one may elect to assign the old value of 'i' to 'j' by writing

```
j = i++;
j = i--;
```

1.6 Functions

The above assignment and arithmetic operations are actually functions that are built into the C language. However, programmers are not limited to these functions; they can also write their own functions.

Functions written in C have a standard format. In the first code line (the heading) of a function, the type of returned value (also referred to as the type of function) appears first. If no type is specified, then it is assumed that the function returns an integer. A function may also be of type "void", which indicates that it returns no value.

The second word in the heading is the function name, given to it by the programmer. It is with this name that the function can later be called (invoked).

The function name is followed by parentheses containing the list of arguments taken by the function, separated by commas and preceded by their types. This completes the function heading.

This heading is followed by the function block, which starts with '{' and ends with '}' and contains the body of the function: the list of instructions to be carried out when it is called. It is good practice to write the '}' symbol that closes the block right under the first character in the heading that contains the '{' symbol at its end and then write the body of the block in the lines in between these symbols, shifted two blank spaces to the right. This way, one is less likely to forget to close the block. Here is a simple example:

```
int add(int i, int j){
  return i+j;
}
```

This code defines a function named "add()" that returns the sum of two integers. Because the sum is also an integer, the type of function is integer as well, as is indeed indicated by the reserved word "int" before the function name. The integers 'i' and 'j' that are added in the body of the function are referred to as local (or dummy) arguments (or variables), because they exist only within the function block and vanish as soon as the function terminates. The local arguments are initialized with the values of the corresponding concrete arguments that are passed to the function when it is actually called.

The command that starts with the reserved word "return" creates an unnamed variable to store the value returned by the function. The type of this variable is specified in the first word in the function heading ("int" in the above example). This new variable is temporary: it exists only in the command line in which the function is actually called and vanishes soon after.

The "return" command also terminates the execution of the function, regardless of whether or not the end of the function block has been reached. For this reason, even functions of type "void", which return no value, may use a "return" command to halt the execution whenever required. In this case, the "return" command is followed by no value, since nothing is returned.

When the C compiler encounters the definition of a function, it creates a finite state machine or automaton (a process that takes input and executes commands to produce output) that implements the function in machine language. This automaton has input lines to take concrete arguments and an output line to return a value. When the compiler encounters a call to the function, it passes the concrete arguments used in this call as input to the automaton and invokes it. The automaton is thus produced once and for all when the function block is compiled and can be used in many calls to the function, with no need to recompile.

Here is how the function "add()" is called:

```
int k=3, m=5, n;
n = add(k,m);
```

Here 'k' and 'm' are well-defined and initialized integer variables that are passed to the function as concrete arguments. When the function is called, its local arguments 'i' and 'j' are initialized to have the same values as 'k' and 'm'. The "add()" function then calculates the sum of its local variables 'i' and 'j', which is actually the required sum of 'k' and 'm'. This sum must be assigned to the well-defined variable 'n', or it will vanish with no trace.

The "add()" function can also be called as follows:

```
int i=3, j=5, n;
n = add(i,j);
```

The integer variables 'i' and 'j' in this code are not the same as the local variables in the definition of the "add()" function. In fact, they are well-defined variables that exist before, during, and after the function is called. These external variables are passed as concrete arguments to the function, where their values are used to initialize the local variables. These local variables, although they have the same names as the external variables, are totally different from them: they only exist inside the function block and vanish once it has ended. Thus, there is no ambiguity in the names 'i' and 'j': in the code in which the function is called, they refer to the external variables that are passed as concrete arguments, whereas, in the function block, they refer to the local variables. It is thus allowed and indeed recommended to use the same names for the external and local variables, as in the latter code, to reflect the fact that they play the same roles in the mathematical formula that calculates the sum.

1.7 Printing Output

Every C program must contain a function named "main()". The commands in this function are executed when the program runs. The "main()" function returns an integer value that is never used in the present program, say 0.

In this section, we present a short program that executes several arithmetic operations and prints the results onto the screen. The "include" command at the beginning of the program gives access to the standard I/O library that supports reading and writing data. In particular, it allows one to call the "printf()" function to print output onto the screen.

The "printf()" function requires several arguments. The first argument is the string to be printed onto the screen. The string appears in double quotation marks, and often ends with the character '\n', which stands for "end of line." The string may also contain the symbols "%d" (integer number)and "%f" (real number). These numbers are specified in the rest of the arguments passed to the "printf()" function.

In the following program, the "printf()" function is used to show the difference between integer and real division. First, it prints the result and residual in the integer division 10/3. Then, it prints the result of the real division 10./3, in which the integer 3 is converted implicitly to the real number 3. before being used to divide the real number 10.:

```
#include<stdio.h>
int main(){
  printf("10/3=%d,10 mod 3=%d,10./3.=%f\n",10/3,10%3,10./3);
  return 0;
}
```

Since the assignment operator is also actually a function that returns the assigned value as output, it is legal to write

```
  int i;
  printf("10/3=%d.\n",i = 10/3);
```

Here, the output of the assignment "i = 10/3", which is actually 3, is printed onto the screen.

Initialization, on the other hand, returns no value. Therefore, it is illegal to write

```
  printf("10/3=%d.\n",int i = 10/3);
               /* wrong!!! no value returned */
```

Here is a simple user-defined function used frequently in what follows. This function just prints a "double" variable onto the screen:

```
void print(double d){
  printf("%f; ",d);
} /* print a double variable */
```

With this function, a "double" variable 'x' can be printed simply by writing "print(x)".

1.8 Conditions

Every programming language supports the logical "if" question. This means that, if a certain condition is met, then the computer is instructed to do one thing, and if the condition is not met, then the computer is instructed to do something else. C supports the logical if in two possible ways. The first way is limited to determining the value of a particular variable. There are two possible values; the choice between them is made according to whether or not the condition is met. For example:

```
double a = 3., b = 5.;
double max = a > b ? a : b;
```

Here, the "double" variable "max" is initialized by the value returned from the "? :" operator. This operator may return two possible values, according to the condition before the question mark. If the condition holds ('a' is indeed greater than 'b'), then the value that appears right after the question mark ('a' in the above example) is returned and used to initialize "max". Otherwise, the value that appears after the symbol ':' ('b' in the above example) is used instead.

The condition "a > b" in the above example may also be viewed as an integer number, which assumes a nonzero value if and only if 'a' is indeed greater than 'b'. In fact, the '>' symbol represents a function that returns a nonzero value, say 1, if and only if its first argument is indeed greater than the second. The value returned from the '>' operator (which is passed to the "? :" operator to decide what its outcome may be) is stored only temporarily in the above code. It could, however, be stored more properly in a well-defined integer variable before being passed to the "? :" operator:

```
double a = 3., b = 5.;
int condition = a > b;
double max = condition ? a : b;
```

The "? :" operator is actually a function of three arguments. The first argument ("condition" in the above example) is an integer. The second and third arguments are the potential outcomes 'a' and 'b', which are separated by the ':' symbol. The operator returns either 'a' or 'b', depending on the value of its first argument, "condition".

The '>' operator used in the above condition returns nonzero if and only if its first argument is greater than the second. The '<' operator, on the other hand, returns nonzero if and only if its first argument is smaller than the second. In addition, the "==" operator returns nonzero if and only if its two arguments are equal: "a == b" assumes a nonzero value if and only if 'a' and 'b' are indeed the same. Conversely, "a != b" assumes a nonzero value if and only if 'a' and 'b' are not equal.

Note that the single '=' character has a totally different meaning than the "==" string. Indeed, '=' means assignment, whereas "==" is the logical operator that checks whether two numbers are equal.

The logical operators "not," "and," and "or" are also available. In particular, the symbol '!' stands for the unary "not" operator: "!cond" is nonzero if and only if the integer "cond" is zero. Furthermore, the symbol "&&" stands for logical "and": "cond1&&cond2" is nonzero if and only if both the integers "cond1" and "cond2" are nonzero. Finally, the symbol "||"

stands for logical "or": "cond1| |cond2" is zero if and only if both "cond1" and "cond2" are zero.

The priority order of these logical operators is as usual: "&&" is prior to "| |", and '!' is prior to "&&". (Parentheses can be used to change these priorities.)

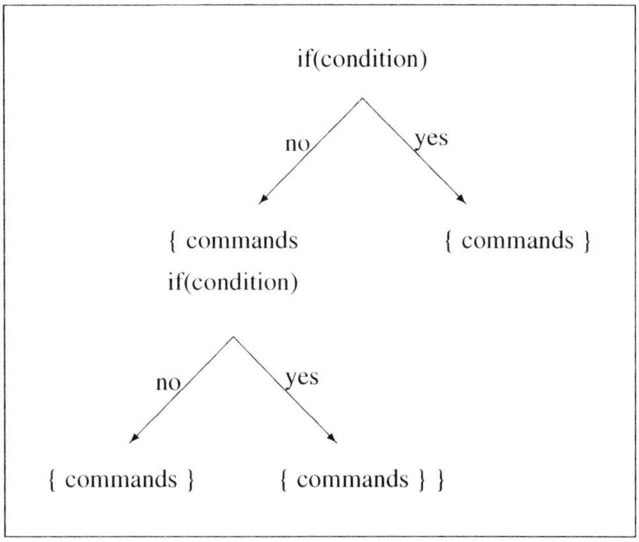

Figure 1.1. *An if-else structure. First, the condition at the top is checked. If it holds, then the commands on the right are executed. If not, then the commands on the left are executed, including another if-else question.*

The second way to give conditional instructions in C is through the "if-else" format (see Figure 1.1). For example, the above code can also be implemented as follows:

```
double a = 3., b = 5.;
int condition = a > b;
double max,min;
if(condition)
  max = a;
else
  max = b;
```

In this code, if the integer "condition" is nonzero, then the instruction that follows the "if" question is executed. Otherwise, the instruction that follows the "else" is executed instead. The "else" part is optional; if it is missing, then nothing is done if "condition" is zero.

1.9 Scope of Variables

The if-else format is more general than the "? :" format, as it allows not only a conditional value but also conditional instructions or even lists of instructions. In fact, the "if" question may be followed by a block of instructions (the "if" block) to be carried out if the condition holds, whereas the "else" can be followed by a completely different block of instructions (the "else" block) to be carried out if it doesn't hold. However, one must keep in mind that variables that are defined within a block exist only in that block and disappear at its end. For example, in the following code, useless local variables are defined inside the "if" and "else" blocks:

```
double a = 3., b = 5.;
if(a>b){
  double max = a;/* bad programming!!!  */
  double min = b;/*  local variables    */
}
else{
  double max = b; /* bad programming!!! */
  double min = a; /* local variables    */
}
```

This code is absolutely useless: since the variable exists only in the block in which it is defined, the variables "max" and "min" no longer exist after the if-else blocks end. The right implementation is as follows:

```
double a = 3., b = 5., max, min;
int condition = a > b;
if(condition){
  max = a;   /*  good  programming!!! */
  min = b;
}
else{
  max = b;
  min = a;
}
```

Here, the variables "max" and "min" are defined in the first code line, before the start of the if-else blocks, so they still exist after these blocks end and are available for further use.

The above variables "min" and "max" contain the minimum and maximum (respectively) of the two variables 'a' and 'b'. It is a better idea to define functions that return the minimum and maximum of their arguments. The following functions take integer arguments and calculate their minimum and maximum:

```
int max(int a, int b){
  return a>b ? a : b;
}  /* maximum of two integers */

int min(int a, int b){
  return a<b ? a : b;
}  /* minimum of two integers */
```

Here, the local variables 'a' and 'b' exist only in the function block in which they are defined. The functions are actually called with concrete arguments that exist before, during, and after the call to the function. The names of these concrete variables are not necessarily 'a' and 'b'. For example, "min(c,d)" returns the minimum of some external integer variables 'c' and 'd'. Still, it is a good idea to use concrete arguments with the same names as the corresponding dummy arguments. For example, "min(a,b)" returns the minimum of the external variables 'a' and 'b'.

One can also define analogous functions that take "double" arguments:

```
double max(double a, double b){
  return a>b ? a : b;
}  /* maximum of real numbers */

double min(double a, double b){
  return a<b ? a : b;
}  /* minimum of real numbers */
```

When the "max()" or "min()" function is called, the compiler invokes the version that fits the concrete arguments with which the call is made. If the concrete arguments are of type "int", then the "int" version is invoked. If, on the other hand, the concrete arguments are of type "double", then the "double" version is invoked.

Another useful function returns the absolute value of a real number:

```
double abs(double d){
  return d > 0. ? d : -d;
}  /* absolute value */
```

This function is actually available in the standard "math.h" library with the slightly different name "fabs()".

1.10 Loops

The best way to execute an instruction many times is in a loop (see Figure 1.2). The loop may be thought of as a single instruction and multiple data (SIMD) pattern, in which the same instruction is repeated over and over again with different data. The data used in the loop are usually indicated by the index of the loop, say 'i'. For example, if one wishes to print to the screen the integer numbers from 1 to 100 (each on a separate line), then one may write

```
int i = 1;
while(i<=100){
  printf("%d\n",i);
  i++;
}
```

The above "while" command consists of two parts: the heading, which contains a condition in parentheses, and the instruction or block of instructions, which is repeated over and over again as long as the condition holds. In the above example, the block contains two instructions: to print 'i' and to increment it by 1. These commands are executed as long as

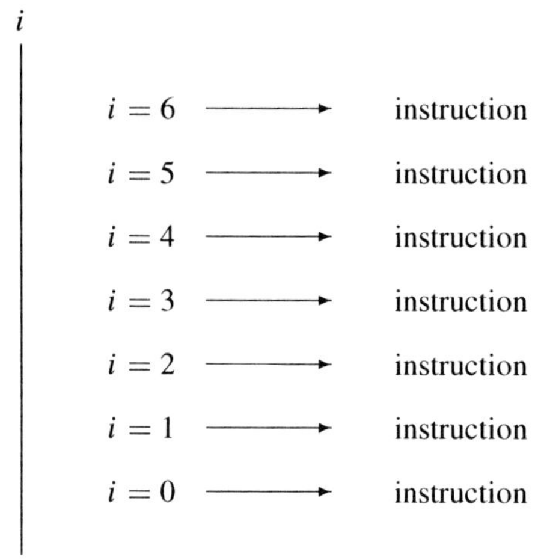

Figure 1.2. *A loop: the same instruction is repeated for* $i = 0, 1, 2, 3, 4, 5, 6$ *(SIMD).*

'i' is not larger than 100. Once 'i' reaches the value 101, the condition no longer holds, the loop terminates, and the execution proceeds to the code line that follows the "while" block.

In fact, since "i++" returns the old value of 'i' before being incremented, the above "while" block could be reduced to one command only:

```
int i = 1;
while(i<=100)
  printf("%d\n",i++);
```

The above "while" loop can also be written as a "do-while" loop:

```
int i = 1;
do
  printf("%d\n",i++);
while(i<=100);
```

This is only different style; the loop is equivalent to the "while" loop used before.

The same results can also be produced by a "for" loop as follows:

```
int i;
for(i=1;i<=100;i++)
  printf("%d\n",i);
```

In the heading of the "for" line, there are three items, separated by ';' characters. The first specifies the initial value of the index 'i' when the loop starts. The second specifies the condition whether to execute the instruction (or block of instructions) or skip it and exit the loop. Finally, the third item specifies what changes should be made every time the loop is repeated, here, how 'i' should change. In the above example, 'i' is initially 1. The instruction is executed for this value, and '1' is printed to the screen. Then, 'i' is successively incremented by 1 and printed, until it reaches the value 101, when it is no longer printed because the condition no longer holds, and the loop is complete.

The change to 'i' in the third item in the heading is made after the instruction has already been executed. Therefore, this item can also be placed at the end of the block of instructions:

```
int i=1;
for(;i<=100;){
  printf("%d\n",i);
  i++;
}
```

Here, only the second item appears in the heading. The first item (initialization) is placed before the loop, and the third item (change) is placed at the end of the block of instructions. This code is equivalent to the previous code.

The first item in the heading can be used not only to initialize but also to actually define 'i':

```
for(int i=1;i<=100;i++)
  printf("%d\n",i);
```

This code does exactly the same as before; the style, though, is slightly improved, because the code line that defines 'i' before the loop is avoided. On some compilers, 'i' exists even after the loop has terminated, and the value 101 contained in it can be used in further code lines. This, however, is no longer the case when the block of instructions (starting with '{' and ending with '}') is used as in the following loop:

```
for(int i=1;i<=100;i++){
  printf("%d\n",i);
}          /*  i disappeared  */
```

This loop is different from the previous one. Although it produces the same results as before, that is, the integer numbers from 1 to 100 are printed onto the screen, 'i' no longer exists after the loop has terminated. In fact, 'i' is a local variable that exists only inside the loop but is unavailable outside it.

If 'i' is never used outside the loop, then there is no reason to store it. Making it a local variable defined in the heading of the "for" loop is thus considered better programming.

1.11 Examples with Loops

Here is an example that shows how a loop can be used to calculate the power base^{exp}, where "base" is an integer number and "exp" is a positive integer number (of type "unsigned"):

```
int power(int base, unsigned exp){
  int result = 1;
  for(int i=0; i<exp; i++)
    result *= base;
  return result;
}  /*  "base" to the "exp"  */
```

Another nice example is the following "log()" function, which returns $\lfloor \log_2 n \rfloor$ (the largest integer that is smaller than or equal to $\log_2 n$), where n is a given positive integer:

```
int log(int n){
  int log2 = 0;
  while(n>1){
    n /= 2;
    log2++;
  }
  return log2;
} /* compute log(n) */
```

A loop is also useful in the implementation of the factorial function, defined by

$$n! = 1 \cdot 2 \cdot 3 \cdots n,$$

where n is a positive integer. (The factorial is also defined for 0: $0! = 1$.) The implementation is as follows:

```
int factorial(int n){
  int result = 1;
  for(int i=1; i<=n; i++)
    result *= i;
  return result;
} /*  compute n! using a loop  */
```

The instruction that follows the loop heading can by itself be a loop. This structure is known as nested loops (Figure 1.3). Here is a program that uses nested loops to print a checkerboard:

```
#include<stdio.h>
int main(){
  for(int i=0;i<8;i++){
    for(int j=0;j<8;j++)
      printf("%c ",((i+j)%2)?'*':'o');
    printf("\n");
  }
  return 0;
} /*  print a checkerboard */
```

1.12 Example: Reversed Integer

Here we show how a loop can be used in a function that reverses the order of the digits in an arbitrarily long integer number. For example, for the concrete argument 123, it returns the output 321:

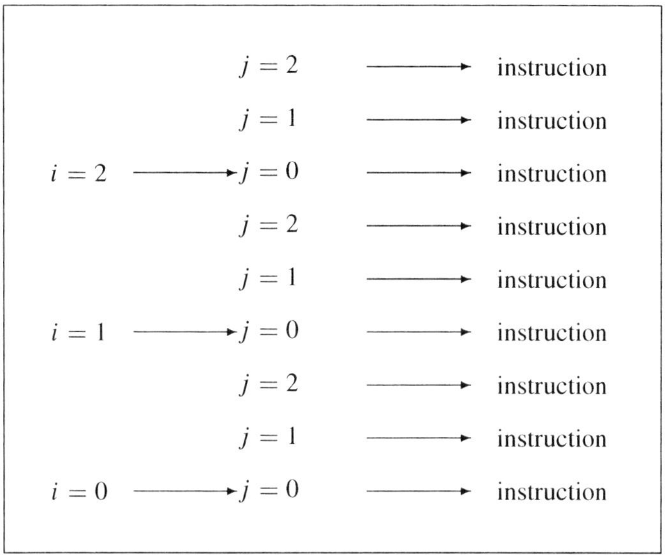

Figure 1.3. *Nested loops: the outer loop goes over* $i = 0, 1, 2$; *for each particular* i, *the inner loop goes over* $j = 0, 1, 2$.

```
int reversed(int number){
  int result=0;
  while (number){
    result *= 10;
    result += number % 10;
    number /= 10;
  }
  return result;
}  /*  reversing an integer number  */
```

This function can be called as follows:

```
int n = 778;
int m = reversed(n);
```

This call initializes 'm' with the value 877. The variable 'n', on the other hand, is never changed by this call, and its value remains 778. Indeed, the "reversed()" function creates its own local copy of the concrete argument 'n'. It is only this copy that is changed during the execution of the function, not the original variable 'n'.

The reversed representation produced above is in the usual decimal base. The "reversed()" function can be slightly modified to produce the reversed representation of an integer number in any base. This is done by introducing the extra argument "base" to denote the base used in the output number:

```
int reversed(int number, int base){
  int result=0;
```

```
    while (number){
      result *= 10;
      result += number % base;
      number /= base;
    }
    return result;
} /*  reversed number in any base */
```

This version is useful for obtaining the binary representation of an integer number. For this purpose, the latter version of "reversed()" is called twice: once to obtain the reversed representation of the original number in base 2, and the second time to reverse this representation and obtain the well-ordered binary representation. The output is, thus, the decimal number that contains the same binary digits (0 or 1) as in the required binary representation.

Note that, when the concrete argument of "reversed()" ends with zeroes, they get lost in the reversed number. For example, 3400 is reversed to 43. This is why 1 should be added to an even argument 'n' to make it odd. This extra 1 is subtracted at the end of the function:

```
int binary(int n){
  int last = 1;
  if(!(n%2)){
    last = 0;
    n += 1;
  }
  return reversed(reversed(n,2),10) - (1-last);
} /* binary representation */
```

1.13 Pointers

A pointer is an integer variable that may contain only the address of a variable of certain type. For example, pointer-to-"double" may contain only the address of a "double" variable, pointer-to-integer may contain only the address of an integer variable, and so on. We say that the pointer "points" to the variable whose address is in it. Here is how pointers are defined:

```
double *p;
int *q;
```

Here '*' stands not for multiplication but rather for dereferencing. The dereferencing operator is applied to a pointer and returns its content, namely, the variable pointed to by it. In the above code, the content of 'p' is defined as "double", which actually defines 'p' as pointer-to-"double", and the content of 'q' is defined as integer, which actually defines 'q' as pointer-to-integer.

The '*' symbol can also be shifted one space to the left with no change to the meaning of the code:

```
double* p;
int* q;
char* w;
```

With this style, "double*" is the pointer-to-"double" type, and "int*" is the pointer-to-integer type. Similarly, "char*" is the pointer-to-character type, and so on.

Since 'p' and 'q' above are not initialized, they take the zero value initially, which means they point to nothing.

One can also define a pointer-to-constant variable:

```
const int* p;
```

Here, the content of 'p' is a constant integer, or, in other words, 'p' points to a constant integer. Thus, once the content of 'p' is initialized to have some value, it can never change throughout the block where 'p' is defined. For this reason, 'p' can never be assigned to a pointer-to-nonconstant-integer or be used to initialize it. Indeed, 'p' contains the address of a constant integer. If one wrote

```
int* q = p;   /* error!!! p points to read-only integer */
```

then the new pointer 'q' would contain the same address. However, the variable in this address could then change inadvertently through 'q', in violation of the fact that it must be constant, because it is also pointed to by 'p'. For this reason, the compiler would refuse to assign the address in 'p' to 'q' and would issue a compilation error.

As we've seen above, the dereferencing operator '*' is applied to a pointer and returns its content. The referencing operator '&' operates in the reverse way: it is applied to a variable and returns its address:

```
double* p;
double v;
p = &v;
```

Here, the address of 'v' is assigned to the pointer 'p'. Both referencing and dereferencing operators are used often in what follows.

1.14 Arrays

An array in C is actually a pointer to the first variable in a long sequence of variables, all of the same type, and placed continuously in the computer memory. For example, the command

```
double a[10];
```

defines an array of 10 variables of type "double", which are allocated consecutive memory and referred to as the components of 'a' or "a[0]", "a[1]", ..., "a[9]". The corresponding addresses of these variables are 'a', "a+1", ..., "a+9", respectively.

The above code defines the array and allocates memory for it in compilation time. In some applications, however, the required size of the array is not yet known in compilation time. In such cases, the array can be initially defined as a pointer, and the precise number of components is left to be determined later. For example, the command

```
char* s;
```

defines a pointer-to-character 's'. In run time, a sequence of character variables may be allocated consecutive memory, with the address of the first one assigned to 's'. This way, 's' becomes an array of characters or a string.

In the above, we've seen an array whose components are "double" variables and an array whose components are "char" variables. It is impossible to define an array with components of mixed types. Still, it is possible to define an array whose components are by themselves arrays, provided that they are all of the same size. This structure is known as a two-dimensional array. For example, one can define an array of five components, each of which by itself is an array of ten "double" components, as follows:

```
double a[5][10];
```

The "double" variables in this array are ordered row by row in the computer memory. More specifically, the first row, "a[0][0]", "a[0][1]", ..., "a[0][9]", is stored first, then the second row, "a[1][0]", "a[1][1]", ..., "a[1][9]", and so on. It is most efficient to use this order when the array is scanned in nested loops: the outer loop jumps from row to row by advancing the first index in the two-dimensional array, whereas the inner loop scans the individual row by advancing the second index. This way, the variables in the array are scanned in their physical order in the computer memory, which increases efficiency.

In the above two-dimensional array, 'a' points to the first "double" variable, "a[0][0]". More generally, "a+10*i+j" points to the "double" variable "a[i][j]", $0 \leq i < 5, 0 \leq j < 10$.

When the size of the two-dimensional array is not known in advance in compilation time, it can still be defined as a pointer-to-pointer-to-double:

```
double** a;
```

Once the size of the array is determined in run time, the required memory is allocated, and the pointer-to-pointers 'a' becomes an array-of-arrays or a two-dimensional array "a[][]".

1.15 Passing Arguments to Functions

As we've seen above, programmers may define their own functions. In this section, we study strategies to pass arguments to such functions.

Let's start with the following simple function, which takes an integer argument and returns its value plus one:

```
int addOne(int i){
  return ++i;
}  /* return value plus one */
```

The argument, however, remains unchanged. For example, if the function is called as

```
int main(){
  int k=0;
  addOne(k);   /*  k remains unchanged  */
```

then the value of 'k' remains zero. This is because 'k' is passed to the function by value (or by name). In other words, when the function is called with 'k' as a concrete argument, a local integer variable, named 'i', is defined and initialized to the same value as 'k'. During the course of the function, it is only 'i' that changes, not 'k'.

The concrete argument can be changed only when it is passed by its address rather than by its value:

```
int addOne(int *q){
  return ++(*q);
}  /* add one */

int main(){
  int k=0;
  addOne(&k);    /* k is indeed incremented by one */
```

This way, the "addOne" function takes an argument of type pointer-to-integer rather than integer. When called, the function increments the contents of this pointer by one, with a real effect that remains after the function has terminated.

In the call "addOne(&k)", the address of 'k' is passed as a concrete argument. The function then creates a local variable of type pointer-to-integer, named 'q', which is initialized with the address of 'k'. Then, 'k' is incremented through its address in 'q'. Although 'q' is destroyed when the function terminates, the effect on 'k' remains valid, as required.

1.16 I/O

So far, we have used the standard "printf" function, which prints a string of characters to the screen. In the code below, we also use the standard "scanf" function, which reads a string of characters from the screen. The first argument of the "scanf" function is this string; the rest of the arguments are the addresses of the variables in which the numbers in the string are placed. These arguments must be passed by address, so that they store the values that are read into them:

```
#include<stdio.h>
#include<stdlib.h>
int main(){
  int i=0;
  double x=0.;
  scanf("%d %f\n",&i,&x);
  printf("i=%d, x=%f\n",i,x);
```

One may also use standard functions that read from and write to a file rather than the screen: the "fscanf" and "fprintf" functions. For this purpose, however, one must first have access to the file and a way to refer to it. The "fopen" standard function opens a file and returns its address. This address is stored in a variable (named "fp") of type pointer-to-file or "FILE*", where "FILE" is a reserved word for a file variable.

The "fopen" function takes two string arguments: the first is the name of the file that is being opened, and the second is either 'r' (if the file is opened for reading from it) or 'w' (if the file is opened for writing in it):

```
    FILE* fp = fopen("readFile","r");
    fscanf(fp,"%d %f\n",&i,&x);
```

The pointer-to-file "fp" (which contains the address of the file) is now passed to the "fscanf" function as its first argument. In other words, the concrete file is passed by its address. The "fscanf" function operates in the same way as "scanf", except that it reads from the file rather than the screen. In the present example, the data are read from the file named "readFile", which is stored in the directory in which the program runs.

Next, another file is opened for the purpose of printing to it. The file is then passed by its address to the "fprintf" function:

```
    fp = fopen("writeFile","w");
    fprintf(fp,"i=%d, x=%f\n",i,x);
    return 0;
  }
```

In this example, the data are printed to the file named "writeFile", which is stored in the directory in which the program runs.

1.17 Recursion

Recursion is a process in which a function is called in its very definition. The recursive call may use new data structures, which must be allocated additional memory in run time. Therefore, recursion cannot be used in structured programming languages such as Fortran, which require memory allocation in compilation time. C, on the other hand, allows memory allocation in run time and supports recursion.

Here is a simple example that shows how useful recursion can be. The "power" function in Section 1.11 is written recursively as follows:

```
  int power(int base, unsigned exp){
    return exp ? base * power(base,exp-1) : 1;
  }  /*  "base" to the "exp"  (with recursion)  */
```

This implementation actually uses the formula

$$\text{base}^{\text{exp}} = \text{base} \cdot \text{base}^{\text{exp}-1}.$$

Indeed, the definition of the "power" function uses a call to the function itself. When the computer encounters this call in run time, it uses the very definition of the "power" function to complete it. The computer automatically allocates the required memory for the arguments and the returned value in this recursive call.

The recursive call can by itself use a further recursive call with an even smaller argument "exp". This nested process continues until the final recursive call, which uses a zero "exp" argument, is reached, in which case no further recursive calls are made.

The extra memory allocations for arguments and output make the above recursive implementation of the "power" function less efficient than the original one. This overhead, however, is negligible when large complex objects are implemented. Recursion is invaluable in complex unstructured applications, as we'll see later on.

Next, we use recursion to rewrite the "log()" function in Section 1.11 as follows:

```
int log(int n){
  return n>1 ? log(n/2)+1 : 0;
}  /* compute log(n) recursively */
```

The "factorial" function in Section 1.11 can also be implemented most elegantly using recursion. This implementation uses the definition of the factorial function:

$$n! = \begin{cases} 1 & \text{if } n = 0, \\ n((n-1)!) & \text{if } n > 0. \end{cases}$$

This recursive (or inductive) definition is used in the following code:

```
int factorial(int n){
  return n>1 ? n * factorial(n-1) : 1;
}  /* compute n! using recursion */
```

Again, this implementation may be slightly less efficient than the original one in Section 1.11 because of the extra memory allocations for the arguments and output in each recursive call. Still, it is particularly short and elegant, and it follows the spirit of the above mathematical definition.

1.18 Example: Binary Representation

Here we show how recursion can be used to compute and print the binary representation of a nonnegative integer number n. The mathematical definition of this representation is

$$n = \sum_{i=0}^{\lfloor \log_2 n \rfloor} a_i 2^i,$$

where a_i is either 0 or 1. This formula is not easy to implement using loops; indeed, the sequence of coefficients a_i must be reversed and reversed again in the code in Section 1.12 before the binary representation can be obtained. Recursion, on the other hand, gives the required binary representation in the correct order. Actually, the recursion is based on a much more straightforward and useful formula:

$$n = 2(n/2) + (n\%2),$$

where $n/2$ means integer division with no residual, and $n\%2$ contains the residual. This formula leads to the following function, which prints the binary representation of a nonnegative integer:

```
void printBinary(int n){
  if(n>1) printBinary(n/2);
  printf("%d",n%2);
}  /* print binary representation */
```

Here, first the binary representation of $n/2$ is printed to the screen by the recursive call, and then the last binary digit $n\%2$ is also printed right after it. This indeed prints the entire binary representation of n in the right order.

The following function uses a similar recursion to reimplement the "binary()" function at the end of Section 1.12 above. This function returns the decimal number that contains the same binary digits (0 or 1) as the binary representation of the original number, in their proper order:

```
int binary(int n){
  return n>1 ? 10*binary(n/2)+n%2 : n%2;
} /* binary representation */
```

Clearly, this implementation is much more reader-friendly and efficient than that in Section 1.12. The output of this function can be printed to the screen as

```
int n = 778;
printf("%d",binary(n));
```

to produce the same result as the call "printBinary(n)".

Let us present another interesting problem with binary numbers. Let i and j be nonnegative integers such that $i \leq j$. Assume that the binary representation of i is extended (if necessary) by leading zeroes, so it contains the same number of binary digits as that of j. Assume also that these binary representations differ from each other by exactly k binary digits. A path leading from i to j is a sequence of integer numbers obtained by starting from i and changing these k digits one by one in some order, until i coincides with j. In this process, i is successively incremented or decremented by powers of 2 until j is reached.

How many such paths exist? Well, it depends on the number of different possible ways to order these k digits. Let's count: there are k possibilities for choosing the first digit to be changed. For each choice, there are $k-1$ possibilities for choosing the second digit to be changed, and so on. By continuing this process, one gets a total of $k!$ possible orderings for these k digits. Since each ordering defines a different path, the total number of paths leading from i to j is also $k!$.

In the well-ordered path leading from i to j, the above k digits are changed one by one in a natural order from the least significant digit to the most significant digit. The following function prints this path to the screen. If $i > j$, then it is called recursively, with i and j interchanged:

```
void path(int i, int j){
  if(i <= j){
    int go = i;
    int power = 1;
    int increment = 0;
    printf("%d\n",i);
    while(j){
if(increment = (j%2 - i%2) * power)
  printf("%d\n", go += increment);
```

```
i /= 2;
j /= 2;
power *= 2;
      }
    }
    else
      path(j,i);
  }  /* well-ordered path from i to j */
```

1.19 Example: Pascal's Triangle

A good example of an application that uses a two-dimensional array and a nested loop is Pascal's triangle. This triangle can be embedded in the lower-left part of a square mesh (see Figure 1.4). In every cell of the mesh, there is room for a single integer number. In what follows, we describe the rules to fill the cells in Pascal's triangle.

Let $i = 0, 1, 2, \ldots$ be the row index (numbered bottom to top) and $j = 0, 1, 2, \ldots$ be the column index (numbered left to right) in the mesh. Assume that a particle initially lies in the origin cell $(0, 0)$. Now, it starts to move. In every move, the particle can advance one cell either to the right or upward.

Consider a cell (k, l) for some fixed nonnegative integers k and l. Along how many different paths can the particle reach (k, l)?

This problem is solved recursively. The particle can approach (k, l) either from $(k, l - 1)$ on the left or $(k - 1, l)$ just below it. Hence, the particle must reach one of these cells before it can reach (k, l). Hence, the number of different paths leading to (k, l) is the number of different paths leading to $(k, l - 1)$ plus the number of different paths leading to $(k - 1, l)$.

The code that implements this algorithm to fill Pascal's triangle uses a two-dimensional array and a nested loop as follows:

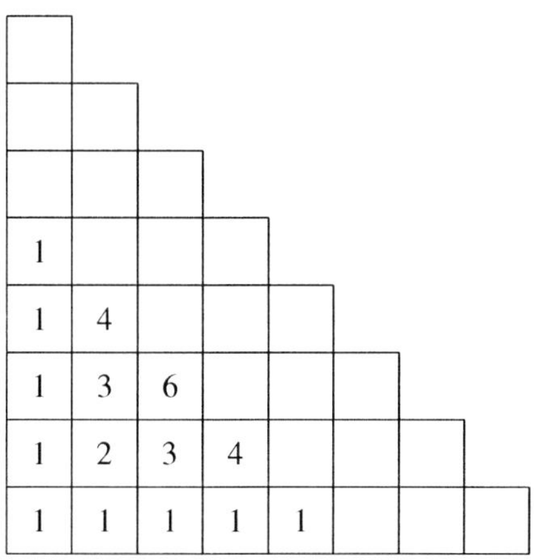

Figure 1.4. *Pascal's triangle.*

```
#include<stdio.h>
int main(){
  const int n=8;
  int triangle[n][n];
  for(int i=0; i<n; i++)
    triangle[i][0]=triangle[0][i]=1;
  for(int i=1; i<n-1; i++)
    for(int j=1; j<=n-1-i; j++)
      triangle[i][j] = triangle[i-1][j]+triangle[i][j-1];
  return 0;
} /* filling Pascal's triangle */
```

There is another formula for calculating the number of different paths leading from the origin cell $(0, 0)$ to the cell (k, l). Obviously, the total number of moves required to reach (k, l) is $k + l$. Of these moves, k are upward and the rest are to the right. In order to focus on a particular path, one should specify which moves are upward. In how many different ways can this be done? In other words, in how many ways can k items be picked from $k + l$ items? The answer is, of course, Newton's binomial coefficient

$$\binom{k+l}{k} = \frac{(k+l)!}{k! \cdot l!}.$$

Indeed, let's count how many different ways there are to pick k items out of the entire set of $k + l$ items. (Two ways are different from each other if there is at least one item that is picked in one way but not in the other.) There are $k + l$ candidates for the first pick; on top of that, there are $k + l - 1$ candidates left for the second pick, and so on, until the kth pick, where there are only $l + 1$ candidates left. The total number of different ways to pick k items is the product of these numbers, that is, $(k + l)!/l!$. Here, however, we are interested in only the final result of the k picks, not the order in which they are made. Therefore, there is some repetition in the above count. In fact, each final result is counted $k!$ times, which is the total number of different ways the k picks can be ordered. Thus, the total number of genuinely different ways to pick k items is indeed $(k + l)!/(k!l!)$.

The direct calculation of binomial coefficients is, however, rather expensive because of the large number of multiplications. The above code provides a much more efficient way to calculate and store them. The binomial coefficients will be used in several applications in what follows.

1.20 Example: Local Maximum

Here we use loops and recursion to implement algorithms that find a local maximum of a given function. The pros and cons of loops vs. recursion are then apparent.

The problem is to find a local maximum of a function $f(x)$ in an interval $[a, b]$. The solution x is required with accuracy, say, of six digits after the decimal point.

The bisection algorithm solves the problem iteratively. Because the required solution is in the interval $[a, b]$, it would help if we replaced this original interval by a smaller one. By repeating this procedure, we subsequently obtain smaller and smaller intervals in which the solution must lie.

More specifically, in each iteration, a and b are updated in such a way that $b - a$ is halved. When $b-a$ is sufficiently small, the iteration terminates, and the midpoint $(a+b)/2$ is accepted as the sufficiently accurate solution.

Here is the algorithm in detail.

Algorithm 1.1.

1. *Compute* $f(a)$, $f(b)$, *and* $f((a+b)/2)$.
2. *Compute* $f((3a+b)/4)$ *and* $f((a+3b)/4)$.
3. *If* $f((a+b)/2)$ *is greater than both* $f((3a+b)/4)$ *and* $f((a+3b)/4)$, *then substitute*
$$a \leftarrow (3a+b)/4 \text{ and } b \leftarrow (a+3b)/4.$$
Otherwise, substitute
$$b \leftarrow (a+b)/2 \text{ if } f((3a+b)/4) \geq f((a+3b)/4),$$
$$a \leftarrow (a+b)/2 \text{ if } f((3a+b)/4) < f((a+3b)/4).$$
4. *If* $b - a < 10^{-6}$, *then accept* $(a+b)/2$ *as the solution; otherwise, go back to step* 2 *above.*

An iteration in the bisection algorithm is displayed in Figure 1.5. Here is how a loop can be used to implement this algorithm:

```
double bisection(double a, double b){
  double Fa = f(a);
  double Fb = f(b);
  double midpoint = (a+b)/2;
  double Fmidpoint = f(midpoint);
  while(b-a>1.e-6){
    double left = (a+midpoint)/2;
    double Fleft = f(left);
    double right = (midpoint+b)/2;
    double Fright = f(right);
    if(Fmidpoint>max(Fleft,Fright)){
      a = left;
      Fa = Fleft;
      b = right;
      Fb = Fright;
    }
    else{
      if(Fleft>Fright){
        b = midpoint;
        Fb = Fmidpoint;
        midpoint = left;
```

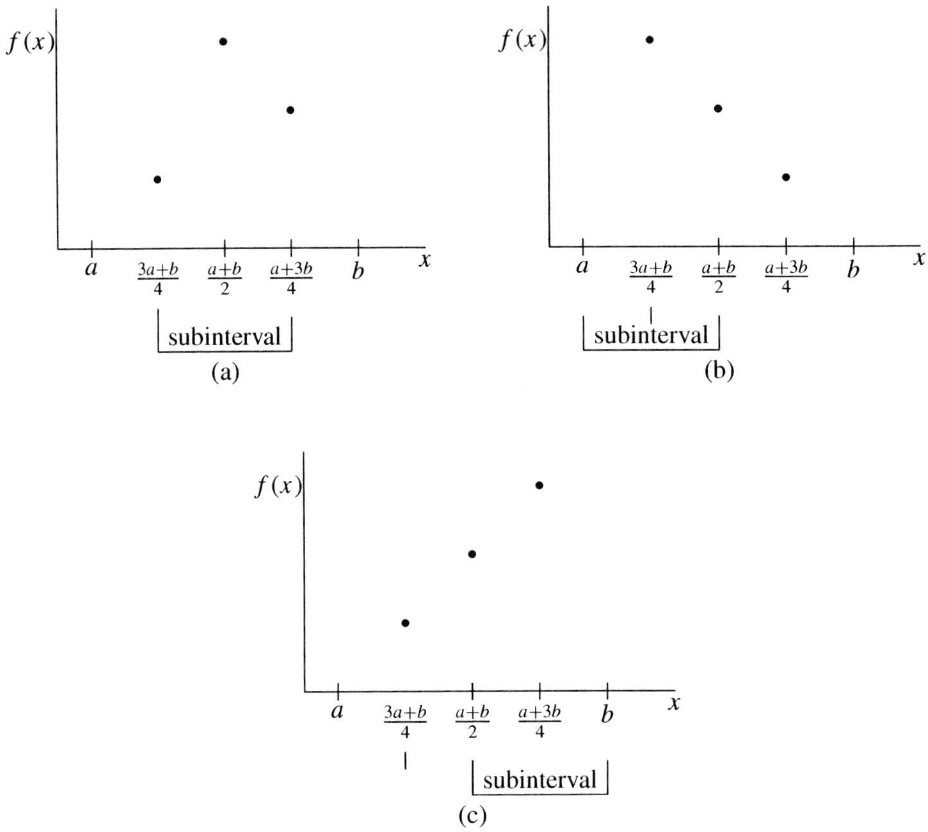

Figure 1.5. *An iteration of the bisection algorithm for choosing a subinterval in which the local maximum of f lies. f is calculated at three intermediate points. If the maximum at these points is obtained at the midpoint, then the middle subinterval is chosen* (a). *If it is obtained at the left point, then the left subinterval is chosen* (b). *If it is obtained at the right point, then the right subinterval is chosen* (c).

```
        Fmidpoint = Fleft;
      }
      else{
        a = midpoint;
        Fa = Fmidpoint;
        midpoint = right;
        Fmidpoint = Fright;
      }
    }
  }
  return midpoint;
}  /* local maximum by bisection algorithm */
```

This is a rather long code. It must use many local variables to avoid potentially expensive calls to the function "f()".

Can the code be shortened? Fortunately, it can. This is done by observing that the bisection algorithm is recursive in nature. Indeed, the reduction from the original interval $[a, b]$ to the subinterval in Figure 1.5 is the main step; the rest can be done recursively by applying the same algorithm itself to the subinterval. This idea is used in the following recursive implementation:

```
double bisection(double a, double b,
                 double midpoint, double Fa,
                 double Fb, double Fmidpoint){
  double left = (a+midpoint)/2;
  double Fleft = f(left);
  double right = (midpoint+b)/2;
  double Fright = f(right);
  return b-a < 1.e-6 ? midpoint
     : Fmidpoint > max(Fleft,Fright) ?
     bisection(left,right,midpoint,Fleft,Fright,Fmidpoint)
     : Fleft > Fright ?
     bisection(a,midpoint,left,Fa,Fmidpoint,Fleft)
     : bisection(midpoint,b,right,Fmidpoint,Fb,Fright);
}  /* local maximum by recursive bisection */
```

In this code, the three possibilities in Figure 1.5 are listed in the "return" command. This is more efficient than the “if-else” blocks used before. Furthermore, this implementation also requires fewer local variables. Still, it requires six arguments, which are allocated memory implicitly in each recursive call.

It is apparent from both implementations that the bisection algorithm requires two calculations of the function "f()" per iteration. Can this number be reduced? Fortunately, it can. The golden-ratio algorithm described below is not only more efficient but also easier to implement.

The golden ratio is the number

$$g = \frac{-1+\sqrt{5}}{2},$$

which satisfies

$$\frac{1-g}{g} = g.$$

This property guarantees that the iteration in Figure 1.6 can repeat in the subinterval as well, with only one calculation of f per iteration. Here is the detailed definition of the algorithm.

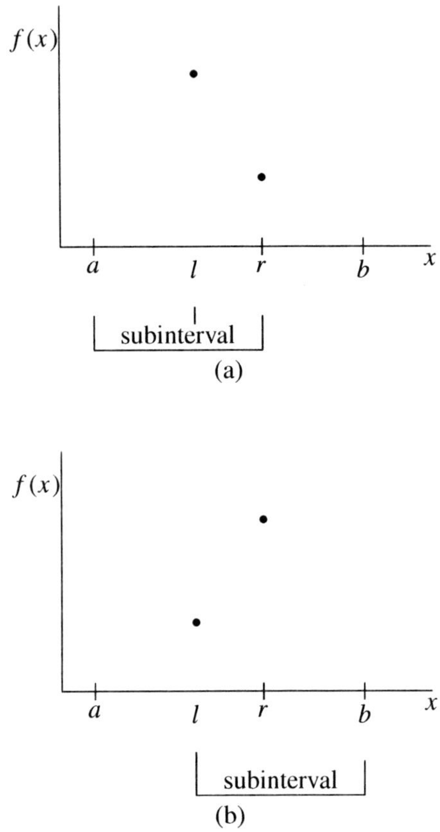

Figure 1.6. *An iteration of the golden-ratio algorithm for choosing a subinterval in which the local maximum of f lies. f is calculated at the intermediate points l and r. If $f(l) > f(r)$, then the left subinterval is chosen* (a)*; otherwise, the right one is chosen* (b).

Algorithm 1.2.

1. *Define*

$$l = b - g(b - a) \quad and \quad r = a + g(b - a).$$

2. *Substitute*

$$b \leftarrow r \text{ if } f(l) > f(r),$$
$$a \leftarrow l \text{ if } f(l) \leq f(r).$$

3. *If* $b - a < 10^{-6}$, *then* $(a + b)/2$ *is accepted as the required solution; otherwise, go back to step* 1.

The special property of the golden ratio g as stated in the above formula implies that the intermediate point (l or r in Figure 1.6) can also be used in the next iteration for the subinterval. Therefore, there is no need to recalculate f at this point. Thus, the cost has been reduced to only one calculation of f per iteration.

The golden-ratio algorithm is not only less expensive but also more straightforward and easier to implement. Here is how a loop can be used for this purpose:

```
double goldenRatio(double a, double b){
  double Fa = f(a);
  double Fb = f(b);
  double right = a + (-0.5+sqrt(5.)/2.) * (b-a);
  double Fright = f(right);
  double left = a + b - right;
  double Fleft = f(left);
  while(b-a>1.e-6){
    if(Fleft>Fright){
      b = right;
      Fb = Fright;
      right = left;
      Fright = Fleft;
      left = a + b - right;
      Fleft = f(left);
    }
    else{
      a = left;
      Fa = Fleft;
      left = right;
      Fleft = Fright;
      right = b - (left - a);
      Fright = f(right);
    }
  }
  return (a+b)/2;
} /* local maximum by golden-ratio algorithm */
```

This implementation can be further simplified and shortened by using recursion as follows:

```
double goldenRatio(double a, double left,
        double right, double b, double Fa,
        double Fleft, double Fright, double Fb){
  return b-a < 1.e-6 ? (a+b)/2
         : Fleft > Fright ?
         goldenRatio(a,a+right-left,left,right,
             Fa,f(a+right-left),Fleft,Fright)
         : goldenRatio(left,right,b+left-right,b,
             Fleft,Fright,f(b+left-right),Fb);
} /* golden-ratio algorithm (recursion) */
```

This implementation uses no local variables at all. Although it uses eight arguments, which must be allocated memory in each recursive call, they are implicit and hidden from the programmer, who can enjoy an elegant and easily debugged code.

It is apparent from both implementations that the golden-ratio algorithm indeed requires only one calculation of f per iteration. This improvement is essential when f is expensive to calculate, as in Chapter 6, Section 15.

Here is how the above four implementations are used to find a local maximum of $f(x) = \cos(x)$ in a particularly long interval:

```
#include<stdio.h>
#include<math.h>

double f(double x){
  return cos(x);
}

int main(){
  double a = -100.;
  double b = 10.;
  printf("%f\n",bisection(a,b));
  printf("%f\n",bisection(a,b,(a+b)/2,
                        f(a),f(b),f((a+b)/2)));
  printf("%f\n",goldenRatio(a,b));
  double right = a + (-0.5+sqrt(5.)/2.) * (b-a);
  double left = a + b - right;
  printf("%f\n",goldenRatio(a,left,right,b,
            f(a),f(left),f(right),f(b)));
  return 0;
}
```

1.21 Example: Arithmetic Expression

Here we use recursion in a program that reads an arbitrarily long arithmetic expression, prints it in the postfix and prefix formats, and calculates it. The postfix (respectively, prefix) format of a binary arithmetic expression puts the symbol of the arithmetic operation before (respectively, after) the arguments. For example, the arithmetic expression $2 + 3$ has the postfix format $+23$ and the prefix format $23+$. (No parentheses are used.)

The above tasks are carried out recursively as follows. The arithmetic expression is stored in a string or an array of digits and arithmetic symbols like '+', '-', etc. This string is fed as input into the function "fix()" that carries out the required task. In this function, the string is scanned in reverse (from right to left). Once a character that corresponds to an arithmetic operation is found, the string is split into two substrings, and the function is called recursively for each of them (see Figure 1.7).

To implement this procedure, we need to define some elementary functions. We start with a function that copies the first 'n' characters from a string 's' to a string 't':

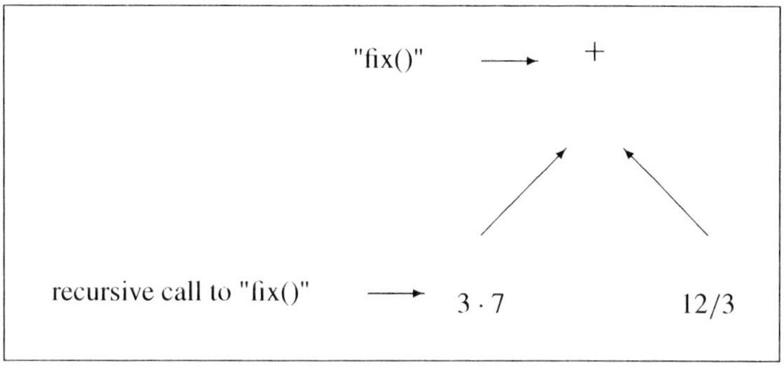

Figure 1.7. *The "fix()" function calculates* $3 \cdot 7 + 12/3$ *by scanning it backward until the '+' is found and splitting it into two subexpressions,* $3 \cdot 7$ *and* $12/3$*, which are calculated recursively and added.*

```
#include<stdio.h>
void copy(char* t, char* s, int n){
  for(int i=0;i<n;i++)
    t[i]=s[i];
  t[n]='\n';
}  /* copy n first characters from s to t  */
```

The function "fix()" defined below carries out one of three possible tasks: writing in postfix format, writing in prefix format, or computing the value of the arithmetic expression. The particular task that is carried out is determined by the value of its last argument, the integer named "task". The arithmetic expression is passed to the "fix()" function as its first argument, and its length is specified in the second argument. The third argument, the integer "task", takes three possible values that specify the required task: if it is zero, then the arithmetic expression is printed in the postfix form; if it is one, then the arithmetic expression is printed in the prefix form; and if it is two, then the arithmetic expression is calculated and its value is returned.

The standard "printf" function that prints the output onto the screen uses the symbol "%c" to denote a variable of type character. Here is the entire implementation of the "fix()" function:

```
int fix(char* s, int length, int task){
  for(int i=length-1;i>=0;i--)
    if((s[i]=='+')||(s[i]=='-')){
      char s1[i+1];
      char s2[length-i];
      copy(s1,s,i);              /* create first substring */
      copy(s2,s+i+1,length-i-1);/* create second substring */
      if(task==2){  /*calculate the arithmetic expression*/
        if(s[i]=='+')
          return fix(s1,i,task) + fix(s2,length-i-1,task);
        else
```

```
        return fix(s1,i,task) - fix(s2,length-i-1,task);
      }
      if(task==0)printf("%c",s[i]);  /* print symbol (postfix) */
      fix(s1,i,task);    /* recursive call (first substring) */
      fix(s2,length-i-1,task);/*recursive call (second substring)*/
      if(task==1)printf("%c",s[i]);  /* print symbol (prefix) */
      return 0;
    }
```

In the above loop, the string is scanned in the reverse order, and we look for the '+' or '-' character. These characters correspond to arithmetic operations of the least priority, and hence are considered first. If a '+' or '-' character is found, then the string is split into two substrings that are added or subtracted (if the task is to calculate) or printed in the postfix or prefix form (if the task is to do so). The next loop does the same for the '%' character, which stands for the modulus operation, which is of the next higher priority:

```
for(int i=length-1;i>=0;i--)
  if(s[i]=='%'){
    char s1[i+1];
    char s2[length-i];
    copy(s1,s,i);          /* create the first substring */
    copy(s2,s+i+1,length-i-1);/* create the second substring */
    if(task==2)return fix(s1,i,task) % fix(s2,length-i-1,task);
    if(task==0)printf("%c",s[i]);/* "mod" symbol (postfix) */
    fix(s1,i,task);    /* recursive call (first substring) */
    fix(s2,length-i-1,task);/*recursive call (second substring)*/
    if(task==1)printf("%c",s[i]);/* "mod" symbol (prefix) */
    return 0;
  }
```

The next loop does the same for the '*' (or '/') character, which corresponds to multiplication (or division), which is of the highest priority, and hence is considered last:

```
for(int i=length-1;i>=0;i--)
  if((s[i]=='*')||(s[i]=='/')){
    char s1[i+1];
    char s2[length-i];
    copy(s1,s,i);          /* create first substring */
    copy(s2,s+i+1,length-i-1);/* create second substring */
    if(task==2){  /* calculate arithmetic expression */
      if(s[i]=='*')
        return fix(s1,i,task) * fix(s2,length-i-1,task);
      else
        return fix(s1,i,task) / fix(s2,length-i-1,task);
    }
    if(task==0)printf("%c",s[i]);  /* print symbol (postfix) */
    fix(s1,i,task);    /* recursive call (first substring) */
```

```
      fix(s2,length-i-1,task);/*recursive call (second substring)*/
      if(task==1)printf("%c",s[i]);  /* print symbol (prefix) */
      return 0;
    }
```

Finally, the string is scanned once again, and this time we look for digits. If these are found, then the value of the integer number they form is calculated (if this is the required task) or printed to the screen (if this is the required task):

```
  if(*s == '\n'){
    printf("error");
    return 0;
  }
  if(task==2){  /* calculation of an integer number */
    int sum=0;
    int exp=1;
    for(int i=length-1;i>=0;i--){
      if((s[i]>='0')&&(s[i]<='9')){
        sum += (s[i]-'0') * exp;
        exp *= 10;
      }
      else{
        printf("error");
        return 0;
      }
    }
    return sum;
  }
  for(int i=0;i<length;i++){  /* print an integer number */
    if((s[i]>='0')&&(s[i]<='9'))
      printf("%c",s[i]);
    else{
      printf("error");
      return 0;
    }
  }
  return 0;
}  /*  calculate or print in prefix/postfix format
```

This completes the "fix()" function that carries out the required task. Here is the "main()" function that actually reads the arithmetic expression (using the standard "getchar()" function to read each individual character) and carries out the three required tasks: printing it in prefix and postfix formats and calculating its value.

```
  int main(){
    char s[80];
    int i;
    for(i=0; (s[i]=getchar()) != '\n'; i++);/*read expression*/
```

This loop (with empty body) reads the entire arithmetic expression character by character from the screen into the string 's' and sets the integer 'i' to the number of characters that have been read. The function "fix()" can now be applied:

```
    fix(s,i,0);                    /* print in postfix form */
    printf("\n");
    fix(s,i,1);                    /* print in prefix form */
    printf("\n");
    printf("%d\n",fix(s,i,2));     /* print expression value */
    return 0;
}
```

As mentioned above, a variable that is defined inside the block of a function (local variable) is destroyed when the function ends. Still, it may be saved by declaring it as "static". For example, if we wrote in the above "fix()" function

```
static FILE* fp = fopen("writeFile","w");
fprintf(fp,"length of subexpression=%d\n",length);
```

then the length of the original arithmetic expression and the subsequent subexpressions would be printed to the file "writeFile". Here, "fp" is a pointer-to-static-file variable. This file is created and initialized the first time the function starts and exists even after it terminates. Thus, writing to the file can continue in further recursive calls to the function.

1.22 Example: The Exponent Function

In this section, we define the exponent function exp(x) (e^x). Although this function is available in the standard "math.h" library, we implement it here explicitly as a good exercise in using loops. Furthermore, the present implementation can be extended to compute the exponent of a square matrix (Chapter 2, Section 22).

The exponent function is defined as an infinite series:

$$\exp(x) = 1 + x + \frac{x^2}{2!} + \frac{x^3}{3!} + \cdots = \sum_{n=0}^{\infty} \frac{x^n}{n!}.$$

This is also the Taylor expansion of the exponent function around zero. We approximate this function by the truncated series

$$T_K(x) = \sum_{n=0}^{K} \frac{x^n}{n!}.$$

Here, K is some predetermined integer, say $K = 10$.

The Taylor polynomial T_K is a good approximation to the exponent function when x is rather small in magnitude. When x is large in magnitude, exp(x) can still be approximated by picking a sufficiently large integer m in such a way that $x/2^m$ is sufficiently small in magnitude and approximating

$$\exp(x) = \exp(x/2^m)^{2^m}$$

by

$$\exp(x) \doteq T_K(x/2^m)^{2^m}.$$

This formula is followed in the function "expTaylor()":

```
double expTaylor(double arg){
  const int K=10;
  double x=arg;
```

First, one needs to find an appropriate integer m. This is done in a loop in which x is successively divided by 2 until its magnitude is sufficiently small, say smaller than 0.5. The total number of times x has been divided by 2 is the value assigned to m:

```
  int m=0;
  while(abs(x)>0.5){
    x /= 2.;
    m++;
  }
```

Now, we return to the evaluation of the Taylor polynomial $T_K(x)$. This can be done most efficiently in the spirit of Horner's algorithm (Chapter 5, Section 11):

$$T_K(x) = \left(\cdots\left(\left(\left(\frac{x}{K}+1\right)\frac{x}{K-1}+1\right)\frac{x}{K-2}+1\right)\cdots\right)x+1.$$

This calculation uses a loop with a decreasing index n in which T_K takes the initial value 1 before being successively multiplied by x/n and incremented by 1:

```
  double sum=1.;
  for(int n=K; n>00; n--){
    sum *= x/n;
    sum += 1.;
  }
```

At this stage, the local variable "sum" has the value $T_K(x/2^m)$. The required output, the 2^m-power of $T_K(x/2^m)$, is obtained from a third loop of length m, in which "sum" is successively replaced by its square:

```
  for(int i=0; i<m; i++)
    sum *= sum;
  return sum;
} /* calculate exp(arg) using Taylor series */
```

The exponent function $\exp(x)$ can also be computed using the diagonal Pade approximation. This approach is considered more stable than the previous one [47], although in our experiments (with x being either scalar or matrix) we have found no evidence for this assertion. In fact, we have observed exactly the same results with both approaches.

The diagonal Pade polynomial of degree K is defined by

$$P_K(x) = \sum_{n=0}^{K} \frac{\binom{K}{n} x^n}{n! \binom{2K}{n}},$$

where the binomial coefficient is given by

$$\binom{k}{n} = \frac{k!}{n! \cdot (k-n)!} \qquad (k \geq n \geq 0).$$

The computation of $P_K(x)$ can be done in the spirit of Horner's algorithm as follows:

$$P_K(x) = \left(\cdots \left(\left(\left(\frac{x}{K(K+1)} + 1 \right) \frac{2x}{(K-1)(K+2)} + 1 \right) \frac{3x}{(K-2)(K+3)} + 1 \right) \cdots \right) \cdot \frac{Kx}{1 \cdot 2K} + 1.$$

The diagonal Pade polynomial P_K is now used to approximate the exponent function by

$$\exp(x) \doteq \left(\frac{P_K(x/2^m)}{P_K(-x/2^m)} \right)^{2^m}.$$

As in the Taylor approximation, the accuracy of the approximation improves as K increases. In our experiments, we have found that $K = 5$ gives sufficient accuracy.

The complete implementation of the Pade approximation is, thus, as follows:

```
double expPade(double arg){
  const int K=5;
  double x=arg;
  int m=0;
  while(abs(x)>0.5){
    x /= 2.;
    m++;
  }
```

So far, we have calculated the integer m as in the Taylor approximation. Next, we calculate the numerator $P_K(x/2^m)$ using the above Horner-like formula:

```
  double nom=1.;
  for(int n=K; n>00; n--){
    nom *= x*(K-n+1)/(n*(2*K-n+1));
    nom += 1.;
  }
```

Next, we calculate the denominator $P_K(-x/2^m)$ using the above Horner-like formula once again:

```
    double denom=1.;
    for(int n=K; n>00; n--){
      denom *= -x*(K-n+1)/(n*(2*K-n+1));
      denom += 1.;
    }
```

Finally, we calculate the 2^m-power of $P_K(x/2^m)/P_K(-x/2^m)$:

```
    double sum = nom/denom;
    for(int i=0; i<m; i++)
      sum *= sum;
    return sum;
  } /* calculate exp(arg) using diagonal Pade method */
```

1.23 Exercises

1. The standard function "sizeof()" takes a name of some type and returns the number of bytes used to store a variable of that type in the computer memory. For example, on most computers, "sizeof(float)" returns 4, indicating that four bytes are used to store a "float" number. Since each byte stores two decimal digits, the precision of type "float" is eight digits. On the other hand, "sizeof(double)" is usually 8, indicating that "double" numbers are stored with a precision of sixteen decimal digits. Write a code that prints "sizeof(float)" and "sizeof(double)" to find out what the precision is on your computer.

2. Verify that arithmetic operations with "double" numbers are indeed more precise than those with "float" numbers by printing the difference $x_1 - x_2$, where $x_1 = 10^{10} + 1$ and $x_2 = 10^{10}$. If x_1 and x_2 are defined as "double" variables, then the result is 1, as it should be. If, however, they are defined as "float" numbers, then the result is 0, due to finite machine precision.

3. It is well known that the harmonic series $\sum 1/n$ diverges. Indeed,

$$\sum_{n=2}^{\infty} \frac{1}{n} = \sum_{k=0}^{\infty} \sum_{n=2^k+1}^{2^{k+1}} \frac{1}{n} \geq \sum_{k=0}^{\infty} \frac{2^k}{2^{k+1}} = \sum_{k=0}^{\infty} \frac{1}{2} = \infty.$$

   Write a function "harmonic(N)" that returns the sum of the first N terms in the harmonic series. (Make sure to use "1./n" in your code rather than "1/n", so that the division is interpreted as division of real numbers rather than integers.) Verify that the result of this function grows indefinitely with N.

4. On the other hand, the series $\sum 1/n^2$ converges. Indeed,

$$\sum_{n=1}^{\infty} \frac{1}{n^2} \leq \sum_{n=1}^{\infty} \frac{2}{n(n+1)} = 2\sum_{n=1}^{\infty} \left(\frac{1}{n} - \frac{1}{n+1}\right) = 2.$$

Write the function "series(N)" that calculates the sum of the first N terms in this series. Verify that the result of this function indeed converges as N increases.

5. Write a function "board(N)" that prints a checkerboard of size $N \times N$, where N is an integer argument. Use '+' to denote red cells and '-' to denote black cells on the board.

6. The Taylor series of the sine function $\sin(x)$ and the cosine function $\cos(x)$ are given by

$$\sin(x) = x - \frac{x^3}{3!} + \frac{x^5}{5!} - \cdots = \sum_{n=0}^{\infty} (-1)^n \frac{x^{2n+1}}{(2n+1)!}$$

and

$$\cos(x) = 1 - \frac{x^2}{2!} + \frac{x^4}{4!} - \cdots = \sum_{n=0}^{\infty} (-1)^n \frac{x^{2n}}{(2n)!},$$

respectively. Write functions "sinTaylor(N,x)" and "cosTaylor(N,x)" that calculate the sum of the first N terms in the series. Include the mathematical library "math.h" at the beginning of your code, so that the standard functions "sin(x)" and "cos(x)" are also available. Verify that the errors

$$|\sin(x) - sinTaylor(N, x)|$$

and

$$|\cos(x) - cosTaylor(N, x)|$$

are bounded by the $(N+1)$th term (the first dropped term) in the corresponding Taylor series.

7. Run the code segments of the examples in Section 1.11 and verify that they indeed produce the required results.

8. Define a two-dimensional array that stores the checkerboard in Section 1.11. Scan it in a nested loop and print it to the screen row by row. Verify that the output is indeed the same as in Section 1.11.

9. Modify the code in Section 1.18 to produce the representation of an integer number in base 3.

10. Modify the functions in Section 1.20 to find the local minimum rather than the local maximum of a function.

11. Run the code in Section 1.19 that constructs Pascal's triangle and verify that the number in cell (k, l) is indeed Newton's binomial coefficient

$$\binom{k+l}{k}.$$

Furthermore, calculate the sum of the entries along the nth diagonal $\{(k, l) \mid k + l = n\}$, and verify that it is equal to

$$\sum_{k=0}^{n} \binom{n}{k} = \sum_{k=0}^{n} \binom{n}{k} 1^k 1^{n-k} = (1+1)^n = 2^n.$$

12. Modify the code in Section 1.21 so that the results are printed to a static file defined in the "fix()" function.

13. Modify the code in Section 1.21 to read arithmetic expressions with parentheses and also print them in the prefix and postfix forms with parentheses.

14. Compare the results of the functions in Section 1.22 to the result of the "exp(x)" function available in the "math.h" library.

Chapter 2

Introduction to C++

In this chapter, we give a concise description of C++ and illustrate its power as an object-oriented programming language. In particular, we show how to construct and use abstract mathematical objects such as vectors and matrices. We also explain the notion of a template class, which can be filled with a concrete type later on in compilation time. We also discuss inheritance and illustrate its potential.

2.1 Objects

As we have seen above, C is a language based on functions. Every command is also a function that returns a value that can be further used or abandoned, according to the wish of the programmer. Furthermore, programmers can write their own functions, which may also return variables of the type specified just before the function name. When the function is called, a temporary, unnamed variable is created to store the returned value until it has been used.

C++, on the other hand, is an object-oriented programming language. In this kind of language, the major concern is not the functions that can be executed but rather the objects upon which they operate. Although C++ supports all the operations and features available in C, its point of view is different. C++ allows users to create not only their own functions but also their own objects and types, which can then be used in much the same way as the "int", "float", and "double" types that are built into C. The new types defined by C++ programmers can be used not only in the specific application for which they have been implemented but also in many other potential applications. Thus, C++ may be viewed as a dynamic extension of C, which constantly develops and grows with the definition of more and more useful objects.

The major advantage of object-oriented programming is the clear separation of the abstract (mathematical) concepts from their detailed implementation. Let us consider a team of two C++ programmers: Programmer A, who implements the objects, and Programmer B, who uses them in the actual application. Just as a C programmer is not interested in the actual implementation of integer and real numbers in the hardware of the computer, Programmer B is not interested in the detailed implementation of the objects prepared by

Programmer A. All that Programmer B needs are the objects themselves and the functions that operate upon them. Once these objects are available, Programmer B is free to use them in his particular application, regardless of their internal structure.

In fact, we could also consider a larger team of C++ programmers, in which Programmer A implements the objects required by all other programmers in their particular applications. Of course, Programmer A should also use feedback from these programmers to develop new functions that may be useful to potential users.

It is important that these programmers (or users) have a convenient interface to use the objects. This interface should not be changed, because any change in it would require changing every application that uses it. The actual implementation of objects, however, can be modified by Programmer A if necessary, so long as the modification doesn't affect the interface. The users should have no access to the internal implementation of objects, or they could change it inadvertently and spoil the entire framework.

Thus, just as C programmers cannot change the properties of integer and real types, users of the objects prepared by Programmer A cannot change their properties and functions. The data structures used in the objects should be accessible to external users only through public interface functions implemented and maintained by Programmer A. In fact, even Programmer A cannot modify the implementation unless he/she makes sure that the public interface functions are called and used in the same way.

The minds of the users of the objects prepared by Programmer A are thus free to develop the algorithms required in their own applications. They can use these objects through the interface functions, with no concern about any detail of implementation. They can treat the objects as perfect abstract concepts, which is particularly useful in the development of new methods and approaches. This is particularly important in numerical modeling, where complex mathematical objects are often involved.

2.2 Classes

As discussed above, C++ is particularly suitable for implementing abstract mathematical structures, which can then be used by other programmers as new objects or types. Assume, for example, that one wants to implement a point in the two-dimensional Cartesian plane. The well-implemented point should then be used by external users as if it were a standard type such as "int" or "double", leaving the users completely unaware of how it is stored and manipulated. This would free the minds of the users to concentrate on their particular application without being distracted by the details of the implementation of the points.

In particular, users should be able to write commands like

```
point P;
point Q=P;
```

to define a point 'P' and use it to define and initialize another point 'Q'. As we'll see below, this objective is achieved by defining the "point" class with its interface functions.

A new object in C++ is defined in a class as follows:

```
class point{
  public:
```

```
    double x;
    double y;   //  not object oriented
};
```

This is called a class block. The symbol "//" indicates the start of a comment line; the words that follow it are intended to describe or explain the code and are ignored by the C++ compiler.

The above block defines a new class called "point", which contains two data fields of type "double". The class can now be used to define variables of type "point". Once a concrete "point" variable 'P' is defined, "P.x" refers to its first field (representing the 'x'-coordinate) and "P.y" refers to its second field (representing the 'y'-coordinate). The reserved word "public:" in the above code indicates that these fields are accessible by every user of the class.

This implementation, however, is not in the spirit of object-oriented programming. Because the data fields 'x' and 'y' are accessible, the "point" object cannot be viewed as a complete "black box" implementation of the point. Because users are not familiar with the internal structure, they could change it inadvertently and spoil the object. Furthermore, if Programmer A, who wrote the "point" class, wanted to change the implementation at some stage, then he/she would have to tell all the users, who would have to change their own codes accordingly. This problem is avoided in the following object-oriented implementation:

```
class point{
    double x;
    double y;   //  an object-oriented implementation
  public:
    double X() const{
      return x
    } //  read x

    double Y() const{
      return y
    } //  read y

    void zero(){
      x=y=0.;
    } //  set to zero
};
```

In this version, the data fields 'x' and 'y' are no longer accessible to external users. Because they appear before the reserved word "public", they are considered by default "private": users who define a "point" object 'P' cannot access its coordinates simply by "P.x" and "P.y" as before. The data in the "point" object are thus safe from inadvertent change by careless users. Still, users can read (but not change) the data fields in 'P' only through the public interface functions "X()" and "Y()". For instance, "P.X()" returns the 'x'-coordinate of 'P', and "P.Y()" returns the 'y'-coordinate of 'P'. We refer to 'P' as the current object or variable with which the functions "X()" and "Y()" are called. The calls "P.X()" and "P.Y()" are no more expensive than the corresponding calls "P.x" and "P.y" in the previous implementation, because the functions contain only one code line each and create no new objects.

The rule is that fields and functions in the class block that have not been declared "public" are by default private. This is why the above 'x' and 'y' fields are private. One may also declare some more private fields and functions at the end of the class block by writing the reserved word "private:" before their declarations.

The functions "X()" and "Y()" read the coordinates of the current "point" object without changing them. This is indicated by the reserved word "const" that precedes the symbol '{' that opens the function block. This reserved word guarantees that the current "point" object can never be changed by the function, and every attempt to change it will lead to a compilation error.

The public function "zero" in the above code lacks the word "const" before its block, because it changes the current "point" object and sets it to zero. Thus, users of "class point" may change the data in a "point" object only through nonconstant functions like "zero".

Interface functions like "X()", "Y()", and "zero" can be modified at any time by Programmer A, who is responsible for the implementation, provided that they still take the same number of arguments and return the same type as before, so they can be used by other users in the same way as before. This way, the users are completely unaware of the actual implementation or any change in it. All they need to know is how to use the interface functions. In fact, they can think of a "point" variable like 'P' as a point in the two-dimensional Cartesian plane. The interface functions associated with it can also be thought of as operations on actual points.

As we've seen above, interface functions are placed inside the class block, right after the definitions of data fields. This style is suitable for short functions that contain no more than a few code lines. These functions are then recompiled every time the function is called. A more efficient style, which is suitable for longer functions as well, declares the function inside the class block, leaving its actual definition until later. The definition is placed outside the class block, and the function name in it is preceded by a prefix containing the class name followed by the symbol "::", to indicate that this is indeed an interface function in this class. This way, it is treated as if it had been placed inside the class block. For example, the "point" class could have been written equivalently as follows:

```
class point{
    double x;
    double y;  //  an object-oriented implementation
  public:
    double X() const;
    double Y() const;  //  declarations only
    void zero();
};

double point::X() const{
  return x;
}  //  definition of X()

double point::Y() const{
  return y;
}  //  definition of Y()
```

```
void point::zero(){
  x=y=0.;
} //  definition of "zero()"
```

Here, the interface functions are only declared in the class block, while their actual definitions are left outside it. Each definition is compiled only once, which creates a finite state machine (automaton). This machine is then invoked every time the function is called, with the concrete arguments that are passed to the function as input and the returned value as output.

The prefix "point::" that precedes the function names in the above code may actually be considered as an operator that "transfers" the definition back into the class block. This format, however, is unnecessary in the present "point" example, which uses very short definitions. The original style, in which the complete definitions appear inside the class block, is preferable in this case.

2.3 Constructors

When the user defines a point object by writing

```
point P;
```

the computer executes this command by allocating memory for the data fields of the new "point" variable 'P'. These data fields, the "double" variables "P.x" and "P.y", are not yet initialized by any meaningful value; in fact, they initially contain random, meaningless values. Only interface functions of the "point" class have access to these variables and can assign meaningful values to them.

C++ has a default constructor that allocates memory for the data fields and initializes them with random values. However, this is not always good enough. Programmer A might have a different idea about what should happen when the user writes code lines like "point P", and he/she might also want users to be able to write more sophisticated code lines like "point P = 0", in which 'P' is constructed and initialized to zero. Indeed, if the "point" object is to behave like standard types, then such commands should be available to users.

It is, thus, good practice to write an explicit constructor in the class. The constructor is a public interface function that tells the computer how to construct a new object and allocate memory for its data fields. The name of the constructor function must be the same as the name of the class itself. For example, if one writes in the block of the "point" class

```
point(){
} //  default constructor for the "point" class
```

then this constructor is invoked every time the compiler encounters a command of the form "point P;". The body of the constructor function contains no code lines at all. However, it still does something: as soon as it is called, the constructor implicitly invokes the default constructor built into C++, which allocates storage for the data fields "P.x" and "P.y" and fills them with random numbers. The above constructor is also called the default constructor for the "point" class, because it is invoked by commands like "point P;", with no initialization at all.

It is possible to write more than one constructor in the class block. In particular, one may also define a more informative constructor, which not only allocates memory for the data fields in the constructed object but also initializes them with meaningful values. For example, if one writes in the block of the "point" class

```
point(double xx,double yy){
  x=xx;
  y=yy;
}
```

then this constructor is invoked every time the compiler encounters a command of the form

```
point P(3.,5.);
```

to create a new "point" object 'P' with the value 3 in its 'x'-field and the value 5 in its 'y'-field.

Here is how this constructor works. When it is called, it first implicitly invokes the default constructor built into the C++ compiler, which allocates memory to the 'x'- and 'y'-fields and fills them with random numbers. Then, the 'x'- and 'y'-fields are reassigned their correct values from the corresponding arguments "xx" and "yy", respectively.

This process is somewhat inefficient. Why not initialize the 'x'- and 'y'-fields immediately with their correct values? This can indeed be done by using an initialization list as follows:

```
point(double xx,double yy):x(xx),y(yy){
} //  constructor with initialization list
```

The initialization list that follows the character ':' is a list of data fields in the constructed object, separated by commas. Each data field is followed by its initial value in parentheses. When this constructor is called, the fields are allocated memory and initialized properly in the order in which they appear (are declared) in the class block, and not necessarily in the order in the initialization list. The construction and initialization are then complete, so there is nothing left to do in the function block, and it remains empty.

Better yet, one may rewrite the above constructor as follows:

```
point(double xx=0.,double yy=0.):x(xx),y(yy){
} //  arguments with default values
```

This way, the arguments "xx" and "yy" take the default value 0, which is used whenever they are not specified explicitly. This constructor also serves as a default constructor in the "point" class, so there is no need to write any other constructor. When the compiler encounters commands like

```
point P(3.,5.);
point Q(3.);  //  or point Q=3.;
point W;
```

it completes the missing arguments with zeroes and constructs three point objects: 'P' with 'x'-field initialized to 3 and 'y'-field initialized to 5, 'Q' with 'x'-field initialized to 3 and 'y'-field initialized to 0 (because the second argument "yy" is not specified and therefore takes the default value 0), and 'W' with both fields initialized to 0 (because both arguments "xx" and "yy" are unspecified).

2.4 Explicit Conversion

The above constructor also provides an explicit conversion operator from type "double" to type "point". As in C, where "(double)n" produces a temporary "double" variable with the same value as the integer variable 'n', "(point)a" produces a temporary "point" object whose first field contains the same value as the "double" variable 'a' and whose second field is zero. This is done by invoking the above constructor, with 'a' serving as the first argument and no second argument. Thanks to the default values in the above code, the second argument implicitly takes the zero value, which is then assigned to the second field in the temporary variable returned by the "(point)" function. This is called explicit conversion.

The term "conversion" is somewhat confusing. Actually, 'a' is never converted or changed in any way. It is only used as an argument for the "(point)" function. In fact, one can write equivalently "point(a)" and obtain the same result: a temporary "point" object with first field equal to 'a' and second field zero. This object can be used only in the very code line in which it was constructed and disappears soon after. Although the term "conversion" is inaccurate, it is convenient and commonly used to refer to this function.

2.5 Implicit Conversion

The above constructor also provides implicit conversion from type "double" to type "point". In code where a variable of type "point" is expected and a variable of type "double" appears instead, the above constructor is invoked implicitly to convert the "double" variable into the required "point" variable. This feature may be particularly useful in functions that take "point" arguments. When such a function is called with a "double" argument, it is converted implicitly into the required "point" argument. This feature is analogous to type conversion in C. On one hand, it may make codes more transparent and straightforward; on the other hand, it may also be too expensive, as we'll see below.

Implicit conversion requires an extra call to the constructor. The memory allocation in this construction requires extra computer time to complete. Although this overhead may be negligible for small objects such as points, it may be significant for larger objects, particularly if the implicit conversion is repeated many times in long loops. One should thus consider carefully whether or not to use implicit conversion.

2.6 The Default Copy Constructor

The above constructor can be used to construct new "point" objects and initialize them with prescribed values. However, users of the "point" class should also be able to use existing objects to initialize new ones. For example, they would surely like to write code such as

```
point P(3.,5.);
point Q(P); //  or point Q=P;
```

where 'P' is first constructed as before, and then 'Q' is constructed and initialized to have the same value as 'P'. This is done by the copy constructor.

The copy constructor constructs (allocates memory for) a new object and initializes it with the value of the object passed to it as an argument. The construction and initialization are done in the same order in which the fields appear in the class block. In the above example, memory is allocated for the new fields "Q.x" and "Q.y", which are initialized with the values "P.x" and "P.y", respectively.

If no copy constructor is defined explicitly in the class block, then the construction is executed by the default copy constructor, available in the C++ compiler. This constructor just copies the data from the fields of the object passed to it as an argument to the corresponding fields in the constructed object. In the "point" class, this is exactly what we want, so there is actually no need to write an explicit copy constructor. Still, it is good practice to write an explicit copy constructor, because the default copy constructor may do the wrong thing. We'll return to this subject in Section 2.10.

The copy constructor is invoked every time an argument is passed to a function by value. In this case, a local copy of the concrete argument is constructed. Consider, for example, the following ordinary (noninterface) function, written outside the class block:

```
const point negative(const point p){
  return point(-p.X(),-p.Y());
}
```

This function returns the negative (minus) of a point in the Cartesian plane. However, its current implementation is rather expensive, because a constructor is used three times in each call to it. First, the concrete argument passed to the function is copied to a local variable 'p' by the copy constructor. Then, the constructor with two "double" arguments is used to create the negative of 'p'. Finally, the copy constructor is used once again to construct the object returned by the function. (Some compilers support a compilation option that avoids this third construction.) In Section 2.9, we'll show how the argument can be passed to the function by address rather than value, avoiding the first call to the copy constructor.

The "point" object returned by the "negative" function has no name and is stored only temporarily. It disappears soon after it is used in the code line in which the function is called. For this reason, it is a good idea to declare it as a constant, as is indeed done in the above code by putting the reserved word "const" before the definition of the function. This way, the returned variable cannot be changed by further calls to other functions. Usually, temporary returned objects have no need to change, because they disappear anyway at the end of the current code line. Declaring them as constants guarantees that they cannot be changed inadvertently.

When an object that is returned from a function is not declared constant, it can further change in the same code line in which it is created. For example, it can be used as a current object in interface functions. However, it cannot be passed by address to serve as an argument of any other function. The C++ compiler would refuse to create a local pointer that points to a temporary object, out of fear that it would change further in the function. The compiler would suspect that this wasn't the real intention of the programmer and would issue a compilation error.

For example, the temporary object "point(1.)", although nonconstant, cannot be passed by address to any function with a pointer-to-(nonconstant)-point argument. However, it can be used as a current object in interface functions such as "zero()":

```
Q = point(1.).zero();   // or Q=((point)1.).zero();
```

The "negative()" function can also be called with a "double" argument, e.g., "negative(1.)" or "negative(a)", where 'a' is a "double" variable. In this call, the "double" argument is first converted implicitly to a "point" object, which is then used as a concrete argument in the "negative" function.

2.7 Destructor

At the end of the block of a function, the local variables are destroyed, and the memory allocated for them is freed for future use. This is done by the destructor invoked implicitly by the computer. If no destructor is defined explicitly in the class block, then the default destructor available in the C++ compiler is used. This destructor goes over the data fields in the object and destroys them one by one. This is done in reverse order: "point" objects, for example, are destroyed by freeing their 'y'-field and then their 'x'-field.

The default destructor, however, does not always do a proper job. It is thus good practice to write an explicit destructor in the class block:

```
~point(){
}  //  destructor
```

Here, the actual destruction is done by the default destructor, which is invoked implicitly at the end of this destructor. This is why the body of this destructor is empty.

The default destructor, however, cannot properly destroy more complicated objects with data fields that are themselves pointers. Indeed, when the default destructor encounters such a field, it only destroys the address in it, not its content. The object in this address, although inaccessible because its address is no longer available, still occupies valuable memory. This is why an explicit destructor is required to delete this field properly using the reserved "delete" command. This command not only deletes the address in the field but also implicitly invokes the appropriate destructor to destroy the object in it and free the memory it occupies.

2.8 Member and Friend Functions

Interface functions may be of two possible kinds: member functions and friend functions. In what follows, we'll describe the features of these kinds of functions.

Constructors, destructors, and assignment operators must be member functions. The above "X()", "Y()", and "zero" functions are also member functions: they are defined inside the class block and act upon the current object with which they are called. For example, the call "P.X()" applies the function "X()" to the "point" object 'P' and returns its 'x'-field, "P.x".

Since member functions are defined inside the class block, their definitions can use (call) only other interface functions declared in the class block; they cannot use ordinary (noninterface) functions defined outside the class block unless they are declared friends of the class.

Member functions are called with a current object, e.g., 'P' in "P.X()". When a member function such as "X()" is executed, it is assumed that the fields 'x' and 'y' mentioned in its definition refer to the corresponding fields in the current object; 'x' is interpreted as "P.x", and 'y' is interpreted as "P.y".

Friend functions, on the other hand, have no current object and can only take arguments such as ordinary functions.

The most important property of member functions is that they have access to all the fields of the current object and objects passed to them as concrete arguments, including private fields. In what follows, we'll explain how this access is granted.

When the user calls "P.X()" for some "point" variable 'P', the address of 'P' is passed to the function "X()", which stores it in a local variable named "this" of type constant-pointer-to-constant-point. (The word "this" is reserved in C++ for this purpose.) The type of "this" guarantees that neither "this" nor its content may change. Indeed, "X()" is a constant function that never changes its current object, as is indicated by the reserved word "const" before the function block.

Now, the member function "X()" can access the private members of 'P', "P.x", and "P.y", through the address of 'P', contained in "this". In the definition of "X()", 'x' and 'y' are just short for "this->x" (or "(*this).x") and "this->y" (or "(*this).y"), respectively. In fact, the definition of "X()" can be rewritten with the command:

```
return this->x;   //  same as (*this).x
```

In nonconstant functions like "zero", "this" is of a slightly different type: it is constant-pointer-to-point, rather than constant-pointer-to-constant-point. This allows changes to the current object through it. Indeed, "zero" is a nonconstant function, which lacks the reserved word "const" before its block. When "P.zero()" is called by the user, "this->x" and "this->y" are set to zero, which actually means that "P.x" and "P.y" are set to zero, as required.

The "this" variable is also useful for returning a value. For example, if we want the function "zero" to return a pointer to the current "point" object with which it is called, then we should rewrite it as follows:

```
point* zero(){
  x=y=0.;
  return this;
}  //  returns pointer-to-current-point
```

This way, a temporary, unnamed variable of type pointer-to-point is created at the end of the block of the function and initialized to the value in "this". This unnamed variable can be further used in the same code line in which the "zero" function is called and can serve as an argument for another function.

Because the pointer returned from the above "zero" function exists only temporarily, it is not good practice to use it to change its content. Usually, contents should change only through permanent, well-defined pointers, not temporary, unnamed pointers returned from functions as output. A better style is, therefore, the following, in which the returned pointer is a pointer-to-constant-point, so it cannot be used to change its content further. This is indicated by the reserved word "const" before the type of the function:

```
const point* zero(){
  x=y=0.;
  return this;
}  //  returns pointer-to-constant-point
```

The pointer returned by the "zero" function can be used, e.g., in the "printf" function, as follows:

```
int main(){
  point P;
  printf("P.x=%f\n",P.zero()->X());
  return 0;
}  //  print P.x after P has been set to zero
```

Here, the function "printf" prints the 'x'-field of the "point" object 'P' whose address is returned by the "zero()" function.

Later on, we'll show how the "zero" function can also be rewritten as a "friend" function. The reserved word "friend" should then precede the function name in the declaration in the class block. No current object or "this" pointer is available; objects must be passed explicitly as arguments, as in ordinary functions. Next, we'll see that arguments should be passed not by name (value) but rather by reference.

2.9 References

In C++, one can define a reference to a variable. A reference is actually another name for the same variable. For example,

```
point p;
point& q = p;
```

defines a variable 'q' of type reference-to-point, initialized to refer to the "point" object 'p'. (Because 'q' is a reference, it must be initialized.) Every change to 'q' affects 'p' as well, and vice versa.

In the previous chapter, we saw that if a function is supposed to change a variable, then this variable must be passed to it by address, that is, by passing a pointer argument that points to it. A more transparent method is to pass a reference to this variable, which allows the function to refer to it and change it. For example, the above "zero" function can also be implemented as a friend function, which sets its argument to zero. This is done as follows. First, it should be declared as a friend in the block of the point class:

```
friend const point* zero(point&);
```

The actual definition can be made outside the block of the point class:

```
const point* zero(point&p){
  p.x=p.y=0.;
  return &p;
}  //  set "point" argument to zero
```

With this implementation, the function "zero" can be declared as a friend of other classes as well, so it can also use their private fields, if necessary. In the present example, this is not needed, so one can actually declare and define the function at the same time inside the block of the "point" class as follows:

```
friend const point* zero(point&p){
  p.x=p.y=0.;
  return &p;
} //  declare as friend and define
```

This way, the "zero" function can be called from "main()" as follows:

```
printf("P.x=%f\n",zero(P)->X());
```

In the above definition, the point argument is passed to the "zero" function by reference. Therefore, when the function is called, no local "point" variable is created; instead, a local variable of type reference-to-point that refers to the same object is created. Every change to this local reference in the function affects the concrete argument as well, as required.

References are not really created: the C++ compiler actually creates a copy of the address of the object and refers to it through this copy. This is why changes to the reference also affect the original object. It is just more convenient for programmers to refer to an object by reference than by address. Both ways are equivalent and give the same result; the syntax, however, is more transparent when references are used.

The syntax used in C++ in general, and in objects passed by reference to functions in particular, may look slightly complicated to newcomers, but soon becomes as easy and natural as ever.

The style of the "zero" function may further improve by returning a reference to its argument rather than a pointer to it. For this purpose, the definition should read

```
friend const point& zero(point&p){
  p.x=p.y=0.;
  return p;
}
```

The type of function is now "const point&" rather than "const point*", indicating that a reference is returned rather than a mere address. The function can then be used as follows:

```
printf("P.x=%f\n",zero(P).X());
```

Here, "zero(P)" returns a reference to 'P', which is further used in conjunction with the function "X()" to print the 'x'-field in 'P'.

Although it is possible to implement the "zero" function as a friend of the "point" class, it is not considered very elegant. Friend functions are usually used to read data from private fields of a class, and their advantages and disadvantages in doing this are discussed in Section 2.14 below. Friend functions are also useful for accessing private data in more than one class. However, when the private data in the class are not only read but also changed, it is more natural to use member functions. The above "zero" function can be written equivalently as a member function in the block of the "point" class as follows:

```
const point& zero(){
  x=y=0.;
  return *this;
}
```

This way, the "zero" member function is of type constant-reference-to-point, as is indeed indicated by the reserved words "const point&" before its name. The function returns a reference to the current object contained in the address "this". The returned reference can be further used in the same code line as follows:

```
printf("P.x=%f\n",P.zero().X());
```

Here, the reference to 'P' returned by "P.zero()" serves as the current object in a further call to the "X()" function.

Passing a reference to a function as an argument is attractive not only when there is a need to change the referenced object but also to reduce costs. Indeed, when the argument is passed by reference rather than by value, the need to invoke the copy constructor to construct a local copy is avoided. For example, if the "negative" function in Section 2.6 were rewritten as

```
const point negative(const point& p){
  return point(-p.X(),-p.Y());
} // passing argument by reference
```

then no local "point" object would be constructed, only a local reference to the concrete argument. Creating this reference requires only copying the address of the concrete argument rather than copying the entire concrete argument physically. The total number of constructions in the call to the "negative" function would then decrease from three to two.

The "negative" function still requires two calls to constructors of "point" objects: one to construct the local negative and the other to return a copy of it. This number cannot be reduced further. Look what happens if one attempts to avoid the second construction by writing

```
const point& negative(const point& p){
  return point(-p.X(),-p.Y());
} // wrong!!! returns reference to nothing
```

This version returns by reference rather than by value. (Indeed, the type of function is "const point&" rather than "const point".) It returns a reference to the local variable that contains the negative of the "point" argument. However, the negative of the "point" argument is a temporary local variable, which no longer exists at the end of the function, so it actually returns a reference to nothing. One should therefore drop this version and stick to the previous one.

2.10 Copy Constructor

As mentioned in Section 2.6, it is good practice to define an explicit copy constructor. For example, a suitable copy constructor can be written in the block of the "point" class as follows:

```
point(const point& p):x(p.x),y(p.y){
} // copy constructor
```

Here, the copied "point" object is passed to the constructor by reference, and its fields are used to initialize the corresponding fields in the constructed "point" object.

Actually, it is not necessary to write this constructor, because the default copy constructor available in the C++ compiler does exactly the same thing. Still, it is good practice to write your own copy constructor, because in many cases the default one does the wrong thing, as we'll see below.

The above copy constructor is invoked whenever the compiler encounters an explicit copying such as

```
point Q = P; // same as point Q(P);
```

or an implicit copying such as passing an argument to a function or returning an object from it by value.

2.11 Assignment Operators

Users of the "point" class may want to assign values in a natural way as follows:

```
point P,W,Q(1.,2.);
P=W=Q;
```

Here, the point objects 'P', 'W', and 'Q' are created in the first code line by the constructor in Section 2.3, which uses "double" arguments with default value 0. This way, the fields in 'Q' are initialized with the specified values 1 and 2, whereas the fields in 'P' and 'W' take the default value 0. In the second code line, the default assignment operator built into the C++ compiler is invoked to assign the value of fields in 'Q' to the corresponding fields in 'W' and 'P'. This is done from right to left as follows. First, the values of fields in 'Q' are assigned to the corresponding fields in 'W' one by one in the order in which they are declared in the class block. In other words, first "W.x" is assigned with "Q.x", and then "W.y" is assigned with "Q.y". This assignment operation also returns a reference to 'W'. This reference is used further to assign the updated 'W' object to 'P', so eventually all three point objects have the same value, as required.

Although the default assignment operator does the right thing here, it is good practice to define your own assignment operator. According to the rules of C++, it must be a member function. Here is how it is defined in the class block:

```
const point& operator=(const point& p){
```

This is the heading of a function, in which the type of argument is declared in parentheses and the type of returned object is declared before the function name, "operator=". Note that the argument is passed and the output is returned by reference rather than by value to avoid unnecessary calls to the copy constructor. Furthermore, the argument and the returned object are also declared as constants, so they cannot change inadvertently. Indeed, both the argument and the function name are preceded by the words "const point&", which stand for reference-to-constant-point.

Look what happens if the argument is declared nonconstant by dropping the word "const" from the parentheses. The compiler refuses to take any constant concrete argument, out of fear that it will change through its nonconstant local reference. Furthermore, the compiler refuses to take even a nonconstant concrete argument that is returned from some other function as a temporary object, out of fear that will change during the execution of the assignment operator. Because it makes no sense to change a temporary object that will disappear soon anyway, the compiler assumes that the call is mistaken and issues a compilation error. Declaring the argument as constant as in the above code line prevents all these problems. The function can now be called with either a constant or a nonconstant argument, as required.

The body of the function is now ready to start. The following "if" question is used to make sure that the compiler hasn't encountered a trivial assignment like "P = P". Once it is made clear that the assignment is nontrivial, it can proceed:

```
    if(this != &p){
      x = p.x;
      y = p.y;
    }
```

Finally, a reference to the current object is also returned for further use:

```
    return *this;
  } //  point-to-point assignment
```

We refer to this operator as a point-to-point assignment operator.

It is also possible to assign values of type "double" to "point" objects. For example, one can write

```
P=W=1.;
```

When this command is compiled, the constructor is first invoked to convert implicitly the "double" number "1." into a temporary unnamed point object with 'x'- and 'y'-fields containing the values 1 and 0, respectively. This object is then assigned to 'W' and 'P' as before. The "zero()" function of Section 2.2 is, thus, no longer necessary, because one can set 'P' to zero simply by writing "P = 0.".

As discussed in Section 2.5 above, implicit conversion may be rather expensive, as it requires the construction of an extra "point" object. This issue is of special importance when assignment is used many times in long loops. In order to avoid this extra construction, one may write an assignment operator that takes a "double" argument:

```
  const point& operator=(double xx){
    x = xx;
    y = 0.;
    return *this;
  } //  double-to-point assignment
```

We refer to this operator as a double-to-point assignment operator. It must also be a member function that is at least declared (or even defined) inside the class block.

When the compiler encounters a command of the form "P = 1.", it first looks for a double-to-point assignment operator. If such an operator exists, then it can be used here, avoiding implicit conversion. More specifically, the double-to-point assignment operator assigns 1 to "P.x" and 0 to "P.y" as required, avoiding any construction of a new object. A reference to the current object is also returned to allow commands of the form "W = P = 1.".

The assignment operator allows compact elegant code lines like "P = Q" and "P = 1.". Still, it can also be called as a regular function:

```
P.operator=(W.operator=(0.));  //  same as P=W=0.;
```

This is exactly the same as "P = W = 0." used above. The original form is, of course, preferable.

Below we'll see many more useful operators that can be written by the programmer of the class. These operators use the same symbols as standard operators in C, e.g., the '=' symbol in the above assignment operators. However, the operators written by the programmer are not necessarily related to the corresponding operators in C. The symbols used only reflect the standard priority order.

2.12 Operators

The programmer of the "point" class may also define other operators for convenient manipulation of objects. The symbols used to denote arithmetic and logical operators may be given new meaning in the context of the present class. The new interpretation of an operator is made clear in its definition.

Although the new operator may have a completely different meaning, it must still have the same structure as in C; that is, it must take the same number of arguments as in C. The type of these arguments and returned object, however, as well as what the operator actually does, is up to the programmer of the class. For example, the programmer may give the symbol "&&" the meaning of a vector product as follows:

```
double operator&&(const point&p, const point&q){
  return p.X() * q.Y() - p.Y() * q.X();
}  //  vector product
```

This way, although the "&&" operator in C has nothing to do with the vector product, it is suitable to serve as a vector-product operator in the context of "point" objects because it takes two arguments, as required. One should keep in mind, though, that the "&&" operator in C is a logical operator, with priority weaker than that of arithmetic operators. Therefore, whenever the "&&" symbol is used in the context of "point" objects, it must be put in parentheses if it should be activated first.

Note that the above operator is implemented as an ordinary (nonmember, nonfriend) function, because it needs no access to any private member of the "point" class. In fact, it accesses the data fields in 'p' and 'q' through the public member functions "X()" and "Y()". The arguments 'p' and 'q' are passed to it by reference to avoid unnecessary copying. These arguments are also declared as constant, so the function can take either constant or nonconstant concrete arguments. The user can now call this function simply by writing "P&&Q", where 'P' and 'Q' are some "point" variables.

2.13 Inverse Conversion

Another optional operator is inverse conversion. This operator is special, because its name is not a symbol but rather a reserved word that represents the type to which the object is converted.

In the context of the "point" class, this operator converts a "point" object to a "double" object. Exactly how this is done is up to the programmer of the class. However, here the programmer has no freedom to choose the status of the function or its name: it must be a member function with the same name as the type to which the object is converted, that is, "double". Here is how this operator can be defined in the block of the "point" class:

```
operator double() const{
  return x;
} //  inverse conversion
```

With this operator, users can write "(double)P" or "double(P)" to read the 'x'-coordinate of a "point" object 'P'. Of course, 'P' never changes in this operation, as is indeed indicated by the reserved word "const" before the '{' character that opens the function block. The term "conversion" is inexact and is used only to visualize the process. Actually, nothing is converted; the only thing that happens is that the first coordinate is read, with absolutely no change to the current object.

Inverse conversion can also be invoked implicitly. If the "point" object 'P' is passed as a concrete argument to a function that takes a "double" argument, then the compiler invokes the above operator implicitly to convert 'P' into the required "double" argument with the value "P.x".

Implicit calls to the inverse-conversion operator are also risky. The programmer is not always aware of them and or able to decide whether or not they should be used. It seems to be better practice not to define inverse conversion and to let the compiler issue an error whenever an argument of the wrong type is passed to a function. This way, the programmer becomes aware of the problem and can decide whether to convert the argument explicitly. In the present classes, we indeed define no inverse conversion.

2.14 Unary Operators

The "negative" function in Section 2.6 can actually be implemented as an operator that takes one argument only (unary operator). For this purpose, one only needs to change the function name from "negative" to "operator-". With this new name, the function can be called more elegantly by simply writing "-Q" for some "point" object 'Q':

```
point W,Q=1.;
W=-Q;  //  same as W=operator-(Q);
```

In this code, the "point" object 'W' is assigned the value $(-1, 0)$.

The "operator-" may also be more efficient than the original "negative" function. For example, the code

```
point W=negative(1.);
```

uses implicit conversion to convert the "double" argument 1 into the "point" argument (1, 0) before applying the "negative" function to it. Once the "negative" function has been renamed "operator-", the above code is rewritten as

```
point W=-1.;
```

which interprets -1 as a "double" number and uses it to initialize 'W' with no conversion whatsoever.

One may define other optional operators on "point" objects. In the definition, the reserved word "operator" in the function name is followed by the symbol that should be used to call the operator.

For example, we show below how the "+=" operator can be defined as a member function of the "point" class. The definition allows users to write the elegant "P+=Q" to add the "point" argument 'Q' to the current "point" object 'P':

```
const point& operator+=(const point& p){
  x += p.x;
  y += p.y;
  return *this;
} //  adding a point to the current point
```

Here, the "point" argument is passed to the "+=" operator by reference to avoid unnecessary copying. Then, the values of its 'x'- and 'y'-coordinates are added to the corresponding fields in the current object stored in "this". The updated current object is also returned by reference as output. This output can be further used in the same code line as follows:

```
P=W+=Q;
```

This code line is executed right to left: first, 'W' is incremented by 'Q', and the resulting value of 'W' is then assigned to 'P'. This assignment can take place thanks to the fact that the point-to-point assignment operator in Section 2.11 takes a reference-to-*constant*-point argument, so there is no fear that it will change the temporary object returned from "W+=Q", which is passed to it as a concrete argument.

The above code is actually equivalent to the following (less elegant) code:

```
P.operator=(W.operator+=(Q));  //  same as P=W+=Q;
```

Like the assignment operator in Section 2.11, the "+=" operator also accepts a "double" argument through implicit conversion. For example, the call

```
W+=1.;
```

first implicitly converts the "double" argument 1 to the temporary "point" object (1, 0), which in turn is used to increment 'W'. As discussed above, implicit conversion can be used here only thanks to the fact that "operator+=" takes a reference-to-*constant*-point argument rather than a mere reference-to-point, so it has no problem accepting as argument the temporary "point" object returned from the implicit conversion.

The above implicit conversion is used only if there is no explicit version of the "+=" operator that takes the "double" argument. If such a version also exists, then the compiler will invoke it, because it matches the type of arguments in the call "W+=1.". This version of "operator+=" can be defined in the class block as follows:

```
const point& operator+=(double xx){
  x += xx;
  return *this;
} //  add real number to the current point
```

This version increases the efficiency by avoiding implicit conversion. This property is particularly important in complex applications, where it may be used many times in long loops.

The above "+=" operators are implemented as member functions, which is the natural way to do it. However, they can also be implemented as friend functions as follows:

```
friend const point&
operator+=(point&P,const point& p){
  P.x += p.x;
  P.y += p.y;
  return P;
}
```

In this style, the function takes two arguments. The first one is nonconstant, because it represents the object that is incremented by the second, constant, argument. Similarly, a "friend" version can also be written for the "+=" operator with a "double" argument. The call to the "+=" operators is done in the same way as before.

The "friend" implementation, although correct, is somewhat unnatural in the context of object-oriented programming. Indeed, it has the format of a C function that changes its argument. In object-oriented programming, however, we think in terms of objects that have functions to express their features, rather than functions that act upon objects. This concept is better expressed in the original implementation of the "+=" operators as member functions.

The "friend" version has another drawback. Although it correctly increments well-defined "point" variables, it refuses to increment temporary, unnamed "point" objects that have been returned from some other function. Indeed, since the incremented argument in the "friend" version is of type reference-to-(nonconstant)-point, it rejects any temporary concrete argument because it sees no sense in changing an object that is going to vanish soon and assumes that this must be a human error. Of course, the compiler might not know that this was intentional. In the original, "member", version of the "+=" operator, on the other hand, even temporary objects can be incremented, because they serve as a nonconstant current object. This greater flexibility of the "member" implementation is also helpful in the implementation of the '+' operator below.

2.15 Binary Operators

In this section, we define binary operators that take two arguments to produce the returned object. In this respect, these operators have the same structure as ordinary C functions. The difference is, however, that they use objects rather than just integer and real numbers as in C. Furthermore, the operators can be called conveniently and produce code that imitates the original mathematical formula.

The '+' operator that adds two "point" objects can be written as an ordinary (nonmember, nonfriend) function that requires no access to private data fields of point objects and hence needs no declaration in the class block:

```
const point
operator+(const point& p, const point& q){
  return point(p.X()+q.X(),p.Y()+q.Y());
} // add two points
```

Unlike the "+=" operator, this operator doesn't change its arguments, which are both passed as reference-to-constant-point, but merely uses them to produce and return their sum. Note that the returned variable cannot be declared as a reference, or it would refer to a local "point" object that vanishes at the end of the function. It must be a new "point" object constructed automatically in the "return" command by the copy constructor to store a copy of that local variable. This is indicated by the words "const point" (rather than "const point&") that precede the function name.

The above '+' operator can be called most naturally as

```
P=W+Q; // the same as P=operator+(W,Q);
```

As mentioned at the end of Section 2.13, we assume that no inverse conversion is available, because the "operator double()" that converts "point" to "double" is dropped. Therefore, since both arguments in "operator+" are of type reference-to-constant-point, "operator+" can be called not only with two "point" arguments (like "W+Q") but also with one "point" argument and one "double" argument (like "Q+1." or "1.+Q"). Indeed, thanks to the implicit double-to-point conversion in Section 2.5, the "double" number "1." is converted to the point $(1, 0)$ before being added to 'W'. Furthermore, thanks to the lack of inverse conversion, there is no ambiguity, because it is impossible to convert 'W' to "double" and add it to "1." as "double" numbers.

If one is not interested in implicit conversion because of its extra cost and risks and wants the compiler to announce an error whenever it encounters an attempt to add a "double" number to a "point" object, then one can drop conversion altogether by not specifying default values for the "double" arguments in the constructor in Section 2.3. This way, the constructor expects two "double" arguments rather than one and will not convert a single "double" number to a "point" object.

In the above implementation, the '+' operator is defined as an ordinary function outside the block of the "point" class. This way, however, it is unavailable in the class block, unless it is declared there explicitly as a "friend":

```
friend const point operator+(const point&,const point&);
```

With this declaration, the '+' operator can also be called from inside the class block. Furthermore, it has access to the data fields of its "point" arguments. In fact, it can be defined inside the class block as follows:

```
friend const point operator+(
        const point& p, const point& q){
  return point(p.x+q.x,p.y+q.y);
} // defined as "friend" in the class block
```

The '+' operator can also be implemented as a member function inside the class block as follows:

```
const point operator+(const point& p) const{
  return point(x+p.x,y+p.y);
}  //  defined as "member" in the class block
```

With this implementation, the '+' operator is still called in the same way (e.g., "W+Q"). The first argument ('W') serves as the current object in the above code, whereas the second argument ('Q') is the concrete argument passed by reference to the above function.

This, however, is a rather nonsymmetric implementation. Indeed, implicit conversion can take place only for the second argument, which is a reference-to-constant-point, but not for the first argument, the current object. Therefore, mixed calls such as "W+1." are allowed, but not "1.+W". This nonsymmetry makes no apparent sense.

The original implementation of "operator+" as an ordinary function outside the class block is more in the spirit of object-oriented programming. Indeed, it avoids direct access to the private data fields 'x' and 'y' in "point" objects and uses only the public member functions "X()" and "Y()" to read them. This way, the '+' operator is independent of the internal implementation of "point" objects.

One could also write two more versions of "operator+" to add a "double" number and a "point" object explicitly. These versions increase efficiency by avoiding the implicit conversion used above. They are also implemented as ordinary functions outside the class block:

```
const point operator+(const point& p, double xx){
  return point(p.X()+xx,p.Y());
}  //  point plus real number

const point operator+(double xx, const point& p){
  return point(p.X()+xx,p.Y());
}  //  real number plus point
```

These versions are invoked by the compiler whenever a call such as "W+1." or "1.+W" is encountered, avoiding implicit conversion.

The original implementation of the '+' operator as an ordinary function can also be written in a more elegant way, using the "operator+=" member function defined in Section 2.14:

```
const point
operator+(const point& p, const point& q){
  return point(p) += q;
}  //  point plus point
```

Thanks to the fact that the "+=" operator is defined in Section 2.14 as a member (rather than a mere friend) of the "point" class, it accepts even temporary "point" objects as concrete arguments. In particular, even the first argument (the current object that is incremented in the "+=" operator) may be a temporary object. This property is used in the above code, where the temporary object "point(p)" returned from the copy constructor is incremented by the "+=" operator before being returned as the output of the entire '+' operator. This elegant style of programming will be used in what follows.

2.16 Example: Complex Numbers

In Fortran, the "complex" type is built in and available along with the required arithmetic operations. The programmer can define a complex variable 'c' by writing simply "complex c" and apply arithmetic operations and some other elementary functions to it.

The C and C++ compilers, on the other hand, don't support the "complex" type. If one wants to define and use complex objects, then one must first define the "complex" class that implements this type. Naturally, the block of the class should contain two "double" fields to store the real and imaginary parts of the complex number and some member operators to implement elementary arithmetic operations. Once the implementation is complete, one can define a "complex" object 'c' simply by writing "complex c" and then apply arithmetic operations to it as if it were built into the programming language.

Some standard C libraries do support complex numbers. Still, the "complex" object implemented here is a good example of object-oriented programming in C++.

This example illustrates clearly how the object-oriented approach works: it provides new objects that can then be used as if they were built into the programming language. These new objects can then be viewed as an integral part of the programming language and can add new dimensions and possibilities to it. The programming language develops dynamically by adding more and more objects at higher and higher levels of programming.

Here is the detailed implementation of the "complex" class:

```
#include<stdio.h>
class complex{
    double real;    //  the real part
    double image;   //  the imaginary part
  public:
    complex(double r=0.,double i=0.):real(r), image(i){
    } //  constructor

    complex(const complex&c):real(c.real),image(c.image){
    }  //  copy constructor

    ~complex(){
    }  //  destructor
```

In the above constructors, all the work is done in the initialization lists, where memory is allocated for the data fields "real" and "image" with the right values. The bodies of these functions remain, therefore, empty. The destructor above also needs to do nothing, because the default destructor called implicitly at the end of it destroys the data fields automatically.

Because the data fields "real" and "image" are declared before the reserved word "public:", they are private members of the class. Thus, only members and friends of the class can access and change data fields of an object in the class. Still, we'd like other users to be able to read these fields from ordinary functions that are neither members nor friends of the class. For this purpose, we define the following two public member functions that can only read (but not change) the data fields:

```
double re() const{
  return real;
}  //  read real part

double im() const{
  return image;
} //  read imaginary part
```

Next, we define the assignment operator. This operator will enable users to assign the value of a "complex" object 'd' to a "complex" object 'c' simply by writing "c = d":

```
const complex&operator=(const complex&c){
  real = c.real;
  image = c.image;
  return *this;
}  //  assignment operator
```

Next, we define some member arithmetic operators that change the current "complex" object. For example, the "+=" operator allows users to write "c += d" to add 'd' to 'c':

```
const complex&operator+=(const complex&c){
  real += c.real;
  image += c.image;
  return *this;
}  //  add complex to the current complex

const complex&operator-=(const complex&c){
  real -= c.real;
  image -= c.image;
  return *this;
}  //  subtract complex from the current complex

const complex&operator*=(const complex&c){
  double keepreal = real;
  real = real*c.real-image*c.image;
  image = keepreal*c.image+image*c.real;
  return *this;
}  //  multiply the current complex by a complex

const complex&operator/=(double d){
  real /= d;
  image /= d;
  return *this;
}  //  divide the current complex by a real number
```

In the latter function, the current complex number is divided by a real number. This operator will be used later in the more general version of the "/=" operator that divides the current complex number by another complex number.

In the end, we'll have two "/=" operators: one that takes a real argument and one that takes a complex argument. When the C++ compiler encounters a command of the form "c /= d", it invokes the first version if 'd' is real and the second if 'd' is complex.

The division of a complex number by a complex number will be implemented later in another version of "operator/=". As a member function, this function will not be able to recognize any ordinary function that is defined outside the class block and is neither a member nor a friend of the class. This is why the two following functions are defined as friends of the "complex" class: they have to be called from the "operator/=" member function defined afterward.

The first of these two functions returns the complex conjugate of a complex number. The name of this function, "operator!", has nothing to do with the "logical not" operator used in C. In fact, this name is chosen here only because it represents a unary operator, which can later be called as "!c" to return the complex conjugate of 'c'.

Note that the returned object must be complex rather than reference-to-complex ("complex&"), or it would refer to a local variable that disappears at the end of the function. The word "complex" before the function name in the above code indicates that the local variable is copied to a temporary unnamed variable to store the returned value. Since this variable is not declared "constant", it can be further changed in the "/=" operator below:

```
    friend complex operator!(const complex&c){
      return complex(c.re(),-c.im());
    }  //  conjugate of a complex
```

The second function defined below, "abs2()", returns the square of the absolute value of a complex number:

```
    friend double abs2(const complex&c){
      return c.re() * c.re() + c.im() * c.im();
    }  //  square of the absolute value of a complex
```

Because they are declared as friends, these two functions can now be called from the "operator/=" member function that divides the current "complex" object by another one:

```
    const complex&operator/=(const complex&c){
      return *this *= (!c) /= abs2(c);
    }  //  divide the current complex by a complex
};
```

In this "/=" operator, the current complex number is divided by the complex argument 'c' by multiplying it by the complex conjugate of 'c' (returned from "operator!") divided by the square of the absolute value of 'c' (returned from the "abs2()" function). Because its argument is real, this division is carried out by the early version of the "/=" operator with a real argument. Thanks to the fact that this operator is a member function, it can change its nonconstant unnamed current object "!c" and divide it by "abs2(c)". The result is used to multiply the current complex number, which is equivalent to dividing it by 'c', as required. This completes the block of the "complex" class.

The following functions are ordinary noninterface (nonmember, nonfriend) functions that implement basic operations on complex numbers. Note that there are two different '-' operators: a binary one for subtraction and a unary one for returning the negative of a complex number. When the C++ compiler encounters the '-' symbol in the program, it invokes the version that suits the number of arguments: if there are two arguments, then the binary subtraction operator is invoked, whereas if there is only one argument, then the unary negative operator is invoked:

```
const complex
operator-(const complex&c){
  return complex(-c.re(),-c.im());
}  //  negative of a complex number

const complex
operator-(const complex&c,const complex&d){
  return complex(c.re()-d.re(),c.im()-d.im());
}  //  subtraction of two complex numbers
```

Here are more binary operators:

```
const complex
operator+(const complex&c,const complex&d){
  return complex(c.re()+d.re(),c.im()+d.im());
}  //  addition of two complex numbers

const complex
operator*(const complex&c,const complex&d){
  return complex(c) *= d;
}  //  multiplication of two complex numbers

const complex
operator/(const complex&c,const complex&d){
  return complex(c) /= d;
}  //  division of two complex numbers
```

In the above functions, the returned object cannot be of type reference-to-complex ("const complex&"), or it would refer to a local variable that no longer exists. It must be of type "const complex", which means that the local variable is copied by the copy constructor to a temporary "complex" object that is constructed at the end of the function and also exists after it terminates.

This concludes the arithmetic operations with complex numbers. Finally, we define a function that prints a complex number:

```
void print(const complex&c){
  printf("(%f,%f)\n",c.re(),c.im());
}  //  printing a complex number
```

Here is how complex numbers are actually used in a program:

```
int main(){
  complex c=1.,d(3.,4.);
  print(c-d);
  print(c/d);
  return 0;
}
```

2.17 Templates

Above, we have implemented the "point" object and the "zero()" function that sets its value to zero. Now, suppose that we need to implement not only points in the two-dimensional Cartesian plane but also points in the three-dimensional Cartesian space. One possible implementation is as follows:

```
class point3d{
    double x;
    double y;
    double z;
  public:
    void zero(){ x=y=z=0.; }
};
```

This implementation, however, is neither elegant nor efficient in terms of human resources, because functions that have already been written and debugged in the "point" class (such as arithmetic operators) will now be written and debugged again in the "point3d" class. A much better approach is to use templates.

A template can be viewed as an object with a parameter that has not yet been specified. This parameter must be specified in compilation time, so that functions that use the object can compile. When the compiler encounters a call to a function that uses this object, it interprets it with the specified parameter and compiles it like regular objects. The compiled function is actually a finite state machine that can be further used in every future call to the function with the same parameter.

The "point" class can actually be defined as a template class with an integer parameter 'N' to determine the dimension of the space under consideration. This way, the class is defined only once for 'N'-dimensional vectors; "point" and "point3d" are obtained automatically as special cases by specifying 'N' to be 2 or 3.

The template class is called "point<N>" instead of "point", where 'N' stands for the dimension. The reserved words "template<int N>" that precede the class block indicate that this is indeed a template class that depends on a yet unspecified parameter 'N':

```
#include<stdio.h>
template<int N> class point{
    double coordinate[N];
  public:
    point(const point&);
```

The copy constructor that is declared here will be defined explicitly later.

The default destructor provided by the C++ compiler is insufficient here. Indeed, this destructor only destroys the data fields in the object to be destroyed. Since the only data field in the "point" object is the array (pointer-to-double) "coordinate", only the address of the coordinates is destroyed and not the actual "double" variables in the array. In the destructor implemented below, on the other hand, the array is deleted by the "delete" command, which automatically invokes the destructor of the "double" variable to destroy every individual coordinate and free the memory occupied by it:

```
      ~point(){
        delete [] coordinate;
      }  //  destructor
```

Because the "coordinate" field appears before the reserved word "public:", it is by default a private member of the "point" class, with no access to anyone but members and friends of the class. Still, we want users to be able to read (although not change) the 'i'th coordinate in a "point" object from ordinary functions written outside the class block. For this purpose, we define the following public member operator:

```
      double operator[](int i) const{
        return coordinate[i];
      }  //  read ith coordinate
  };
```

This completes the block of the "point" template class.

The above "operator[]" allows users to read the 'i'th coordinate in a "point" object 'P' simply as "P[i]". This is a read-only implementation; that is, the 'i'th coordinate can be read but not changed. This property is obtained by returning a copy of the 'i'th coordinate rather than a reference to it, as is indicated by the word "double" (rather than "double&") that precedes the function name.

Another possible strategy to implement the read-only operator is to define the returned variable as a constant reference to double:

```
      const double& operator[](int i) const{
        return coordinate[i];
      }  //  read-only ith coordinate
```

This approach is preferable if the coordinates are themselves big objects that are expensive to copy; here, however, since they are only "double" objects, it makes no difference whether they are returned by value or by constant reference.

The "operator[]" function can also be implemented as a “read-write” operator as follows:

```
      double& operator[](int i){
        return coordinate[i];
      }  //  read/write ith coordinate (risky)
```

This version returns a nonconstant reference to the 'i'th coordinate, which can be further changed even from ordinary functions that are neither members nor friends of the "point" class. With this version, a call of the form "P[i]" can be made only if the "point" object 'P' is nonconstant; otherwise, its 'i'th coordinate is constant and cannot be referred to as nonconstant, for fear that it will be changed through it. For this reason, the reserved word "const" before the '{' character that opens the function block is missing: this indicates that the current object can be changed by the function and, therefore, cannot be constant.

The latter version, however, is somewhat risky, because the coordinates can be changed inadvertently, which will spoil the original "point" object. This version must, therefore, be used with caution and only when necessary.

It is also possible to define "operator()" rather than "operator[]" by using parentheses. The 'i'th coordinate in the "point" object 'P' is then read as "P(i)" rather than "P[i]". Here, there is no real difference between the two styles. In other cases, however, "operator()" may be more useful because it may take any number of arguments, whereas "operator[]" must take exactly one argument.

The only thing left to do in the "point" template class is to define explicitly the copy constructor declared above. This task is discussed and completed below.

The "point" template class must have a copy constructor, because the default copy constructor provided by the compiler does the wrong thing. Indeed, this copy constructor just copies every data field from the copied object to the constructed object. Since the field in the "point<>" template class is an array, it contains the address of a "double" variable. As a result, the default copy constructor just copies this address to the corresponding field in the constructed object, and no new array is created, and both objects have the same array in their data field. The result is that the constructed object is not really a new object but merely a reference to the old one, which is definitely not what is required. The copy constructor in the code below, on the other hand, really creates a new object with the same data as in the old one, as required.

The words "template<int N>" before the definition indicate that this is a template function. The prefix "point<N>::" before the function name indicates that this is a definition of a member function:

```
template<int N> point<N>::point(const point&P){
    for(int i = 0; i < N; i++)
      coordinate[i] = P.coordinate[i];
} //  copy constructor
```

The default constructor available in the C++ compiler is used here to construct new "point" objects. In fact, it allocates memory for an array of length 'N' with components of type "double". This is why 'N' must be known in compilation time. The components in this array are not yet assigned values and are initialized with meaningless, random values.

This constructor is also invoked automatically at the start of the above copy constructor, since no initialization list is available in it. The correct values are then copied from the coordinates in the copied object to the corresponding coordinates in the constructed object.

Here is how the "point<N>" template class is actually used in a program:

```
int main(){
  point<2> P2;
```

```
    point<3> P3;
    printf("P2=(%f,%f)\n",P2[0],P2[1]);
    return 0;
}
```

When a concrete "point<N>" object is created, the parameter 'N' must be specified numerically and used in every function that uses the object as a current object or a concrete argument. For example, "P2" and "P3" in the above code are constructed as points in the two-dimensional and three-dimensional spaces, respectively, and the functions that are applied to them also use 'N' = 2 and 'N' = 3, respectively.

The above template class uses only a single parameter, the integer 'N'. More advanced template classes may use several parameters of different types. In particular, a parameter may specify not only the value but also the class that is used within the template class. For example, the above "point" template class can use a coordinate of type 'T', where 'T' is a parameter that should be specified in compilation time as integer, double, complex, or any other type or class. This gives the user greater freedom in choosing the type of coordinate in the "point" object.

When the template class is defined, the type 'T' is not yet specified. It is only specified at the call to the constructor of the class. The template class "point<T,N>" that implements 'N'-dimensional points with coordinates of type 'T' is written similarly to the "point<N>" class above, except that "<int N>" is replaced by "<class T, int N>"; "<N>" is replaced by "<T,N>"; and "double" is replaced by 'T':

```
template<class T, int N> class point{
    T coordinate[N];
};

int main(){
    point<double,2> P2;
    return 0;
}
```

In the next section we provide a more complete version of this template class with many useful functions, from which both two-dimensional and three-dimensional points can be obtained as special cases.

2.18 Example: The Vector Object

In this section, we present the "vector<T,N>" template class that implements an N-dimensional vector space, in which each vector has N components of type 'T'. The arithmetic operators implemented in this class provide a useful and convenient framework to handle vectors. The two-dimensional and three-dimensional point objects are also obtained as a special case.

Most operators and functions, except very short ones, are only declared inside the class block below, and their actual definition is placed outside it later on. The prefix "vector<T,N>::" that precede a function name indicates that a member function is defined. The words "template<class T, int N>" that precede the definition of a function indicate that this is a template function that uses the as yet unspecified type 'T' and integer 'N':

```
#include<stdio.h>
template<class T, int N> class vector{
    T component[N];
  public:
    vector(const T&);
    vector(const vector&);
    const vector& operator=(const vector&);
    const vector& operator=(const T&);
```

So far, we have only declared the constructor, copy constructor, and assignment operators with scalar and vector arguments. The actual definitions will be provided later. Next, we define the destructor. Actually, the destructor contains no command. The actual destruction is done by the default destructor invoked at the end of the function:

```
    ~vector(){
    } //  destructor
```

Because the components in the vector are private class members, we need public functions to access them. These functions can then be called even from nonmember and nonfriend functions:

```
    const T& operator[](int i) const{
      return component[i];
    } //read ith component

    void set(int i,const T& a){
      component[i] = a;
    } //  change ith component
```

Here we declare more member arithmetic operators, to be defined later:

```
    const vector& operator+=(const vector&);
    const vector& operator-=(const vector&);
    const vector& operator*=(const T&);
    const vector& operator/=(const T&);
};
```

This concludes the block of the "vector" class. Now, we define the member functions that were only declared in the class block: the constructor, copy constructor, and assignment operators with vector and scalar arguments:

```
template<class T, int N>
vector<T,N>::vector(const T& a = 0){
   for(int i = 0; i < N; i++)
     component[i] = a;
} //  constructor
```

```
template<class T, int N>
vector<T,N>::vector(const vector<T,N>& v){
  for(int i = 0; i < N; i++)
    component[i] = v.component[i];
}  //  copy constructor

template<class T, int N>
const vector<T,N>& vector<T,N>::operator=(
          const vector<T,N>& v){
  if(this != &v)
    for(int i = 0; i < N; i++)
      component[i] = v.component[i];
  return *this;
}  //  assignment operator

template<class T, int N>
const vector<T,N>& vector<T,N>::operator=(const T& a){
  for(int i = 0; i < N; i++)
    component[i] = a;
  return *this;
}  //  assignment operator with a scalar argument
```

Next, we define some useful arithmetic operators:

```
template<class T, int N>
const vector<T,N>&
vector<T,N>::operator+=(const vector<T,N>&v){
    for(int i = 0; i < N; i++)
      component[i] += v[i];
    return *this;
}  //  adding a vector to the current vector

template<class T, int N>
const vector<T,N>
operator+(const vector<T,N>&u, const vector<T,N>&v){
  return vector<T,N>(u) += v;
}  //  vector plus vector
```

The following are unary operators that act upon a vector as in common mathematical formulas: $+v$ is the same as v, and $-v$ is the negative of v.

```
template<class T, int N>
const vector<T,N>&
operator+(const vector<T,N>&u){
  return u;
}  //  positive of a vector
```

```
template<class T, int N>
const vector<T,N>
operator-(const vector<T,N>&u){
  return vector<T,N>(u) *= -1;
} // negative of a vector
```

The following functions return the inner product of two vectors and the sum of squares of a vector:

```
template<class T, int N>
const T
operator*(const vector<T,N>&u, const vector<T,N>&v){
    T sum = 0;
    for(int i = 0; i < N; i++)
      sum += u[i] * +v[i];
    return sum;
} // vector times vector (inner product)

template<class T, int N>
T squaredNorm(const vector<T,N>&u){
    return u*u;
} // sum of squares
```

Finally, here is a function that prints a vector to the screen:

```
template<class T, int N>
void print(const vector<T,N>&v){
  printf("(");
  for(int i = 0;i < N; i++){
    printf("v[%d]=",i);
    print(v[i]);
  }
  printf(")\n");
} // printing a vector
```

The template class "vector<T,N>" and its functions are now complete. The two-dimensional and three-dimensional point classes can be obtained from it as special cases:

```
typedef vector<double,2> point;
typedef vector<double,3> point3d;
```

The "typedef" command gives a short and convenient name to a type with a long and complicated name. This way, "point" is short for a two-dimensional vector, and "point3d" is short for a three-dimensional vector. In what follows, we show how the "point" class can be further used to derive an alternative implementation for the "complex" class.

2.19 Inheritance

In Section 2.16, we defined the "complex" class that implements complex numbers. The data hidden in a "complex" object are two "double" fields to store its real and imaginary parts. In this section we introduce a slightly different implementation, using the geometric interpretation of complex numbers as points in the two-dimensional Cartesian plane, with the x-coordinate representing the real part and the y-coordinate representing the imaginary part. For example, one may write

```
class complex{
      point p;
    public:
      complex(const point&P):p(P){}

      complex(const complex&c):p(c.p){}

      const complex&operator=(const complex&c){
        p=c.p;
        return *this;
      }

      const complex&operator+=(const complex&c){
        p+=c.p;
        return *this;
      }
};
```

and so on. This implementation uses the "has a" approach: the "complex" object has a field "point" to contain the data. In this approach, one must explicitly write the functions required in the "complex" class.

The above approach is not quite natural. Mathematically, the complex number doesn't "have" any point, but rather "is" a point in the two-dimensional Cartesian plane. This leads us to the "is a" approach available in C++. In this approach, the "complex" object is actually a "point" object with some extra features or functions. This approach is more in the spirit of object-oriented programming in this case, because it allows one to implement complex numbers precisely as they are in mathematical terms: points in the Cartesian plane with some extra algebraic features.

The "is a" concept in C++ is implemented by inheritance or derivation. The new class is derived from the "base" class and inherits its properties (see Figure 2.1). In the present example, the "complex" class is derived from the "point" class and inherits its properties as a point in the two-dimensional Cartesian plane (see Figure 2.2). On top of these properties, more arithmetic operations that are special to complex numbers can be defined in the derived "complex" class. These algebraic operations (multiplication and division of two complex numbers) complete the Cartesian plane from a mere vector space into a complete mathematical field.

The definition of the derived class (derivation) is similar to the standard class definition, except that the name of the derived class is followed by the character ':' followed by the

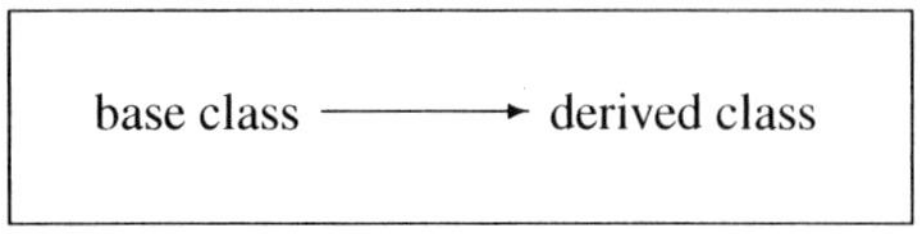

Figure 2.1. *Schematic representation of inheritance.*

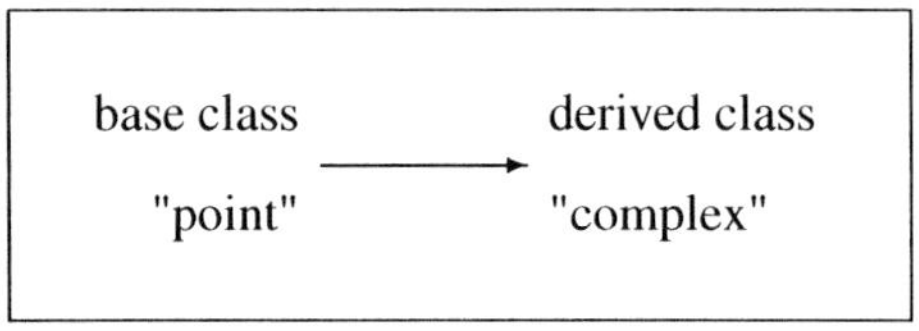

Figure 2.2. *Schematic representation of inheritance from the base class "point" to the derived class "complex".*

reserved word "public" and the name of the base class from which it is derived. All these words precede the '{' character that opens the block of the derived class:

```
class complex : public point{
```

The word "public" before the name of the base class is optional. However, without it the derivation is private in the sense that the nonpublic functions in the base class are unavailable from objects or functions of the derived class. In other words, the derived class has the same access rights as ordinary classes. This way, the users are completely "unaware" of the derivation in the sense that they are unable to use objects from the derived class in functions written in terms of the base class. This is definitely not what we want here: in fact, we definitely want users of the "complex" class to be able to add and subtract complex numbers as if they were mere "point" objects. Therefore, we use here public derivation by writing the word "public" before the name of the base class as in the above code line.

Next, we implement the constructors in the body of the derived "complex" class:

```
public:
  complex(const point&p){
    set(0,p[0]);
    set(1,p[1]);
  } //  constructor with "point" argument

  complex(double re=0., double im=0.){
    set(0,re);
    set(1,im);
  } //  constructor with "double" arguments
```

When a "complex" object is constructed, the underlying "point" object is first constructed by its own default constructor, which sets the values of its fields to 0. These values are then reset to their correct values obtained from the argument. This resetting must use the public "set" function in the "vector" template class in Section 2.18. It is impossible to change the components of a vector object directly because they are declared as private in the "vector" class and are thus inaccessible not only by users but also by derivers.

Next, we define a friend function that calls the constructor to return the complex conjugate:

```
    friend complex operator+(const complex&c){
      return complex(c[0], -c[1]);
    } // complex conjugate
```

Next, we declare member arithmetic operations that do not exist in the base "point" class or should be rewritten. The actual definition will be given later:

```
    const complex&operator+=(double);
    const complex&operator-=(double);
    const complex&operator*=(const complex&);
    const complex&operator/=(const complex&);
};
```

This completes the block of the derived "complex" class.

The derived class has no access to the private fields of the base class. However, it has access to “half private” fields: fields that are declared as "protected" in the block of the base class by simply writing the reserved word "protected:" before their names. These fields are accessible by derivers only, not by other users. In fact, if the "component" field is declared "protected" in the base "vector" class, then it can be accessed from the derived "complex" class and set in its constructors directly. With the present implementation of the "vector" class in Section 2.18, however, the "component" field is private, so the constructors of "complex" objects must access it indirectly through the public "set" function as above.

In summary, the members of a class are of three possible kinds: (a) public members that can be used by everyone; (b) private members that are accessible only to members and friends; and (c) protected members that are accessible to members and friends of derived classes (or when called in conjunction with current objects that have been derived by public derivation), but not to ordinary functions, even if they take arguments from the class or any other class derived from it (see Figure 2.3).

When the derived object is constructed, the data fields that belong to the underlying base class are constructed first by the default constructor of the base class. Thus, the constructors in the derived class cannot use an initialization list for these fields, as they are already initialized to their default values. If these data fields are declared protected in the base class, then they can be reset in the body of the constructors of the derived class. If, on the other hand, they are private members of the base class, then they can be reset only indirectly through public member functions like "set" in the above example. In this example, the "set" function is applied to the "complex" object that is currently constructed. Since no such function is available in the derived "complex" class, the "complex" object is interpreted as a "point" object, and the "set" function of Section 2.18 is invoked to reset the data fields that represent the real and imaginary parts of the complex number to their correct values.

Similarly, when a "complex" object is passed as a concrete argument to a function, the compiler first looks for this function among the functions that take a "complex" argument.

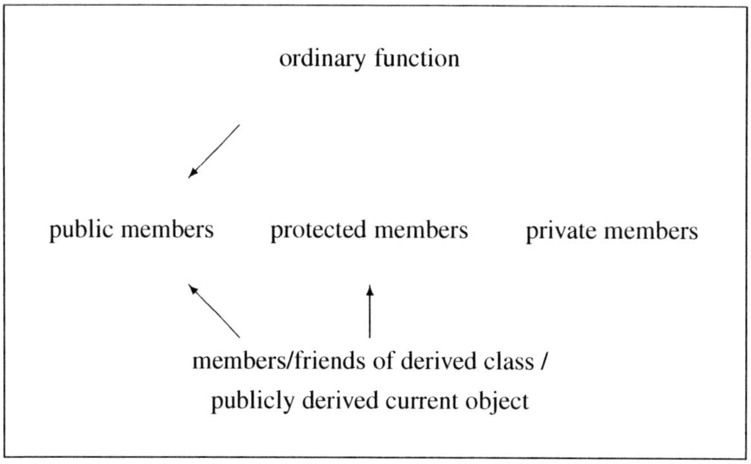

Figure 2.3. *The three kinds of members of a class (public, protected, and private) and their access pattern.*

Only if no such function exists does it interpret the passed "complex" object as a "point" object and look for a function that takes a "point" argument.

When a derived object is destroyed, the data members inherited from the base class are destroyed last. This is done by the default destructor of the base class, which is invoked automatically at the end of the destruction and destroys the members inherited from the base class in reverse order to the order in which they are declared in it. For this reason, the destructor in the "complex" class needs to do nothing: the destructor in the base "point" class does the actual destruction implicitly. To invoke it, the default destructor available automatically in the "complex" class is sufficient, and no explicit destructor needs to be written.

The above discussion indicates that the process of inheritance suffers from slight overhead in terms of both time and storage due to the base object hidden behind every derived object. Still, it may be well worth it for the sake of elegant and transparent programming and for using objects that are already implemented properly to derive various kinds of more advanced objects. Furthermore, inheritance gives us the opportunity to follow the true concept of mathematical objects and the relation between them. It is particularly useful in high-level programming, where the programmer can concentrate on the special properties of the new object derived from more technical elementary objects.

The members of the base class also function as members of the derived class and can thus be used in the definition of its own members. Furthermore, since the derivation is public, the public members of the base class remain public in the derived class and can thus be used by its users.

Members of the base class can also be rewritten in the derived class. In this case, the version in the derived class overrides the version in the base class. For example, the "*=" and "/=" operators in the "complex" class override the corresponding operators in the base "point" class and are therefore used for "complex" objects. One can still call the version in the base class by adding a prefix of the form "base::" (where "base" stands for the name of the base class) before the function name to indicate that the old version is called. For example, "point::operator*=" invokes the "*=" operator in the base "point" class.

The unary '+' operator defined above also overrides the corresponding operator in the "point" class and returns the complex conjugate of a complex number. The '+' symbol is chosen for this purpose because it represents the only operator that leaves real numbers unchanged, as indeed does the complex-conjugate operator when applied to complex numbers with zero imaginary part. This operator is used in the division of complex numbers:

```
const complex&
complex::operator*=(const complex&c){
  double keep0 = (*this)[0];
  set(0,(*this)[0]*c[0]-(*this)[1]*c[1]);
  set(1,keep0*c[1]+(*this)[1]*c[0]);
  return *this;
} // multiplying by complex

const complex&
complex::operator/=(const complex&c){
  return *this *= (complex)((point)(+c)/=squaredNorm(c));
} // dividing by complex
```

The above "/=" operator works as follows. Dividing by the complex number c is the same as multiplying by $\bar{c}/|c|^2$. First, $|c|^2$ is calculated by the "squaredNorm" function. Since the derived "complex" class has no such function, the argument 'c' is interpreted as a "point" object, and the "squaredNorm" function of the base "vector" class in Section 2.18 is invoked. The unary '+' operator, on the other hand, is available in both base and derived classes, so the version in the derived "complex" class overrides the version in the base "vector" class. As a result, "+c" is interpreted as the required complex conjugate $\bar{c}$. Now, the compiler needs to invoke a "/=" operator to divide the "complex" object $\bar{c}$ by the "double" number $|c|^2$. However, if one attempts to do this naively, then, since no division by "double" is available in the "complex" class, the compiler will implicitly convert the number $|c|^2$ from "double" to "complex" and invoke the "/=" operator recursively. Of course, this will lead to infinitely many recursive calls, with no result.

The cure to the above problem is to convert $\bar{c}$ explicitly from "complex" to "point" by adding the prefix "(point)", which converts a derived object to the underlying base object, while preserving its value. The "point" object $\bar{c}$ is then divided by the scalar $|c|^2$ unambiguously by the "/=" operator of the "vector" class in Section 2.18, as required.

The resulting "point" object $\bar{c}/|c|^2$ is then converted back to a "complex" object by the prefix "(complex)", which invokes the constructor that takes a "point" argument and constructs a "complex" object with the same value. Since $\bar{c}/|c|^2$ is now interpreted as a complex number, it can multiply the current "complex" object stored in "this", which completes the required division by the complex number c.

The above implementation involves two extra conversions, which may cause considerable overhead in applications that divide many times. These conversions are avoided in the following alternative implementation:

```
const complex&
complex::operator/=(const complex&c){
  return *this *= (+c).point::operator/=(squaredNorm(c));
} // dividing by complex
```

In this code, the prefix "point::" before the inner call to the "/=" operator indicates that the "/=" operator of the base "point" class is to be used. This implementation produces the correct result with no conversion at all.

There is no need to rewrite the "+=" and "−=" operators with a "complex" argument, because the corresponding operators in the base "point" class work just fine. However, there is a need to write explicitly the "+=" and "−=" operators with a "double" argument. Otherwise, the compiler would implicitly convert the "double" argument a to the "point" object (a, a), as in the constructor that takes a double argument in Section 2.18. Of course, this object has the wrong value; the correct value should be $(a, 0)$. Therefore, these operators must be implemented explicitly as follows:

```
const complex& complex::operator+=(double a){
    set(0,(*this)[0] + a);
    return *this;
} // adding a real number

const complex& complex::operator-=(double a){
    set(0,(*this)[0] - a);
    return *this;
} // subtracting a real number
```

The same applies to the (nonmember) binary '+' and '−' operators. These should be rewritten for mixed "complex" and "double" arguments as follows:

```
const complex
operator+(double a, const complex&c){
    return complex(c) += a;
} // double plus complex

const complex
operator+(const complex&c, double a){
    return complex(c) += a;
} // complex plus double

const complex
operator-(double a, const complex&c){
    return complex(a) - c;
} // double minus complex

const complex
operator-(const complex&c, double a){
    return complex(c) -= a;
} // complex minus double
```

The above member operators "*=" and "/=" are now used as in Section 2.16 to implement binary multiplication and division operators:

```
const complex
operator*(const complex&c, const complex&d){
  return complex(c) *= d;
} // complex times complex

const complex
operator/(const complex&c, const complex&d){
  return complex(c) /= d;
} // complex divided by complex
```

The present implementation clearly shows the risks taken in implicit conversion. For example, when the compiler encounters an expression like "3. * c", where 'c' is a complex number, it doesn't know whether to convert implicitly the real number '3.' into a complex number and invoke a complex-times-complex multiplication, or treat 'c' as a "point" object and invoke the scalar-times-point operator of the base "point" class. Although the mathematical result is the same, the compiler doesn't know that in advance and issues a compilation error due to ambiguity.

In order to avoid this problem, operators with mixed "complex" and "double" arguments must be defined explicitly as follows:

```
const complex
operator*(double a, const complex&c){
  return (point)c *= a;
} // double times complex

const complex
operator*(const complex&c, double a){
  return (point)c *= a;
} // complex times double

const complex
operator/(const complex&c, double a){
  return (point)c /= a;
} // complex divided by double
```

In some cases, it is also recommended to implement explicitly the binary addition and subtraction operators to avoid ambiguity:

```
const complex
operator+(const complex&c, const complex&d){
  return complex(c) += d;
} // complex plus complex
```

```
const complex
operator-(const complex&c, const complex&d){
  return complex(c) -= d;
} // complex minus complex
```

Here is how complex numbers are actually used in a program:

```
int main(){
  complex c=1.,i(0.,1.);
  complex d=c*3+4.*i;
  print(c+1);
  print(c/d);
  return 0;
}
```

It seems that the original implementation of the "complex" object in Section 2.16 is simpler and cheaper than the latter one, because it avoids constructing a base "point" object and converting implicitly to and from it whenever a base-class function is called. Still, inheritance may be invaluable in many applications, as we'll see below. The present implementation of the "complex" class can be viewed as a good exercise in using inheritance.

2.20 Example: The Matrix Object

Here we use inheritance to implement the "matrix" object and useful arithmetic operations with it. First, we describe briefly the object and its main functions.

Suppose that we want to implement an $N \times M$ matrix, that is, a matrix with M columns of N elements each:

$$A \equiv \left(A_{i,j}\right)_{0<i<N,\ 0<j<M} = (c_0 \mid c_1 \mid \cdots \mid c_{M-1}).$$

Here the matrix A is represented as a set of M columns $c_0, c_1, \ldots, c_{M-1}$, each of which contains N elements.

For an N-dimensional vector u, the product u times A is the M-dimensional vector

$$uA \equiv ((u, c_0), (u, c_1), \ldots, (u, c_{M-1})),$$

where (u, c_j) is the inner product of u and c_j.

For an M-dimensional vector $v = (v_0, v_1, \ldots, v_{M-1})$, the product A times v is the N-dimensional vector

$$Av \equiv \sum_{j=0}^{M-1} v_j c_j.$$

Let B be a $K \times N$ matrix. The product B times A is the $K \times M$ matrix

$$BA \equiv (Bc_0 \mid Bc_1 \mid \cdots \mid Bc_{M-1}).$$

In order to implement the "matrix" object, we use the vector as in Section 2.18, with components that are themselves vectors, representing the columns in the matrix. In other

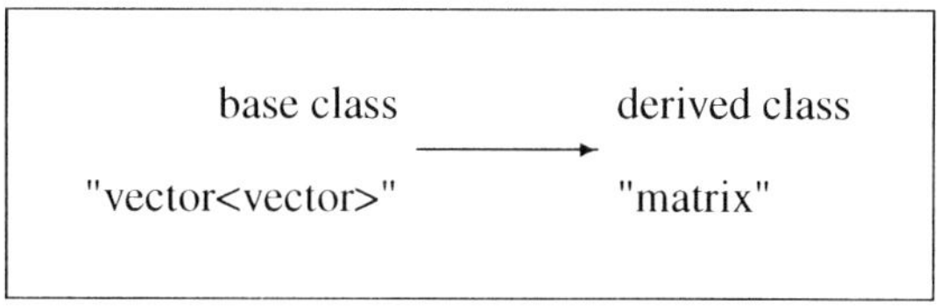

Figure 2.4. *Schematic representation of inheritance from the base class "vector<vector>" to the derived class "matrix".*

words, the matrix is implemented as a vector of vectors. For this purpose, the "matrix" class is derived from the "vector<vector>" class (see Figure 2.4). The three different kinds of products (vector-matrix, matrix-vector, and matrix-matrix) are then implemented exactly as in the above mathematical definitions.

The implementation of the template class uses three parameters, to be specified later in compilation time: 'T' to specify the type of elements, 'N' to specify the number of rows, and 'M' to specify the number of columns. This way, the definition is general, with unspecified type and dimensions:

```
template<class T, int N, int M>
class matrix : public vector<vector<T,N>,M>{
  public:
      matrix(){}

      matrix(const vector<T,N>&u, const vector<T,N>&v){
        set(0,u);
        set(1,v);
      } // constructor

      const T& operator()(int i,int j) const{
        return (*this)[j][i];
      } // read the (i,j)th matrix element

      const matrix& operator*=(const T&);
      const matrix& operator/=(const T&);
};
```

This concludes the block of the "matrix" class. The copy constructor, assignment operator, and destructor don't have to be defined here, because the corresponding functions in the base "vector<vector>" class work just fine. The operators that multiply and divide by a scalar, on the other hand, must be rewritten, because the corresponding operators in the base "vector" class return base "vector<vector>" objects rather than the required "matrix" objects. The actual definition of these operators is left as an exercise.

We can now define new types "matrix2" and "matrix3" of square matrices of orders 2 and 3, respectively, as special cases of the above template class:

```
typedef matrix<double,2,2> matrix2;
typedef matrix<double,3,3> matrix3;
```

The addition and subtraction of two "matrix" objects are done by the corresponding operators inherited from the base "vector" class, so they don't have to be rewritten here. The actual implementation of the above products of vector times matrix, matrix times vector, and matrix times matrix is left as and exercise, with detailed solution in Section A.2 of the Appendix. Assuming that these functions are available, one can define and manipulate matrices as follows:

```
int main(){
  matrix2 m(point(2,1),point(1,1));
  print(m+m);
  print(m-m);
  print(m*m);
  return 0;
}
```

Because there are no addition and subtraction operators in the derived "matrix" class, the corresponding operators of the base "vector" class are invoked. Multiplication operators, on the other hand, do exist in both the base and derived classes. Fortunately, since derived-class functions override base-class functions, they are the ones that are invoked, as required.

2.21 Determinant and Inverse of a Square Matrix

Two difficult and important tasks are the calculations of the determinant $\det(A)$ and the inverse A^{-1} of a square matrix A (of order N). The best way to do this is by using the LU decomposition of A described below.

Let $e^{(j)}$ be the jth standard unit vector, namely, the vector whose jth component is equal to 1, and all other components are equal to 0. The LU decomposition of a nonsingular square matrix A is given by

$$A = LUP,$$

where L, U, and P are square matrices (of order N) with the following properties.

1. L is lower triangular with main-diagonal elements that are equal to 1:

$$L_{i,j} = 0, \quad 0 \le i < j < N,$$

and

$$L_{i,i} = 1, \quad 0 \le i < N.$$

2. U is upper triangular with nonzero main-diagonal elements (pivots):

$$U_{i,j} = 0, \quad 0 \le j < i < N,$$

and

$$U_{i,i} \ne 0, \quad 0 \le i < N.$$

3. P is a permutation matrix, namely, a matrix whose columns are distinct standard unit vectors.

The above LU decomposition is obtained from Gaussian elimination, with suitable permutations of columns to avoid zero pivots. Here, we present a simplified version of this algorithm, in which it is assumed that the pivots are not too small in magnitude, so no permutation is needed, and P is just the identity matrix I.

Algorithm 2.1.

1. *Initialize* $L = \left(L_{i,j}\right)_{0 \le i,j < N}$ *to be the identity matrix* I.
2. *Initialize* $U = \left(U_{i,j}\right)_{0 \le i,j < N}$ *to be the same matrix as* A.
3. *For* $i = 1, 2, 3, \ldots, N-1$, *do the following:*
   - *For* $j = 0, 1, 2, \ldots, i-1$, *do the following:*
     (a) *Define*
     $$factor = U_{i,j}/U_{j,j}.$$
     (b) *For* $k = j, j+1, \ldots, N-1$, *set*
     $$U_{i,k} \leftarrow U_{i,k} - factor \cdot U_{j,k}.$$
     (c) *Set*
     $$L_{i,j} \leftarrow factor.$$

The determinant of A can be calculated most efficiently as follows:

$$\det(A) = \det(L)\det(U)\det(P) = \pm\det(U) = \pm U_{0,0}U_{1,1}U_{2,2}\cdots U_{N-1,N-1}.$$

In other words, the determinant of A is just the product of pivots $U_{i,i}$ obtained during Gaussian elimination.

The LU decomposition is also useful in calculating the inverse of A, A^{-1}. Indeed, the jth column in A^{-1} is just the solution x of the vector equation

$$Ax = e^{(j)}.$$

This equation can be solved by substituting the above LU decomposition for A:

$$LUPx = e^{(j)}.$$

This equation is solved in three steps. First, the equation

$$Lz = e^{(j)}$$

is solved for the unknown vector z (forward elimination in L). Then, the equation

$$Uy = z$$

is solved for the unknown vector y (back substitution in U). Finally, since P is orthogonal, its inverse is the same as its transpose: $P^{-1} = P^t$. Therefore, the required vector x is obtained as

$$x = P^t y.$$

This way, neither L nor U needs to be inverted explicitly, which saves a lot of computation. Furthermore, the triangular factors L and U calculated during Gaussian elimination can be stored for further use. In fact, if A is no longer required, then they can occupy the same array occupied previously by A, to save valuable computer memory.

2.22 Exponent of a Square Matrix

The exponent of a square matrix A of order N is defined by the converging infinite series

$$\exp(A) = I + A + \frac{A^2}{2!} + \frac{A^3}{3!} + \cdots = \sum_{n=0}^{\infty} \frac{A^n}{n!},$$

where I is the identity matrix of order N.

This function can be approximated by either the Taylor or the Pade approximation in Chapter 1, Section 22 above, by just replacing the scalar x used there by the matrix A. For this purpose, the present "matrix" class is most helpful.

As in Chapter 1, Section 22, one must first find a sufficiently large integer m such that the l_2-norm of $A/2^m$ is sufficiently small (say, smaller than $1/2$). Since the l_2-norm is not available, we estimate it in terms of the l_1- and l_∞-norms:

$$\|A\|_2 \le \sqrt{\|A\|_1 \|A\|_\infty},$$

where the l_1- and l_∞-norms are given by

$$\|A\|_1 = \max_{0 \le j < N} \sum_{i=0}^{N-1} |A_{i,j}|,$$

$$\|A\|_\infty = \max_{0 \le i < N} \sum_{j=0}^{N-1} |A_{i,j}|.$$

Thus, by finding an integer m so large that

$$2\sqrt{\|A\|_1 \|A\|_\infty} < 2^m,$$

we also guarantee that the l_2-norm of $A/2^m$ is smaller than $1/2$, as required.

The algorithm to approximate $\exp(A)$ proceeds as in Chapter 1, Section 22, with either Taylor or Pade polynomials. The scalar x used in Chapter 1, Section 22, is replaced by the square matrix A. The codes in Chapter 1, Section 22, are easily adapted to apply also to square matrices, provided that the required arithmetic operations are well defined.

The power of templates is apparent here. Indeed, if the functions in Chapter 1, Section 22 were written as template functions, then they could be used immediately to calculate the exponent of a matrix by simply specifying their template to be of the "matrix" class. This approach is left as an exercise below.

2.23 Exercises

1. Implement complex numbers as a template class "complex<T>", where 'T' is the type of the real and imaginary parts. Define the required arithmetic operations and test them on objects of type "complex<float>" and "complex<double>".

2. Implement complex numbers in polar coordinates: a "complex" object contains two fields, 'r' and "theta", to store the parameters $r \geq 0$ and $0 \leq \theta < 2\pi$ used in the polar representation $r \exp(i\theta)$. Define and test the required arithmetic operations.

3. Do users of the "complex" class have to be informed about the modification made above? Why?

4. Complete the missing operators in Section 2.18, such as subtraction of vectors and multiplication and division by a scalar. The solutions are given in Section A.1 of the Appendix.

5. Implement the vector-matrix, matrix-vector, matrix-matrix, and scalar-matrix products that are missing in the code in Section 2.20. The solutions are given in Section A.2 of the Appendix.

6. Write functions that return the transpose, inverse, and determinant of 2×2 matrices ("matrix2" objects in Section 2.20). The solution can be found in, Section A.2 of the Appendix.

7. Rewrite the "expTaylor" and "expPade" functions in Chapter 1, Section 22, as template functions that take an argument of type 'T', and apply them to an argument A of type "matrix2" to compute $\exp(A)$. Make sure that all the required arithmetic operations between matrices are available in your code.

8. Apply the above functions to objects of type "matrix<complex,4,4>", and verify that, for a complex parameter λ,

$$\exp\left(\begin{pmatrix} \lambda & & & \\ 1 & \lambda & & \\ & 1 & \lambda & \\ & & 1 & \lambda \end{pmatrix}\right) = \exp(\lambda)\begin{pmatrix} 1 & & & \\ 1/1! & 1 & & \\ 1/2! & 1/1! & 1 & \\ 1/3! & 1/2! & 1/1! & 1 \end{pmatrix}$$

   (the blank spaces in the above matrices indicate zero elements).

9. Compare your code to the Fortran code (461 lines) that appears on the Web page http://www.siam.org/books/cs01. Are the numerical results the same? Which code is easier to read and use?

Chapter 3

Data Structures

In this chapter, we describe several useful data structures, along with their C++ implementation. In particular, we present dynamic data structures, with a size that is specified in run time rather than compilation time. Furthermore, we implement lists of objects with variable size. Finally, we implement flexible connected lists, with algorithms to merge and order them. These data structures are most useful throughout the book.

3.1 Data Structures

As discussed above, a variable defined and used during the execution of a program is just a particular place in the computer memory where the corresponding datum is stored. The name of a variable is actually a way to refer to it and access the datum stored in the computer memory.

In most applications, one must use not only individual variables but also large structures that may contain many variables. The structures observe some pattern, according to which the variables are ordered or related. These structures are called data structures.

The most common data structure is the array. In the array, the variables are stored one by one continuously in the computer memory. Thus, they must all be of the same type, e.g., integer, float, double, etc.

The array is considered a particularly efficient data structure. Because the variables are stored one by one in the computer memory, it is particularly easy to manipulate them in loops. However, the array is often not sufficiently flexible; its size and type are fixed and cannot be changed, and it is also impossible to add more items to it or drop items from it after it is constructed.

The array is particularly useful in implementing algorithms in linear algebra. Indeed, the matrix and vector objects that are often used in linear algebra can be naturally implemented in arrays. The natural implementation of these mathematical objects provides the opportunity to implement numerical algorithms in the same spirit in which they were developed originally.

This approach is further developed in C++, where arrays are used not only to store vectors and matrices efficiently but also to actually provide complete matrix and vector objects, with all the required arithmetic operations between them. These objects can then be used exactly as in the original mathematical algorithm; the code is free of any detail about storage issues and is transparent and easy to debug.

In summary, in C++, the mathematical algorithm is completely separate from the details of storage. The arrays are hidden inside the vector and matrix classes and are manipulated only through interface functions. Once the vector and matrix objects are ready, they are used as complete mathematical objects in the required numerical algorithm.

3.2 Templates in Data Structures

As we have seen above, an array must be homogeneous: it can contain variables of only one type. This feature leads naturally to the use of templates in C++. When vectors and matrices are defined as template classes, the type of variable in them is not determined in advance but rather remains implicit until a concrete object is actually constructed. The opportunity to use templates thus saves a lot of programming effort, because it allows the definition of data structures with every possible type of variable. For example, the vector and matrix classes can be used with integer, float, double, or complex type, or any other variable with constant size.

Templates, however, provide not only an efficient and economic programming style but also a suitable approach to the implementation of data structures. Indeed, data structures are independent of the particular type of variable in use. They can be better defined with a general type, to be determined later when used in a concrete application.

Using templates in the definition of data structures is thus most natural; it takes out of the way any distracting detail about the particular type of variable and lets the programmer concentrate on the optimal implementation of the pure data structure and its mathematical features.

The mathematical features of a data structure include also the way in which variables are organized in it. These features can be rather complex: for example, they can describe the way a particular variable can be accessed from another variable. The relations between variables are best implemented in the template class, from which the particular type of variable is eliminated.

So far, we have dealt only with the simplest data structure: the vector. In this data structure, the relation between the variables is determined by their order in the vector: a particular variable is related to the previous one and the next one in the vector. This is why vectors are particularly suitable for loops. In matrices, variables are also related to variables that lie above and below them in the matrix. Therefore, matrices are also suitable for nested loops.

In the next section, we introduce an implementation of vectors in which the dimension is determined in run time rather than compilation time. This property is most important in practical applications.

3.3 Dynamic Vectors

The implementation of the "vector" object in Chapter 2, Section 18, requires that its dimension 'N' be specified in compilation time. In many cases, however, the dimension is known only in run time. For example, the dimension may be a parameter read from an external file, specified in a recursive process, or passed as an argument to a function. In such cases, the dimension is not yet known in compilation time, and the "vector" class of Chapter 2, Section 18, cannot be used. The need for dynamic vectors whose dimension is determined in run time is clear.

In this section, we provide the required dynamic implementation. In this implementation, the dimension of the "dynamicVector" object is stored in a private data member. Thus, the "dynamicVector" object contains not only the components of the vector but also an extra integer to specify the dimension of the vector. The value of this integer can be set in run time, as required.

The memory used to store a "dynamicVector" object cannot be allocated in compilation time, because the actual dimension is not yet known then. Instead, the memory allocation is done in run time, using the reserved word "new". The "new" function allocates memory for a certain object and returns the address of this memory. For example,

```
double* p = new double;
```

allocates memory for a "double" variable and uses its address to initialize a pointer-to-double named 'p'.

The present implementation is flexible and dynamic, as required. Templates are used only to specify the type of component, not the number of components. This number is indeed determined dynamically during run time.

The data fields in the "dynamicVector" class are declared "protected" to make them accessible from derived classes to be defined later. Two data fields are used: the integer "dimension" that indicates the dimension of the vector and the pointer "component" that points to the components of the vector.

Because the dimension is not yet available, the "component" field must be declared as pointer-to-T rather than array-of-T's as in the "vector" class in Chapter 2, Section 18. It is only at the actual call to the constructor of the "dynamicVector" class that the required memory is allocated and the "component" points to the concrete array of components:

```
#include<stdio.h>
template<class T> class dynamicVector{
  protected:
    int dimension;
    T* component;
  public:
    dynamicVector(int, const T&);
    dynamicVector(const dynamicVector&);
    const dynamicVector& operator=(const dynamicVector&);
    const dynamicVector& operator=(const T&);
```

The constructors and assignment operators are only declared above. The actual definition will be given later on. Next, we define the destructor:

```
    ~dynamicVector(){
      delete [] component;
    } //  destructor
```

The "delete[]" command used in the destructor deletes the entire array "component" and frees the memory occupied by it for future use. Note that the "dimension" field doesn't have to be removed explicitly. Because it is not a pointer, it is deleted implicitly by the default destructor built in the C++ compiler, which is invoked automatically at the end of every call to the above destructor.

Because the "dimension" field is protected, we need a public function to read it from ordinary classes that are not derived from the "dynamicVector" class:

```
    int dim() const{
      return dimension;
    } //  return the dimension
```

The "operator()" defined below returns a reference to the 'i'th component in the current "dynamicVector" object. Because this reference is nonconstant, it can be changed by the user. The "operator[]", on the other hand, returns a constant reference, so the 'i'th component can only be read by it:

```
    T& operator()(int i){
      return component[i];
    } //  read/write ith component

    const T& operator[](int i) const{
      return component[i];
    } //  read only ith component
```

We also declare some member arithmetic operators that operate on the current "dynamic Vector" object. The complete definition will be given later on.

```
    const dynamicVector& operator+=(const dynamicVector&);
    const dynamicVector& operator-=(const dynamicVector&);
    const dynamicVector& operator*=(const T&);
    const dynamicVector& operator/=(const T&);
};
```

This concludes the block of the "dynamicVector" class.

We now define the functions that are only declared above. In the constructor defined below, memory for the array of components is allocated in the initialization list, provided that the dimension is nonzero: this dynamic memory allocation is done by the "new" command. The address returned by this command is placed in the "component" field for future access:

```
template<class T>
dynamicVector<T>::dynamicVector(
        int dim = 0, const T& a = 0)
        : dimension(dim), component(dim ? new T[dim] : 0){
    for(int i = 0; i < dim; i++)
      component[i] = a;
} //  constructor
```

The same approach is used in the copy constructor:

```
template<class T>
dynamicVector<T>::dynamicVector(const dynamicVector<T>& v)
  : dimension(v.dimension),
   component(v.dimension ? new T[v.dimension] : 0){
  for(int i = 0; i < v.dimension; i++)
    component[i] = v.component[i];
} // copy constructor
```

Next, we define the assignment operator:

```
template<class T>
const dynamicVector<T>&
dynamicVector<T>::operator=(const dynamicVector<T>& v){
  if(this != &v){
```

We have just entered the "if" block that makes sure that the assignment operator is not called by a trivial call of the form "u = u". Now, we need to make the dimension of the current vector the same as that of the argument vector 'v':

```
    if(dimension > v.dimension)
     delete [] (component + v.dimension);
    if(dimension < v.dimension){
       delete [] component;
       component = new T[v.dimension];
    }
```

This way, the array "component[]" has the same dimension as the array "v.component[]" and can be filled with the corresponding values in a standard loop:

```
    for(int i = 0; i < v.dimension; i++)
      component[i] = v.component[i];
    dimension = v.dimension;
  }
  return *this;
} // assignment operator
```

This completes the definition of the assignment operator that takes a vector argument. Next, we define another assignment operator that takes a scalar argument. This scalar is simply assigned to all the components in the current dynamic vector:

```
template<class T>
const dynamicVector<T>&
dynamicVector<T>::operator=(const T& a){
  for(int i = 0; i < dimension; i++)
    component[i] = a;
  return *this;
} // assignment operator with a scalar argument
```

Next, we implement some useful arithmetic operators:

```
template<class T>
const dynamicVector<T>&
dynamicVector<T>::operator+=( const dynamicVector<T>&v){
    for(int i = 0; i < dimension; i++)
      component[i] += v[i];
    return *this;
}// adding a dynamicVector to the current one

template<class T>
const dynamicVector<T>
operator+(const dynamicVector<T>&u,
          const dynamicVector<T>&v){
  return dynamicVector<T>(u) += v;
} // dynamicVector plus dynamicVector

template<class T>
const dynamicVector<T>
operator-(const dynamicVector<T>&u){
  return dynamicVector<T>(u) *= -1.;
} // negative of a dynamicVector
```

Finally, here is a function that prints a dynamic vector to the screen:

```
template<class T>
void print(const dynamicVector<T>&v){
  print("(");
  for(int i = 0;i < v.dim(); i++){
    printf("v[%d]=",i);
    print(v[i]);
  }
  print(")\n");
} // printing a dynamicVector
```

The implementation of some arithmetic operations such as subtraction, multiplication, division by scalar, and inner product is left as an exercise. Assuming that these operators are also available, one can write all sorts of vector operations as follows:

```
int main(){
  dynamicVector<double> v(3,1.);
  dynamicVector<double> u;
  u=2.*v;
  printf("v:\n");
  print(v);
  printf("u:\n");
  print(u);
```

```
    printf("u+v:\n");
    print(u+v);
    printf("u-v:\n");
    print(u-v);
    printf("u*v=%f\n",u*v);
    return 0;
}
```

3.4 Lists

The vector and dynamic vector above are implemented as arrays of components of type 'T'. By definition, an array in C (as well as other programming languages) must contain components that are all of the same type and size. No array can contain components that occupy different amounts of memory.

In many applications, however, one needs to use sequences of objects that differ in size from each other. For example, one might need to use sequences of vectors whose dimensions are not yet known in compilation time. This kind of application can no longer use a standard array to store the vectors because of the different sizes they may take in the end. A more flexible data structure is needed.

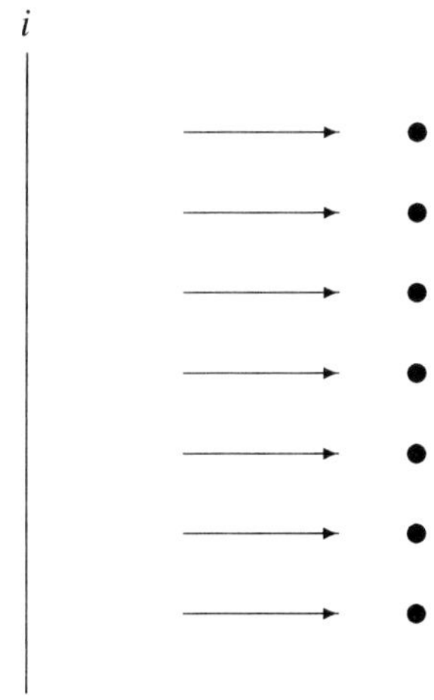

Figure 3.1. *Schematic representation of a list of objects. The arrows represent pointers, and the bullets represent constructed objects. i stands for the index in the array of pointers.*

The required data structure is implemented in the "list" class, which has an array of pointer-to-'T' objects (see Figure 3.1). Although objects of type 'T' may have different sizes (e.g., when 'T' is a "dynamicVector"), their addresses are just integer numbers that occupy a fixed amount of memory and, hence, can be placed in an array. During run time, concrete objects of type 'T' are placed in the addresses in the array. For example, if 'T' is specified in compilation time as "dynamicVector", then the dimensions of the "dynamicVector" objects in the list are specified during run time using the constructor of the "dynamicVector" class and then placed in the addresses contained in the array in the list.

The length of the list, that is, the number of pointers-to-'T' in the array in it, can also be determined dynamically in run time. As in the "dynamicVector" class, this is done by using an extra integer field to store the number of items. The "list" class thus contains two protected data fields: "number" to indicate the number of items and "item" to store their addresses. The detailed implementation is as follows:

```
template<class T> class list{
  protected:
    int number;
    T** item;
  public:
    list(int n=0):number(n), item(n ? new T*[n]:0){
    } // constructor

    list(int n, const T&t)
      : number(n), item(n ? new T*[n] : 0){
      for(int i=0; i<number; i++)
        item[i] = new T(t);
    } // constructor with T argument

    list(const list<T>&);
    const list<T>& operator=(const list<T>&);
```

The copy constructor and assignment operator are only declared above and will be defined later. The destructor defined below deletes first the pointers in the array in the "list" object and then the entire array itself:

```
    ~list(){
      for(int i=0; i<number; i++)
        delete item[i];
      delete [] item;
    } // destructor
```

Because the "number" field is protected, it cannot be read from ordinary functions. The only way to read it is by using the "size()" function:

```
    int size() const{
      return number;
    } // list size
```

Similarly, because the "item" field is protected, it cannot be accessed by ordinary functions. The only way to access its items is through "operator()" (read/write) or "operator[]" (read only):

```
    T& operator()(int i){
      if(item[i])return *(item[i]);
    } // read/write ith item

    const T& operator[](int i)const{
      if(item[i])return *(item[i]);
    } // read only ith item
};
```

This concludes the block of the "list" class, including the definitions of the constructor, destructor, and functions to read and access individual items in the list. Note that one should be careful to call "l(i)" or "l[i]" only for a list 'l' that contains a well-defined 'i'th item.

The copy constructor and assignment operator are only declared in the class block above. Here is the actual definition:

```
template<class T>
list<T>::list(const list<T>&l):number(l.number),
      item(l.number ? new T*[l.number] : 0){
  for(int i=0; i<l.number; i++)
    if(l.item[i]) item[i] = new T(*l.item[i]);
} // copy constructor
```

Here is the definition of the assignment operator:

```
template<class T>
const list<T>&
list<T>::operator=(const list<T>& l){
   if(this != &l){
```

We have just entered the "if" block that makes sure that the assignment operator has not been called by a trivial call of the form "l = l". Now, we make sure that the current list contains the same number of items as the list 'l' that is passed as an argument (in other words, the array "item[]" has the same dimension as the array "l.item[]"):

```
     if(number > l.number)
       delete [] (item + l.number);
     if(number < l.number){
       delete [] item;
       item = new T*[l.number];
     }
```

We can now go ahead and copy the items in 'l' to the current "list" object:

```
     for(int i = 0; i < l.number; i++)
       if(l.item[i]) item[i] = new T(*l.item[i]);
     number = l.number;
   }
   return *this;
 } // assignment operator
```

Finally, we implement the function that prints the items in the list to the screen:

```
template<class T>
void print(const list<T>&l){
  for(int i=0; i<l.size(); i++){
    printf("i=%d:\n",i);
    print(l[i]);
  }
} // printing a list
```

3.5 Connected Lists

The "list" object in Section 3.4 is implemented as an array of pointers or addresses of objects. This array, however, is too structured: it is not easy to add new items or remove old ones. Furthermore, the number of items is determined once and for all when the list is constructed and cannot be changed afterward. These drawbacks make the "list" data structure unsuitable for many applications. We need to have a more flexible, easily manipulated, and unrestrained kind of list.

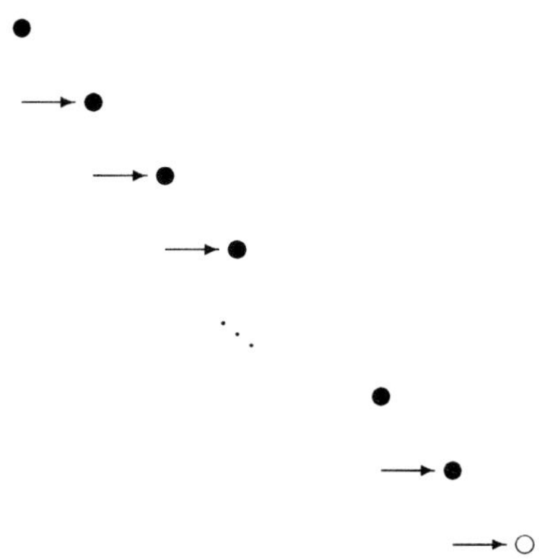

Figure 3.2. *Schematic representation of a connected list: each item (denoted by a bullet) contains a pointer (denoted by an arrow) to the next item (except the last item, which contains the null (or zero) pointer).*

Connected lists, also known as linked lists, don't use arrays at all. Instead, each item also has a pointer that points to the next item in the list (see Figure 3.2). This structure allows the addition of an unlimited number of items to the connected list if necessary and also allows inserting and dropping items at every location in it.

Accessing items through pointers as above is also called "indirect indexing." Clearly, it is less efficient than the direct indexing used in arrays, because the items are no longer stored continuously in the computer memory. Nevertheless, its advantages far exceed this disadvantage. Indeed, the freedom to insert and remove items is essential in implementing many useful objects (see Chapter 4). Furthermore, the slight inefficiency in computer resources may be well worth it for the sake of better use of human resources: programmers who use connected lists can benefit from their special recursive structure to implement complex objects, functions, and algorithms.

The connected list is implemented in the template class "connectedList" defined below. (The items in the connected list are of type 'T', to be defined later in compilation time.) The "connectedList" class contains two data fields: "item", which contains the first item in the connected list, and "next", which contains the address of the rest of the connected list. Both data fields are declared "protected", so they can be accessed from derived classes later on.

The definition of the "connectedList" object is recursive: the shorter connected list that contains all items but the first one is defined recursively as a "connectedList" object as well and is placed in the address in the field "next". This recursive pattern is useful in many basic operations.

Here is the full implementation of the "connectedList" class:

```
template<class T> class connectedList{
  protected:
    T item;
    connectedList* next;
  public:
    connectedList():next(0){
    } // default constructor

    connectedList(T&t, connectedList* N=0)
      : item(t),next(N){
    } // constructor
```

The data fields "item" and "next" can also be read (although not changed) from ordinary functions by the public member functions "operator()" and "readNext()" defined below:

```
    const T& operator()() const{
      return item;
    } // read item field

    const connectedList* readNext() const{
      return next;
    } // read next
```

The recursive pattern of the connected list is particularly useful in the copy constructor. In fact, it needs only to copy the "item" field and then be applied recursively to the shorter connected list that contains the rest of the items:

```
    const connectedList& operator=(const connectedList&);
    connectedList(const connectedList&l):item(l()),
        next(l.next ? new connectedList(*l.next):0){
    } // copy constructor
```

Here, all the work is done in the initialization list: the "item" field in the constructed object is initialized to be "l()", which is just the first item in 'l' (accessed by the "operator()" defined above). The "next" field in the constructed object is then initialized by the "new" command and a recursive call to the copy constructor itself.

The recursive structure is also useful in defining the destructor. In fact, the destructor should contain only one command that deletes the field "next", which points to the shorter connected list that contains all but the first item. When this pointer is destroyed, the same destructor is invoked automatically to destroy its content (the shorter connected list) and free it for further use. This way, the rest of the items in the connected list are destroyed as well, with no need to write any explicit code:

```
~connectedList(){
  delete next;
  next = 0;
} // destructor
```

The first field in the "connectedList" object, "item", needs no explicit command to destroy it, because it is not a pointer. It is destroyed automatically right after "next" is destroyed.

Further, we define recursive functions that return a reference to the last item in the connected list and the number of items in it. These functions are then used to append a new item at the end of the connected list:

```
connectedList& last(){
  return next ? next->last() : *this;
} // last item

int length() const{
  return next ? next->length() + 1 : 1;
} // number of items

void append(T&t){
  last().next = new connectedList(t);
} // append item
```

The following functions insert new items at different locations in the connected list. The function "insertNextItem()" places the new item right after the first item in the connected list:

```
void insertNextItem(T&t){
  next = new connectedList(t,next);
} // insert item in second place
```

The function "insertFirstItem()" places the new item just before the first item in the connected list:

```
void insertFirstItem(T&t){
  next = new connectedList(item,next);
  item = t;
} // insert item at the beginning
```

We also declare some more functions that drop items from different places in the connected list. (The full definition will be given later.)

```
    void dropNextItem();
    void dropFirstItem();
    void truncateItems(double);
    const connectedList& operator+=(connectedList&);
    connectedList& order(int);
};
```

This concludes the block of the "connectedList" class; the functions that are only declared above will be defined below.

One may rightly ask why the 'T' argument in the above constructor and other functions like "insertNextItem" and "insertFirstItem" has not been declared constant. After all, the current "connectedList" object is the one that is being changed in these functions, and surely there is no need to change the 'T' argument as well. Declaring it as a constant could protect the code from compilation- and run-time errors, couldn't it?

The answer is that usually it is indeed a good idea to declare the argument as a constant. Here, however, we plan to derive from the "connectedList" class an unusual class with a constructor that changes its argument as well. This unusual class is described in Chapter 13.

Next, we introduce the assignment operator. This operator also benefits from the recursive pattern of the connected list. In fact, after the first item has been assigned a value, the assignment operator is applied recursively to the rest of the connected list:

```
template<class T>
const connectedList<T>&
connectedList<T>::operator=(const connectedList<T>&L){
  if(this != &L){
    item = L();
    if(next){
      if(L.next)
        *next = *L.next;
      else{
        delete next;
        next = 0;
      }
    }
    else
      if(L.next)next = new connectedList(*L.next);
  }
  return *this;
} // assignment operator
```

The main advantage of connected lists is the opportunity to insert and drop items. Here is the function that drops the second item in the connected list. (The next item can be dropped by applying the same function to the shorter connected list of which it is the second item.)

```
template<class T>
void connectedList<T>::dropNextItem(){
    if(next){
      if(next->next){
        connectedList<T>* keep = next;
        next = next->next;
        keep->item.~T();
      }
      else{
        delete next;
        next = 0;
      }
    }
    else
      printf("error: cannot drop nonexisting next item\n");
} //  drop the second item from the connected list
```

The above function is also used in the following function to drop the first item in the connected list:

```
template<class T>
void connectedList<T>::dropFirstItem(){
      if(next){
        item = next->item;
        dropNextItem();
      }
      else
        printf("error: cannot drop first item; no next.\n");
} //  drop the first item in the connected list
```

Note that when a connected list contains only one item, it is never dropped. If an attempt is made to drop it, then an error message is printed to the screen.

The next function drops items that are smaller (in magnitude) than some prescribed threshold. It is assumed that the 'T' class has a "getValue()" function that returns the value of the 'T' object. If this value is smaller (in absolute value) than the prescribed threshold, then the item is dropped from the connected list. If, on the other hand, the 'T' class has no "getValue()" function, then the present function cannot be called.

The detailed implementation is as follows:

```
template<class T>
void connectedList<T>::truncateItems(double threshold){
   if(next){
     if(abs(next->item.getValue()) <= threshold){
       dropNextItem();
       truncateItems(threshold);
     }
```

In the above "if" block, the second item in the connected list is considered for dropping. If it is indeed dropped, then the third item replaces it as the new second item and is then considered for dropping in the recursive call. If, on the other hand, it is not dropped, then the third item remains in its original place and is considered for dropping by a recursive call applied to the shorter connected list that starts from the second item:

```
      else
        next->truncateItems(threshold);
    }
```

Finally, the first item is also considered for dropping, provided that it is not the only item in the connected list:

```
    if(next&&(abs(item.getValue()) <= threshold))
        dropFirstItem();
  }  //  truncate certain items
```

Note how the recursive pattern of the connected list is used in the above implementation. First, the second item in the connected list is considered for dropping. Then, the function is called recursively for the remaining items. This approach is taken because the last item in the connected list that is dropped (if it is small enough) by the innermost recursive call can never be dropped by the "dropFirstItem" function, but rather by the "dropNextItem" function called from the previous item. Therefore, the recursion must always look ahead and truncate the next item, rather than the current item.

Actually, the latter code segment can be removed from this function and placed in another function that drops only the first item, if appropriate. This might increase efficiency, because as it stands there are some unnecessary repetitive checks.

The recursive structure of the connected list is also useful to print it. In fact, after the first item has been printed, the "print" function is applied recursively for the rest of the items in the connected list:

```
template<class T>
void print(const connectedList<T>&l){
  printf("item:\n");
  print(l());
  if(l.readNext())print(*l.readNext());
}  //  print a connected list
```

Here is how connected lists are actually used in a program:

```
int main(){
  connectedList<double> c(3.);
  c.append(5.);
  c.append(6.);
  c.dropFirstItem();
  print(c);
  return 0;
}
```

3.6 The Merging Problem

Users of connected lists may want to manipulate them in many ways. For example, they may want to insert items into them or drop items from them. Furthermore, they would like to be able to do this without dealing with storage issues such as pointers and addresses. The functions that perform these tasks are described above.

Users of connected lists may want to do another operation: merge two ordered connected lists with each other while preserving the order. Completing this task efficiently is called the merging problem.

The power of templates is apparent here. Indeed, it is particularly convenient to deal with an unspecified type 'T', provided that it supports some order. This way, we can concentrate on the data structure under consideration rather than the type used in the variables in it. It is only when the merging function is called that the concrete type is specified.

In the following, we describe a function that merges two connected lists into a single connected list while preserving the order. It is assumed that the type 'T' of the items in the connected list supports a complete priority order; that is, every two 'T' objects can be compared by the '<', '>', and "==" binary operators. It is also assumed that the current "connectedList" object and the "connectedList" argument that is merged into it are well ordered in the sense that each item is smaller than the next item.

The purpose is to merge the connected list that is passed as an argument into the current connected list while preserving the correct order. This operation is most useful, particularly in the sparse matrix implemented in Chapter 16.

It is also assumed that the 'T' class supports a "+=" operator. With these assumptions, the "+=" operator that merges a connected list into the current connected list can also be defined.

The code uses two "runners" to scan the items in the connected lists (see Figure 3.3). The main runner scans the items in the current connected list. The room between the item pointed at by this runner and the item that follows it should be filled by items from the second connected list that is passed as an argument, provided that they indeed belong there in terms of order. For this purpose, a secondary runner, called "Lrunner", is used to scan the items in the second connected list that indeed belong in the location pointed at by the main runner. These items are then inserted one by one into the current connected list in their correct places.

In case an item in the second connected list has the same priority order '<' as an existing item in the current connected list, that is, it is equal to it in terms of the "==" operator of the 'T' class, then it is added to it using the "+=" operator of the 'T' class:

```
template<class T>
const connectedList<T>&
connectedList<T>::operator+=(connectedList<T>&L){
  connectedList<T>* runner = this;
  connectedList<T>* Lrunner = &L;
```

Initially, "Lrunner" points to the connected list 'L' that is passed to the function as an argument. However, in order to start the merging process, we must first make sure that the first item in the current connected list is prior (according to the priority order '<') to the

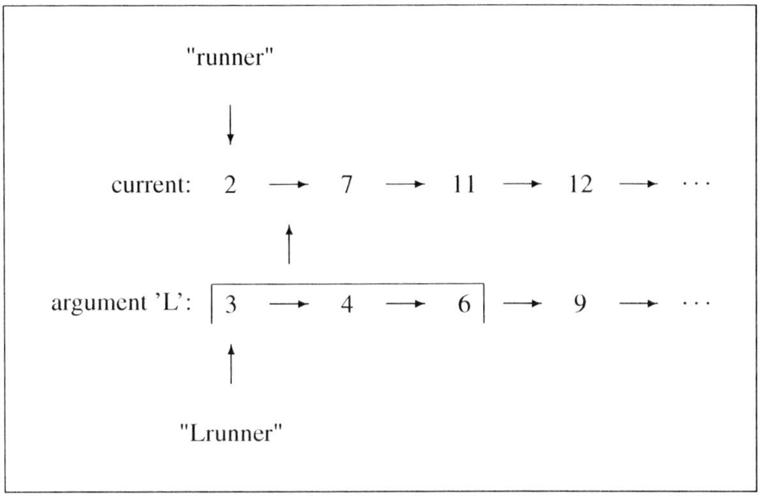

Figure 3.3. *Merging two connected lists while preserving order. The items in the top connected list (the current object) are scanned by the pointer "runner" in the outer loop. The items in the bottom connected list 'L' (the argument) are scanned by the pointer "Lrunner" in the inner loop and inserted in the right place.*

first item in 'L', "L.item". If this is not the case, then "L.item" must first be placed at the beginning of the current connected list and "Lrunner" advanced to the next item in 'L':

```
if(L.item < item){
  insertFirstItem(L.item);
  Lrunner = L.next;
}
for(; runner->next; runner=runner->next){
```

Here we enter the main loop. Initially, "runner" points to the entire current connected list. Then, it is advanced gradually to point to subsequent sublists that contain fewer and fewer items. The loop terminates when "runner" points to the sublist that contains only the last item in the original connected list.

We are now ready to add an item from 'L' to the current connected list. If the item pointed at by "Lrunner" has the same priority as the item pointed at by "runner", then it is added to it using the "+=" operator available in the 'T' template class:

```
if(Lrunner&&(Lrunner->item == runner->item)){
  runner->item += Lrunner->item;
  Lrunner = Lrunner->next;
}
for(; Lrunner&&(Lrunner->item < runner->next->item);
        Lrunner = Lrunner->next){
```

We now enter the inner loop, in which the items in 'L' pointed at by "Lrunner" are added one by one to the current connected list and placed in between the item pointed at by "runner" and the item that follows it. Once an item from 'L' has been added to the current connected list, the "runner" pointer must be advanced to skip it:

```
        runner->insertNextItem(Lrunner->item);
        runner = runner->next;
      }
    }
```

The inner and outer loops are now complete. However, 'L' may still contain more items that should be placed at the end of the current connected list. Fortunately, "runner" and "Lrunner" were defined before the loops started, so they still exist. In fact, at this stage, "runner" points to the last item in the current connected list, and "Lrunner" points to the remaining items in 'L', if any. These items are appended to the current connected list as follows:

```
    if(Lrunner&&(Lrunner->item == runner->item)){
      runner->item += Lrunner->item;
      Lrunner = Lrunner->next;
    }
    if(Lrunner)runner->next=new connectedList<T>(*Lrunner);
    return *this;
  }  //  merge two connected lists while preserving order
```

This completes the merging of 'L' into the current connected list while preserving order, as required.

3.7 The Ordering Problem

The ordering problem is as follows: find an efficient algorithm to order a given disordered list of items. Naturally, this algorithm requires changing the order in the original list. Thus, the connected list implemented above, which easily accepts every required change, is the obvious candidate for the data structure of the required algorithm.

Here, we use the "+=" operator that merges two ordered connected lists to define the "order()" template function that orders a disordered connected list (see Figure 3.4). The algorithm for doing this uses recursion as follows. Divide the disordered connected list into two sublists. Apply the "order()" function separately to each of these sublists. Then, merge them into a single, well-ordered list using the above "+=" operator.

The complete code is as follows:

```
template<class T>
connectedList<T>&
connectedList<T>::order(int length){
  if(length>1){
    connectedList<T>* runner = this;
    for(int i=0; i<length/2-1; i++)
      runner = runner->next;
```

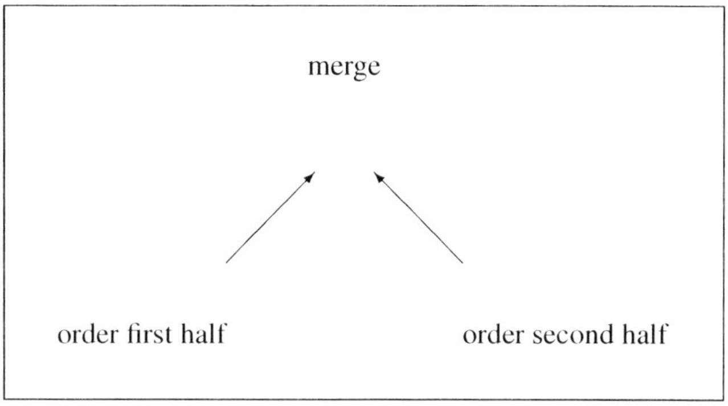

Figure 3.4. *The "order()" function that orders a disordered connected list: the original list is split into two sublists, which are ordered recursively and merged (while preserving order).*

At the end of this short loop, "runner" points to the item that lies at the middle of the current connected list. The "runner" pointer is now being used to define the pointer "second", which points to the second half of the original connected list:

```
        connectedList<T>* second = runner->next;
        runner->next = 0;
```

We are now ready for the recursion. The "order()" function is applied recursively to its first and second halves, before they are merged by the "+=" operator:

```
        order(length/2);
        *this += second->order(length-length/2);
      }
      return *this;
    }  //  order a disordered connected list
```

This completes the ordering of the current connected list.

3.8 Vectors vs. Lists

So far, we have discussed four different kinds of vectors and lists: vector, dynamic vector, list, and connected list. The pros and cons of these objects are summarized in Table 3.1.

The "vector" object uses a standard array, which allows the most efficient loops. However, the components stored in the array must all have the same size, and the dimension of the array must be determined in compilation time. The "dynamicVector" object improves on "vector" by allowing the dimension to be set at run time rather than compilation time. However, the components in the array must still have the same size. The "list" object improves the situation further by allowing the items in the list to have different sizes. However, this comes at the price of using indirect indexing, which may slow down loops over the list. The "connectedList" object improves on all the others by having extra flexibility

Table 3.1. *Different kinds of vectors and lists and their pros and cons.*

	vector	dynamicVector	list	connectedList
Storage	fixed array	dynamic array	array of pointers	recursive addressing
Pros	efficient storage	efficient storage	variable-size items	variable-size items
	efficient loops	efficient loops		high flexibility
Cons	same-size items	same-size items	indirect indexing	indirect indexing
	fixed dimension			expensive recursion

in inserting new items and removing old ones. However, this comes at the price of using not only indirect indexing but also expensive recursion to manipulate the recursively defined connected list. Nevertheless, the flexible nature of the connected list makes it most useful in this book and elsewhere (see Chapter 4).

In the rest of this chapter, we show how the "connectedList" object can be improved further to have other desirable features. Naturally, this comes at the price of using extra storage and more complicated definitions and functions. In this book, the "connectedList" object is sufficient, and the more complex data structures are actually never used. They are discussed here briefly only for the sake of completeness.

3.9 Two-Sided Connected Lists

A data structure is characterized by the way one can access data about variables in it. For example, in a connected list, each item has access to the next item only. An item cannot access the item that points to it or the previous item in the connected list.

In some algorithms, this may cause a problem. There may be a need to loop on the connected list in the reverse direction or to know what the previous item in standard loops is. In such cases, one must use a two-sided connected list.

The "twoSidedConnectedList" class can be derived from the "connectedList" class by adding one field of type pointer-to-"twoSidedConnectedList". This field, named "previous", should contain the address of the previous item in the connected list. Because we don't actually use this class in this book, we omit the details.

3.10 Trees

In connected lists, each item contains a pointer to the next item. But what if it contained two pointers? We would then have a binary tree (see Figure 3.5):

```
template<class T> class binaryTree{
  protected:
    T item;
    binaryTree* left;
    binaryTree* right;
  public:

      ...

};
```

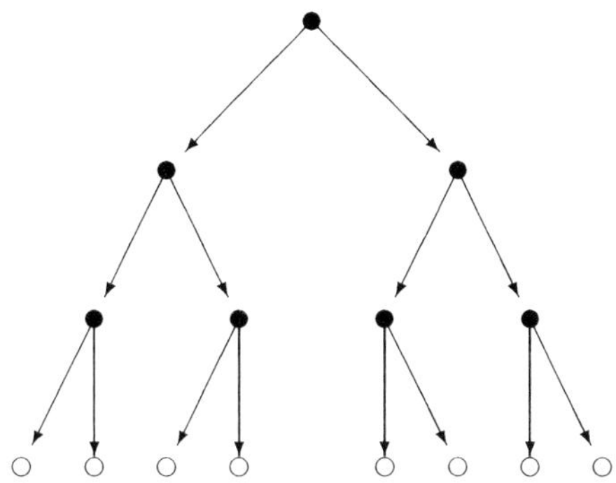

Figure 3.5. *Schematic representation of a binary tree with three levels. The arrows represent pointers, and the bullets represent constructed objects. The circles at the lowest level stand for the null (zero) pointer.*

In a connected list, access to the individual items is not always easy. In fact, accessing the last item requires stepping along a path whose length is the length of the entire list. In a tree, on the other hand, this task is much easier and requires a path whose length is at most the logarithm of the number of items.

In object-oriented programming, however, there is little need to access individual items in the connected list. One treats the connected list as a whole and tries to avoid accessing individual items in it. Thus, we use connected lists rather than trees in this book.

Furthermore, binary trees lack the natural order of items available in connected lists. Although one can order the items recursively (e.g., left, middle, right), this order can be much more complicated than the natural order in connected lists.

One can also define more general trees by replacing the "left" and "right" pointers by a pointer to a connected list of trees:

```
template<class T> class tree{
  protected:
    T item;
    connectedList<tree<T> >* branch;
  public:

    ...

};
```

This tree may contain any number of subtrees connected to each other in a connected list. This connected list is pointed at by the private member "branch". Note that the '>' symbols in the definition of "branch" are separated by a blank space to distinguish them from the string ">>", which has a completely different meaning in the "iostream.h" standard library.

In the above data structure, it is particularly easy to remove existing subtrees and add new ones. Two-sided trees can also be derived by adding a pointer field named "parent" to contain the address of the parent tree.

3.11 Graphs

In the above trees, an item can point only to new items; it cannot point to parent or ancestor trees or even sibling subtrees. Thus, a tree cannot contain circles.

In a graph, on the other hand, an item can point not only to new items but also to existing ones; be they ancestors, siblings, or even itself.

In fact, the mathematical definition of a graph is a set of nodes numbered $1, 2, 3, \ldots, N$ and a set of edges (pairs of nodes). The set of edges is commonly denoted by E.

There are two kinds of graphs. In oriented graphs, each edge is an ordered pair of nodes, so the pair (i, j) (also denoted by $i \to j$) is not the same as (j, i) (also denoted by $j \to i$). It may well be that $(i, j) \in E$ but $(j, i) \notin E$, or vice versa (see Figure 3.6).

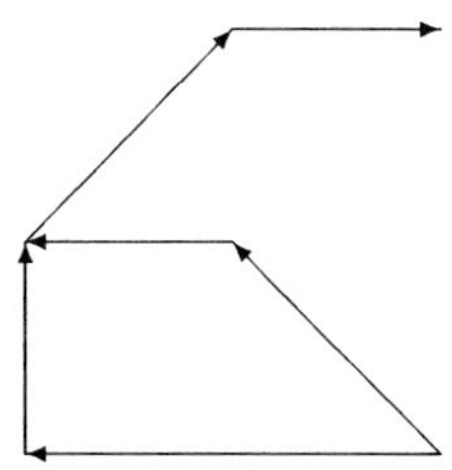

Figure 3.6. *Schematic representation of an oriented graph.*

In nonoriented graphs, on the other hand, a pair has no order in it: (i, j) is the same as (j, i), and either both are in E or both are not in E (see Figure 3.7).

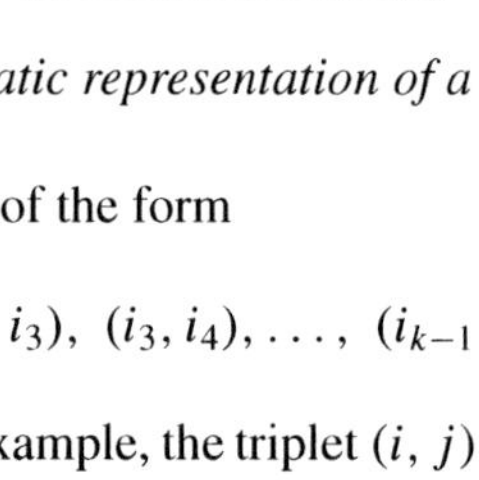

Figure 3.7. *Schematic representation of a nonoriented graph.*

In a graph, a set of k edges of the form

$$(i_1, i_2), \ (i_2, i_3), \ (i_3, i_4), \ldots, \ (i_{k-1}, i_k), \ (i_k, i_1)$$

is called a circle of k edges. For example, the triplet (i, j), (j, k), and (k, i) forms a circle of three edges, the pair of edges (i, j) and (j, i) forms a circle of two edges, and even the single edge (i, i) by itself forms a circle of one edge only. These circles, although not allowed in trees, must be allowed in the implementation of graphs.

Nevertheless, the above implementation of trees is also suitable for graphs, provided that the rules are changed so that circles are allowed. This means that the constructor in the "graph" class must be different from that in the "tree" class. The tree constructor must use the "new" command in every subtree in the "*branch" connected list, whereas the graph constructor may also use existing addresses in this connected list.

The fact that circles are allowed means that there is no natural order in the items in the graph. This is a drawback in the above implementation, because there is no natural way to loop over the nodes. In the next chapter, we'll see more practical implementations for graphs in which the original order $1, 2, 3, \ldots, N$ is preserved.

3.12 Exercises

1. Complete the missing arithmetic operators in the implementation of the dynamic vector in Section 3.3, such as subtraction and multiplication and division by scalar. The solution is given in Section A.3 of the Appendix.

2. Using the "list" object in Section 3.4, write the function "Taylor()" that takes as arguments the real number h and the list of N derivatives of a function $f(x)$ (calculated at some point x) and produces as output the Taylor approximation to $f(x+h)$:

$$f(x+h) \doteq \sum_{n=0}^{N} \frac{f^{(n)}(x)h^n}{n!},$$

   where $f^{(n)}$ denotes the nth derivative of $f(x)$. The solution can be found in Chapter 5, Section 11.

3. Write the function "deriveProduct()" that takes as input two lists that contain the derivatives of two functions $f(x)$ and $g(x)$ at some point x and returns the list of derivatives of the product $f(x)g(x)$, using the formula

$$(f(x)g(x))^{(n)} = \sum_{k=0}^{n} \binom{n}{k} f^{(k)}(x)g^{(n-k)}(x).$$

   The solution can be found in Chapter 5, Section 12.

4. Implement Pascal's triangle in Chapter 1, Section 19, as a list of diagonals. The diagonals are implemented as dynamic vectors of increasing dimension. (The first diagonal is of length 1, the second is of length 2, and so on.) The components in these vectors are integer numbers. Verify that the sum of the entries along the nth diagonal is indeed 2^n and that the sum of the entries in the first, second, $\ldots$, nth diagonals is indeed $2^{n+1}-1$.

5. Define the template class "triangle<T>" that is derived from a list of dynamic vectors of increasing dimension, as above. The components in these vectors are of the unspecified type 'T'. Implement Pascal's triangle as a "triangle<int>" object. Verify that the sum of the entries in the nth diagonal is indeed 2^n.

6. Use the above "triangle<double>" object to store the (mixed) partial derivatives of a function of two variables $f(x, y)$. In particular, use the (k, l)th cell in the triangle to

store the value

$$\frac{\partial^{k+l} f}{\partial^k x \partial^l y}(x, y)$$

at some point (x, y). This way, derivatives of order up to n are stored in the first n diagonals in the triangle.

7. Write a function "Taylor2()" that takes the above triangle and two small "double" parameters h_x and h_y as arguments and produces the two-dimensional Taylor approximation of $f(x + h_x, y + h_y)$ according to the formula

$$f(x + h_x, y + h_y) = \sum_{n=0}^{\infty} \frac{1}{n!} \sum_{k=0}^{n} \binom{n}{k} \frac{\partial^n f}{\partial^k x \partial^{n-k} y}(x, y) h_x^k h_y^{n-k}.$$

   The solution is indicated in the exercises at the end of Chapter 5.

8. Apply the "order()" function in Section 3.7 to a connected list of integer numbers and order it with respect to absolute value. For example, verify that the list

$$(-5, 2, -3, 0, \ldots)$$

   is reordered as

$$(0, 2, -3, -5, \ldots).$$

9. Complete the implementation of the "binaryTree" class, with constructors, destructor, and assignment operator.

10. Complete the implementation of the "tree" class, with constructors, destructor, and assignment operator.

11. Modify your "tree" class to obtain the "graph" class.

Part II

The Object-Oriented Approach

Objects are abstract concepts that one can operate on. In C, only numerical objects such as integer and real numbers and characters are available. These objects merely serve as arguments passed to functions. In C++, on the other hand, programmers may construct their own objects, which can be much more complex and have many useful functions. Thus, in C++, objects play a much more active role. They are the center of the implementation, which focuses on them and their properties. They not only wait until some function uses them as arguments but actively call functions that reflect their potential.

In C++, the programmer implements not only isolated algorithms but also complete mathematical objects (such as vectors and matrices) that can be used in many algorithms and applications. This makes the programming language dynamic: the new objects may serve as new types, which make the programming language richer and more suitable for new applications.

This part introduces the object-oriented approach that is in the very heart of C++. In the first chapter in this part (Chapter 4), a unified approach from graph theory is used to introduce the present object-oriented framework, with potential applications in unstructured meshes (Part IV) and sparse matrices (Part V). The second chapter in this part (Chapter 5) introduces algorithms that use mathematical objects in their implementation. In particular, it implements fully the "polynomial" object and associated functions, to be used later in Part IV. Finally, the third chapter in this part (Chapter 6) shows how the object-oriented approach can be used in mathematical analysis.

Chapter 4

Object-Oriented Programming

In this chapter, we present the concept of object-oriented programming and use it in the graph-coloring and triangle-coloring problems. This concept proves to be most helpful in the implementation of mathematical objects required in numerical methods, such as unstructured meshes and sparse matrices.

4.1 Object-Oriented Language

C may be considered as a process-oriented programming language. Indeed, the basic element in C is the process or function (with one or more input arguments), which calculates and produces the required output. In fact, every command in C is actually a function, which also returns a value.

The input data used in processes in C are not considered an integral part of the process. In fact, the process can be defined before the data have been specified. For example, functions can be defined with dummy arguments, to be given concrete values only when the function is actually called. In fact, when the function is compiled, a finite state machine (automaton) is created, and each dummy argument is represented by an input line, to be filled by a concrete argument later on when the function is actually called.

The concrete arguments may also be thought of as objects upon which the function operates. Thus, in C, the focus is on functions, which operate upon objects.

One of the most important functions in C is the referencing operator '&'. When this operator is applied to a variable, it returns its address. For example, "&x" is the address of the variable 'x'. The inverse of the referencing operator is the dereferencing operator '*'. When applied to an address, it returns the variable stored in it. For example, "*p" is the content of the address 'p'. These two operators allow the use of pointers, which are so important in the implementation of data structures.

Fortran, on the other hand, may be considered as a storage-oriented programming language. Because pointers are not available, every data structure must be explicitly allocated memory by the user. In fact, every data structure must be embedded in an array, with the index in the array serving as its virtual address. This way, the user can simulate pointers by indices in arrays. This approach, however, can be extremely complicated and hard to use and read. This shows clearly why C is considered as a more flexible and advanced language than Fortran.

C++ is even more advanced than both C and Fortran in the sense that it offers the opportunity to define and implement new objects, along with their own functions. Unlike in C, functions that operate on the object are considered an integral part of its implementation. Well-prepared mathematical objects can then be further used in complicated algorithms, applications, and definitions of even more complex objects. In fact, the objects can be used in further code much like standard types, provided that the required functions are well implemented. For example, if the multiplication operator of two objects is well defined, then they can be multiplied like two integer or real numbers.

The opportunity to define and use new objects is not only convenient but also efficient, particularly in terms of human resources. Indeed, complex mathematical structures and algorithms can be implemented in their original spirit. The transparent and modular implementation allows easy debugging, modification, and further improvements. These advantages are illustrated next.

4.2 Example: The Graph-Coloring Problem

In order to illustrate the advantage of object-oriented programming, we consider the following graph-coloring problem: in a graph as in Chapter 3, Section 11, use the minimal possible number of colors to color the edges in such a way that two node-sharing edges have distinct colors. In other words, find the minimal positive integer C such that every edge in E can be assigned an integer number between 1 and C, and every two node-sharing edges are assigned distinct numbers. The numbers between 1 and C are then called "colors," and C is the number of colors.

It is sometimes easier to view a graph as a matrix. In fact, a graph of N nodes is equivalent to an $N \times N$ matrix (square matrix of order N) whose elements are either -1 or 0. More specifically, the matrix A corresponding to the graph is defined by

$$A_{i,j} \equiv \begin{cases} -1 & \text{if } (i,j) \in E, \\ 0 & \text{if } (i,j) \notin E \end{cases}$$

(where E is the set of edges in the graph).

For oriented graphs, A is defined uniquely by the above definition. Furthermore, the inverse direction is also true: every square matrix A with elements that are either 0 or -1 uniquely defines an oriented graph.

For nonoriented graphs, on the other hand, the above definition is not sufficiently precise. Indeed, for a nonoriented graph, $A_{i,j}$ must be the same as $A_{j,i}$, so the matrix A is symmetric. Still, strictly speaking, $A_{i,j}$ and $A_{j,i}$ are distinct elements in the matrix A, whereas, in the corresponding graph, (i, j) and (j, i) are actually the same edge. A more precise (and economic) method for representing a nonoriented graph is by only the upper triangular part of A, which contains only elements $A_{i,j}$ with $i \le j$. The matrix A associated with a nonoriented graph is, thus, defined by

$$A_{i,j} \equiv \begin{cases} -1 & \text{if } (i,j) \in E \text{ and } i \le j, \\ 0 & \text{otherwise} \end{cases}$$

(where E is the set of edges in the graph).

Oriented and nonoriented graphs can now be treated in a uniform way by looking at the matrices representing them: an oriented graph is represented by a general (symmetric or nonsymmetric) matrix, whereas a nonoriented graph is represented by an upper triangular matrix. Using these representations, the graph-coloring problem can be reformulated in terms of matrices rather than graphs: find the minimal positive integer C such that the nonzero elements in A can be replaced by positive integers between 1 and C in such a way that no positive number appears more than once in the same row or column in A. More precisely, every nonzero element $A_{i,j}$ is replaced by a positive integer $c_{i,j}$ ($1 \leq c_{i,j} \leq C$) in such a way that, for oriented graphs,

$$c_{i,j} \neq c_{k,l} \text{ if } i = k \text{ or } j = l$$

or, for nonoriented graphs,

$$c_{i,j} \neq c_{k,l} \text{ if } i = k \text{ or } j = l \text{ or } i = l \text{ or } j = k.$$

The numbers $1, 2, 3, \ldots, C$ are also called colors, so one can also say that the elements in A are colored in such a way that a color appears at most once in each row or column.

Let us present an algorithm to color the matrix [36]. For oriented graphs, the algorithm is as follows. For $i = 1, 2, 3, \ldots, N$, do the following. For $j = 1, 2, 3, \ldots, N$, do the following: color $A_{i,j}$ by the first color $c_{i,j}$ that has not yet been used in the ith row and jth column; in other words, $c_{i,j}$ is the minimal positive integer satisfying

$$c_{i,j} \neq c_{i,l}, \qquad 1 \leq l < j,$$

and

$$c_{i,j} \neq c_{k,j}, \qquad 1 \leq k < i.$$

This completes the algorithm for oriented graphs.

For nonoriented graphs, the above algorithm is slightly modified as follows. For $i = 1, 2, 3, \ldots, N$, do the following. For $j = i, i+1, i+2, \ldots, N$, do the following: assign to $A_{i,j}$ the minimal positive integer $c_{i,j}$ satisfying

$$c_{i,j} \neq c_{i,l}, \qquad i \leq l < j,$$

and

$$c_{i,j} \neq c_{k,j}, \qquad 1 \leq k < i,$$

and

$$c_{i,j} \neq c_{k,i}, \qquad 1 \leq k < i.$$

This completes the algorithm for nonoriented graphs.

Let us now estimate the complexity of the algorithm, namely, the amount of storage and computation time it requires. Clearly, the storage requirement is proportional to N^2, the number of elements in A. The number of calculations required is proportional to $N^2\bar{e}$, where $\bar{e}$ is the average number of nonzero elements per row.

The reason for these rather large complexity estimates is that valuable resources are wasted on the zero elements in A. More precisely, the drawback is not in the algorithm but rather in the implementation. Indeed, when $\bar{e}$ is much smaller than N, A is sparse, namely, most of its elements vanish. It is thus pointless to store the zero elements in A or use them in the above algorithm. It makes more sense to store only the nonzero elements in A, using flexible data structures. The above algorithm can then be implemented much more efficiently, because only nonzero elements are used.

In what follows, we consider a suitable implementation for the sparse matrix A and the coloring algorithm. When talking about complexity, we often omit the words "proportional to" and assume that the complexity estimates are up to multiplication by a constant independent of N and $\bar{e}$.

4.3 Downward Implementation

The above implementation of the coloring algorithm may be viewed as an "upward" implementation, in which the basic objects are implemented first and then used to implement more complex objects (Figure 4.1). Indeed, the basic objects in the graphs, the nodes, are stored in an N-dimensional array. This way, the ith entry in the array may contain information about the ith node, such as its geometric location. The index $1 \le i \le N$ is then used to refer to the ith node; in fact, it can be considered as the "virtual address" of this node in the array of nodes. This virtual address may be different from the physical address of the ith entry in the array in the computer memory. Still, it is good enough for referring to the node in the coloring problem and algorithm.

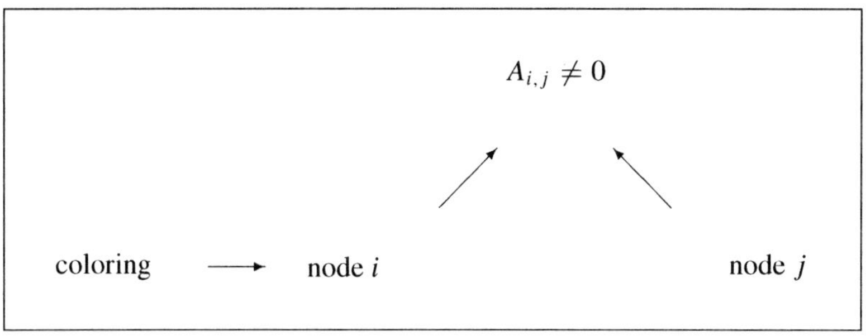

Figure 4.1. *Upward implementation of a graph. Nodes are implemented as objects and stored in an array. Edges are not implemented but only indicated by the function $A_{i,j}$, which takes a nonzero value if and only if i and j are connected by an edge.*

The more complex objects in the graph are the edges, which are actually pairs of nodes. The edges, however, are never actually stored as physical objects; instead, the information about them is provided in terms of a function of two variables, i and j: this function, named $A_{i,j}$, assumes a nonzero value if and only if i and j are connected by an edge in the graph. The information about the existence of the edge can be accessed only through the nodes i and j, the arguments of the function $A_{i,j}$.

The function $A_{i,j}$ may often assume the value 0, which is never used in the coloring problem or algorithm. In fact, the function $A_{i,j}$ must be stored in an $N \times N$ array, where entries may often vanish. This is not a good use of computer resources.

Let us now consider a "downward" implementation, in which the more complex objects are considered before the basic ones (Figure 4.2). Of course, this approach assumes

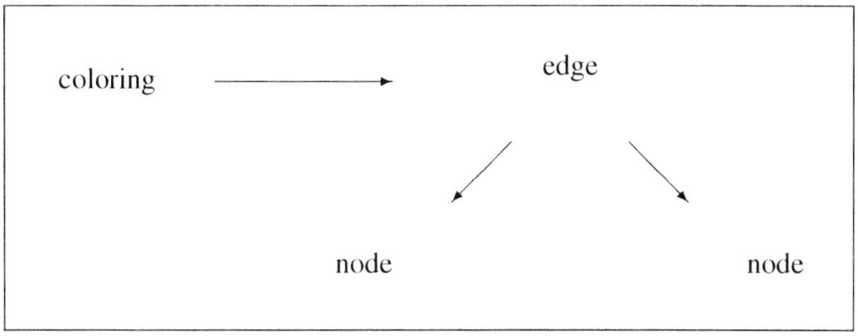

Figure 4.2. *Downward implementation of a graph. Edges are implemented and stored in an array. Each edge points to its two endpoints, which provides edge-to-node (but not node-to-edge) data access.*

that the required basic objects are available; only their actual implementation is delayed to a later stage. This approach gives one the opportunity to concentrate on the complex object, which is usually the more important one in the entire problem area, rather than the basic object, which usually contains only technical details, which may be rather irrelevant to the problem.

More concretely, let us start by implementing the more complex objects in the graph: the edges. After all, the edges are what are colored in the coloring problem. We assume that the nodes are available, with their virtual addresses $1, 2, 3, \ldots, N$. Naturally, each edge is implemented as a pair of nodes (i, j). This time, however, the edges are stored physically in a long array of edges whose length is $N\bar{e}$. Each edge in this array contains the two virtual addresses of the nodes that are its endpoints in the graph. In other words, each edge has access to its endpoints through their virtual addresses.

In the downward implementation, the edges are implemented physically and actually stored in the array of edges. Each edge has its own virtual address: its index in the array of edges. Nodes can now be accessed not only directly through their virtual addresses but also indirectly through edges that use them as endpoints.

The above downward implementation can be also viewed as a method to store sparse matrices. In fact, the edges can be viewed as nonzero matrix elements. Assume that these elements are ordered row by row and listed in a long array of length $N\bar{e}$. Each entry in this array actually contains two numbers: i and j. (This structure can be implemented physically in two arrays of integers.) This array is thus equivalent to the above array of edges, except that here the order of the nonzero matrix elements is specified to be row by row. Because the zero matrix elements are not stored, the storage requirement is reduced from N^2 to only $N\bar{e}$.

Let us now define a prototype coloring algorithm that is particularly suitable for the above downward implementation. The algorithm reads as follows. Loop on the edges in the array of edges. For each edge encountered, color it in the first color that has not yet been used in former edges that share a node with it; in other words, an inner loop on the former edges in the array of edges must be used to find the node-sharing former edges and avoid using their color. This completes the definition of the prototype coloring algorithm.

Clearly, this algorithm is mathematically equivalent to the previous one. The only difference is that here nonzero matrix elements are ignored, because they are never needed when coloring is considered. The algorithm is thus more efficient in terms of storage requirements.

Unfortunately, the prototype algorithm is still expensive in terms of computation time. Indeed, because it uses nested loops on the long array of edges, it requires $N^2\bar{e}^2$ calculations. The source of this drawback is the one-sided data access available in the downward implementation. In fact, we only have edge-to-node access, as the edge object contains the virtual addresses of its endpoints, but not node-to-edge access, as a node "doesn't know" which edges issue from or to it. This is why the inner loop is so large: it must scan all the former edges, because we don't know which edges exactly share a node with the current edge (scanned in the outer loop).

The above discussion leads to an improved implementation with both edge-to-node and node-to-edge data access (Figure 4.3). In this implementation, a node also "knows" which edges issue from or to it. This is done as follows. For every $1 \le i \le N$, let $1 \le n(i) \le N\bar{e}$ be the place in the array of edges where the nonzero elements in the ith row in A start to be listed. The entries in the array of edges that are indexed $n(i), n(i)+1, \ldots, n(i+1)-1$ correspond to edges issuing from the node i. Thus, the ith node has access to the continuous list of edges issuing from it, or the nonzero elements in the ith row in A. If the transpose of A, A^t is stored in the same way, then the ith node also has access to the continuous list of edges that point to it, or the nonzero elements in the ith column in A. (Compare with the column pointers used in the Harwell–Boeing method for storing sparse matrices in Chapter 18, Section 17.)

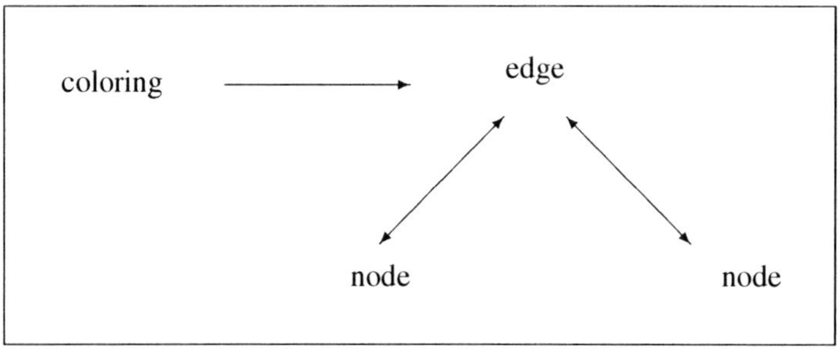

Figure 4.3. *Downward implementation with complete data access. Edges have access to their endpoints, and nodes have access to edges that use them as endpoints.*

Thus, the expensive inner loop in the prototype algorithm can be replaced by two short loops over the nonzero elements in the ith row and jth column in A (where i and j are the indices of the nonzero element $A_{i,j}$ scanned in the outer loop). With this implementation, the number of calculations in the prototype algorithm is reduced from $N^2\bar{e}^2$ to $N\bar{e}^2$, which is rather efficient.

In what follows, we'll see how this implementation can be carried out nicely in C++.

4.4 The C++ Implementation

The downward implementation requires memory allocation for the individual edges. The edges are stored in a long array, and the virtual address of an edge is just its index in this array. This implementation is equivalent to storing the nonzero elements in a sparse matrix row by row in an array of length $N\bar{e}$.

This is how the downward implementation is carried out in a storage-oriented programming language like Fortran, in which no pointers are available, and virtual addresses must be used instead. Clearly, the implementation is involved and heavy with details of storage, which may obscure the global picture.

Fortunately, this is unnecessary in C++, because pointers can be used to form the required data structures and objects (Figure 4.4). In fact, the nonzero elements in the ith row in A can be stored in a connected list, avoiding storage of unnecessary zero elements. More specifically, A can be implemented as a list of N items, each of which is by itself a connected list of integers. These integers are just the column indices j of the corresponding nonzero elements $A_{i,j}$. In other words, an edge (i, j) has access to its second endpoint j, so we have one half of the required edge-to-node data access. Furthermore, a node i has access to the ith item in the list, which contains the nonzeroes in the ith rows in A, or the edges issuing from i. Thus, we also have one half of the required node-to-edge data access. If A^t is stored in the same way, then we have complete edge-to-node and node-to-edge data access, yielding the economic implementation of the prototype coloring algorithm with two short inner loops over the connected lists that store the ith row in A and the jth row in A^t (where i and j are the endpoints of the current edge in the outer loop, or the indices of $A_{i,j}$, the nonzero element scanned in the outer loop). With this implementation, the total number of calculations in the entire prototype algorithm is reduced from $N^2\bar{e}^2$ to only $N\bar{e}^2$.

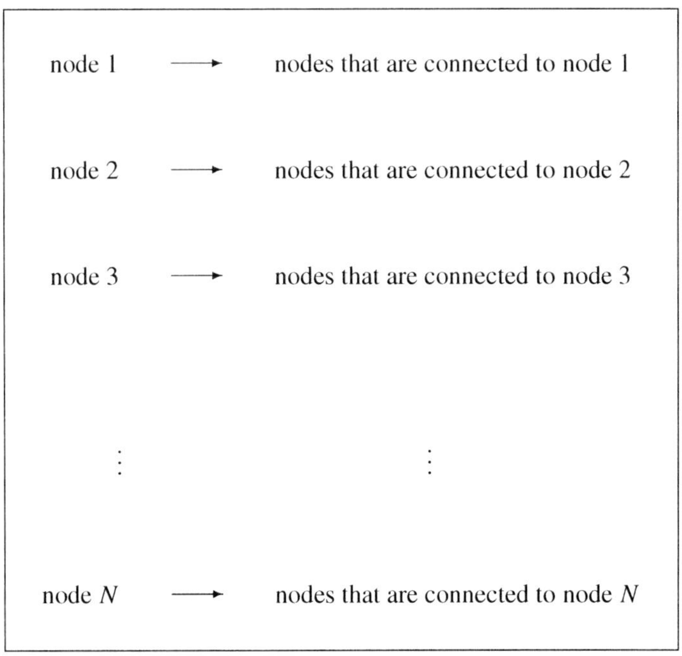

Figure 4.4. *Implementation of a graph in C++. Each node points to the nodes that are connected to it. The edges are implemented implicitly through these pointers.*

The above implementation of the transpose matrix A^t is actually equivalent to the Harwell–Boeing method for storing sparse matrices (Chapter 18, Section 17). Of course, the present implementation is much more transparent because it uses real rather than virtual addresses. In fact, because matrix rows are stored in connected lists, there is no need to store the nonzero matrix elements in the long array of edges as before. After all, their physical address in the computer memory is available in the pointers in the connected lists, so no virtual address is needed. The complete C++ implementation is available in Chapter 16.

Note that, in the C++ implementation, the nonzero matrix elements are no longer stored in a continuous array. Thus, the outer loop must be carried out by looping over the nodes $i = 1, 2, 3, \ldots, N$ and, for the ith node, over the connected list in which the ith row is stored. This way, the outer loop is carried out using indirect rather than direct indexing.

The inner loop can also be carried out efficiently. Because the items in the connected list contain the integer j, the column index of $A_{i,j}$, both i and j are available to carry out the two short inner loops over the ith row in A and the jth row in A^t to check whether a particular color has already been used or not.

The C++ implementation has another important advantage: because the nonzero elements are stored in connected lists, it is particularly easy to modify the matrix by adding new nonzero elements or dropping old ones (or, equivalently, introducing new edges to the graph or dropping old ones from it). This extra flexibility of the C++ implementation is particularly useful in the context of triangulation, discussed next.

4.5 Triangulation

Let's use the downward approach to implement triangulation. In order to define a triangulation, we first need to define a shape in a graph.

A shape of k nodes in a graph is a set of k nodes

$$i_1,\ i_2,\ i_3,\ldots,\ i_k$$

that are connected to each other by k edges that together form a circle:

$$(i_1, i_2),\ (i_2, i_3),\ (i_3, i_4),\ldots,\ (i_{k-1}, i_k),\ (i_k, i_1).$$

A subshape of this shape is a subset of the above set of nodes that is also a shape in its own right. For example, if i_1, i_2, and i_4 are connected by the edges (i_1, i_2), (i_2, i_4), and (i_4, i_1), then they form a subshape of the above shape.

A triangle is a shape of three nodes. In the above example, i_1, i_2, and i_4 form a triangle.

We are now ready to define a triangulation. A triangulation is a nonoriented graph that is embedded in the Cartesian plane (so each node can be described in terms of the Cartesian coordinates (x, y), and an edge is a straight line connecting two nodes in the Cartesian plane) and satisfies the following conditions:

1. Edges cannot cross each other.

2. Every shape of $k > 3$ nodes contains a subshape of three nodes (triangle).

3. Each node is shared by at least two edges as their joint endpoint.

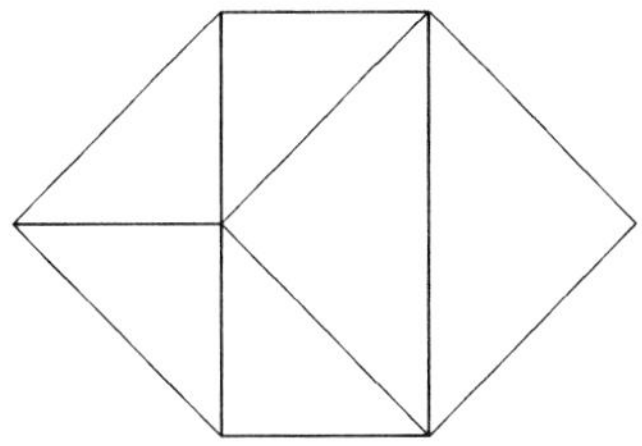

Figure 4.5. *A triangulation, or conformal mesh of triangles.*

4. A node cannot lie in between the endpoints of an edge unless it coincides with one of them.

Note that the first condition guarantees that the triangulation is a planar graph, the second guarantees that it contains only triangles, the third guarantees that it contains no dangling nodes or edges, and the fourth guarantees that it is conformal. Thus, a triangulation is actually a conformal mesh of triangles (see Figure 4.5).

In the next section, we'll see how the downward implementation is particularly suitable for triangulation.

4.6 Example: The Triangle-Coloring Problem

The triangle-coloring problem is defined as follows: use the minimal possible number of colors (denoted by $1, 2, 3, \ldots, \mathcal{C}$) to color the triangles in the triangulation in such a way that adjacent (edge-sharing) triangles are colored by different colors.

The triangle-coloring problem can also be formulated in terms of matrices. For this purpose, it is particularly convenient to represent the triangulation as an $N \times N \times N$ cube of integers. In this representation, we avoid the edge object, which is irrelevant in the triangle-coloring problem, and use nodes and triangles only. In fact, the triangulation is equivalent to a cube (three-dimensional matrix) A defined by

$$A_{i,j,k} \equiv \begin{cases} -1 & \text{if } i < j < k \text{ and } (i, j), (j, k), (i, k) \in E, \\ 0 & \text{otherwise} \end{cases}$$

(where E is the set of edges in the triangulation). In this representation, A is actually a function of three integer variables: i, j, and k. Because the triangulation is in particular a nonoriented graph, it is sufficient to consider the case $i < j < k$ (otherwise, i, j, and k could be interchanged to satisfy these inequalities). In this case, $A_{i,j,k} = -1$ if and only if i, j, and k belong to the same triangle in the triangulation. Thus, the cube A is indeed equivalent to the triangulation.

The above implementation can be considered as an upward implementation (Figure 4.6). Indeed, the nodes have the virtual addresses $1, 2, 3, \ldots, N$ to refer to them, but the triangles are not implemented physically and have no physical or virtual address. The function that indicates the existence of triangle, $A_{i,j,k}$, often returns the zero value, hence requires the large storage complexity of N^3.

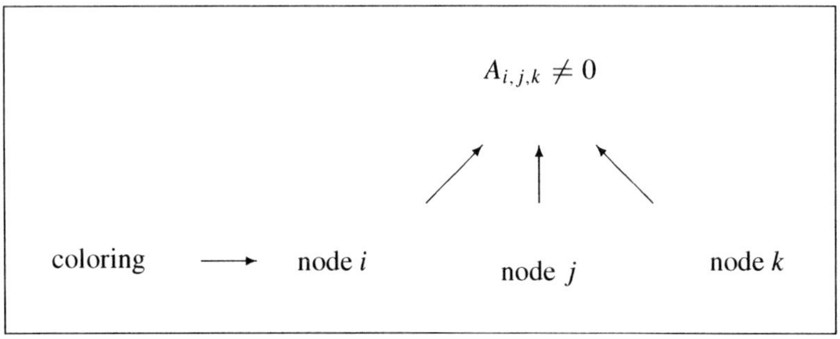

Figure 4.6. *Naive upward implementation of triangulation. Nodes are implemented and stored in an array. Triangles are not implemented but only indicated in the function $A_{i,j,k}$, which takes a nonzero value if and only if i, j, and k form a triangle (cubic storage and time).*

Although the coloring algorithm in Section 4.2 may be extended to a triangle-coloring algorithm, it would require a large computational complexity of N^3. Indeed, every cell in the cube A must be entered at least once to check whether or not it contains a nonzero integer. This complexity is prohibitively large; the only cure is to use the downward approach.

4.7 Downward Implementation

The basic mathematical objects in the triangulation are the nodes. We assume that the nodes are already implemented and numbered by the index $i = 1, 2, 3, \ldots, N$.

The more complex objects in the triangulation are the triangles, each of which consists of three vertices. The triangles have not yet been implemented. In fact, in the above implementation, they are not implemented at all. This is why this implementation is called upward implementation: it focuses on the basic node objects, which are used to access the information about the existence of the triangle through the function $A_{i,j,k}$. As we have seen above, this implementation requires too much storage and is also too slow for the triangle-coloring problem.

This is why we switch here to the downward implementation. This approach focuses on the more complex mathematical objects in the problem area: the triangles (Figure 4.7). Each triangle is implemented as a triplet of pointers to the nodes that are used in it as vertices. The nonexisting triangles (for which $A_{i,j,k} = 0$) are now avoided and not implemented at all. Thus, the implementation is much more economic in terms of storage: it requires only $N\bar{t}$ storage units, where $\bar{t}$ is the average number of triangles per node, which is usually pretty close to 1.

The downward implementation is also most suitable for the triangle-coloring problem. After all, the triangles are the objects to be colored in it, so they must be implemented properly. The prototype triangle-coloring algorithm is similar to the prototype graph-coloring algorithm in Section 4.3. It reads as follows. Loop over the individual triangles in the triangulation. For each one, use an inner loop to scan the former triangles to find the adjacent ones and avoid using their color. This completes the definition of the prototype triangle-coloring algorithm.

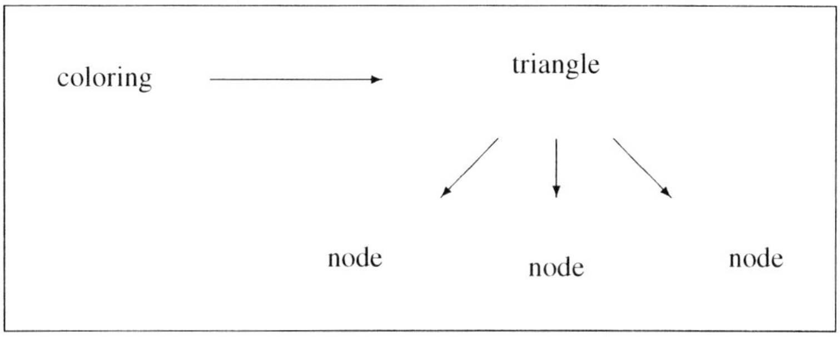

Figure 4.7. *Downward implementation of triangulation. Each triangle is implemented as an object that points to its three vertices, providing triangle-to-node (but not node-to-triangle) data access (linear storage, quadratic time).*

The prototype algorithm is still rather expensive in terms of time. Indeed, because of the nested loops over the triangles, it requires $N^2\bar{t}^2$ time units. This is because the downward implementation uses only triangle-to-node data access, by which a triangle "knows" the addresses of its vertices, but not node-to-triangle data access, so a node doesn't "know" to which triangles it belongs. This is why the inner loop is so long.

The above algorithm can be improved in much the same way as in Section 4.4 above. Since we also want to have node-to-triangle data access, the triangulation should be implemented as a list of connected lists. More specifically, the list contains N items, each of which is by itself a connected list. For $1 \leq i \leq N$, the ith connected list stores the addresses of the triangles that use the ith node as a vertex.

The long inner loop in the above prototype algorithm can now be avoided. Since the vertices of the current triangle in the outer loop have access to the triangles that use them as a vertex, the adjacent triangles can be found easily as the triangles that share two vertices with the current triangle. The current triangle can then be colored by the first color that has not yet been used to color the adjacent triangles. The complexity of the algorithm is thus reduced to $N\bar{t}^2$.

Note that the items in the above connected lists are pointers-to-triangles rather than triangles. This is because a triangle appears in three different connected lists. Because the triangle object cannot be constructed more than once, its address must be used in these three connected lists to allow node-to-triangle data access.

Because the above connected lists don't store the physical triangles, it still remains to decide what is the best data structure to store the triangles themselves physically in the computer memory. This question is discussed next.

4.8 Separation of Information

The downward approach focuses on the more complex mathematical objects in the problem area and implements them first, assuming that the more elementary objects are already available. In triangulation, this means that triangles are implemented as triplets of the addresses of their vertices. The triangle object is then available for further use in the coloring algorithm and other useful algorithms.

One question, however, is still unanswered: what is the best data structure to store the triangle objects? After all, the triangles in the triangulation are related to each other and surely shouldn't be separated from each other in the computer memory. The data structure in which they are stored must reflect the relation between the triangles.

At first glance, a suitable data structure to store the triangles is a nonoriented graph. Each node in this graph would represent a triangle, and two nodes in the graph would be connected by an edge if and only if the corresponding triangles were adjacent to each other in the triangulation.

The above approach, however, has a major drawback: it is not sufficiently flexible. Indeed, adding new triangles to an existing triangulation is too complicated, since it requires removing existing edges and introducing new ones in the graph. A more easily modified data structure is required.

Furthermore, it turns out that the edges in the above graph are completely unnecessary. Indeed, information about the adjacency of triangles can be obtained from their vertices: two triangles are adjacent if and only if they share two vertices. There is absolutely no need to store this information once again in the form of a graph edge.

The above discussion leads to the concept of separation of information. This principle means that, in a system with both elementary and complex objects, it is advisable to store as much technical information as possible in the elementary objects and to hide it from the complex objects. The complex (high-level) objects are then free to help implement the mathematical algorithm in its original spirit, while the elementary (low-level) objects take care of the technical details that are irrelevant or secondary in the algorithms and applications.

In our case, the nodes are the elementary (low-level) objects, and the triangles are the complex (high-level) objects. Thus, the node objects are responsible for storing technical data (such as their geometric location in the Cartesian plane). Indeed, this information is completely irrelevant in the triangle-coloring problem and algorithm, which are actually defined in terms of triangles and the adjacency relation only. Therefore, an adjacency function is needed that takes two triangle arguments and checks whether or not they share two vertices. The coloring algorithm should use this function in the inner loop as a black box, without knowing or caring how it actually works. This is in the spirit of the principle of separation of information: the algorithm deals with triangles and their relations only and should be independent of any geometric consideration.

The concept of separation of information is kept particularly well in C++. Indeed, in C++ the geometric information can be kept private in the "node" class and unknown to the outer world, including triangle objects. This information is never used in the coloring algorithm. Indeed, because triangles are implemented as triplets of pointers-to-nodes, the adjacency function needs only to compare the addresses of nodes in two triangle objects to find out whether or not these addresses coincide. The triangle objects themselves are thus free from any geometric data and ready to be colored in the coloring algorithm.

Thus, a suitable data structure for storing the triangles should disregard their location in the mesh and focus on its main task: to provide a flexible storage mechanism. The suitable data structure for this job is not a graph but rather a connected list. Indeed, in a connected list, the outer loop in the triangle-coloring algorithm can be carried out easily. Furthermore, the connected list is highly flexible and allows inserting new triangles in the triangulation and removing old ones from it if necessary. The connected list only stores the triangles and disregards any relation between them (including the relation of adjacency, which is checked by other means). This is why it provides the most suitable storage method for the individual triangles.

In the next section, we illustrate the suitability of the present implementation not only for the coloring algorithm but also for numerical algorithms, such as those implemented later in this book.

4.9 Application in Numerical Schemes

Realistic applications in applied science and engineering are often formulated in terms of partial differential equations (PDEs). The solution to the PDE is a function defined in the domain in which the PDE is originally given. Unfortunately, PDEs can rarely be solved analytically; they can often be solved only approximately, using a numerical scheme. In the numerical scheme, the domain is approximated by a mesh, the solution is a discrete grid function defined only at the discrete nodes in the mesh, and the original PDE is replaced by a suitable difference equation in terms of values at the nodes in the mesh. The solution of this discrete system of equations is called the numerical solution. If the numerical scheme approximates the original problem sufficiently accurately and adequately, then the numerical solution is a good approximation to the solution of the original PDE.

In most applications, the domain in which the PDE is defined is complicated and may have a curved and irregular boundary. It is thus particularly important to have a sufficiently fine mesh that approximates it as accurately as possible. The natural candidate for this task is triangulation.

The triangulation may contain triangles of variable size. In particular, it may use small triangles near curved or irregular boundary segments to provide the high resolution required there and bigger triangles where only lower resolution is required. This flexibility is particularly attractive in realistic applications in numerical modeling. This is why triangulation is used often in numerical schemes.

In order to construct the discrete system of equations that approximates the original PDE, one must loop over the triangles and assemble their contributions (see Chapter 12, Section 5). For this purpose, the relevant object is the triangle, not the node. The downward implementation used above is thus indeed the appropriate approach, because it focuses on the triangles rather than the nodes.

In numerical schemes, one often needs to refine the triangulation by adding smaller and smaller triangles, until a sufficiently fine triangulation that approximates the original domain well is obtained. It is thus most important to store the triangulation in a flexible data structure that supports adding new triangles and removing old ones. The original downward implementation in Section 4.7 is most suitable for this purpose. In this approach, the triangles may be stored in a connected list, which gives the opportunity to add or remove triangles easily and efficiently. Because the more complex objects, the triangles, are implemented explicitly, triangle-to-node data access is available. Node-to-triangle data access, on the other hand, is unavailable, because it requires extra data structures that are not easy to modify when new triangles are added to the triangulation or old ones are removed from it. Although this may require nested loops over the entire list of triangles, this is a price worth paying for the sake of having an easily refined mesh. In fact, because in the downward implementation nodes are never stored in any array or list, it is particularly easy to introduce new nodes or remove old ones. In fact, a new node object can be created by the "new" command and its address placed in the triangles that use it as a vertex.

Thus, the downward approach is a particularly useful tool in the implementation of triangulation. The required hierarchy of objects (node, triangle, triangulation) is particularly well implemented in C++, as is indeed done in Chapter 13.

4.10 Exercises

1. Use the downward approach in Section 4.4 to implement oriented graphs efficiently. The solution can be found in Chapter 16 in terms of sparse matrices.

2. Use your code from the previous exercise to implement the improved version of the graph-coloring algorithm in Section 4.3 for oriented graphs.

3. Use the downward approach in Section 4.4 to implement nonoriented graphs efficiently. The solution can be found in Chapter 16 in terms of upper triangular sparse matrices.

4. Use your code from the previous exercise to implement the improved version of the graph-coloring algorithm in Section 4.3 for nonoriented graphs.

5. Use the downward approach in Section 4.7 to implement triangulation efficiently. The solution can be found in Chapter 13.

6. Use your code from the previous exercise to implement the prototype triangle-coloring algorithm in Section 4.7. What is the complexity in your code? Can it be reduced?

Chapter 5

Algorithms and Their Object-Oriented Implementation

In this chapter, we discuss computational problems, solution methods, and their efficient implementation. We describe different approaches to writing algorithms to solve a particular problem and compare their storage and computation requirements. The abstract objects used here help not only to implement algorithms but also to develop them in the first place and modify them later on if necessary. We illustrate these points in the implementation of the "polynomial" object, along with useful arithmetic operations and functions. This object is particularly useful in high-order finite elements, discussed later in the book.

5.1 Ideas and Their Implementation

The programmer has an idea in mind of how to complete a particular task. This idea can usually be expressed in words in a natural language such as English. This formulation makes the idea clearer and more practical; indeed, when you explain your idea to a friend or colleague, it becomes clearer and more concrete to you too.

Describing the idea in words, however, is usually insufficient from a practical point of view. A more useful description must take the form of a list of operations that can be carried out one by one to produce the required solution to the problem under consideration. This is called an algorithm.

The individual operations in the algorithm are still written in a natural language, which may be somewhat vague and ambiguous. In fact, the interpretation of the terms in the natural language may well depend on the context in which they are used. Therefore, the algorithm must contain only unambiguous (context-free) instructions that can be carried out by a human being or a machine.

When the idea is about how to solve a computational problem, it can often be written in context-free formal language, using mathematical symbols, structures, and objects. This is really the best way to formulate the idea, because then the algorithm derived from it can also be written in terms of unambiguous mathematical instructions.

The original idea is useless unless it can be communicated to people and machines. While humans can usually understand it in (context-dependent) natural language, machines must have an explicit algorithm, written in context-free formal language. Translating the idea into a formal algorithm is also helpful for the developers themselves, because it gives them the opportunity to debug it and check whether or not it indeed does what it is supposed to do.

The most important step in using the idea is, thus, to formulate it in mathematical language. This is where C++ comes to our aid: it provides the framework for defining the required mathematical objects. These objects may serve as words in the formal language in which the algorithm is written. Writing the original idea and algorithm in a high-level programming language like C++ is called "implementation."

Actually, the object-oriented approach not only helps to implement the idea and algorithm in their original spirit but also provides the terms and objects required to think about them and develop them in the first place. By introducing useful terms and objects, it provides the required vocabulary to develop, express, and reshape the original raw idea, until it ripens to its final form.

5.2 Multilevel Programming

In a high-level programming language, objects such as characters and numbers are available. One can define variables that may take different values and manipulate them with unary and binary operations and functions. These elementary objects, however, are usually insufficient for implementing practical algorithms. In fact, even for elementary algorithms, abstract mathematical objects are needed. Although these objects can in theory be implemented using characters only, as in the original Turing machine, this approach is, of course, highly impractical. Traditional programming languages that use arrays are of course better, but the implementation of complex data structures in them may still be too complicated and hard to read, use, and modify if necessary. A programming language with a high level of abstraction is clearly necessary to implement complex mathematical structures.

With an object-oriented language like C++, one can implement sophisticated algorithms using objects that are unavailable in the standard language. This is called "high-level programming." In this approach, the programmer assumes that the required objects and functions that manipulate them are available. The actual implementation of these objects and functions can be delayed to a later stage, called "low-level programming."

High-level programming requires a good understanding of the algorithm and concepts and ideas behind it. The programmer who writes the high-level code should write footnotes about what objects exactly are needed and what functions should be available to manipulate them. These requirements should then be passed to a colleague who is experienced in low-level programming. The programmer of the high-level code can now continue implementing the algorithm in its original spirit, without being distracted by the details of the implementation of the objects used for this purpose.

Low-level programming may require knowledge and experience in computer hardware and memory. The programmer who does it should implement the objects as efficiently as possible according to the requirements passed on from the programmer of the high-level code.

The two programmers can thus work rather independently. The programmer of the high-level code can concentrate on realizing the true meaning of the original algorithm, having every abstract object available, while the programmer of the low-level code can concentrate on the optimal storage-allocation strategy to implement the objects.

The above workplan is ideal. In practice, interaction between the two programmers is required, especially in complex numerical applications, where the efficiency requirements in the low-level programming may put constraints on the high-level programming. Still, dividing the project into high-level and low-level tasks is helpful as a starting point to help organize the work on the entire project.

This method of work is called two-level programming. It is suitable not only for a team with members who have different levels of knowledge and expertise but also for a single programmer who must do the entire project. This programmer can still benefit from dividing the job into high-level and low-level tasks and working on each of them separately with full attention.

If the low-level programming is done properly and the required objects are well implemented and sufficiently general, then they can be used in many algorithms and applications. Actually, the programmer of the low-level code can start implementing objects that are commonly used even before having specific requirements, thus creating a library of objects for future use. The objects should be well documented to make them useful for potential users, to implement more complex objects and create more advanced libraries. The process may continue, with higher and higher levels of programming that use objects from lower levels to define new ones. This is called multilevel programming; it actually contributes to the development of the standard programming language by introducing more and more objects ready for future use.

5.3 Information and Storage

Each step in the computational algorithm requires data to complete a particular instruction. These data can be either calculated or fetched from the computer memory. Obviously, data that have already been calculated should be stored for future use unless recalculation is cheaper than fetching.

Surprisingly, it often is. Storing and fetching can be so expensive and slow that it is no longer worth it. Furthermore, it involves the extra effort of allocating sufficient memory for new variables and giving them meaningful names to remind the user what kind of data they contain. All this can make the code less elegant and harder to read and debug.

One of the great advantages of C is the opportunity to use functions that return the required result, be it a number or a pointer to a sequence of numbers. This way, one can use a function to recalculate data rather than fetch it from memory, provided that this recalculation is not too expensive.

This feature is made yet more elegant in C++, where functions may take and return actual objects rather than mere pointers. The high-level programming that uses such functions is thus much more transparent and clear, because it avoids dealing with pointers or addresses in the computer memory. Surely, a function that takes and returns objects is far more useful and transparent than a function that takes and returns arrays of numbers.

C++ functions, however, may be slightly slower than the corresponding C functions, because the returned objects may be constructed by an extra call to the constructor of the class. For example, a C++ function may define and use a local object that contains the required result, but then, in order to be returned as output, this object must be copied (by the copy constructor of the class) to a temporary external object to store the result after the local object has vanished. Still, this slight overhead is well worth it for the sake of transparent and useful code. Furthermore, some C++ compilers support versions that reduce this overhead to a minimum.

In what follows, we'll illustrate the effectiveness of C++ in implementing the polynomial object and related algorithms.

5.4 Example: The Polynomial Object

Here, we show how convenient it is to implement polynomials as objects in C++. The object-oriented approach gives one the opportunity to define functions that take and return objects rather than pointers or arrays, thus avoiding the need to deal with details of storage. Furthermore, this approach enables the definition of useful arithmetic operations between polynomials, such as addition, multiplication, etc.

We mainly consider two common problems: the multiplication of two polynomials and the calculation of the value of a polynomial at a given point.

Consider the polynomial

$$p(x) \equiv \sum_{i=0}^{n} a_i x^i,$$

where x is the independent variable, n is the degree of the polynomial (maximal power of x), and $a_0, a_1, \ldots, a_n$ are the coefficients.

The first question is how to store the polynomial. To answer this question, observe that the polynomial $p(x)$ is characterized by its coefficients $a_0, a_1, \ldots, a_n$. Thus, to store a polynomial, it is sufficient to store its coefficients.

In C, one would naturally store the coefficients in an array; but then again, passing them to a function involves getting into details of storage and distracts the programmer from the mathematical algorithms.

It is far more efficient to do this in C++, using, e.g., the "list" object in Chapter 3, Section 4. Indeed, the "polynomial" class can be derived from a list of numbers, so the "polynomial" object is actually a list of coefficients. This object can then be passed easily to functions by reference as usual, regardless of its internal implementation. Furthermore, "polynomial" objects can also be returned as output from functions and used further as input in subsequent calls to other functions.

Because the type of x and the a_i's is not yet specified, we use another powerful tool available in C++: templates. This way, the parameter 'T' in the "polynomial" template class stands for the type of independent variable and coefficients and can be used in the definition of the function. Because the particular type is immaterial in the algorithm used in the function, the template also means that the type can be disregarded and the mathematical concepts in the algorithm can be focused on. The concrete type substituted for 'T' will be specified later, when the compiler encounters calls to functions that use "polynomial" objects. These functions can then be called for polynomials of the specified type: integer, real, complex, etc.

Here is the code that derives the "polynomial" class from the base "list" class (Figure 5.1):

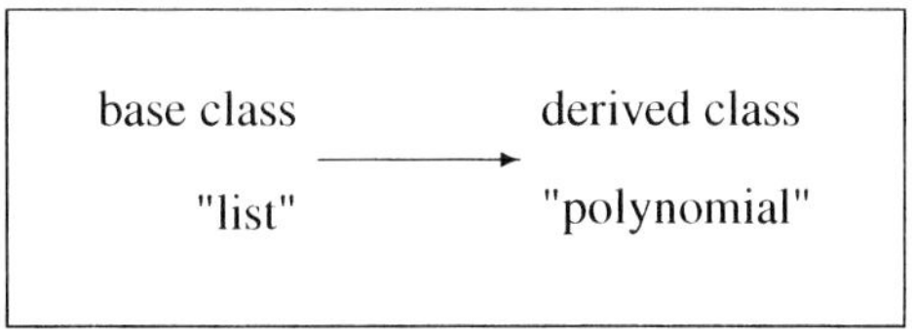

Figure 5.1. *Schematic representation of inheritance from the base class "list" to the derived class "polynomial".*

```
template<class T> class polynomial:public list<T>{
  public:
    polynomial(int n=0){
      number = n;
      item = n ? new T*[n] : 0;
      for(int i=0; i<n; i++)
        item[i] = 0;
    } // constructor

    polynomial(int n, const T&a){
      number = n;
      item = n ? new T*[n] : 0;
      for(int i=0; i<n; i++)
        item[i] = new T(a);
    } // constructor with 'T' argument
```

These constructors first implicitly invoke the default constructor of the base "list" class, which constructs a trivial list with no items in it. Thanks to the fact that the "item" field in the base "list" class is declared "protected" rather than private, the above constructors can access and reconstruct it to contain meaningful coefficients.

The copy constructor and assignment operator don't have to be defined, because the corresponding operators in the base "list" class do the right thing, that is, copy or assign the items in the argument one by one to the current object.

The following destructor also needs to do nothing, because the destructor of the base "list" class (invoked implicitly at the end of it) destroys first the individual items in the underlying list and then the list itself, as required:

```
    ~polynomial(){
    } // destructor
```

The following member function returns the degree of the polynomial:

```
    int degree() const{
      return number-1;
    } // degree of polynomial
};
```

This concludes the block of the "polynomial" class.

Next, we also implement operators that add two polynomials. In particular, the "+=" operator takes two "polynomial" arguments and adds the second one to the first one:

```
template<class T>
const polynomial<T>&
operator+=(polynomial<T>& p, const polynomial<T>&q){
  if(p.size() >= q.size())
    for(int i=0; i<q.size(); i++)
      p(i) += q[i];
  else{
    polynomial<T> keepQ = q;
    p = keepQ += p;
  }
  return p;
}  //  add polynomial
```

This operator works as follows. If the degree of 'p' is larger than or equal to the degree of 'q', then the coefficients in 'q' are added one by one to the corresponding coefficients in 'p'. In the addition of individual components, the "operator()" inherited from the base "list" class is used to change the item "p(i)" on the left, whereas the "operator[]" is used to read the item "q[i]" on the right.

If, on the other hand, the degree of 'p' is smaller than that of 'q', then the above algorithm is no longer applicable. Instead, an inner call to the "+=" operator is made, with the roles of 'p' and 'q' interchanged.

Because the "+=" operator is implemented here as a nonmember function, its first (nonconstant) argument cannot be a temporary variable returned as output from another function. Indeed, the compiler won't accept such a call, because it makes no sense to change a temporary object that is going to disappear soon anyway. This is why the extra "polynomial" object "keepQ" is defined and passed as the first argument to the inner call to the "+=" operator.

The above "+=" operator is now further used in the '+' operator, which returns the sum of two polynomials:

```
template<class T>
const polynomial<T>
operator+(const polynomial<T>& p,
        const polynomial<T>&q){
  polynomial<T> keep = p;
  return keep += q;
}  //  add two polynomials
```

Next, we implement the multiplication of a polynomial by a scalar. The "*=" operator takes two arguments, a polynomial and a scalar, and multiplies the first by the second:

```
template<class T>
const polynomial<T>&
operator*=(polynomial<T>& p, const T&a){
  for(int i=0; i<p.size(); i++)
    p(i) *= a;
  return p;
} // multiplication by scalar
```

The above "*=" operator is now used in the '*' operator that returns the product of a scalar and a polynomial:

```
template<class T>
const polynomial<T>
operator*(const T&a, const polynomial<T>&p){
  polynomial<T> keep = p;
  return keep *= a;
} // scalar times polynomial
```

Once the "polynomial" object, along with its arithmetic operations and other useful functions, is properly implemented as above, it can be placed in a library of objects for further use in high-level programming, such as the implementation of high-order finite elements (Chapter 15). Because the details of implementation are well hidden in the private part of the "polynomial" class, the implementation can later be changed if necessary, without affecting the high-level codes and with no need to debug them again. For example, if one deals mostly with sparse polynomials with only a few nonzero coefficients a_i, then it makes sense to store only these nonzero coefficients in a connected list (as in Chapter 16, Section 3) rather than a list. The low-level programming required for this change has absolutely no effect on codes that use "polynomial" objects, provided that the interface remains the same; that is, the reimplemented functions must still take and return the same arguments as before.

Here, we are not particularly interested in sparse polynomials, so we stick to our original implementation of a polynomial as a list of coefficients. The reason for this will become clear in Section 5.13.

5.5 Multiplication of Polynomials

Let us now consider the problem of multiplying two polynomials. Let $q(x)$ be the polynomial

$$q(x) \equiv \sum_{i=0}^{k} b_i x^i .$$

The product polynomial pq is given by

$$
\begin{aligned}
(pq)(x) &\equiv p(x)q(x) \\
&= \sum_{i=0}^{n}\sum_{j=0}^{k} a_i b_j x^{i+j} \\
&= \sum_{m=0}^{n+k}\left(\sum_{i+j=m} a_i b_j\right) x^m \\
&= \sum_{m=0}^{n+k}\left(\sum_{j=\max(0,m-n)}^{\min(m,k)} a_{m-j} b_j\right) x^m.
\end{aligned}
$$

Thus, the required polynomial pq is of degree $n+k$, and its coefficients are given in the above formula in terms of the coefficients of p and q.

Once the polynomials are implemented as objects, the above formula can be used to define the "operator*()" that takes two polynomials p and q and produces their product pq:

```
template<class T>
polynomial<T>
operator*(const polynomial<T>&p,const polynomial<T>&q){
  polynomial<T> result(p.degree()+q.degree()+1,0);
  for(int i=0; i<result.size(); i++)
    for(int j=max(0,i-q.degree());
        j<=min(i,p.degree()); j++){
      if(j == max(0,i-q.degree()))
        result(i) = p[j] * q[i-j];
      else
        result(i) += p[j] * q[i-j];
    }
  return result;
} // multiply two polynomials
```

The above '*' operator is also used to define the "*=" operator:

```
template<class T>
polynomial<T>&
operator*=(polynomial<T>&p, const polynomial<T>&q){
  return p = p * q;
} // multiply by polynomial
```

The following program defines a polynomial with three coefficients of value 1, namely, $p(x) = 1 + x + x^2$. Then, it calls the '*' operator to produce the polynomial p^2 and uses the "print()" function in the base "list" class to print it onto the screen:

```
int main(){
  polynomial<double> p(3,1);
  print(p * p);
  return 0;
}
```

5.6 Calculation of a Polynomial

Another common task that involves the polynomial p is to calculate its value $p(x)$ at a given point x. Here, we also benefit from the present implementation of the polynomial object, which gives us the opportunity to pass it to the function as a whole, without bothering with storage details.

Here is the function that takes the polynomial p and the point x and returns $p(x)$:

```
template<class T>
const T
calculatePolynomial(const polynomial<T>&p, const T&x){
  T powerOfX = 1;
  T sum=0;
  for(int i=0; i<p.size(); i++){
    sum += p[i] * powerOfX;
    powerOfX *= x;
  }
  return sum;
} // calculate a polynomial
```

Note that we have used here the local variable "powerOfX" to store the powers x^i used in the polynomial. This extra variable slightly reduces the elegance of the code. In what follows, we'll see an improved algorithm that is not only more efficient but also more elegantly implemented.

5.7 Algorithms and Their Implementation

So far, we have discussed mostly implementation issues. We mentioned that object-oriented languages such as C++ give us the opportunity to divide the entire project into two parts: a low-level part, where elementary objects are implemented, and a high-level part, where these objects are actually used to implement the algorithm in its original spirit. This two-level approach also has the advantage that the well-implemented objects can be used not only in the present algorithm but also in other algorithms and applications. Actually, the low-level part of the project could extend to create an entire library of objects, ready for future use. Furthermore, the hierarchy of the libraries could be built one on top of the other, each of which uses objects from lower libraries to form more sophisticated objects. This is called multilevel programming.

The low-level part of the code, where frequently used objects are implemented, should be particularly efficient in terms of memory allocation and data access. The high-level part, on the other hand, where the mathematical algorithm is implemented, should be as efficient as possible in terms of operation count. Even more importantly, it should be modular, transparent, and reader friendly, to aid not only potential readers but also the programmer in the process of writing, debugging, and modifying if necessary.

Although efficiency is an important property, transparency and clarity may be even more important to guarantee the correctness and usefulness of the code. The objects, in particular, should be complete and ready to use in every future application. In particular, storage details should be hidden from the users, who should remain completely unaware of the internal structure of the objects and know them only by their mathematical properties, available through interface functions.

Moreover, we would even recommend compromising efficiency for the sake of transparency and clarity if necessary. For example, a function that returns an object uses an extra call to the copy constructor of the class to construct the temporary output object. Although this extra call reduces efficiency, it may be well worth it to make the code clearer. Indeed, the returned object may be further used as input in other calls to other functions in the same code line. Returning an object is far more appropriate than returning an array of numbers with no apparent interpretation as a complete object.

Most often, there is mutual interaction between the high-level and low-level programmers. The high-level programmer may pass requirements to the low-level programmer, but also receive feedback about how realistic the requirements are, and what the limits and drawbacks are in the required objects. This feedback may lead the high-level programmer to modify the original implementation and adapt it to hardware issues and limits.

Actually, even the developer of a mathematical algorithm is not entirely independent. An algorithm may look efficient in terms of theoretical operation count but require objects that are too hard to implement. Thus, although the algorithm is usually developed in a purely mathematical environment, more practical implementation issues shouldn't be ignored. (See also Chapter 18, Section 14.)

The first concern of the algorithm developer is, of course, to reduce the theoretical operation count. Fortunately, it turns out that algorithms that are efficient in this theoretical sense are often also straightforward and efficient in terms of practical implementation.

One of the common principles of efficient algorithm is to avoid recalculating data that can be easily fetched from the memory. In the next section, we'll see an algorithm that uses this principle in the present problem of calculating the value of a polynomial at a given point. We'll also see that this algorithm indeed has a straightforward and efficient implementation.

5.8 Horner's Algorithm

The first rule in efficient calculation is as follows:

Don't open parentheses unless absolutely necessary!

Indeed, the distributive law says:

$$A(B + C) = AB + AC,$$

where A, B, and C are members of some mathematical field. Now, the right-hand side, where no parentheses are used, requires two multiplications and one addition to calculate, whereas the left-hand side, where parentheses are used, requires only one addition and one multiplication. This is also the idea behind Horner's algorithm for calculating the value of a polynomial. In fact, this algorithm introduces in the polynomial as many parentheses as possible.

The polynomial in Section 5.6 is actually calculated directly from the formula

$$p(x) = \sum_{i=0}^{n} a_i x^i.$$

Because the right-hand side contains no parentheses, it requires $2n$ multiplications (n multiplications to calculate the powers x^i, and another n multiplications to multiply them by the coefficients a_i) and n additions. Can this number be reduced?

Yes, it can. The Horner algorithm uses the following formula:

$$p(x) = (\cdots(((a_n x + a_{n-1})x + a_{n-2})x + a_{n-3})\cdots)x + a_0.$$

The process starts from the innermost parentheses, where the term with coefficient a_n is calculated, and progresses gradually to the outer ones, to which the free coefficient a_0 is added at the end. Because of the large number of parentheses, the algorithm requires only a total of n multiplications and n additions and is also implemented nicely as follows:

```
template<class T>
const T
HornerPolynomial(const polynomial<T>&p, const T&x){
  T result = p[p.degree()];
  for(int i=p.degree(); i>0; i--){
    result *= x;
    result += p[i-1];
  }
  return result;
} //  Horner algorithm to calculate a polynomial
```

The algorithm is not only more efficient but also more efficiently implemented. Indeed, the code is slightly shorter than the code in Section 5.6 and avoids the extra variable "powerOfX" used there.

In the next section, we use the above algorithms to efficiently calculate the value of the single power x^n at a given point x. It turns out that, for this purpose, although Horner's algorithm is slightly less efficient, it is far more elegant and easy to understand and debug. This tradeoff between efficiency and transparency arises often not only in simple examples such as this one but also in complex applications, where elegance and transparency are crucial for the sake of well-debugged and useful code.

5.9 Calculation of a Power

Usually, a polynomial is calculated for a real or complex argument x. Here, however, we show that the concept of a polynomial with an integer argument is also helpful. Actually, this polynomial is never calculated explicitly, because its value is already available. Still, the actual representation of an integer number as a polynomial (as in Chapter 1, Section 18) helps to solve the present problem.

Consider the following problem: for a given x and a large integer n, calculate x^n efficiently. Of course, this can be done using n multiplications in the recursive formula

$$x^n = x \cdot x^{n-1}$$

(see Chapter 1, Section 17), but can it be done more efficiently?

The answer is yes, it can be calculated in $2\log_2 n$ multiplications only. For this purpose, we use the binary representation of the integer n as a polynomial in the number 2:

$$n = \sum_{i=0}^{k} a_i 2^i,$$

where the coefficients a_i are either 0 or 1. With this representation, we have

$$x^n = \Pi_{i=0}^{k} x^{a_i 2^i},$$

which is the product of all the $a_i 2^i$-powers of x. Now, the 2^i-power of x can be calculated by i applications of the substitution

$$x \leftarrow x^2.$$

The total cost of the calculation is, thus, at most $2\log_2 n$ multiplications.

The algorithm is implemented nicely as follows:

```
template<class T>
const T
power(const T&x, int n){
  T result = 1;
  T powerOfX = x;
  while(n){
    if(n % 2) result *= powerOfX;
    powerOfX *= powerOfX;
    n /= 2;
  }
  return result;
} // compute a power
```

In the body of the "while" loop above, the last digit in the binary representation of n is found by the modulus operation n % 2. Once this digit has been used, it is dropped by dividing n by 2 without residual. With this approach, the code is particularly elegant.

There is, however, an even better approach to calculating the power x^n. This approach is based on Horner's polynomial. Recall that Horner's representation of a polynomial $p(x)$ is actually based on the recursion

$$p(x) = \sum_{i=0}^{n} a_i x^i = a_0 + x p_1(x),$$

where

$$p_1(x) \equiv \sum_{i=0}^{n-1} a_{i+1} x^i$$

is a polynomial of degree $n-1$. In fact, the polynomial $p_1(x)$ can be reformulated recursively in the same way, leading eventually to the representation in Section 5.8.

Of course, the above recursive formulation is never used explicitly due to the large cost of constructing $p_1(x)$ as a "polynomial" object. This is why the code in Section 5.8

actually "opens up" the recursion and starts from the end of it (the innermost parentheses), going back to the beginning in an ordinary loop.

In the present problem, however, where the polynomial $p(2) = n$ is just the binary representation of the integer n, $p_1(2) = n/2$ is immediately available by dividing n by 2 without residual. In fact, we have

$$n = (n \% 2) + 2(n/2),$$

where $n/2$ means integer division without residual and $n\%2$ contains the residual (see Chapter 1, Section 18). This implies that

$$x^n = x^{n \% 2}(x^2)^{n/2}.$$

This leads to the following recursive implementation of the "power()" function:

```
template<class T>
const T
power(const T&x, int n){
  return n ? (n%2 ? x * power(x * x,n/2)
               : power(x * x,n/2)) : 1;
} // compute a power recursively
```

This way, the "power()" function contains only one command. This style is particularly efficient, because it avoids explicit definition of local variables. Because the "power()" function is called recursively $\log_2 n$ times, the total cost in terms of operation count is still at most $2\log_2 n$ multiplications, as before. However, one should also take into account the extra cost involved in the recursive calls. In fact, in each recursive call, the computer needs to allocate memory for the local variables 'x' and 'n'. Although 'x' is passed by reference, its address must still be stored locally in each recursive call. Nevertheless, this extra cost may be well worth it for the sake of short and elegant code. Indeed, the recursive reformulation of n is much shorter and simpler than the original explicit formulation, hence also easier to debug. Although debugging is unnecessary for the present well-tested codes, it may pose a crucial problem in complex applications. This is why writing short and elegant code is a most attractive skill.

5.10 Calculation of Derivatives

In the above discussion, it is assumed that one needs to calculate only the nth power of x, x^n. But what if all powers $x^2, x^3, \ldots, x^n$ are required? In this case, of course, it makes no sense to use the above algorithms. The original approach is much more sensible:

$$x^k = x \cdot x^{k-1}, \quad k = 2, 3, \ldots, n.$$

Indeed, this method computes all the required powers in n multiplications only.

The question raised now is where to store all these calculated powers. In C, one is forced to use an array. But then again, an array is not a meaningful object, and passing it to a function or returning it from a function makes little sense. The "list" object in Chapter 3,

Section 4, is far more suitable for this purpose. Although the advantage of the "list" object over the standard array may look tiny, it becomes essential in complex applications, where many functions are applied to objects returned from other functions. Thus, using suitable objects rather than standard arrays is most useful.

Consider, for example, the following problem: let $f \equiv f(x)$ be a function of the independent variable x. At a given point x, calculate the derivatives of f up to order n, denoted by

$$\begin{aligned} f^{(0)}(x) &= f(x), \\ f^{(1)}(x) &= f'(x), \\ f^{(2)}(x) &= f''(x), \\ f^{(k)}(x) &= f^{(k-1)'}(x), \end{aligned}$$

and so on. A suitable candidate for storing these derivatives for future use is the "list" object. For example, when

$$f(x) = \frac{1}{x},$$

we have

$$f^{(k)}(x) = -\frac{k}{x} f^{(k-1)}(x).$$

The code that calculates and stores these derivatives is as follows:

```
template<class T>
const list<T>
deriveRinverse(const T&r, int n){
  list<T> Rinverse(n+1,0);
  Rinverse(0) = 1/r;
  for(int i=0; i<n; i++)
    Rinverse(i+1) = -double(i+1)/r * Rinverse[i];
  return Rinverse;
} // derivatives of 1/r
```

This function returns the list of derivatives of $1/x$ up to and including order n. In the next section, we'll see how lists can also be used in the Taylor expansion of a function.

5.11 The Taylor Expansion

In this section, we discuss efficient ways to calculate the Taylor expansion of a function $f(x)$. Let x be fixed, and let h be a small parameter. Assume that f has sufficiently many derivatives in the closed interval $[x, x+h]$. The Taylor expansion of order n at $x+h$ gives the value of the function at $x+h$ in terms of the values of the function and its derivatives at x, plus an error term that involves the $(n+1)$th derivative at an intermediate point $x \le \xi \le x+h$:

$$f(x+h) = \sum_{i=0}^{n} \frac{f^{(i)}(x)h^i}{i!} + \frac{f^{(n+1)}(\xi)h^{n+1}}{(n+1)!}.$$

When f is sufficiently smooth, one can assume that the $(n+1)$th derivative of f is bounded in $[x, x+h]$, so for sufficiently large n the error term is negligible. In this case, a good approximation to $f(x+h)$ is given by

$$f(x+h) \doteq \sum_{i=0}^{n} \frac{f^{(i)}(x)h^i}{i!}.$$

The computational problem is to calculate this sum efficiently. This problem contains two parts: first, to find an efficient algorithm, and then to implement it efficiently and elegantly on a computer.

The elementary task in the implementation is to pass the required information about f and its derivatives at x to the computer. The function "Taylor()" in the code below must have as input the numbers $f(x)$, $f'(x)$, $f''(x)$, ..., $f^{(n)}(x)$ before it can start calculating the Taylor approximation to $f(x+h)$. As discussed above, these numbers are placed and passed to the function in a single "list" object. This way, the programmer who writes the function can disregard storage issues and concentrate on the mathematical algorithm.

Since the above Taylor approximation is actually a polynomial of degree n in h, one may use a version of the algorithm in Section 5.6. In this version, the terms in the polynomial are calculated recursively by

$$\frac{h^i}{i!} = \frac{h}{i} \cdot \frac{h^{i-1}}{(i-1)!}.$$

These terms are then multiplied by $f^{(i)}(x)$ (available in the input "list" object) and added on to the sum. We refer to this version as the standard algorithm; it is implemented as follows:

```
template<class T>
const T
Taylor(const list<T>&f, const T&h){
  T powerOfHoverIfactorial = 1;
  T sum=0;
  for(int i=0; i<f.size()-1; i++){
    sum += f[i] * powerOfHoverIfactorial;
    powerOfHoverIfactorial *= h/(i+1);
  }
  return sum;
} // Taylor approximation to f(x+h)
```

Note that the last item in the input list, which contains the $(n+1)$th derivative, is not actually used here; it is reserved for the purpose of estimating the error (see Chapter 6, Section 10).

The above standard algorithm requires a total of $3n$ multiplications and n additions. A more efficient algorithm is a version of the Horner algorithm in Section 5.8. This version is based on the observation that the Taylor approximation can be written in the form

$$\left(\cdots\left(\left(f^{(n)}(x)\frac{h}{n} + f^{(n-1)}(x)\right)\frac{h}{n-1} + f^{(n-2)}(x)\right)\frac{h}{n-2}\cdots\right)\frac{h}{1} + f^{(0)}(x).$$

The following code implements this formula:

```
template<class T>
const T
HornerTaylor(const list<T>&f, const T&h){
  T result = f[f.size()-2];
  for(int i=f.size()-2; i>0; i--){
    result *= h/i;
    result += f[i-1];
  }
  return result;
} // Horner algorithm for Taylor approximation
```

This code requires only $2n$ multiplications and n additions. Furthermore, its implementation is slightly shorter and more elegant.

The question is, though, is it worth it? The standard algorithm has one major advantage: it adds the terms in the natural order, from 0 to n. This is not only more in the spirit of the original formula but also more economic in terms of storage in some cases.

Consider, for example, a "short-memory" process, in which $f^{(i)}(x)$ depends only on the previous number $f^{(i-1)}(x)$, but not on the yet previous numbers $f(x)$, $f'(x)$, $f''(x)$, ..., $f^{(i-2)}(x)$ (Figure 5.2). In this case, it makes more sense to calculate $f^{(i)}(x)$ inside the loop and drop it once it has contributed to the sum and has been used to calculate the next number $f^{(i+1)}(x)$. Actually, the new number $f^{(i+1)}(x)$ can be stored in the same variable used previously to store the old number $f^{(i)}(x)$. This strategy, however, is possible only in the standard algorithm, where terms are added in the natural order, but not in Horner's algorithm, where they are added in the reverse order (Figure 5.3). Thus, the standard algorithm becomes in this case much more attractive, because it can avoid storing and passing input to the "Taylor()" function.

$$f(x) \longrightarrow f'(x) \longrightarrow f''(x) \longrightarrow f'''(x)$$

Figure 5.2. *Short-memory process: for fixed x, each derivative can be calculated from data about the previous one only.*

In our applications, however, we are mainly interested in "long-memory" processes, in which $f^{(i)}(x)$ depends not only on the previous number $f^{(i-1)}(x)$ but also on all the yet previous numbers $f(x)$, $f'(x)$, $f''(x)$, ..., $f^{(i-2)}(x)$ (Figure 5.4). In this case, the list of derivatives must be stored in its entirety as in the above codes, so the Horner algorithm is preferable thanks to its lower cost in terms of operation count.

We'll see examples of the short-memory process in Chapter 6, Section 9, and the long-memory process in Chapter 6, Section 14. In the next section, we illustrate how useful it is to pass entire lists to a function, particularly when all the items in them are combined to produce the required result.

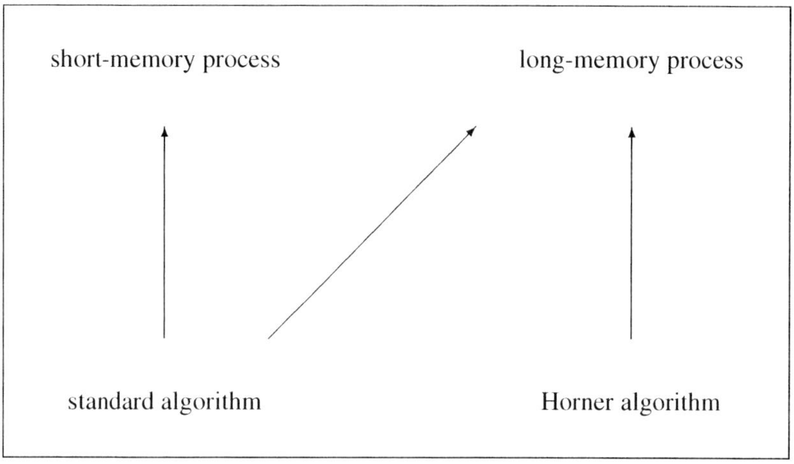

Figure 5.3. *Horner's algorithm is preferable in a long-memory process, where the derivatives must be calculated and stored in advance anyway, but not in a short-memory process, where they are better used and dropped, as can be done only in the standard algorithm.*

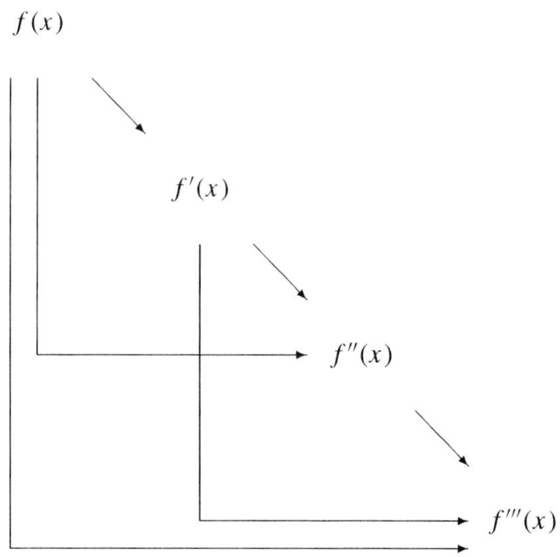

Figure 5.4. *Long-memory process: for fixed x, the calculation of each derivative requires data about all the previous ones.*

5.12 Derivatives of a Product

In this section, we show why it is particularly important to have the opportunity to store the list of derivatives as an object. For this purpose, we present a function that takes list arguments and combines all the items in them in the calculation. This function will be particularly useful in the long-memory process in Chapter 6, Section 14.

Assume that the derivatives of the functions $f(x)$ and $g(x)$ at a given point x up to and including order n are available. Compute the nth derivative of the product fg at x, denoted by

$$(fg)^{(n)}(x).$$

Note that this quantity has the same algebraic structure as Newton's binomial:

$$(f+g)^n = \sum_{i=0}^{n} \binom{n}{i} f^i g^{n-i}.$$

The only difference is that here the sum is replaced by a product, and the power is replaced by a derivative. Therefore, we have the formula

$$(fg)^{(n)} = \sum_{i=0}^{n} \binom{n}{i} f^{(i)} g^{(n-i)}$$

at the given point x.

This formula is implemented in the code below. The function "deriveProduct()" takes as input two "list" objects that contain the derivatives of f and g up to and including order n at the fixed point x and returns the nth derivative of fg at x. It is assumed that a global array of integers that contains Pascal's triangle is available. This array, named "triangle", is formed in advance, as in Chapter 1, Section 19, and placed in a global domain accessible to the "deriveProduct()" function. This seems to be a better strategy than recalculating the required binomial coefficients each time the function is called:

```
template<class T>
const T
deriveProduct(const list<T>&f,const list<T>g,int n){
  T sum = 0;
  for(int i=0; i<=n; i++)
    sum += triangle[n-i][i] * f[i] * g[n-i];
  return sum;
} //  nth derivative of a product
```

5.13 Polynomial of Two Variables

The polynomial $\sum a_i x^i$ in Section 5.4 is implemented as the list of coefficients a_0, a_1, $a_2, \ldots, a_n$. This implementation is appropriate because the polynomial is defined uniquely by its coefficients.

Still, one may ask, why use a list rather than a vector? After all, the coefficients are all of the same type, so they can be safely stored in a vector of dimension $n+1$. Furthermore, a vector is more efficient than a list thanks to direct indexing, that is, using an array to store the coefficients themselves rather than their addresses. This saves not only storage but also valuable time, by using efficient loops over the coefficients stored continuously in the computer memory. Why then use a list?

The answer is that, in some cases, the coefficients in the polynomial occupy different amounts of memory, and hence cannot be stored in an array. Consider, for example, the polynomial of two independent variables x and y:

$$p(x, y) \equiv \sum_{i+j \le n} a_{i,j} x^i y^j,$$

where i and j are nonnegative indices, $a_{i,j}$ are the given coefficients, and n is the degree of the polynomial. This polynomial may also be written in the form

$$\begin{aligned} p(x, y) &= \sum_{k=0}^{n} \sum_{i+j=k} a_{i,j} x^i y^j \\ &= \sum_{k=0}^{n} \left(\sum_{j=0}^{k} a_{k-j,j} \left(\frac{y}{x} \right)^j \right) x^k. \end{aligned}$$

In this form, $p(x, y)$ can be viewed as a polynomial of degree n in x, with coefficients that are no longer scalars but rather polynomials in y/x. In fact, the kth coefficient is by itself the polynomial of degree k in y/x, given by

$$a_k(y/x) \equiv \sum_{j=0}^{k} a_{k-j,j} \left(\frac{y}{x} \right)^j.$$

Thus, the original polynomial $p(x, y)$ can be implemented as the list $a_0(y/x)$, $a_1(y/x)$, $a_2(y/x), \ldots, a_n(y/x)$. Each item in this list is a polynomial in its own right, thus also implemented as a "polynomial" object. More specifically, the original polynomial $p(x, y)$ is implemented as a polynomial of polynomials, or a "polynomial<polynomial<T> >" object, whose kth item is the polynomial $a_k(y/x)$. The polynomial $a_k(y/x)$ is defined and stored in terms of its own coefficients $a_{k,0}, a_{k-1,1}, a_{k-2,2}, \ldots, a_{0,k}$.

Clearly, this implementation is possible thanks to the fact that the "polynomial" class is derived from the "list" class, which may contain items of different sizes (such as polynomials of different degrees). With this implementation, it is particularly easy to multiply two polynomials of two variables. (This operation is particularly useful in high-order finite elements in Chapter 15.) Here is how this is done.

Let $q(x, y)$ be another polynomial of two variables:

$$q(x, y) \equiv \sum_{k=0}^{m} b_k(y/x) x^k,$$

where $b_k(y/x)$ is by itself a polynomial of degree k in y/x. Then, the product of p and q is

$$\begin{aligned} (pq)(x, y) &\equiv p(x, y) q(x, y) \\ &= \sum_{k=0}^{m+n} \left(\sum_{j=\max(0,k-n)}^{\min(k,m)} a_{k-j}(y/x) b_j(y/x) \right) x^k. \end{aligned}$$

Note that the product of polynomials $a_{k-j}(y/x)b_j(y/x)$ in the above parentheses is by itself a polynomial of degree k in y/x, so the sum in these parentheses is also a polynomial of degree k in y/x. This sum of products can be calculated using arithmetic operations between polynomials of a single variable (Sections 5.4 and 5.5). Thus, the required product of p and q can be carried out in the same way as in Section 5.5; the only difference is that the coefficients a_{k-j} and b_j are polynomials rather than scalars, and hence the arithmetic operations between them are interpreted as arithmetic operations between polynomials rather than scalars. This interpretation is used automatically once the compiler encounters polynomials rather than scalars.

The multiplication operator in Section 5.5 can thus be used for polynomials of two variables as well. Indeed, it works just as before, with the template 'T' in it (denoting the type of coefficient) specified to be "polynomial". For example, the following code computes and prints to the screen the coefficients of the polynomial $(1 + x + y)^2$:

```
int main(){
  polynomial<polynomial<double> >
      p2(2,polynomial<double>(1,1));
  p2(1) = polynomial<double>(2,1);
  print(p2 * p2);
  return 0;
}
```

5.14 Integration of a Polynomial

Here we show how convenient it is to use the "polynomial" object to calculate integrals in certain domains. (This task is required, e.g., in high-order finite elements in Chapter 15.)

We start with a polynomial of one independent variable:

$$p(x) = \sum_{i=0}^{n} a_i x^i .$$

The integral of this polynomial in the unit interval [0, 1] is

$$\int_0^1 p(x)dx = \sum_{i=0}^{n} a_i \int_0^1 x^i dx = \sum_{i=0}^{n} \frac{a_i}{i+1}.$$

This formula is implemented in the following code:

```
template<class T>
const T
integral(const polynomial<T>&p){
  T sum = 0;
  for(int i=0; i<p.size(); i++)
    sum += (1./(i+1)) * p[i];
  return sum;
} // integral in the unit interval
```

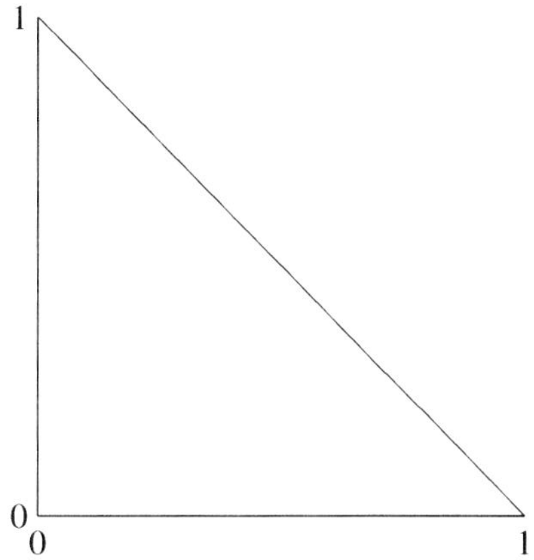

Figure 5.5. *The triangle in which the polynomial $p(x, y)$ is integrated.*

Next, we consider the problem of integrating a polynomial of two independent variables x and y in the right-angle triangle in Figure 5.5. Consider the polynomial $p(x, y)$ given by

$$\begin{aligned} p(x, y) &= \sum_{k=0}^{n} \sum_{i+j=k} a_{i,j} x^i y^j \\ &= \sum_{k=0}^{n} \sum_{j=0}^{k} a_{k-j,j} y^j x^{k-j}. \end{aligned}$$

The integral of this polynomial in the triangle in Figure 5.5 takes the form

$$\begin{aligned} \int p(x, y) dx dy &= \sum_{k=0}^{n} \sum_{j=0}^{k} a_{k-j,j} \int_0^1 \left(\int_0^{1-x} y^j dy \right) x^{k-j} dx \\ &= \sum_{k=0}^{n} \sum_{j=0}^{k} a_{k-j,j} \int_0^1 \frac{(1-x)^{j+1}}{j+1} x^{k-j} dx. \end{aligned}$$

Thus, we have every tool required for calculating this integral. Indeed, since $1 - x$ is a polynomial in x, and we already know how to multiply polynomials, we can compute the polynomials

$$(1-x)^{j+1} = (1-x)^j (1-x) \quad (j = 2, 3, 4, \ldots, n).$$

Furthermore, we know how to multiply these polynomials by the polynomials x^{k-j} and scalars $a_{k-j,j}/(j+1)$. Finally, we also know how to sum these polynomials and integrate in the unit interval. This completes the algorithm for integrating $p(x, y)$ in the triangle in Figure 5.5.

Here is the code that implements this algorithm:

```
template<class T>
const T
integral(const polynomial<polynomial<T> >&p){
  polynomial<T> sum(p.size()+1,0);
  polynomial<T> one(1,1);
  polynomial<T> x(2,0);
  x(1) = 1;
  polynomial<T> oneMinusX(2,1);
  oneMinusX(1) = -1;
  list<polynomial<T> > xPowers(p.size(),one);
  list<polynomial<T> > oneMinusXpowers(p.size()+1,one);
  for(int i=1; i<p.size(); i++)
    xPowers(i) = x * xPowers[i-1];
  for(int i=1; i<=p.size(); i++)
    oneMinusXpowers(i) = oneMinusX * oneMinusXpowers[i-1];
  for(int k=p.degree(); k>=0; k--)
    for(int j=0; j<=k; j++)
      sum += (p[k][j]/(j+1))
          * oneMinusXpowers[j+1] * xPowers[k-j];
  return integral(sum);
} // integral in the triangle
```

Although this function bears the same name as the previous function that integrates in the unit interval, no ambiguity occurs. Indeed, when the function is actually called, the compiler looks at its concrete argument. If it is a polynomial of one variable, then the previous version is invoked. If, on the other hand, it is a polynomial of two variables, then this version is invoked.

5.15 Exercises

1. Use the code in Section 5.5 to compute the coefficients of the polynomial
$$(1 + 2x + 2x^2 + x^3)^2.$$

2. Use the above code in a loop to compute the coefficients of the polynomial
$$(a + bx)^n,$$
where a and b are some real numbers and n is a large integer number. Verify that the result agrees with Newton's binomial:
$$(a + bx)^n = \sum_{i=0}^{n} \binom{n}{i} a^{n-i} b^i x^i.$$

3. Compare the efficiency and accuracy of the functions in Sections 5.6 and 5.8 in calculating the value of the polynomial
$$\sum_{n=0}^{N} x^n = \frac{x^{N+1} - 1}{x - 1}$$
at the points $x = 2$, $x = 3$, and $x = 4$.

4. It would be more in the spirit of object-oriented programming to rename the functions in Sections 5.6 and 5.8 as "operator()". This function should then be a member of the "polynomial" class and take only one argument 'x' to return the value of the current "polynomial" object at 'x'. This way, the function can be simply called as "p(x)" to return the value of the polynomial 'p' at 'x'. This way, the function doesn't act on 'p' but rather reflects its property to return different values for different values of 'x'. Why is this impossible here? Is it because the base "list" class already uses an "operator()" that shouldn't be overridden? Can you get around this problem?

5. Use the code segments in Section 5.9 to calculate
$$2^{10}, 2^{20}, 3^{17}, 7^{16}, \ldots .$$
Compare the results with those of the recursive "power()" function in Chapter 1, Section 17, in terms of efficiency and correctness.

6. Which code segment in Section 5.9 do you find particularly easy to read, understand, and modify? What writing style and programming strategy is most suitable for you?

7. Use the code in Section 5.10 to compute the lists of derivatives of the function $f(x) = 1/x$ at the points $x = 2$, $x = 3$, and $x = 4$.

8. Use the above lists and the code in Section 5.11 to approximate $1/(2.1)$, $1/(2.9)$, and $1/(3.9)$, using the Taylor expansion around $x = 2$, $x = 3$, and $x = 4$ (respectively). Verify that the error is indeed within the expected error estimate.

9. Which algorithm in Section 5.11 is more efficient in the calculation in the previous exercise?

10. Apply the code in Section 5.12 to construct the list of derivatives of the function $1/x^2 = (1/x)(1/x)$ at $x = 1$, $x = -1$, and $x = 0.5$.

11. Use the lists from the previous exercise in the code in Section 5.11 to obtain the Taylor approximation to $1/(0.9)^2$, $1/(-1.1)^2$, and $1/(0.45)^2$. Are the errors within the expected error estimates?

12. Apply the code in Section 5.12 with the above lists to construct the list of derivatives of the function $1/x^4 = (1/x^2)(1/x^2)$ at $x = 1$, $x = -1$, and $x = 0.5$.

13. Use the lists from the previous exercise in the code in Section 5.11 to obtain the Taylor approximation to $1/(0.9)^4$, $1/(-1.1)^4$, and $1/(0.45)^4$. Are the errors within the expected error estimates?

14. Construct the lists of derivatives of the functions $\sin(x)$, $\cos(x)$, and $\sin(2x)$ at the points $x = 0$, $x = \pi/4$, and $x = \pi/3$.

15. Use the lists of derivatives of $\sin(x)$ and $\cos(x)$ calculated above in the code in Section 5.12 to construct the list of derivatives of the function
$$\sin(2x) = 2\sin(x)\cos(x).$$
Compare the results with those from the previous exercise.

16. Use the lists from the previous exercise in the Taylor expansion around the above points, and verify that the errors are indeed within the expected error estimates.

17. Which code in Section 5.11 do you find particularly easy to read, understand, and modify? What writing style and programming strategy is most suitable for you?

18. Use the code in Section 5.13 to compute the coefficients of the polynomial of two variables
$$(1 + 2x + 2y + x^2 + xy + y^2)^2.$$

19. Use the above code in a loop to compute the coefficients of the polynomial of two variables
$$(ax + by)^n,$$
where a and b are some real numbers and n is a large integer number. Verify that the result agrees with Newton's binomial:
$$(ax + by)^n = \sum_{i=0}^{n} \binom{n}{i} a^i b^{n-i} x^i y^{n-i}.$$

20. Use the code in Section 5.14 to calculate the integral of the above polynomials in the triangle in Figure 5.5. Verify that the result is indeed correct by calculating it manually.

21. Write a function that calculates the value of a polynomial of two variables $p(x, y)$ as follows. First, the polynomial is written in the form
$$p(x, y) = \sum_{k=0}^{n} a_k(y/x)x^k,$$
as in Section 5.13. In this form, $p(x, y)$ can be passed to the function as a "polynomial<polynomial<T> >" object, in which the coefficients are themselves polynomials. In fact, the kth coefficient in $p(x, y)$ is $a_k(y/x)$, which contains the coefficients $a_{k,0}$, $a_{k-1,1}$, $a_{k-2,2}, \ldots, a_{0,k}$. Now, the "HornerPolynomial" function in Section 5.8 is called to compute the individual $a_k(y/x)$. These values are stored in a local "polynomial<T>" object, to which the "HornerPolynomial()" function is applied to produce the required value $p(x, y)$.

22. Rewrite your code from the previous exercise in a short-memory approach, in which the $a_k(y/x)$'s are calculated one by one and deleted right after the term $a_k(y/x)x^k$ is calculated and added to the current sum that will eventually produce $p(x, y)$. In this approach, the $a_k(y/x)$'s don't have to be stored in a local "polynomial<T>" object. However, the final call to the Horner algorithm must be replaced by the standard algorithm in Section 5.6. How does your present code compete with the code in the previous exercise?

23. Let $f(x, y)$ be a given function of two variables, and let x and y be fixed. Consider the polynomial $p(h_x, h_y)$ of the two variables h_x and h_y, whose coefficients are given by

$$a_{i,j}\frac{1}{i!\cdot j!}\cdot\frac{\partial^{i+j} f}{\partial^i x \partial^j y}(x, y).$$

Actually, $p(h_x, h_y)$ is the Taylor approximation of $f(x + h_x, y + h_y)$:

$$f(x + h_x, y + h_y) \doteq p(h_x, h_y).$$

Use the algorithms in the two previous exercises to calculate the Taylor approximation of the function $f(x, y) = \sin(x)\cos(y)$ at $x = \pm\pi/4$, $y = \pm\pi/3$, $h_x = \pm 0.1$, and $h_y = \pm 0.1$. Verify that the approximation indeed improves as n (the degree of p) increases.

24. Compare your code from the previous exercise to your answer to a similar exercise at the end of Chapter 3. Do you obtain the same numerical results? Which code is more efficient? Which code is more transparent?

25. Use the guidelines in Section 5.13 to implement polynomials of three variables x, y, and z.

Chapter 6

Object-Oriented Analysis

In this chapter, we show how useful object-oriented programming can be not only in numerical applications but also in computational error estimates that are relevant in mathematical analysis. For this purpose, we first describe concisely ordinary differential equations (ODEs), along with some stability analysis and numerical schemes and their C++ implementation. The "interval" object implemented in the low-level code is then used in the numerical scheme to obtain the error estimates required to prove the existence of an asymptotic solution to a particular nonlinear ODE.

6.1 ODEs

The main objective of this chapter is to show how useful object-oriented programming can be not only in numerical applications but also in mathematical analysis. Indeed, once the "interval" object is well implemented and used in numerical schemes, it produces not only the required numerical solution but also a computational error estimate, which can be further used to prove the existence of an analytic solution.

We start with a concise introduction to ODEs, including stability and basic numerical schemes. Then, we focus on a particular nonlinear ODE, where object-oriented analysis is used.

In an ODE, one looks for an unknown function $u(t)$ of the single independent variable t $(0 < t < \infty)$ using available data about its derivative $u'(t)$. More specifically, the value of u at the initial time $t = 0$ is available, and the ODE tells us how u changes as t increases.

Consider, for example, the ODE

$$u'(t) = Su(t),$$

where S is a given constant. The ODE is also accompanied by the initial condition

$$u(0) = u_0,$$

where u_0 is a given parameter. The ODE combines with the initial condition to form an initial-value problem.

The solution to this initial-value problem is

$$u(t) = \exp(tS)u_0.$$

The solution can be calculated efficiently as in Chapter 1, Section 22. However, sometimes we are not particularly interested in the explicit solution itself but rather in its mathematical properties. In this case, no computation is needed; the answer may lie in the original ODE itself. This subject is discussed next.

6.2 Stability in the ODE

Stability is an important property of the ODE, which requires no explicit solution. (Stability in the ODE shouldn't be confused with stability of the numerical scheme, studied in Chapter 8.)

A stable solution to the ODE is a solution that remains bounded (in magnitude) as t increases. An unstable solution, on the other hand, is a solution that grows indefinitely as t increases. A stable ODE is an ODE with no unstable solution.

To check stability, no explicit solution is required. It is sufficient to look at the parameter S on the right-hand side of the ODE. Assume that S is a complex number of the form

$$S = \Re(S) + i \cdot \Im(S),$$

where $i = \sqrt{-1}$ is the imaginary number and $\Re(S)$ and $\Im(S)$ are the real and imaginary parts of S, respectively. Then, since

$$\exp(tS) = \exp(t\Re(S))\exp(it\Im(S)),$$

the solution $u(t)$ is unstable if and only if $\Re(S) > 0$. Thus, the ODE is stable if and only if $\Re(S) \leq 0$.

6.3 System of ODEs

In the above ODE, u has been interpreted as a scalar function. It is also possible to interpret it as a vector function:

$$u(t) = (u_1(t), u_2(t), \ldots, u_k(t)).$$

Here, $u(t)$ is a k-dimensional vector, with individual components $u_1(t), u_2(t), \ldots, u_k(t)$ that are scalar functions of t. Similarly, the derivative of $u(t)$, $u'(t)$, is interpreted as the k-dimensional vector

$$u'(t) = (u_1'(t), u_2'(t), \ldots, u_k'(t)).$$

The given parameter u_0 that contains the initial condition is also interpreted as a k-dimensional vector. Finally, the given parameter S on the right-hand side of the ODE is interpreted as a $k \times k$ matrix. With this new interpretation, the ODE is called a vector ODE or a system of ODEs.

The k-dimensional solution $u(t)$ of the vector ODE takes the same form as before:

$$u(t) = \exp(tS)u_0,$$

which is interpreted as the result of applying the $k \times k$ matrix $\exp(tS)$ to the k-dimensional vector u_0.

Computing the exponent of the matrix tS is, thus, the main task in solving the vector ODE. Fortunately, it can be done efficiently, as in Chapter 2, Section 22. Still, the task must be repeated for every time t under consideration, which could be particularly expensive. In what follows, we'll see how the cost can be reduced by using previous calculations.

Often, the solution is required not at every time but only in a mesh of discrete times:

$$t = h, 2h, 3h, \ldots, jh, \ldots,$$

where h is a fixed (and small) parameter called the meshsize. In such cases, it is advisable to compute and store only the exponent of the matrix hS and reuse it recursively to calculate the solution in the above mesh:

$$u(jh) = \exp(jhS)u_0 = \exp(hS)\exp((j-1)hS)u_0 = \exp(hS)u((j-1)h).$$

This way, the solution on the entire mesh requires no matrix-times-matrix operations, but merely matrix-times-vector operations, which are relatively inexpensive.

In some cases, the explicit solution of the system of ODEs is not at all required. All that is required are the properties or behavior of the solution. In such cases, the entire computation may be avoided; the required information may be deduced directly from the original system.

6.4 Stability in a System of ODEs

A system of ODEs is called stable if it has no unstable (unbounded) solution. In order to check whether a system is stable or not, no explicit solution is required; it is sufficient to look at the matrix S in the system. Indeed, let

$$S = J^{-1}\Lambda J$$

be the Jordan form of S, where J is a nonsingular matrix and Λ is a "nearly diagonal" matrix; that is, it contains the eigenvalues of S on its main diagonal, the numbers 0 or 1 in the diagonal just below it, and zeroes elsewhere. For example,

$$\Lambda = \begin{pmatrix} \lambda_1 & & & & \\ 1 & \lambda_1 & & & \\ & & \lambda_2 & & \\ & & 1 & \lambda_2 & \\ & & & 1 & \lambda_2 \end{pmatrix}.$$

In this example, λ_1 and λ_2 are the eigenvalues of S. (The zero elements in the matrix are not indicated.) The matrix Λ in the example contains two blocks (Jordan blocks): the first one (of order 2) corresponds to the eigenvalue λ_1, and the second one (of order 3) corresponds to the eigenvalue λ_2.

Using the Jordan form, the exponent of tS can be written in terms of the exponent of $t\Lambda$:

$$\begin{aligned}
\exp(tS) &= \sum_{n=0}^{infty} \frac{t^n S^n}{n!} \\
&= \sum_{n=0}^{infty} \frac{t^n (J^{-1}\Lambda J)^n}{n!} \\
&= J^{-1} \left(\sum_{n=0}^{infty} \frac{t^n \Lambda^n}{n!} \right) J \\
&= J^{-1} \exp(t\Lambda) J.
\end{aligned}$$

Of course, the Jordan form of S is not available explicitly, so the above formula is not proposed as a practical solution method for the original system. However, it might shed light on the stability question. Indeed, in the above example, we have

$$\exp(t\Lambda) = \begin{pmatrix} \exp(t\lambda_1) & & & & \\ \frac{t}{1!}\exp(t\lambda_1) & \exp(t\lambda_1) & & & \\ & & \exp(t\lambda_2) & & \\ & & \frac{t}{1!}\exp(t\lambda_2) & \exp(t\lambda_2) & \\ & & \frac{t^2}{2!}\exp(t\lambda_2) & \frac{t}{1!}\exp(t\lambda_2) & \exp(t\lambda_2) \end{pmatrix}.$$

(Larger Jordan blocks produce a similar structure.) Thus, stability depends on the real parts of the eigenvalues of S: negative real parts produce no instability, whereas positive real parts produce instability. Zero real parts produce instability if and only if the order of the corresponding Jordan block is greater than one.

Thus, the system is stable (has no unstable solution) if and only if S has no eigenvalue with positive real part or zero real part and nontrivial Jordan block. This property can be checked without solving the system explicitly.

6.5 Stable Invariant Subspace

In this section, we use the above stability condition to define stable invariant subspaces. An invariant subspace is characterized by the property that if the initial-condition vector u_0 lies in it, then the solution $u(t)$ never leaves it at any time t. A stable invariant subspace is an invariant subspace that produces only stable solutions.

Let e denote the k-dimensional vector in which all components except the last one, which is equal to 1, are 0. (In other words, e is the last column in the identity matrix of order k.) Let q denote the last column in J^{-1}. From the Jordan form of S, we have

$$Sq = J^{-1}\Lambda J q = J^{-1}\Lambda e = J^{-1} e \lambda_2 = q\lambda_2.$$

Thus, q is an eigenvector of S, with the eigenvalue λ_2. Similarly, if e stands for the last three columns in the identity matrix of order k, q stands for the last three columns in J^{-1}, and Λ_2 is the lower-right block in Λ, then we have

$$Sq = J^{-1}\Lambda J q = J^{-1}\Lambda e = J^{-1} e \Lambda_2 = q\Lambda_2,$$

which are again three linear combinations of the last three columns in J^{-1}. Thus, the last three columns in J^{-1} span an invariant subspace of S in the sense that S applied to a vector in the subspace produces a vector in it.

From the Jordan form of $\exp(tS)$, it follows that the above subspace is invariant for $\exp(tS)$ as well. In other words, if the initial-condition vector u_0 lies in the subspace, then so also do all the vectors $u(t)$ at every time t. Therefore, this subspace is called an invariant subspace of the system of ODEs.

If the real part of λ_2 is negative, then this subspace is also called stable, because every initial-condition vector u_0 in it produces a stable solution $u(t)$. The union of all stable invariant subspaces forms the stable invariant subspace of the system of ODEs.

6.6 The Inhomogeneous Case

So far, we have considered only homogeneous ODEs, in which only the term $Su(t)$ appears on the right-hand side. Here, we consider also the more general inhomogeneous case, in which a free term $f(t)$ is added:

$$u'(t) = su(t) + f(t),$$

where f is a given function that is integrable (in absolute value) in $[0, \infty)$:

$$\int_0^\infty |f(\tau)|d\tau < \infty.$$

(In systems of ODEs, f is interpreted as a vector function, and the above condition should hold for each of its components.)

The solution to the initial-value problem (with initial condition as before) is

$$\begin{aligned} u(t) &= \exp(tS)u_0 + \exp(tS)\int_0^t \exp(-\tau S)f(\tau)d\tau \\ &= \exp(tS)u_0 + \int_0^t \exp((t-\tau)S)f(\tau)d\tau. \end{aligned}$$

Unfortunately, the above integral is not always available in closed form. In the next section, we'll see how it can be approximated numerically on a discrete mesh.

6.7 Numerical Solution

The numerical solution of an ODE is a numerical approximation to the solution $u(t)$ of the original ODE on a grid of discrete times $t_1, t_2, \ldots$. Here, we describe some basic numerical methods of obtaining a numerical solution. These methods are also called numerical schemes or discretization methods.

The above formula for the solution of an inhomogeneous ODE gives the analytic solution. However, it is useful only when the integral in the formula is easily calculated. Unfortunately, this integral is rarely available in closed form. Furthermore, the function f on the right-hand side of the ODE may be given only as a machine function, which returns the output $f(t)$ for an input t. In such cases, the integral should be approximated numerically using a discrete mesh of meshsize h, as above. For this purpose, it is most helpful to compute $\exp(hS)$ once and for all and store it for future use. For a fixed time $t = jh$, the integral can be approximated by

$$\int_0^t \exp((t-\tau)S)f(\tau)d\tau$$
$$\doteq h\sum_{i=1}^{j}\exp(hS)^{j-i}f(ih)$$
$$= h\sum_{i=0}^{j-1}\exp(hS)^{i}f((j-i)h).$$

The most expensive task here is to calculate the individual powers of $\exp(hS)$. Fortunately, this task can be avoided by observing that this is actually a polynomial of degree $j-1$ in $\exp(hS)$, which can be calculated efficiently by Horner's algorithm in Chapter 5, Section 8. In this algorithm, matrix-matrix operations are completely avoided, because the matrix $\exp(hS)$ is only applied to a vector to produce a new vector, using matrix-vector rather than matrix-matrix products.

This approach is suitable for calculating the numerical solution at a particular time $t = jh$. In the next section, we'll see how the numerical solution can be computed efficiently in the entire mesh, using previously calculated data.

6.8 Difference Schemes

Often, the numerical solution is required not only at the particular time $t = jh$, as above, but also in the entire mesh of discrete times $t = h, 2h, \ldots$. The above approach, which focuses on the isolated time point $t = jh$, is unsuitable for this task. It makes more sense to calculate the numerical solution at jh using its values at the previous time points $h, 2h, \ldots, (j-1)h$. After all, valuable computer resources have been used to obtain this information, so why not use it further?

In order to better use and avoid recalculating information that has already been calculated during the solution process, one can break the original initial-value problem into smaller problems, defined in cells of the form $[jh, (j+1)h]$, with initial conditions at jh obtained from the previous cell $[(j-1)h, jh]$. More specifically, the original initial-value problem is rewritten in each cell as

$$u'(t) = Su(t) + f(t), \quad jh < t \le (j+1)h, \ j = 0, 1, 2, \ldots,$$

with initial condition about $u(jh)$ obtained from u_0 (for $j = 0$) or the previous cell $[(j-1)h, jh]$ (for $j > 0$). The solution at $(j+1)h$ is, thus,

$$u((j+1)h) = \exp(hS)u(jh) + \int_{jh}^{(j+1)h} \exp(((j+1)h-\tau)S)f(\tau)d\tau.$$

Since this integral is in general unavailable in closed form, it is approximated by evaluating the integrand only at $\tau = (j+1)h$. This gives the numerical solution $\tilde{u}(jh)$, which satisfies the initial condition

$$\tilde{u}(0) = u_0$$

and the difference equation

$$\tilde{u}((j+1)h) = \exp(hS)\tilde{u}(jh) + hf((j+1)h).$$

This scheme is mathematically equivalent to the one in Section 6.7, but is much more efficient thanks to the effective use of data from previous time points and the avoidance of unnecessary recalculations. Furthermore, it uses only matrix-vector operations rather than expensive matrix-matrix operations. (Actually, it is just Horner's algorithm, with the intermediate values calculated during the loop being used as the numerical solution at the previous grid points.)

The above method is called finite-difference discretization or difference scheme. Its accuracy is evaluated in terms of the (discretization) error, namely, the difference between the numerical solution and the solution of the original ODE at the grid:

$$|\tilde{u}(jh) - u(jh)|, \quad j = 1, 2, 3, \ldots.$$

A more accurate scheme can be obtained by evaluating the above integrand at the midpoint $\tau = (j + 1/2)h$ rather than at $\tau = (j+1)h$:

$$\tilde{u}((j+1)h) = \exp(hS)\tilde{u}(jh) + h\exp(hS/2)f((j+1/2)h).$$

This difference scheme is more accurate in the sense that the discretization error is smaller than before. Clearly, more accurate numerical integration leads to more accurate difference schemes. In the next section, we will see a scheme that not only is accurate but also gives explicit error estimates.

6.9 The Taylor Difference Scheme

The Taylor scheme is based on the Taylor expansion

$$u((j+1)h) = \sum_{i=0}^{n} \frac{h^i u^{(i)}(jh)}{i!} + \frac{h^{n+1}u^{(n+1)}(\xi)}{(n+1)!},$$

where ξ is some intermediate point between jh and $(j+1)h$. Because the last term in the above formula (the error term) is usually very small, it can be dropped in the numerical scheme.

One may ask how we know the derivatives of u at jh. The answer is that they can be obtained from the ODE itself, provided that the derivatives of f are available. Indeed, the derivatives of u can be calculated by the following recursive formula:

$$\begin{aligned} u'(jh) &= Su(jh) + f(jh), \\ u''(jh) &= Su'(jh) + f'(jh), \\ u^{(i+1)}(jh) &= Su^{(i)}(jh) + f^{(i)}(jh). \end{aligned}$$

Thus, the numerical solution $\tilde{u}$ should satisfy the initial condition

$$\tilde{u}(0) = u_0$$

and the difference equation

$$\tilde{u}((j+1)h) = \sum_{i=0}^{n} \frac{h^i \tilde{u}^{(i)}(jh)}{i!},$$

where the $\tilde{u}^{(i)}(jh)$'s are calculated by the recursive formula

$$\tilde{u}^{(0)}(jh) = \tilde{u}(jh),$$
$$\tilde{u}^{(i+1)}(jh) = S\tilde{u}^{(i)}(jh) + f^{(i)}(jh).$$

This recursion is a short-memory process: $\tilde{u}^{(i+1)}(jh)$ is a function of $\tilde{u}^{(i)}(jh)$ alone and is independent of $\tilde{u}^{(0)}(jh)$, $\tilde{u}^{(1)}(jh)$, ..., $\tilde{u}^{(i-1)}(jh)$. This is why the code below is based on the standard algorithm in Chapter 5, Section 11, rather than Horner's algorithm: it avoids storing the entire list of numbers calculated in the above recursion. Instead, each one of them contributes to the sum and is then replaced by the next one.

The advantage of object-oriented programming is clear here. Indeed, the matrix and vector objects implemented in the low-level code in Chapter 2, Sections 18 and 20, are used in the present high-level code, which preserves the spirit of the original mathematical formula:

```
template<class T, int N>
const vector<T,N>
TaylorScheme(const vector<T,N>&u0,
    const matrix<T,N,N>&S,
    const list<vector<T,N> >&f,
    const T&h){
  T powerOfHoverIfactorial = 1;
  vector<T,N> sum=0;
  vector<T,N> uDerivative = u0;
  for(int i=0; i<f.size(); i++){
    sum += powerOfHoverIfactorial * uDerivative;
    uDerivative = S * uDerivative + f[i];
    powerOfHoverIfactorial *= h/(i+1);
  }
  return sum;
} // Taylor scheme
```

This completes the definition and implementation of the Taylor scheme. Next, we study the discretization error in this scheme.

6.10 Computational Error Estimates

Here we see the main advantage in the Taylor scheme and its C++ implementation: the opportunity to compute error estimates.

The error term in the Taylor approximation is given by

$$\frac{h^{n+1}u^{(n+1)}(\xi)}{(n+1)!},$$

where ξ is somewhere in the interval $[jh, (j+1)h]$. In order to estimate this error, we must estimate the $(n+1)$th derivative of u at ξ. However, since the exact location of ξ is unavailable, we must estimate $u^{(n+1)}$ in the entire interval $[jh, (j+1)h]$. How can this possibly be done? Even the solution u is unavailable in this interval, let alone its derivatives!

To our aid comes again the object-oriented approach. Suppose that we know that every individual component in the vector u is bounded in magnitude in the interval $[jh, (j+1)h]$ by a constant L. Then we can repeat the above recursion, but this time the vector $\tilde{u}^{(0)}(jh)$ is replaced by the vector of intervals

$$[-L, L]^k \equiv ([-L, L], [-L, L], \ldots, [-L, L]),$$

which contains the individual components in u in $[jh, (j+1)h]$. In this recursion, we use interval arithmetics: adding two intervals means adding their corresponding endpoints to produce the sum interval, and multiplying an interval by a scalar means multiplying each endpoint. These definitions are then used in the application of the matrix S to a vector of intervals to produce a vector of intervals. Because f and its derivatives are available in the entire interval $[jh, (j+1)h]$, their corresponding vectors of intervals are also available and can be added. As a result, the recursion gives the vectors of intervals in which the derivatives of u must lie for any point in $[jh, (j+1)h]$.

Thus, the recursion that uses vectors of intervals gives us bounds for the components in the vectors u, u', u'', ..., $u^{(n+1)}$ in the entire interval $[jh, (j+1)h]$. Therefore, for sufficiently small h and large n, the error in the Taylor approximation can be estimated computationally and shown to be indeed negligible.

The "interval" objects should be defined in the low-level code along with the required arithmetic operations (addition, subtraction, multiplication, and division). The output interval should be extended slightly to account for errors due to finite-precision arithmetics.

Once this object is well prepared, it can be used in template classes such as "vector" to produce vectors of intervals. The following code can be used to compute the error estimate once 'T' is interpreted as an interval:

```
template<class T, int N>
const vector<T,N>
error(const vector<T,N>&boundForU,
    const matrix<T,N,N>&S,
    const list<vector<T,N> >&f,
    const T&h){
  T powerOfHoverIfactorial = 1;
  vector<T,N> uDerivative = boundForU;
  for(int i=0; i<f.size(); i++){
    uDerivative = S * uDerivative + f[i];
    powerOfHoverIfactorial *= h/(i+1);
  }
  return powerOfHoverIfactorial * uDerivative;
} // error in Taylor scheme
```

Finally, we need to verify that our original conjecture that u lies in $[-L, L]^k$ for every point in $[jh, (j+1)h]$ is indeed true. For this purpose, we again use interval arithmetics.

More specifically, we apply the above "TaylorScheme()" function to vectors of intervals, with the initial-condition vector being the vector of intervals that contains $u(jh)$, and h replaced by the interval $[0, h]$. This produces the vector of intervals that contains the Taylor approximation for every point in $[jh, (j+1)h]$. Next, we apply the above "error()" function, again with h replaced by $[0, h]$. This gives the vector of intervals that contains the error for every point in $[jh, (j+1)h]$. The sum of outputs from these two functions gives the vector of intervals in which $u(t)$ lies for every $jh \leq t \leq (j+1)h$. Now, if these intervals are contained in $[-L, L]$, then our original conjecture proves to be true; otherwise, we have to restart the error estimate once again with larger L.

6.11 Nonlinear ODEs

So far, we have dealt only with linear ODEs, in which the right-hand side of the equation is a linear function of u. In nonlinear ODEs, where the right-hand side is a nonlinear function of u, the situation is much more complicated. The stability analysis in Section 6.4 no longer holds, and straight invariant subspaces are replaced by curved invariant manifolds. Some initial conditions may produce asymptotic solutions that converge to fixed (steady-state) points in the k-dimensional Cartesian space as $t \to \infty$, whereas others may produce solitons that spiral around some fixed point and seem to converge to it, but then suddenly depart from it and start to spiral around another fixed point, and so on, wandering among fixed points with no apparent order and never converging to any of them. Solving nonlinear ODEs is, thus, a particularly challenging task.

The most important tool for studying nonlinear problems is linearization. Once an ODE has been linearized around a fixed point, the linearized ODE may produce a stable subspace that is tangent to the stable manifold of the original ODE at the fixed point. In fact, if some initial conditions produce a solution that approaches the stable manifold sufficiently close in terms of the unstable direction, then there exist initial conditions that produce a solution that converges to the fixed point in it [19]. The latter solution is called the asymptotic solution.

Computational error estimates can thus be of great importance here. They may show that a particular solution indeed gets so close to a stable manifold (in terms of the unstable dimension of the linearized equation around the fixed point) that there must exist initial conditions that produce an asymptotic solution as well [27].

Using interval arithmetics, one can add the error interval to the numerical solution to obtain the interval in which the solution of the ODE must lie at a particular time t. If the entire interval is sufficiently close to the stable manifold, then one can deduce the existence of an asymptotic solution.

6.12 Object-Oriented Analysis

In what follows, we show how the object-oriented approach can be used in the analysis of nonlinear ODEs. In this analysis, the template function that implements the Taylor scheme is called with the template 'T' being not only a scalar but also an interval. The "interval" object used for this purpose is assumed to be available from the low-level part of the code. This object is most suitable in computational error estimates in complex nonlinear ODEs.

Once the error interval is added to it, the interval that is the output of the Taylor scheme contains the solution to the ODE at the final time under consideration. This output interval accounts for the uncertainty due to finite-precision arithmetics, discretization error, and other possible errors due to numerical approximation. If the output interval is sufficiently small, then the solution of the ODE at the final time under consideration is known with rather good accuracy, and mathematical estimates can be used to check whether it is so close to a stable manifold that the existence of an asymptotic solution can be deduced.

6.13 Application

Let us apply the above plan to a particular nonlinear ODE that arises from the Kuramoto–Sivashinsky equation. This equation describes the phenomenon of combustion. Under certain symmetry assumptions, the equation can be reduced to the nonlinear ODE

$$(u' + u/r)'' = c^2 - u^2/2 - (u' + u/r)'/r - (u' + u/r),$$

where c is a given real constant. Here, we use the independent variable r $(0 < r < \infty)$ rather than t to represent the radius of the flame.

The unknown solution $u \equiv u(r)$ also satisfies the initial conditions

$$u(0) = u''(0) = 0 \quad \text{and} \quad u'(0) = a_0,$$

where a_0 is a parameter of the problem.

In order to make the equation easier to study, we introduce new unknown functions

$$\begin{aligned} v &= u' + u/r, \\ w &= v'. \end{aligned}$$

With these unknowns, the equation can be written as a system of nonlinear ODEs:

$$\begin{aligned} u' &= v - u/r, \\ v' &= w, \\ w' &= c^2 - u^2/2 - w/r - v. \end{aligned}$$

To this system, we would like to apply the Taylor scheme with computational error estimates. Unfortunately, this scheme cannot be used at $r = 0$, since $1/r$ is undefined there. We must find a way to advance to $r > 0$ before the Taylor scheme can be applied.

This is done as follows. From the initial conditions

$$u(0) = u''(0) = 0,$$

one may reasonably assume that $u(r)$ is an odd function around $r = 0$; that is, it can be expanded as a power series that contains only odd powers of r:

$$u(r) = \sum_{i=0}^{\infty} a_i r^{2i+1} \quad (0 \le r \le r_0),$$

where r_0 is the convergence radius of the series. Of course, a_0 is known from the initial conditions. By substituting the power series for u in the original ODE and equating coefficients of r^0, one obtains

$$a_1 = (c^2 - 2a_0)/16.$$

Similarly, for $i = 1, 2, 3, \ldots$, one equates coefficients of r^{2i}; this gives the recursion

$$(2i+2)^2(2i+4)a_{i+1} = -\frac{1}{2}\sum_{m=0}^{i-1} a_m a_{i-1-m} - (2i+2)a_i.$$

From this recursion, it can be seen that all the coefficients a_i are bounded, so the power series indeed has convergence radius $r_0 \geq 1$ and can indeed be derived term by term for every $0 < r < r_0$ to produce $v(r)$ and $w(r)$ as well. (Actually, it can be shown by induction that a_i decreases rapidly with i, so r_0 is actually much larger than 1.)

When error estimates are computed, one can also implement the above recursion in interval arithmetic, adding in the end an error interval due to the truncated tail of the power series. This yields (with little uncertainty) the solution $u(r)$ (and $v(r)$ and $w(r)$ as well) at a suitable starting point, say $r = r_0/2$, where the numerical scheme can be used to compute $u(r+h)$, $v(r+h)$, and $w(r+h)$.

6.14 Taylor Scheme with Error Estimates

The Taylor scheme used to advance from $r = r_0/2$ to $r+h$ requires the derivatives of u, v, and w at r up to and including order $n+1$. From the original ODE, these derivatives can be obtained recursively by

$$\begin{aligned} u^{(i+1)} &= v^{(i)} - (u/r)^{(i)}, \\ v^{(i+1)} &= w^{(i)}, \\ w^{(i+1)} &= -(u^2)^{(i)}/2 - (w/r)^{(i)} - v^{(i)} \end{aligned}$$

at r. This recursion formula can also be used in the numerical scheme to obtain $\tilde{u}^{(i)}$, $\tilde{v}^{(i)}$, and $\tilde{w}^{(i)}$ and in the Taylor scheme to obtain the numerical solution $\tilde{u}$, $\tilde{v}$, and $\tilde{w}$ at $r+h$. Furthermore, if interval arithmetic is available, then one can replace the initial values of $\tilde{u}$, $\tilde{v}$, and $\tilde{w}$ at r by corresponding intervals that contain them and apply the above recursion formula to these intervals to obtain the $(n+1)$th derivatives of u, v, and w in the form of the intervals that contain them. Once these intervals are multiplied by $h^{n+1}/(n+1)!$ and added to the numerical solution at $r+h$, we obtain the intervals in which $u(r+h)$, $v(r+h)$, and $w(r+h)$ must lie.

In the following high-level code, we assume that the "interval" object is already available and use it as a concrete type in the template functions. This lets us concentrate on implementing the above mathematical algorithm.

The function "deriveKS" below uses the above recursion to calculate the lists of derivatives of u, v, and w at the initial point r. For this purpose, we need the derivatives of $1/r$ (computed by the "deriveRinverse" function in Chapter 5, Section 10) and the derivatives of the product of two functions such u/r, u^2, and w/r (computed by the "deriveProduct" function in Chapter 5, Section 12):

```
template<class T>
void deriveKS(const T&c, const T&r,
    list<T>&u, list<T>&v, list<T>&w){
  list<T> Rinverse = deriveRinverse(r,u.size());
  for(int i=0; i<u.size()-1; i++){
    u(i+1)=v[i]-deriveProduct(u,Rinverse,i);
    v(i+1) = w[i];
    w(i+1) = (-0.5)*deriveProduct(u,u,i)
           - deriveProduct(w,Rinverse,i)
           - v[i] + (i ? 0. : (c*c) );
  }
} // derivatives in KS equation
```

Because of the products of functions, the recursion is a long-memory process: all derivatives up to order i are needed to calculate the ith derivative. This is why the derivatives must be stored in lists.

The "TaylorKS" function below calls the "deriveKS" function to produce the lists of derivatives of u, v, and w at r. In order to start the recursion, it uses the argument "u0", which is the vector of intervals that contain $u(r)$, $v(r)$, and $w(r)$ (obtained from the truncated power series and estimate for its tail). These lists are then used in Horner's algorithm to produce the numerical solution at $r+h$. Then, the "deriveKS" function is called once again, this time to calculate recursively the intervals that contain the derivatives of u, v, and w in the entire interval $[r, r+h]$. For this purpose, we assume that the intervals that contain u, v, and w are placed in the vector of intervals named "bound". Once these intervals are used to start the recursion in "deriveKS", it produces intervals that contain the derivatives of u, v, and w in the entire interval $[r, r+h]$. This produces the error interval, which is then added to the numerical solution to yield the interval in which the solution at $r+h$ must lie:

```
template<class T>
const vector<T,3>
TaylorKS(const T&c, const T&r, const T&h, int n,
    const vector<T,3>&u0,
    const vector<T,3>&bound){
  list<T> u(n,0.);
  list<T> v(n,0.);
  list<T> w(n,0.);
  u(0) = u0[0];
  v(0) = u0[1];
  w(0) = u0[2];
  deriveKS(c,r,u,v,w);
  vector<T,3> result(HornerTaylor(u,h),
     HornerTaylor(v,h), HornerTaylor(w,h));
  u(0) = bound[0];
  v(0) = bound[1];
  w(0) = bound[2];
  deriveKS(c,r,u,v,w);
```

```
    vector<T,3> highDerivative(u[n-1],v[n-1],w[n-1]);
    vector<T,3> error =
(power(h,n-1)/factorial(n-1)) * highDerivative;
    return result + error;
  } //  Taylor for KS equation + error estimate
```

In order to verify that the intervals in "bound" indeed contain u, v, and w at every point in $[r, r+h]$, one should call "TaylorKS" once again, but this time with h replaced by the interval $[0, h]$. The output is the intervals that contain u, v, and w for every possible point in $[r, r+h]$. If these intervals are indeed contained in the corresponding intervals in "bound", then our conjecture proves to be true; otherwise, we have to restart the error estimate with larger intervals in "bound" or use smaller h and larger n in Taylor's scheme.

The above algorithm can now be repeated to obtain the intervals in which $u(r+2h)$, $v(r+2h)$, and $w(r+2h)$ must lie, and so on.

6.15 Asymptotic Solution

In [27], the above algorithm is introduced and used to show that, for a particular choice of the parameter a_0, the solution $(u(r), v(r), w(r))$ (for some r) indeed gets sufficiently close to the stable invariant manifold that contains the fixed point. For this purpose, the fixed point is first obtained by setting $r = \infty$, so all the r-derivatives in the ODE vanish. This yields the fixed point

$$\begin{pmatrix} u_\infty \\ v_\infty \\ w_\infty \end{pmatrix} = \begin{pmatrix} -\sqrt{2}c \\ 0 \\ 0 \end{pmatrix}.$$

Because $1/r$ vanishes and $d(u^2)/du = 2u$, the linearized ODE around the above fixed point takes the form

$$\begin{pmatrix} u \\ v \\ w \end{pmatrix}' = \begin{pmatrix} 0 & 1 & 0 \\ 0 & 0 & 1 \\ \sqrt{2}c & -1 & 0 \end{pmatrix} \begin{pmatrix} u \\ v \\ w \end{pmatrix} + \begin{pmatrix} 0 \\ 0 \\ c^2 \end{pmatrix}.$$

The stability analysis of the above matrix is carried out as in Section 6.4. In particular, the stability depends on the real parts of the eigenvalues. Even without calculating them explicitly, one can use results from linear algebra to have some idea about them. First, since the matrix is of order 3, there are three eigenvalues. Second, their sum is equal to the trace of the matrix (sum of main-diagonal elements), which, in this case, vanishes. Third, their product is equal to the determinant of the matrix, which, in our case, is equal to $\sqrt{2}c$. Fourth, because the matrix is real, the complex conjugate of an eigenvalue is also an eigenvalue. All these facts lead us to the conclusion that the above matrix has two eigenvalues with negative real part that are complex conjugates of each other, and a third, real positive eigenvalue. The unstable direction is thus the direction of the eigenvector corresponding to this third eigenvalue. (This eigenvector is the third and last column in J^{-1} in Section 6.4.) Now, Newton's method can be used to find the third eigenvalue as the positive root of the characteristic polynomial of the matrix. Once it has been found, the characteristic polynomial is the product of a polynomial of degree 1 and a polynomial of degree 2, so the other two eigenvalues can be found as the roots of the latter. Using the eigenvalues, the eigenvectors can also be calculated as solutions of the corresponding linear systems. Thus, the entire matrix J^{-1} in Section 6.4 is available, and so is its inverse J.

Clearly, applying J to a three-dimensional vector gives its representation in terms of the above eigenvectors. In particular, the third component corresponds to the unstable direction. Thus, if for sufficiently large r

$$\left(J \begin{pmatrix} u(r) - u_\infty \\ v(r) - v_\infty \\ w(r) - w_\infty \end{pmatrix} \right)_3$$

is sufficiently small in magnitude, then the solution is sufficiently close to the stable manifold in terms of the unstable direction, and the existence of an asymptotic solution (for some initial conditions) is proved.

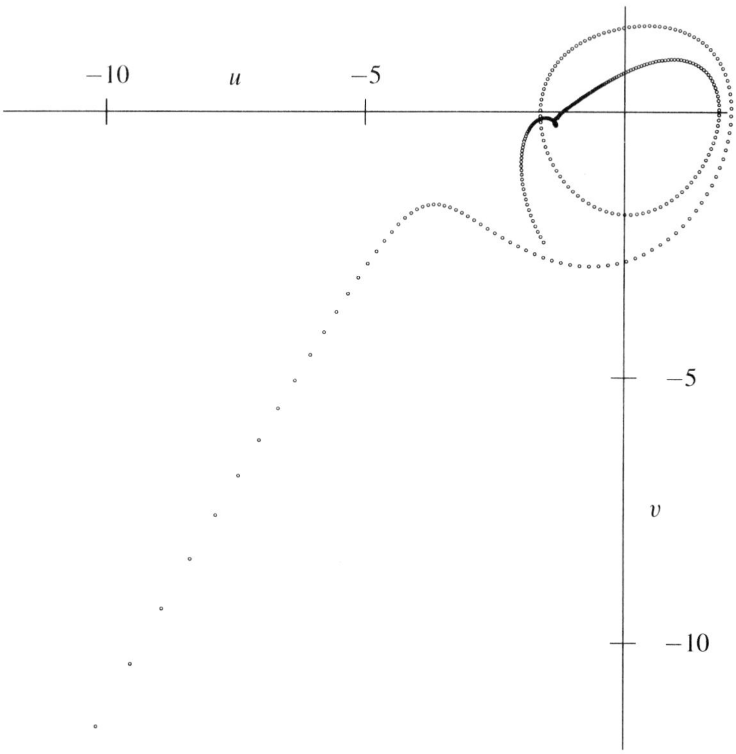

Figure 6.1. *Numerical solution of the Kuramoto–Sivashinsky equation, projected onto the (u, v) plane. Truncated power series are used to start the numerical marching. Before diverging to $(-\infty, -\infty)$, the solution gets sufficiently close to the fixed point, which proves the existence of an asymptotic solution.*

The stable manifold of the original nonlinear system that contains the fixed point can also be approximated by expanding u as a power series in $1/r$ around $r = \infty$. Using the computational error estimate and the theory in [19], one can then show that the maximal possible distance from the solution to the stable manifold (in terms of the unstable direction) is so small that there must exist initial conditions that produce an asymptotic solution.

The computational error estimate can thus be used to prove the existence of an asymptotic solution for complicated systems of nonlinear ODEs. The asymptotic solution issues from some initial conditions, which are not exactly the same as those used to produce the numerical solution. In our numerical solution, we have two zero initial conditions, $u(0) = u''(0) = 0$, but the third one, $u'(0) = a_0$, has yet to be specified. In fact, a_0 should be chosen in such a way that the numerical solution remains reasonably bounded for r as large as possible. More specifically, if $f(a_0)$ is the function that returns the maximal ρ for which $|u(r)| \le 10$ for every $0 < r \le \rho$ in the mesh, then a_0 should be a local maximum of f, calculated as in Chapter 1, Section 20. For $c = 0.85$, for example, the optimal a_0 turns out to be $a_0 = -1.725517$. (Actually, it is calculated with an accuracy of 20 digits after the decimal point.)

It is with this optimal a_0 that the computational error estimate is carried out in [37] and used to prove the existence of an asymptotic solution in [27]. The numerical solution with this particular a_0 is displayed in Figure 6.1. In this figure, the horizontal axis stands for $u(r)$, the vertical axis stands for $v(r)$, and $w(r)$ is not indicated. The independent variable r takes discrete values on a mesh stretched from $r = 1$ to $r = 61$, with meshsize $h = 1/16$. Truncated power series are used to approximate u, v, and w at $r = 1$ to start the numerical marching. Before it diverges to $(-\infty, -\infty)$ as r gets too large, the numerical solution at $r = 15$ approaches the fixed point $(-\sqrt{2c}, 0)$ as closely as 10^{-3} (with the entire interval of the computational error estimate in each spatial direction in the u-v-w space, including the unstable one). This implies that there exist initial conditions that produce an asymptotic solution as well.

6.16 Exercises

1. Solve the ODE
$$u'(t) = Su(t)$$
in the unit interval $0 < t < 1$, with the parameter $S = -1$, $S = 2$, and $S = 10$. Is it stable?

2. Apply the Taylor scheme to the above ODEs. What is the maximal meshsize h that still provides reasonable accuracy? Compare the stable and unstable cases.

3. Solve the ODE
$$u'(t) = Su(t)$$
in the unit interval $0 < t < 1$, where u is a two-dimensional vector and S is the 2×2 matrix
$$S = \begin{pmatrix} -1 & 0 \\ 1 & 1 \end{pmatrix},$$
$$S = \begin{pmatrix} 1 & 1 \\ -1 & 1 \end{pmatrix},$$

or

$$S = -\begin{pmatrix} 1 & 1 \\ -1 & 1 \end{pmatrix}.$$

Use the matrix-exponent function in Chapter 2, Section 22. Find the stable and unstable directions. Compare your solution to the numerical solution obtained from the Taylor scheme. For what h is the error still acceptable? Is it within the expected error estimate?

4. Is the recursion in Section 6.13 a short-memory or a long-memory process (in the terminology of Chapter 5, Section 11)?

5. Use the recursion in Section 6.13 to calculate the coefficients a_i and store them in a "polynomial" object. Remember that the even coefficients in this polynomial are zero.

6. Repeat the previous exercise, only this time use a polynomial in r^2 to avoid storing trivial coefficients.

7. Use the code in Chapter 5, Section 8, to calculate the power series that represents $u(r)$ at $r = 1/2$. Use $N = 50$ and $N = 100$ terms in the series. Is there any change in $u(1/2)$? What is the error estimate in both cases?

8. Derive the above power series term by term, and obtain the power series for $u'(r)$. Add to it the power series of $u(r)/r$ (obtained by shifting the coefficients in the power series of $u(r)$) to produce the power series for $v(r)$. Store it in a "polynomial" object, and calculate $v(1/2)$.

9. Derive the above power series once again term by term, and obtain the power series for $w(r)$. Use the Horner algorithm once again to calculate $w(1/2)$.

10. Write the "interval" class that contains two "double" fields to denote the left and right edges of an interval. Implement the required arithmetic operations on intervals. Remember to add to the output interval the maximal possible error due to finite machine precision.

11. Rewrite the "interval" class as a template class "interval<T>", where 'T' is the type of the fields denoting the left and right edges of an interval.

12. Apply the Taylor scheme to the above ODEs with "interval<double>" objects for the template 'T' rather than "double" scalars. Add the interval of the computational error estimate, and obtain the interval in which the solution of the original ODE must lie. Are the results in agreement with those from previous exercises?

13. For which S and initial conditions in the above system of ODEs does the solution converge to a fixed point? Answer both analytically and numerically, using the above code.

14. Use the "interval<double>" class for the template 'T' in the code in Section 6.14 to estimate a (bounded) solution to the Kuramoto–Sivashinsky equation in Section 6.13. Compare your code with the code in [37], which contains 620 lines in Fortran that appear on the Web page http://www.siam.org/books/cs01. Which code is easier to read and use?

Part III

Partial Differential Equations and Their Discretization

We are now fairly familiar with the concepts and properties of C++ and are ready to use them in numerical applications. As we'll see below, numerical methods are particularly well implemented in C++ using object-oriented programming.

The problems described below and their solution methods are most important in applied science and engineering. Knowing how to develop and program the required numerical algorithms is thus a great advantage.

Many important problems in applied science and engineering can be formulated as differential equations. In scalar differential equations, there is only one unknown function (usually denoted by u) to solve for. In vector (or systems of) differential equations, there are several unknown functions (usually denoted by u, v, w, etc.) to solve for.

The data that we're given to help discover these unknown functions are in the form of a differential equation and initial and boundary conditions. The unknown functions are available at the initial time through the initial conditions and at the boundary of the spatial domain through the boundary conditions. These values propagate in time to the entire time-space domain, using the propagation law contained in the differential equation. Solving the problem means actually finding explicitly the unknown functions that result from this process.

We first consider linear differential equations, in which the propagation law in the differential equation is independent of the solution. Later on, we'll also consider the more difficult case of nonlinear differential equations, in which the propagation law in the differential equation may change according to the particular function used as solution.

In ordinary differential equations (ODEs), the unknown function u is a function of a single variable, say $u \equiv u(x)$. It is assumed that u is defined in a domain, say the unit interval $0 \le x \le 1$. The ODE gives information about the derivative (or derivatives) of u in the interior of the domain. For example,

$$u''(x) + C(x)u'(x) + K(x)u(x) = F(x), \quad 0 < x < 1.$$

This equation combines u, its first derivative u', and its second derivative u'' (multiplied by the known coefficients $C(x)$ and $K(x)$) to produce the known right-hand-side function $F(x)$. This determines the propagation law for u from $x = 0$ onward. The boundary conditions imposed at $x = 0$ and $x = 1$ fix u at the endpoints, while the differential equation shapes it in the interior of the domain. Together, they determine the solution u uniquely.

The coefficients, as well as the right-hand side, are known in advance; it is only the unknown function u that needs to be discovered. In other words, u is the solution of the differential equation.

The above ODE is of variable coefficients, since $C(x)$ and $K(x)$ depend on x and may assume different values for different values of x in the domain. The order of the differential equation is the order of the highest derivative in it. In the above example, the order is 2, since the derivative of highest order in it is the second derivative $u''(x)$.

The presentation of the problem is not yet complete. Boundary conditions that specify the values of u (or its first derivative) at the endpoints of the domain are also necessary. For example,

$$u'(0) = u(1) = 0.$$

A boundary condition that specifies the value of the solution u itself is called a boundary condition of the first kind or a Dirichlet boundary condition. A boundary condition that specifies the value of the derivative of u is called a boundary condition of the second kind or a Neumann boundary condition. A boundary condition that specifies a linear combination of u and u' is called a boundary condition of the third kind or a mixed boundary condition. In the above example, a Dirichlet boundary condition is given at the right endpoint of the x-interval, and a Neumann boundary condition is given at the left endpoint.

The number of boundary conditions should be the same as the order of the ODE. This guarantees that the ODE is well posed in the sense that it has a unique solution $u(x)$. The well-posed differential equation, together with its boundary conditions, is also called a boundary-value problem.

In partial differential equations (PDEs), the unknown function u (as well as the coefficients and right-hand side) are functions of more than one variable. The domain is also of more than one dimension, e.g., the unit square $0 \le x, y \le 1$. The derivatives of u are now partial: $u_x(x, y)$ is the derivative of u with respect to x with y kept fixed (evaluated at the point (x, y)), $u_y(x, y)$ is the derivative of u with respect to y with x kept fixed, $u_{xx}(x, y)$ is the second derivative of u with respect to x with y kept fixed, $u_{yy}(x, y)$ is the second derivative of u with respect to y with x kept fixed, and so on.

The PDE gives information about the behavior of the unknown function u in the interior of the domain in terms of its partial derivatives. For example, the Poisson equation in the unit square is given by

$$-u_{xx}(x, y) - u_{yy}(x, y) = F(x, y), \quad 0 < x, y < 1.$$

To this equation, we also need to add boundary conditions to specify the behavior of u at the boundary of the unit square in terms of its lower derivatives. For example,

$$\alpha u + \beta u_n = G(x, y),$$

where α, β, and G are functions defined on the boundary of the unit square and u_n is the directional derivative of u in the direction of the outer unit normal vector at the boundary. In other words, $u_n = -u_x$ on the $x = 0$ edge, $u_n = u_x$ on the $x = 1$ edge, $u_n = -u_y$ on the $y = 0$ edge, and $u_n = u_y$ on the $y = 1$ edge. Together, the PDE and boundary conditions are called a boundary-value problem. It is well known that the Poisson equation is a well-posed boundary-value problem in the sense that it has a unique solution u (Chapter 11).

The Poisson equation is a typical example of an elliptic equation. Elliptic equations are characterized by the fact that, if the derivatives of lower order are dropped from the PDE and only the derivatives of highest order remain, then the boundary-value problem may be formulated as a minimization problem (Chapter 11, Section 3). For example, the solution of the Poisson equation is also the solution of the following minimization problem:

Find the function $u(x, y)$ that minimizes the functional
$$\frac{1}{2}\int_0^1\int_0^1 (u_x(x, y)^2 + u_y(x, y)^2)dxdy - \int_0^1\int_0^1 F(x, y)u(x, y)dxdy.$$

it also satisfies the above (Dirichlet) boundary conditions. Thus, u is also the steady state in a stable equilibrium problem; that is, there is a time-dependent process $u(t, x, y)$ that wanders around the point of equilibrium $u(x, y)$ and eventually converges to it. The time-dependent process $u(t, x, y)$ is the solution of the corresponding time-dependent PDE

$$u_t(t, x, y) - u_{xx}(t, x, y) - u_{yy}(t, x, y) = F(t, x, y),$$

known as the heat equation.

In order to start this time-dependent process, one must also have the initial conditions that specify the values $u(0, x, y)$ explicitly. More precisely, in order to be well posed, the PDE must be accompanied not only by boundary conditions to specify the solution (or its normal derivative) at the boundary of the spatial domain at each particular time t but also by initial conditions of the form $u(0, x, y) = u^{(0)}(x, y)$ (where $u^{(0)}$ is a given function defined in the spatial domain) to specify the solution at the beginning of the process at time $t = 0$. With the required initial and boundary conditions, the problem is a well-posed initial-boundary-value problem.

This kind of PDE, which governs not only the final steady state at $t = \infty$ but also the entire time-dependent process that leads to it, is called a parabolic PDE. In the above example, the parabolic PDE is the heat equation, and its solution approaches the solution of the Poisson equation as $t \to \infty$.

There is yet another kind of PDE, which models time-dependent processes with no steady state at all. These are known as hyperbolic PDEs. For example, the convection equation in the unit interval is formulated as

$$u_t(t, x) + C(t, x)u_x(t, x) = F(t, x), \quad 0 < t < T, \ 0 < x < 1.$$

Here, C and F are given functions of the two independent variables t and x, and only the unknown function u is solved for. Since the PDE is of first order, only one boundary condition is required, say the boundary condition that specifies the value of u at $x = 0$ (if $C > 0$) or $x = 1$ (if $C < 0$) for each time t. In addition, initial conditions that specify u at the initial time $t = 0$ for each x are also required. With these data, the initial-boundary-value problem is well posed.

The convection equation above is of a different nature from the heat equation, because the data in the initial (and boundary) conditions no longer approach a steady state but rather progress in time along the characteristic lines defined by

$$dx(t)/dt = C(t, x).$$

In fact, the initial and boundary conditions imposed on the lower and left edges of the x-t (time-space) domain (assuming $C > 0$) progress along these characteristic lines (or curves) to determine the value of u in the entire x-t domain. This way, discontinuities in the initial (or boundary) conditions are not smoothed out, as in the heat equation, but rather preserved along the characteristic lines.

When the coefficient C depends not only on t and x but also on u itself, the equation is no longer linear but rather quasi-linear. In this case, the shape of the characteristic lines, along which data propagate to the entire time-space domain, depends on the as yet unknown solution u. For a certain type of initial (and boundary) conditions, two characteristic lines may collide, which may produce a conflict between the data carried along them. The result is a shock wave.

An intermediate case between parabolic and hyperbolic equations is the singularly perturbed convection-diffusion equation:

$$u_t(t, x) - \varepsilon u_{xx}(t, x) + C(t, x)u_x(t, x) = F(t, x), \quad 0 < t < T, \ 0 < x < 1,$$

where ε is a small positive parameter. Formally, the equation is parabolic, because the derivative of highest order is u_{xx}, as in the heat equation. Therefore, two boundary conditions at $x = 0$ and $x = 1$ are required. However, since the coefficient of u_{xx} (the diffusion coefficient) is very small, the solution behaves much like the solution to the convection equation. In fact, the data contained in the initial conditions and the boundary conditions imposed at $x = 0$ progress in much the same way as in the convection equation along the characteristic lines, with only slight dissipation due to the small diffusion term. When these data reach the neighborhood of the right boundary $x = 1$, they must change sharply to fit the boundary condition imposed at $x = 1$. The neighborhood of $x = 1$ in which this sharp variation occurs is called the boundary layer.

Boundary-value and initial-boundary-value problems are most important for modeling real phenomena in applied science and engineering. Unfortunately, only a few of them can be solved analytically in closed form. Most PDEs must be solved numerically using a digital computer.

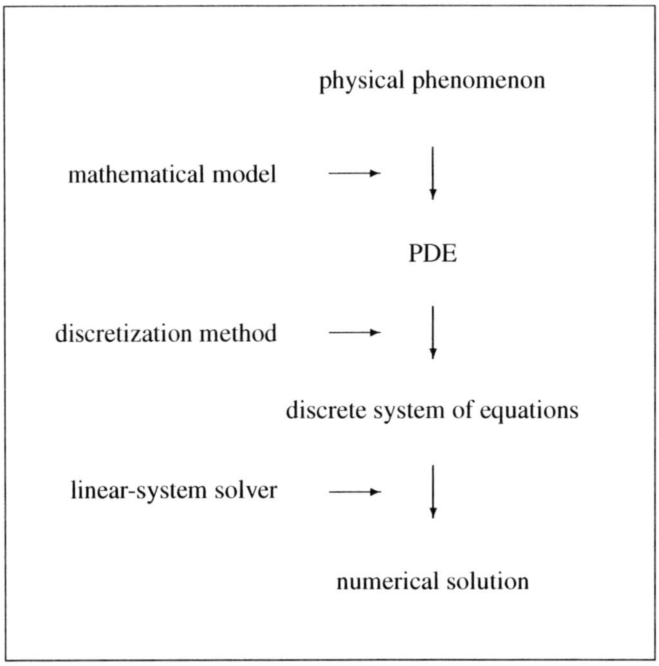

Figure III.1. *The entire process of producing the numerical approximation to the original physical phenomenon.*

The invention of the digital computer in the 1950s transformed the science of applied mathematics completely. Before this, only a few elementary examples could be solved. The analytic methods used for this purpose, although interesting and inspiring, are limited to these examples only, and cannot be used in more realistic cases.

Once the digital computer became available, PDEs no longer had to be solved analytically. Instead, they could be solved numerically on a finite, discrete grid approximating the original domain. The field of numerical analysis provides reliable and accurate discretization methods to approximate PDEs on discrete grids, and the field of scientific computing provides efficient algorithms to solve the difference equations resulting from the discretization.

The entire process of producing the numerical approximation to the original physical phenomenon is described in Figure III.1.

This part of the book describes some linear and nonlinear time-dependent PDEs and their finite-difference discretization methods, along with their C++ implementation. In the first chapter (Chapter 7), the time-dependent convection-diffusion equation is described and discretized using the upwind and time-marching schemes. In the second chapter (Chapter 8), these schemes are analyzed. In the third chapter (Chapter 9), nonlinear hyperbolic PDEs are studied and discretized. In the fourth chapter (Chapter 10), nonlinear PDEs are used in the field of image processing.

Chapter 7
The Convection-Diffusion Equation

In this chapter, we describe finite-difference methods to discretize the convection-diffusion equation in one and two spatial dimensions. In particular, we use explicit, implicit, and semi-implicit schemes to march from the current time step to the next one and the upwind scheme to discretize the spatial derivatives. The entire algorithm is implemented in C++ using a hierarchy of objects: from the time-space grid at the high level to the individual grid lines at the low level, with the difference operators that act upon them.

7.1 Initial-Boundary-Value Problems

We are now ready to use the programming tools developed above in the numerical solution of initial-boundary-value problems. Here, we consider one of the most important parabolic PDEs, which is commonly used to model physical processes: the convection-diffusion equation [29, 32]. In one spatial dimension, this equation takes the form

$$u_t(t,x) - \varepsilon u_{xx}(t,x) + C(t,x)u_x(t,x) = F(t,x), \quad 0 < t < T,\ 0 < x < L.$$

Here, T is the maximal time, L is the length of the x-interval, $F(t,x)$ is the given right-hand side, ε is the small diffusion coefficient, $C(t,x)$ is the given convection coefficient, and $u(t,x)$ is the unknown solution.

We further assume that initial conditions are given in the following form:

$$u(0,x) = u^{(0)}(x), \quad 0 < x < L,$$

where $u^{(0)}(x)$ is a given function. We also assume that Dirichlet boundary conditions are imposed on the right edge:

$$u(t,L) = G(t,L), \quad 0 < t < T,$$

and mixed boundary conditions are imposed on the left edge:

$$\alpha(t,0)u(t,0) + u_n(t,0) = G(t,0), \quad 0 < t < T,$$

where $\alpha()$ and $G()$ are given functions and $n = -x$ is the direction that points away from the interval $[0, L]$ at its left edge 0 (the outer normal direction).

The above initial and boundary conditions must be compatible with each other; that is, they must agree with each other at the corner points $(0, 0)$ and $(0, L)$, so that u can be defined continuously there. With these conditions, the problem is well posed in the sense that it indeed has a unique solution $u(t, x)$.

7.2 Finite Differences

PDEs can be solved analytically only in very few model cases. In general, the PDE is solved numerically on a discrete grid, which is just a finite set of points in the x-interval $[0, L]$. For example, if N is the number of points in a uniform grid with meshsize $h = L/N$, then the grid is just the set of points

$$(0, h, 2h, \ldots, (N-1)h).$$

To be solved numerically, the PDE must be approximated (discretized) on the grid. A common way to do this is by finite differences. In this approach, a derivative is approximated by the corresponding difference, and the entire differential equation is approximated by the corresponding system of difference equations, which can then be solved numerically on a computer.

Let us now describe the finite-difference discretization method in some more detail. Let the time t be fixed, and let u_j be an approximation to the solution at the jth grid point:

$$u_j \doteq u(t, jh), \quad 0 \le j < N$$

(where the symbol $\doteq$ stands for "approximately equal to"). The derivative of u at the midpoint between the jth and $(j+1)$th grid points is approximated by the divided finite difference

$$u_x(t, (j+1/2)h) \doteq h^{-1}(u_{j+1} - u_j).$$

Similarly, the derivative of u at $(j - 1/2)h$ is approximated by

$$u_x(t, (j-1/2)h) \doteq h^{-1}(u_j - u_{j-1}).$$

By subtracting the latter approximation from the former and dividing by h, we obtain the approximation to the second derivative of u at $x = jh$:

$$u_{xx}(t, jh) \doteq h^{-1}(u_x(t, (j+1/2)h) - u_x(t, (j-1/2)h)) \doteq h^{-2}(u_{j+1} - 2u_j + u_{j-1}).$$

This way, the diffusion term u_{xx} can be approximated by the above finite differences on the discrete grid.

The above discretization uses symmetric differencing; that is, the divided difference between two grid points approximates the derivative at their midpoint. This scheme is of second-order accuracy; that is, the discretization error $u_j - u(t, jh)$ is as small as h^2 as $h \to 0$. However, below we'll see that accuracy is not always the most important property of a numerical scheme. In some cases, more important properties such as adequacy should also be verified before the numerical scheme is used.

7.3 The Upwind Scheme

Let us now discretize the convection term $C(t,x)u_x$. The naive way to do this is by symmetric finite differencing as before:

$$C(t, jh)u_x(t, jh) \doteq (2h)^{-1}C(t, jh)(u_{j+1} - u_{j-1}).$$

This approach, however, is inappropriate because it excludes the jth grid point. In fact, it only uses differences between two even-numbered grid points or two odd-numbered grid points, so it is completely unaware of frequent oscillations such as

$$(1, -1, 1, -1, \ldots),$$

which is overlooked by the scheme. Indeed, because the scheme uses only the $(j-1)$th and $(j+1)$th grid points, it produces the same discrete convection regardless of whether or not the discrete solution contains a component of the above oscillation. Because this oscillation depends highly on h, it has nothing to do with the solution of the original PDE, $u(t,x)$, which must be independent of h. It is therefore known as a nonphysical oscillation.

In theory, the above symmetric differencing is of second-order accuracy as $h \to 0$. In practice, however, h must take a small positive value. When the diffusion coefficient ε is very small (particularly, smaller than h), the resulting scheme becomes inadequate in the sense that it produces a numerical solution that has nothing to do with the required solution of the original PDE. A more stable and adequate scheme is needed to discretize the convection term properly.

Such a scheme is the "upwind" scheme. This scheme also uses u_j in the approximation to $u_x(t, jh)$, thus avoiding nonphysical oscillations. More specifically, the scheme uses backward differencing at grid points j for which $C(t, jh) > 0$:

$$C(t, jh)u_x(t, jh) \doteq h^{-1}C(t, jh)(u_j - u_{j-1}),$$

and forward differencing at grid points j for which $C(t, jh) < 0$:

$$C(t, jh)u_x(t, jh) \doteq h^{-1}C(t, jh)(u_{j+1} - u_j).$$

This way, the coefficient of u_j in the discrete approximation of the convection term Cu_x is always positive, as in the discrete approximation to the diffusion term $-\varepsilon u_x x$ in Section 7.2 above. These coefficients add up to produce a substantial positive coefficient to u_j in the difference equation, which guarantees stability and no nonphysical oscillations.

In summary, the difference approximation to the entire convection-diffusion term takes the form

$$D_{j,j-1}u_{j-1} + D_{j,j}u_j + D_{j,j+1}u_{j+1} \doteq -\varepsilon u_{xx}(t, jh) + C(t, jh)u_x(t, jh),$$

where

$$\begin{aligned} D_{j,j-1} &= -\varepsilon h^{-2} - \frac{|C(t, jh)| + C(t, jh)}{2h}, \\ D_{j,j} &= 2\varepsilon h^{-2} + |C(t, jh)|h^{-1}, \\ D_{j,j+1} &= -\varepsilon h^{-2} - \frac{|C(t, jh)| - C(t, jh)}{2h}. \end{aligned}$$

The convection-diffusion equation is solved numerically on a time-space grid. This is a rectangular $M \times N$ grid, containing M rows of N points each. The rows have the running index $i = 1, 2, 3, \ldots, M$ to order them from the bottom row to the top row. (The bottom row is numbered by 1, and the top row is numbered by M.) The ith row corresponds to the ith time step (time level) in the time marching in the scheme.

The above matrix D that contains the discrete convection-diffusion term may change from time step to time step according to the particular convection coefficient $C()$ at the corresponding time. Thus, we also use the superscript i to indicate the relevant time step. This way, $D^{(i)}$ denotes the matrix corresponding to the ith time step (or time level, or row) in the time-space grid ($1 \le i \le M$).

7.4 Discrete Boundary Conditions

At the endpoint of the grid, the above discretization cannot take place because the $(j+1)$th or $(j-1)$th grid point is missing. For example, for $j = N-1$, the $(j+1)$th point lies outside the grid, so the last equation

$$D_{N-1,N-2}u_{N-2} + D_{N-1,N-1}u_{N-1} + D_{N-1,N}u_N = \cdots$$

is not well defined, because it uses the dummy unknown u_N. In order to fix this, one should use the Dirichlet boundary condition available at the right edge of the domain:

$$u_N = G(t, L).$$

Once this equation is multiplied by $D_{N-1,N}$ and subtracted from the previous one, the dummy u_N unknown is eliminated, and the equation is well defined.

Similarly, the dummy unknown u_{-1} is used in the first equation:

$$D_{0,-1}u_{-1} + D_{0,0}u_0 + D_{0,1}u_1 = \cdots.$$

Fortunately, one can still use the discrete mixed boundary conditions to eliminate this unknown. Indeed, the discrete boundary conditions on the left can be written as

$$\alpha(t, 0)u_0 + h^{-1}(u_{-1} - u_0) = G(t, 0).$$

Once this equation is multiplied by $hD_{0,-1}$ and subtracted from the previous one, the dummy u_{-1} unknown is eliminated, and the first equation is also well defined. By the way, the above subtraction also increases the $D_{0,0}$ coefficient by

$$-hD_{0,-1}\alpha(t, 0) + D_{0,-1}.$$

The above discretization of the convection-diffusion terms actually produces the matrix (i.e., difference operator) D. This operator maps any N-dimensional vector $v \equiv (v_0, v_1, \ldots, v_{N-1})$ to another N-dimensional vector Dv, defined by

$$\begin{aligned}
(Dv)_0 &= D_{0,0}v_0 + D_{0,1}v_1, \\
(Dv)_j &= D_{j,j-1}v_{j-1} + D_{j,j}v_j + D_{j,j+1}v_{j+1} \quad (0 < j < N-1), \\
(Dv)_{N-1} &= D_{N-1,N-2}v_{N-2} + D_{N-1,N-1}v_{N-1}.
\end{aligned}$$

Note that the matrix or difference operator D depends on the time t, because the functions $C()$ and $\alpha()$ depend on t. Therefore, it should actually be denoted by $D(t)$ or $D^{(i)}$, where i is the index of the time step. In the above, however, we omit the time indication for the sake of simplicity.

7.5 The Explicit Scheme

The time derivative u_t in the convection-diffusion equation is also discretized by a finite difference. For this purpose, the entire x-t domain should be first discretized or approximated by a two-dimensional uniform grid with M rows of N points each (Figure 7.1). The ith row $(1 \le i \le M)$, also known as the ith time level or time step, approximates the solution of the original PDE at a particular time. More specifically, the numerical solution at the ith time step, denoted by $u^{(i)}$, approximates $u(t, x)$ on the discrete grid:

$$u_j^{(i)} \doteq u(i\Delta t, jh),$$

where $\Delta t = T/M$ is the cell size in the time direction. Clearly, for $i = 0$, the approximate solution is available from the initial conditions

$$u_j^{(0)} = u^{(0)}(jh),$$

where $u^{(0)}$ on the left is the vector of values on the grid at the zeroth time level, and $u^{(0)}$ on the right is the given function that specifies the initial condition at $t = 0$.

For $i > 0$, the numerical solution at the ith time level is computed by marching across time levels (time marching). In other words, the numerical solution at the current time level is obtained from the numerical solution computed before on the previous time level and the boundary conditions.

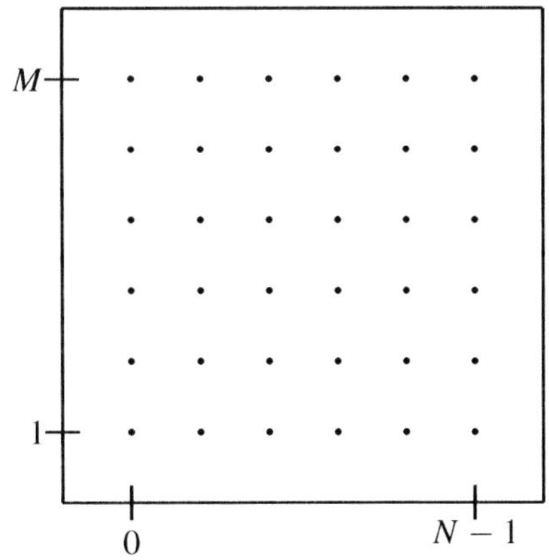

Figure 7.1. *The uniform $M \times N$ time-space grid for the discretization of the convection-diffusion equation in the time-space domain $0 < x < L$, $0 < t < T$.*

Time marching can be done in (at least) three different ways, which depend on the discretization of the time derivative u_t: forward differencing leads to explicit time marching (or the explicit scheme), backward differencing leads to implicit time marching (the implicit scheme), and midpoint differencing leads to semi-implicit time marching (the semi-implicit scheme).

In the explicit scheme, forward differencing is used to discretize u_t. This way, the numerical solution at the current time level is computed from a discrete approximation to the original PDE, in which the convection-diffusion term is evaluated at the previous time level:

$$\begin{aligned} &u_t(t,x) - \varepsilon u_{xx}(t,x) + C(t,x)u_x(t,x) \\ &\doteq (\Delta t)^{-1}\left(u_j^{(i)} - u_j^{(i-1)}\right) + \left(D^{(i-1)}u^{(i-1)}\right)_j = F(t, jh), \end{aligned}$$

where $D^{(i-1)}$ is the discrete convection-diffusion term, evaluated at time $t = (i-1)\Delta t$ as in Section 7.3. This evaluation uses the numerical solution $u^{(i-1)}$ at the previous or $(i-1)$th time level to calculate the numerical solution $u^{(i)}$ at the current or ith time level. This is why this scheme is known as the explicit scheme.

Note that, when $j = 0$ or $j = N-1$, the right-hand side should be incremented by the appropriate contribution from the boundary-condition function $G(t,x)$. In fact, the right-hand side is the N-dimensional vector $f^{(i)}$ defined by

$$\begin{aligned} f_0^{(i)} &= F(i\Delta t, 0) - hD_{0,-1}^{(i)}G(i\Delta t, 0), \\ f_j^{(i)} &= F(i\Delta t, jh) \quad (0 < j < N-1), \\ f_{N-1}^{(i)} &= F(i\Delta t, (N-1)h) - D_{N-1,N}^{(i)}G(i\Delta t, L). \end{aligned}$$

The above formulas lead to the explicit time marching

$$u^{(i)} = u^{(i-1)} - (\Delta t)D^{(i-1)}u^{(i-1)} + (\Delta t)f^{(i-1)}.$$

This equation defines the numerical solution at the current or ith time level in terms of the numerical solution at the previous or $(i-1)$th time level.

The explicit scheme can actually be viewed as the analogue of the scheme in Chapter 6, Section 8, with the first-order Taylor approximation

$$\exp\left(-(\Delta t)D^{(i-1)}\right) \doteq I - (\Delta t)D^{(i-1)},$$

where I is the identity matrix.

The implementation of this marching in C++ requires not only "vector" objects to store and manipulate the data in the time levels but also "difference" objects to handle the difference operators that act upon vectors. This is the subject of Section 7.12 below.

In the above explicit scheme, the numerical solution $u^{(i)}$ is computed by evaluating the discrete convection-diffusion terms at the $(i-1)$th time level, where the numerical solution $u^{(i-1)}$ is already known from the previous step. Clearly, this can be done in a straightforward loop over the time levels, from the first one to the Mth one. Although the cost of each step in this loop is fairly low, its total cost may be rather large. Indeed, as discussed in Chapter 8, Section 3, the parameter Δt used in the explicit scheme must be as small as h^2 to prevent nonphysical oscillations from accumulating during the time marching. Thus, the number of time levels M must be as large as Th^{-2}, leading to a prohibitively lengthy process. More stable schemes, which use fewer time levels, are clearly needed.

7.6 The Implicit Scheme

A more stable scheme, which requires fewer time levels to discretize the same time-space domain, is the implicit scheme. In this scheme, the discrete convection-diffusion terms are evaluated at the current (or ith) time level rather than the previous (or $(i-1)$th) time level:

$$\begin{aligned}&u_t(t,x) - \varepsilon u_{xx}(t,x) + C(t,x)u_x(t,x)\\ &\doteq (\Delta t)^{-1}\left(u_j^{(i)} - u_j^{(i-1)}\right) + \left(D^{(i)}u^{(i)}\right)_j = F(i\Delta t, jh)\end{aligned}$$

where $t = i\Delta t$. Using this formula, the implicit scheme can be written compactly in terms of N-dimensional vectors as:

$$\left(I + (\Delta t)D^{(i)}\right)u^{(i)} = u^{(i-1)} + (\Delta t)f^{(i)},$$

where I is the identity matrix of order N.

The calculation of $u^{(i)}$ requires the "inversion" of the tridiagonal matrix $I+(\Delta t)D^{(i)}$. Of course, the inverse of this matrix is never calculated explicitly, because this would be too expensive. By "inversion" we mean solving the above linear system for the unknown vector $u^{(i)}$. The algorithm for doing this is discussed below. This completes the definition of the implicit time marching. The detailed implementation will be discussed later.

Although the computation of each individual $u^{(i)}$ is more expensive than in the explicit scheme, the entire loop over the entire time-space grid is usually far less expensive, because M can be much smaller. Indeed, since the implicit scheme is unconditionally stable (Chapter 8, Section 3), it can use Δt as large as h. This reduces considerably the number of time levels M and the total computational cost.

The implicit scheme can actually be viewed as the analogue of the scheme in Chapter 6, Section 8, with the first-order Taylor approximation

$$\exp\left(-(\Delta t)D^{(i)}\right) = \exp\left((\Delta t)D^{(i)}\right)^{-1} \doteq \left(I + (\Delta t)D^{(i)}\right)^{-1}.$$

7.7 The Semi-Implicit Scheme

The semi-implicit scheme, also known as the Crank–Nicolson or "midpoint" scheme, can be viewed as a compromise between the explicit and implicit schemes. In this scheme, the discrete convection-diffusion term is the average of the evaluation at the previous time level as in the explicit scheme and the evaluation at the current time level as in the implicit scheme:

$$\begin{aligned}&u_t(t,x) - \varepsilon u_{xx}(t,x) + C(t,x)u_x(t,x)\\ &\doteq (\Delta t)^{-1}\left(u_j^{(i)} - u_j^{(i-1)}\right) + \frac{1}{2}\left(D^{(i)}u^{(i)} + D^{(i-1)}u^{(i-1)}\right)_j\\ &= F((i-1/2)\Delta t, jh)\end{aligned}$$

at the "half" time level $t = (i-1/2)\Delta t$. Because the time derivative u_t is discretized symmetrically around this half time level, the scheme is of second-order accuracy in time wherever the solution to the original PDE is indeed smooth in time.

The semi-implicit scheme can be written compactly as the vector equation

$$\left(I + \frac{\Delta t}{2}D^{(i)}\right)u^{(i)} = \left(I - \frac{\Delta t}{2}D^{(i-1)}\right)u^{(i-1)} + (\Delta t)f^{(i-1/2)}.$$

The algorithm to solve this system for the unknown vector $u^{(i)}$ will be provided later. This completes the definition of the semi-implicit scheme. The complete implementation will be given later.

Like the implicit scheme, the semi-implicit scheme is unconditionally stable (Chapter 8, Section 3). Thus, Δt can be relatively large without producing any unstable nonphysical oscillations. Thus, M and the computational cost are kept rather low.

The semi-implicit scheme can actually be viewed as the analogue of the scheme in Chapter 6, Section 8, with the first-order diagonal Pade approximation (Chapter 1, Section 22)

$$\exp\left(-(\Delta t)D^{(i-1/2)}\right) \doteq \left(I + (\Delta t/2)D^{(i)}\right)^{-1}\left(I - (\Delta t/2)D^{(i-1)}\right).$$

In the next sections, we use the present C++ framework to implement the semi-implicit scheme. The explicit and implicit schemes can be implemented in a similar way.

7.8 The Implementation

In the remainder of this chapter, we introduce the C++ implementation of the semi-implicit scheme for the numerical solution of the convection-diffusion equation. The main advantage of this implementation in comparison with standard Fortran codes is the opportunity to specify the size of the grid in run time rather than compilation time. This flexibility is particularly important in adaptive discretizations, where the size of the grid becomes available only when the program is actually executed.

The method of adaptive mesh refinement is implemented fully in Chapter 14, Section 2. Here, we use no adaptivity, because the time-space grid is uniform. Thus, the present problem can actually be solved in Fortran or C. Nevertheless, the objects implemented here in C++ may be useful not only for this particular application but also for more complicated ones. Furthermore, the multigrid linear-system solver that can be used to solve the linear system of equations in each time step is particularly well implemented in C++, as we'll see in Chapter 10, Section 5, and Chapter 17, Section 10. In fact, in C++, one can actually develop libraries of objects, which can then be used not only in the present application but also in more advanced numerical applications and algorithms.

Let us illustrate how object-oriented languages in general and C++ in particular can be used in practice to develop the library or hierarchy of objects required in numerical applications. Suppose that a user wants to solve the convection-diffusion equation. This user is only interested in the solution of the physical model and not in the particular numerical method used for this purpose. In fact, the user may have little experience in numerical algorithms and prefer to leave this part of the job to another member of the team.

Thus, the user writes code like this:

```
int main(){
  domain D(10.,1.,.1);
  D.solveConvDif();
  return 0;
}
```

That's it, problem solved! All the details will be worked out by the numerical analyst. For this, however, the analyst must know precisely what objects to implement and what their properties or functions should be.

In order to work, the above code must have a class "domain" with a constructor that takes three "double" arguments to specify T (the size of the time interval), L (the size of the x-interval), and the required accuracy. In the above code, for example, 'D' is the time-space domain $0 \le t \le 10$, $0 \le x \le 1$, and the discretization error $|u^{(i)} - u(i\Delta t, jh)|$ should be at most 0.1.

The "domain" class must also have a member function "solveConvDif", which solves the convection-diffusion equation (to the above accuracy) and prints the numerical solution at the final time $t = 10$ onto the screen.

The numerical analyst, in turn, also wants to make his/her life easy, so he/she assumes the existence of a template class named "xtGrid" that implements the time-space discrete grid. This class must have at least three member functions: a constructor "xtGrid(M,N)" to construct the $M \times N$ grid, "timeSteps()" to return the number of time levels, and "width()" to return the number of points in the x-direction. Then, he/she implements the "domain" class as follows:

```
class domain{
  xtGrid<double> g;
  double Time;
  double Width;
```

In this implementation, the "domain" class contains three data fields: "Time" to indicate the size of the time interval, "Width" to indicate the size of the space interval, and 'g' to contain the discrete time-space grid.

The constructor of the "domain" class takes three "double" arguments to specify the size of the time interval, the size of the space interval, and the required accuracy. This third argument is used to calculate the required numbers of time levels and grid points in the x-direction. The constructor has an empty body; all the work is done in the initialization list. In particular, the first field, the grid 'g', is constructed by calculating the required number of time levels and number of grid points in the x-direction and using them to invoke the constructor of the "xtGrid" class (whose existence is assumed):

```
public:
  domain(double T, double L, double accuracy)
    : g((int)(T/accuracy)+1,(int)(L/accuracy)+1),
      Time(T), Width(L){
  } //  constructor
```

The numerical analyst also assumes that there exists an ordinary function named "march(g,h,Δt)", which takes "xtGrid", "double", and "double" arguments, and does the actual time marching in this grid. Then, he/she uses this function to define the function that solves the convection-diffusion equation in the "domain" class:

```
  void solveConvDif(){
    march(g,Width/g.width(),Time/g.timeSteps());
  } //  solve the convection-diffusion equation
};
```

This completes the block of the "domain" class.

By writing the above code, the numerical analyst actually completed his/her part of the job and can pass the rest of it on to a junior numerical analyst to work out the details. The problem of solving the convection-diffusion equation has now been transformed to the numerical problem of marching in time in the time-space grid. The junior numerical analyst is asked to write the ordinary function "march(g,h,Δt)" that does the actual time marching in the time-space grid.

The junior numerical analyst also makes assumptions to make his/her life easier. In particular, he/she assumes that the "xtGrid" class is already implemented with some useful member functions. First, he/she assumes that the 'i'th row in the grid 'g' can be read as "g[i]" and accessed (for reading/writing) as "g(i)". Next, he/she assumes that the value at the 'j'th point in it can be read as "g(i,j,"read")" and accessed (for reading/writing) as "g(i,j)". Finally, he/she assumes that the member functions "timeSteps()" and "width()" return the number of time levels and the number of points in an individual time level, respectively.

The junior numerical analyst also assumes the existence of a template class "difference" that implements difference operators like $D^{(i)}$ above, along with some elementary arithmetic operations, such as "difference" plus "difference", "difference" times "dynamicVector", etc. In fact, the "difference" object 'D' is also interpreted as a tridiagonal matrix, whose "(i,j)"th element can be read as "D(i,j,"read")" and accessed for reading/writing as "D(i,j)". Finally, it is also assumed that "D.width()" returns the order of the matrix 'D'.

With these tools, the junior numerical analyst is ready to implement the semi-implicit time marching in the time-space grid. For this purpose, he/she must first define a function named "convDif()" to set the "difference" object that contains the discrete convection-diffusion term at the particular time under consideration, along with the corresponding right-hand-side vector 'f'. For this purpose, it is assumed that the functions used in the original PDE and initial-boundary conditions (F, G, α, etc.) are available as global external functions. (In the present example, most of them are set to zero for the sake of simplicity; only the initial conditions in the function "Initial()" are nonzero.)

```
double F(double, double){return 0.;}
double C(double, double){return 0.;}
double Alpha(double, double){return 0.;}
double G(double, double){return 0.;}
double Initial(double x){return 1.-x*x;}
const double Epsilon=1.;
```

It is assumed that these functions are global, so they are accessible from the "convDif()" function defined next. The first arguments in this function are the "difference" object 'd', where the discrete convection-diffusion term at time 't' is placed, and the "dynamicVector" object 'f', where the corresponding right-hand side is placed. It is assumed that both 'd' and 'f' are initially zero; in the "convDif()" function, they are set to their correct values:

```
template<class T>
void convDif(difference<T>&d,dynamicVector<T>&f,
        double h,double deltaT,double t){
    for(int j=0; j<d.width(); j++){
      if(t>deltaT/2)f(j)=F(j*h,t-deltaT/2);
      double c=C(j*h,t);
```

The local variable 'c' contains the convection coefficient at the 'j'th grid point at time 't'. The upwind scheme decides whether to use forward or backward differencing in the x-direction according to the sign of 'c':

```
        if(c>0.){
          d(j,j)=c/h;
          d(j,j-1)=-c/h;
          d(j,j+1)=0.;
        }
        else{
          d(j,j)=-c/h;
          d(j,j+1)=c/h;
          d(j,j-1)=0.;
        }
      }
```

So far, we have introduced the discrete convection term into the difference operator 'd'. This has been done in a loop over the grid points in the time level under consideration. Now, we add to 'd' the discrete diffusion term. This is done in one command only, using the constructor and "operator+=" to be defined later in the "difference" class:

```
      d += Epsilon/h/h * difference<T>(d.width(),-1.,2.,-1.);
```

Finally, we introduce the discrete boundary conditions into the difference operator 'd' and the right-hand-side vector 'f':

```
      d(0,0) += d(0,-1);
      d(0,0) -= d(0,-1) * h * Alpha(0,t);
      if(t>deltaT/2){
        f(0) -= d(0,-1) * h * G(0,t-deltaT/2);
        f(d.width()-1) -= d(d.width()-1,d.width())
                * G(d.width()*h,t-deltaT/2);
      }
  } // set the convection-diffusion matrix and right-hand side
```

The "convDif()" function is now used in the "march()" function that implements the semi-implicit time marching. For this purpose, we use a loop over the time levels, with a local integer index named "time" that goes from the first time level at the bottom to the final one at the top of the time-space grid. In this loop, we use two local "difference" objects named "current" and "previous" to store the discrete convection-diffusion terms at the current and previous time levels (the difference operators $D^{(\text{time})}$ and $D^{(\text{time}-1)}$):

```
template<class T>
void march(xtGrid<T>&g, double h, double deltaT){
  difference<T> I(g.width(),0.,1.,0.); // identity matrix
  for(int j=0; j<g.width(); j++)
    g(0,j) = Initial(j*h);
  dynamicVector<T> f(g.width());
  difference<T> previous(g.width());
```

```
    convDif(previous,f,h,deltaT,0);
    for(int time=1; time < g.timeSteps(); time++){
      difference<T> current(g.width());
      convDif(current,f,h,deltaT,time*deltaT);
```

We are now in the loop that does the actual time marching. We now advance from the previous time step to the current one. We have just put the discrete convection-diffusion spatial derivatives (evaluated at the current time) into the difference operator "current". We have also saved the discrete convection-diffusion spatial derivatives from the previous time step in the difference operator "previous". These difference operators are now used in the following command, which is the heart of the semi-implicit scheme. In this command, the '/' symbol below calls the "operator/" (to be defined later) that solves the linear system obtained in the semi-implicit scheme, which can be viewed symbolically as dividing a vector by a tridiagonal matrix or difference operator:

```
      g(time) =
          ((I-.5*deltaT*previous)*g[time-1]+deltaT*f)
          / (I + 0.5 * deltaT * current);
      previous = current;
    }
    print(g[g.timeSteps()-1]);
  } // semi-implicit time marching
```

The actual implementation of the "xtGrid" and "difference" classes and required functions will be discussed later on.

7.9 Hierarchy of Objects

The hierarchy of objects used in the entire workplan is described in Figure 7.2. In the highest level, the "domain" object is placed, along with the function "solveConvDif()" that acts upon it. In the lower level, the "xtGrid" object is placed, along with the function "march()" that acts upon it. In fact, "solveConvDif()" only invokes "march()" to act upon the "xtGrid" object contained in the "domain" object. In the lowest level, the "dynamicVector" and "difference" objects are placed. These objects correspond to a particular time level in the entire "xtGrid" object; they are also connected by arithmetic operations between them, and the function "convDif()" sets them with the discrete convection-diffusion term at the relevant time level or time step.

7.10 List of Vectors

All that is left to do is define the "xtGrid" and "difference" template classes, with the required functions. These tasks can actually be carried out independently by two other members of the team, who are C++ programmers who don't necessarily have a background in numerical methods.

The "xtGrid" and "difference" objects can be viewed as lists of vectors. In an "xtGrid", each vector corresponds to a time level, so the number of vectors is the same as the number of time levels. In a "difference" object, on the other hand, there are only three vectors:

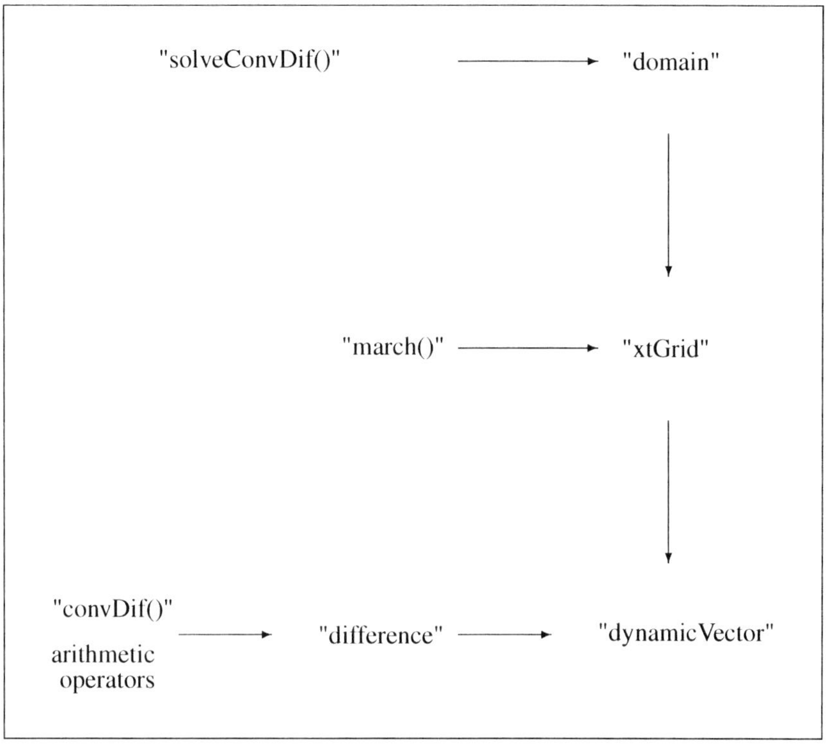

Figure 7.2. *Hierarchy of objects for the convection-diffusion equation: the "domain" object uses an "xtGrid" object, which uses "dynamicVector" and "difference" objects.*

the first contains the $(j, j-1)$ elements in the tridiagonal matrix, the second contains the (j, j) elements (the main diagonal), and the third contains the $(j, j+1)$ elements. It is thus natural to derive the "xtGrid" and "difference" classes from a list of dynamic vectors.

Because the data fields in the "list" template class in Chapter 3, Section 4, are declared "protected" rather than "private", they are accessible from derived classes. This is why it is possible to define constructors in the derived "xtGrid" and "difference" classes with integer argument to specify the dimension of the vectors in the list. Indeed, when the constructors of the "xtGrid" and "difference" classes are called, the data fields of the base "list" class are constructed first by the default constructor of that class, resulting in a trivial list with no items at all. These lists are then reconstructed in the derived-class constructor thanks to their access privilege to the data fields of the base class.

7.11 The Time-Space Grid

Here, we introduce the "xtGrid" template class that implements the time-space grid as a list of time levels, each of which is implemented as a dynamic vector.

Here is the detailed implementation of the "xtGrid" class, derived from the list of dynamic vectors:

```
template<class T>
class xtGrid : public list<dynamicVector<T> >{
  public:
   xtGrid(int,int,const T&);
```

The constructor is only declared here; its full definition will be provided later. Here are the required functions that return the number of time levels and the number of grid points in each time level:

```
    int timeSteps() const{
      return size();
    } // number of time levels

    int width() const{
      return item[0]->dim();
    } // width of grid
```

The individual time levels can be accessed by three different versions of "operator()". Specifically, the 'i'th time level in the "xtGrid" object 'g' can be accessed by either "g[i]" for reading only or "g(i)" for reading or writing, and the value at the 'j'th grid point in it can be accessed by "g(i,j)" for reading or writing. The compiler invokes the version that best fits the number and type of arguments in the call.

The "operator[]" is inherited from the base "list" class in Chapter 3, Section 4. The "operator()", on the other hand, although available in the "list" class, must be rewritten explicitly to prevent confusion with the other version:

```
    dynamicVector<T>& operator()(int i){
      if(item[i])return *item[i];
    } //  ith time level (read/write)

    T& operator()(int i, int j){
      return (*item[i])(j);
    } //  (i,j)th grid point (read/write)
};
```

This concludes the block of the "xtGrid" class. All that is left to do is to define the constructor, which is only declared in the class block above. This constructor takes two integer arguments: the first specifies the number of items in the underlying list of vectors (or the number of time levels), and the second specifies the dimension of these vectors (or the number of points in each individual time level). When the constructor is called, the underlying list of dynamic vectors is initialized automatically by the default constructor of the base "list" class to be an empty list that contains no items at all. This list is then reset in the present constructor to the required nontrivial list. This reconstruction can be done thanks to the fact that the "number" and "item" data fields are declared as "protected" rather than private in the base "list" class.

```
template<class T>
xtGrid<T>::xtGrid(int m=0,int n=0,const T&a=0){
  number = m;
  item = m ? new dynamicVector<T>*[m] : 0;
  for(int i=0; i<m; i++)
    item[i] = new dynamicVector<T>(n,a);
} // constructor
```

Note that the order of the above codes should actually be reversed: since the "xtGrid" class is used in the "domain" class and the "convDif" and "march" functions, it must appear before them in the program. It is only for the sake of clear presentation that the order is reversed in the above discussion.

7.12 Difference Operators

Here we define the "difference" class that implements the difference operator or the tridiagonal matrix. The "difference" class is derived from the "list<dynamicVector<T> >" template class (Figure 7.3). (Note the blank space between the two '>' symbols, which distinguishes them from the ">>" string, which has a totally different meaning in the "iostream.h" library.)

The "difference" object is actually a list of three vectors: the first vector contains the $D_{j,j-1}$ elements in the tridiagonal matrix D, the second vector contains the $D_{j,j}$ elements (the main diagonal), and the third vector contains the $D_{j,j+1}$ elements. Because the items in a "list" object are declared "protected" in Chapter 3, Section 4, they can be accessed from members of the derived "difference" class:

```
template<class T>
class difference : public list<dynamicVector<T> >{
   public:
    difference(int,const T&,const T&,const T&);
    const difference<T>& operator+=(const difference<T>&);
    const difference<T>& operator-=(const difference<T>&);
    const difference& operator*=(const T&);
```

The constructor and arithmetic operators are only declared above; they will be defined explicitly later.

Particularly important operators in the derived "difference" class are the "operator()" member functions defined below. These operators allow one to refer to the elements $D_{j,j-1}$, $D_{j,j}$, and $D_{j,j+1}$ simply as "D(j,j-1)", "D(j,j)", and "D(j,j+1)", respectively. Because this operator returns a nonconstant reference to the indicated element, the above calls can be used to actually change the value in it. Therefore, they must be used with caution, so that values are not changed inadvertently:

```
    T& operator()(int i,int j){
      return (*item[j-i+1])(i);
    } // (i,j)th element (read/write)
```

To read an element of the "difference" object 'D', one can also use the read-only version of "operator()". This version is invoked in calls like "D(j,j,"read")", which also use an extra

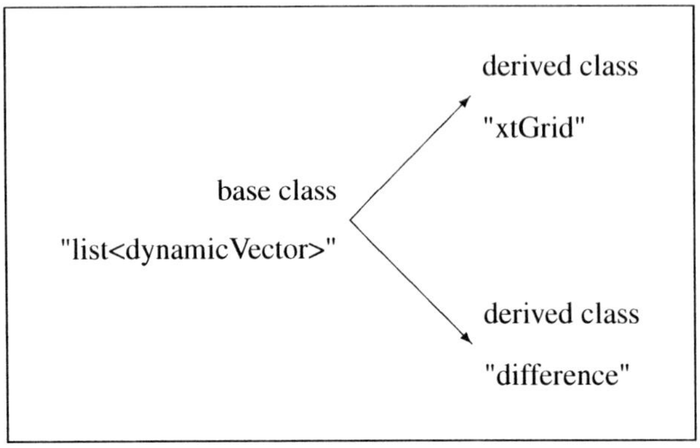

Figure 7.3. *Schematic representation of inheritance from the base class "list" (list of dynamic vectors) to the derived classes "xtGrid" and "difference".*

argument of type "char*" (array of characters). Because this version returns a constant reference, the indicated element can only be read and not changed:

```
      const T& operator()(int i,int j,char*) const{
        return (*item[j-i+1])[i];
      }  //  (i,j)th element (read only)
```

Another required function is the function that returns the order of the tridiagonal matrix:

```
      int width() const{
        return item[0]->dim();
      }  //  width of grid
  };
```

This completes the block of the "difference" class. The copy constructor and assignment operator need not be redefined, because when they are called, the corresponding function in the base "list<dynamicVector<T> >" class is implicitly invoked to set the required values to data fields. Similarly, no destructor needs to be defined, because the implicitly invoked destructor of the base "list" class destroys every data field in the object.

The constructor that takes integer and 'T' arguments, on the other hand, must be redefined explicitly in the derived "difference" class. This is because it implicitly invokes the default constructor of the base "list<dynamicVector<T> >" class, with no arguments at all. This implicit call constructs a trivial list with no items in it, which must then be reset in the body of the constructor below:

```
template<class T>
difference<T>::difference(int n=0,
    const T&a=0,const T&b=0,const T&c=0){
  number = 3;
  item = new dynamicVector<T>*[3];
  item[0] = new dynamicVector<T>(n,a);
```

```
    item[1] = new dynamicVector<T>(n,b);
    item[2] = new dynamicVector<T>(n,c);
  } // constructor
```

Next, we define the required arithmetic operations with "difference" objects. These functions are defined here rather than in the base "list" class in Chapter 3, Section 4. This is because, if they had been defined there, then they would have needed to return a "list<dynamicVector<T> >" object, which would then have needed to be converted to a "difference" object. The present implementation avoids this extra conversion:

```
  template<class T>
  const difference<T>&
  difference<T>::operator+=(const difference<T>&d){
    for(int i=0; i<number; i++)
      *item[i] += d[i];
    return *this;
  } // adding another "difference" to the current one

  template<class T>
  const difference<T>&
  difference<T>::operator-=(const difference<T>&d){
    for(int i=0; i<number; i++)
      *item[i] -= d[i];
    return *this;
  } // subtracting a "difference" from the current one
```

So far, we have implemented addition and subtraction of another difference operator to or from the current one. Next, we define multiplication by a scalar:

```
  template<class T>
  const difference<T>&
  difference<T>::operator*=(const T&t){
    for(int i=0; i<size(); i++)
      *item[i] *= t;
    return *this;
  } // multiplying the current "difference" by a scalar T
```

Next, we implement the above arithmetic operations as (nonmember) binary operators:

```
  template<class T>
  const difference<T>
  operator+(const difference<T>&d1,
            const difference<T>&d2){
    return difference<T>(d1) += d2;
  } // adding two "difference" objects
```

```
template<class T>
const difference<T>
operator-(const difference<T>&d1,
        const difference<T>&d2){
  return difference<T>(d1) -= d2;
} // subtracting two "difference" objects

template<class T>
const difference<T>
operator*(const T&t, const difference<T>&d){
  return difference<T>(d) *= t;
} // scalar times "difference"

template<class T>
const difference<T>
operator*(const difference<T>&d, const T&t){
  return difference<T>(d) *= t;
} // "difference" times scalar
```

Next, we introduce the operator that returns the product of a difference operator and a vector, or the difference operator applied to a vector. Here, the read-only version of the "operator()" of the "difference" class and the read-only "operator[]" of the "dynamicVector" class are used in the calculations, and the read/write "operator()" of the "dynamicVector" class is then used to assign these calculated values to the appropriate components in the output vector. We also use here the "min()" and "max()" functions from Chapter 1, Section 9:

```
template<class T>
const dynamicVector<T>
operator*(const difference<T>&d,
        const dynamicVector<T>&v){
  dynamicVector<T> dv(v.dim(),0.);
  for(int i=0; i<v.dim(); i++)
    for(int j=max(0,i-1); j<=min(v.dim()-1,i+1); j++)
      dv(i) += d(i,j,"read")*v[j];
  return dv;
} // "difference" times vector
```

Next, we implement the operator that “divides” a vector by a difference operator, or, more precisely, solves numerically a tridiagonal linear system of the form $Dx = f$. Because the solution vector x can be expressed as $D^{-1}f$ or f/D, the binary "operator/" seems to be most suitable for denoting this operation.

For simplicity, the solution vector x is computed approximately using the Gauss–Seidel iterative method of Chapter 17, Section 3. Of course, this is not a very efficient linear-system solver; the multigrid algorithm used in Chapter 10, Section 5 and Chapter 17, Section 8, makes a much better solver that is particularly well implemented in C++. Still, the Gauss–Seidel iteration is good enough for our main purpose: writing and debugging the overall algorithm and code for the numerical solution of the convection-diffusion equation:

```
template<class T>
const dynamicVector<T>
operator/(const dynamicVector<T>&f,
        const difference<T>&d){
  dynamicVector<T> x(f);
  for(int iteration=0; iteration < 100; iteration++)
    for(int i=0; i<f.dim(); i++){
      double residual = f[i];
      for(int j=max(0,i-1); j<=min(f.dim()-1,i+1); j++)
        residual -= d(i,j,"read")*x[j];
      x(i) += residual/d(i,i,"read");
    }
  return x;
}  //  solving d*x=f approximately by 100 GS iterations
```

7.13 Two Spatial Dimensions

We turn now to the more complicated case of the convection-diffusion equation in two spatial dimensions (the Cartesian dimensions x and y). This equation has the form

$$\begin{aligned}
&u_t(t,x,y) - \varepsilon(u_{xx}(t,x,y) + u_{yy}(t,x,y)) \\
&+ C_1(t,x,y)u_x(t,x,y) + C_2(t,x,y)u_y(t,x,y) \\
&= F(t,x,y), \quad 0 < t < T,\ 0 < x < L_x,\ 0 < y < L_y,
\end{aligned}$$

where T, L_x, and L_y are positive numbers denoting (respectively) the length of the time interval and the width and length of the two-dimensional spatial domain, and C_1 and C_2 are the given convection coefficients. In order to have a well-posed problem, initial and boundary conditions must also be imposed:

$$\begin{aligned}
u(0,x,y) &= u^{(0)}(x,y), & 0 \le x \le L_x,\ 0 \le y \le L_y, \\
u(t,x,L_y) &= G(t,x,L_y), & 0 \le t \le T,\ 0 \le x \le L_x, \\
u(t,L_x,y) &= G(t,L_x,y), & 0 \le t \le T,\ 0 \le y \le L_y, \\
\alpha(t,x,0)u(t,x,0) + u_n(t,x,0) &= G(t,x,0), & 0 \le t \le T,\ 0 \le x \le L_x, \\
\alpha(t,0,y)u(t,0,y) + u_n(t,0,y) &= G(t,0,y), & 0 \le t \le T,\ 0 \le y \le L_y,
\end{aligned}$$

where F, G, and α are given functions and n is the outer normal vector; that is, $n = -x$ at $y = 0$ and $n = -y$ at $x = 0$. Thus, Dirichlet boundary conditions are imposed at the right and upper edges of the rectangular spatial domain, and mixed boundary conditions are imposed on the other two edges.

The finite-difference discretization is as in Section 7.3 above, except that here both the x- and y-derivatives should be discretized. Note also that here the spatial grid is rectangular as in Figure 7.4 rather than one-dimensional as in Section 7.3.

Let us now describe the finite-difference scheme in some more detail. Let N_x and N_y denote the number of grid points in the x and y spatial directions, respectively. Let $h_x = L_x/N_x$ and $h_y = L_y/N_y$ be the corresponding meshsizes. Then, the rectangular

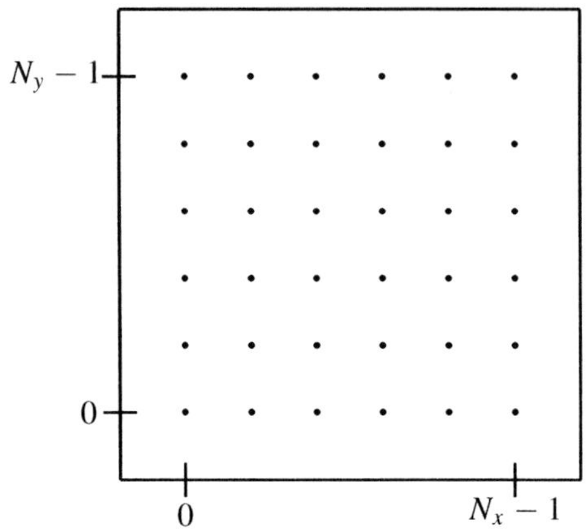

Figure 7.4. *The uniform $N_x \times N_y$ spatial grid for the discretization of the convection-diffusion terms in the unit square $0 \le x, y \le 1$.*

$N_x \times N_y$ grid that approximates the spatial domain is the set of pairs

$$(0, h_x, 2h_x, \ldots, (N_x - 1)h_x) \times (0, h_y, 2h_y, \ldots, (N_y - 1)h_y)$$

(see Figure 7.4). Let the time t be fixed, and let u be the $N_x N_y$-dimensional vector that represents the numerical solution of the discrete approximation to the initial-boundary-value problem at the time level corresponding to t. More specifically, the component in u that corresponds to the (i, j)th point in the grid is denoted by $u_{i,j}$. The difference operator D that approximates the spatial derivatives in the PDE (the convection-diffusion terms) has four indices to refer to the elements in it. More specifically, the discrete approximation to the convection-diffusion terms takes the form

$$\begin{aligned} &D_{i,j,i,j-1}u_{i,j-1} + D_{i,j,i,j+1}u_{i,j+1} \\ &+ D_{i,j,i-1,j}u_{i-1,j} + D_{i,j,i+1,j}u_{i+1,j} \\ &+ D_{i,j,i,j}u_{i,j} \\ &\doteq -\varepsilon(u_{xx} + u_{yy}) + C_1u_x + C_2u_y, \end{aligned}$$

where

$$\begin{aligned} D_{i,j,i,j-1} &= -\varepsilon h_x^{-2} - \frac{|C_1(t, jh_x, ih_y)| + C_1(t, jh_x, ih_y)}{2h_x}, \\ D_{i,j,i,j+1} &= -\varepsilon h_x^{-2} - \frac{|C_1(t, jh_x, ih_y)| - C_1(t, jh_x, ih_y)}{2h_x}, \\ D_{i,j,i-1,j} &= -\varepsilon h_y^{-2} - \frac{|C_2(t, jh_x, ih_y)| + C_2(t, jh_x, ih_y)}{2h_y}, \\ D_{i,j,i+1,j} &= -\varepsilon h_y^{-2} - \frac{|C_2(t, jh_x, ih_y)| - C_2(t, jh_x, ih_y)}{2h_y}, \\ D_{i,j,i,j} &= -\left(D_{i,j,i,j-1} + D_{i,j,i,j+1} + D_{i,j,i-1,j} + D_{i,j,i+1,j}\right). \end{aligned}$$

Of course, points that lie outside the grid must be eliminated using the discrete boundary conditions as in Section 7.4 above. For example, when $i = 0$, $u_{-1,j}$ is eliminated using the discrete mixed boundary conditions at $y = 0$ and $x = jh_x$, and so on.

The rest of the details are in principle the same as in one spatial dimension. This completes the definition of the finite-difference scheme in two spatial dimensions. The actual implementation is left as an exercise; the complete solution can be found in Section A.4 of the Appendix.

7.14 Exercises

1. Modify the code in Section 7.12 above to solve problems with mixed boundary conditions at both the $x = 0$ and $x = L$ edges of the x-interval. The vectors used in this case must be of dimension $N + 1$ rather than N, because an extra unknown at the point $Nh = L$ must also be solved for. The dummy u_{N+1} unknown should be eliminated using the discrete boundary conditions (like the dummy u_{-1} unknown).

2. The code in this chapter uses a long-memory approach, in which the numerical solution in the entire time-space grid is stored. Rewrite it using a short-memory approach, in which the numerical solution at each time step is dropped right after it is used to compute the numerical solution at the next time step, and the output of the code is the numerical solution at the final time step. Note that, with this approach, the "xtGrid" object can be avoided.

3. Modify the time-marching method in the code to use the explicit or implicit scheme rather than the semi-implicit scheme.

4. Define the "dynamicVector2" class that implements a rectangular grid. This object is useful for implementing the individual time steps in the numerical solution of the convection-diffusion equation in two spatial dimensions in Section 7.13 above. The answer can be found in Section A.4 of the Appendix.

5. Define the "difference2" object that implements the discrete spatial derivatives in both the x- and y-directions in Section 7.13 above. The answer can be found in Section A.4 of the Appendix.

6. Define the required arithmetic operations between the "difference2" and "dynamicVector2" objects. (In principle, they are the same as the arithmetic operators between the "difference" and "dynamicVector" objects.) The solution can be found in Section A.4 of the Appendix.

7. The "difference2" object in Section A.4 of the Appendix contains nine numbers per grid point, which is not really necessary here. Improve it to contain only five numbers per grid point.

8. Use the above objects to implement the semi-implicit scheme for the convection-diffusion equation in two spatial dimensions. The solution can be found in Section A.4 of the Appendix.

9. Modify your code to use the implicit scheme rather than the semi-implicit scheme.

10. Modify your code to use the explicit scheme rather than the semi-implicit scheme. Run your code with small Δt. Run it again with larger Δt. What happens when Δt is too large? Does this also happen with the implicit or semi-implicit scheme?

11. Modify your code in such a way that only mixed boundary conditions are used.

Chapter 8
Stability Analysis

In this chapter, we show that the above finite-difference discretization methods are stable in the sense that the possible numerical error in the initial conditions grows at most linearly with the number of time steps used. Stability implies that the discretization method is not only accurate in the sense that the discretization error (difference between the numerical and exact solutions) approaches zero with the meshsize but also adequate in the sense that the discretization error approaches zero when both the meshsize and the diffusion coefficient approach zero at the same time.

8.1 Preliminaries

This chapter contains some theoretical analysis of finite-difference schemes and can be skipped by readers who are interested in the practical aspects only. More advanced readers, on the other hand, may find it particularly useful to understand better the nature of finite-difference methods.

Here, we study the stability of the finite-difference discretization methods used in Chapter 7. The stability of discretization methods should not be confused with the stability of differential equations. The former is a property of the numerical scheme that guarantees that the error due to the numerical approximation is well under control, so the numerical solution is indeed meaningful and approximates well the solution of the original differential equation.

For this study, we need some preliminary definitions and elementary results from matrix theory, particularly about square matrices. These preliminaries will be particularly useful in writing the schemes in matrix form, from which stability soon follows.

The eigenvalue of a square matrix $A = (a_{i,j})$ is a complex number λ for which

$$Av = \lambda v$$

for some nonzero vector v.

Gersgorin's theorem gives us some idea about the location of λ in the complex plane. According to this theorem, λ must lie not too far from one of the main-diagonal elements in A, say $a_{i,i}$. More precisely, the distance between λ and $a_{i,i}$ cannot exceed the sum of the absolute values of the off-diagonal elements in the ith row:

$$|\lambda - a_{i,i}| \le \sum_{j \ne i} |a_{i,j}|.$$

(For the proof, see, e.g., Chapter 1, Section 11, in [39].)

We say that the matrix A is diagonally dominant if, for every row i in it, the sum of the absolute values of the off-diagonal elements does not exceed the corresponding main-diagonal element:

$$a_{i,i} \ge \sum_{j \ne i} |a_{i,j}|.$$

From Gersgorin's theorem, it therefore follows that the eigenvalues of a diagonally dominant matrix lie in the right side of the complex plane; that is, they have nonnegative real parts.

Similarly, we say that A is strictly diagonally dominant if the above inequality can also be written in a sharp version:

$$a_{i,i} > \sum_{j \ne i} |a_{i,j}|.$$

In this case, Gersgorin's theorem implies that the eigenvalues of A lie strictly in the right side of the complex plane; that is, they have positive real parts.

The Jordan form of A is the representation

$$A = J^{-1} \Lambda J,$$

where J is a nonsingular matrix and Λ is an "almost" diagonal matrix with the eigenvalues of A on its main diagonal, arbitrarily small values on the diagonal just below it, and zeroes elsewhere.

The norm of a vector $v = (v_1, v_2, v_3, \ldots)$ is the square root of the sum of the squares of the absolute values of its components:

$$\|v\| = \left(\sum_i |v_i|^2 \right)^{1/2}.$$

The norm of a matrix A is the maximum possible enlargement of a vector by applying A to it:

$$\|A\| = \max_{v \ne 0} \|Av\| / \|v\|.$$

The above definition implies that, for every nonzero vector v,

$$\|Av\| / \|v\| \le \|A\|$$

or

$$\|Av\| \le \|A\| \cdot \|v\|.$$

The triangle inequality holds for matrices as well. Indeed, if A and B are matrices of the same order, then

$$\begin{aligned}
\|A+B\| &= \max_{v\neq 0} \|(A+B)v\|/\|v\| \\
&\le \max_{v\neq 0}(\|Av\| + \|Bv\|)/\|v\| \\
&\le \max_{v\neq 0} \|Av\|/\|v\| + \max_{v\neq 0} \|Bv\|/\|v\| \\
&= \|A\| + \|B\|.
\end{aligned}$$

Furthermore, the norm of the product is smaller than or equal to the product of the norms:

$$\begin{aligned}
\|AB\| &= \max_{v\neq 0} \|ABv\|/\|v\| \\
&\le \|A\| \max_{v\neq 0} \|Bv\|/\|v\| \\
&= \|A\| \cdot \|B\|.
\end{aligned}$$

8.2 Algebraic Representation

We are now ready to rewrite the finite-difference schemes from Chapter 7 in algebraic form. In this form, the entire time-marching process is obtained as the numerical solution of the linear system of equations

$$Au = \mathcal{F},$$

where $\mathcal{F}$ is the vector obtained from the right-hand-side function $F(t, x)$ in the original PDE by restricting it to the discrete time-space grid and also incorporating the discrete boundary and initial data, u is the vector of unknown values at the points in the time-space grid, and A is the square matrix that governs the time-marching process. In fact, A can be written in block form, with rows of blocks rather than rows of numbers. Every block is by itself a square matrix of order N (or order N^2 in the two-dimensional case in Chapter 7, Section 13), and the ith row of blocks corresponds to the ith time level or time step in the time-space grid. Because the numerical solution at each time step depends only on the previous one, most of the blocks in A vanish. The only nonzero blocks lie on the main block diagonal and the next block diagonal just below it.

More explicitly, A can be written as

$$\begin{aligned}
A &= \begin{pmatrix} B_1 & & & \\ -Q_1 & B_2 & & \\ & -Q_2 & B_3 & \\ & & \ddots & \ddots \end{pmatrix} \\
&= \begin{pmatrix} I & & & \\ -Q_1 B_1^{-1} & I & & \\ & -Q_2 B_2^{-1} & I & \\ & & \ddots & \ddots \end{pmatrix} \begin{pmatrix} B_1 & & & \\ & B_2 & & \\ & & B_3 & \\ & & & \ddots \end{pmatrix}.
\end{aligned}$$

Here, only the nonzero blocks are indicated. Blocks that are not indicated are zero blocks. Blocks of the form B_i and Q_i correspond to the ith time step in the time-space grid and depend on the particular strategy of time marching.

More specifically, let A_i be the $N \times N$ matrix containing the discrete convection-diffusion spatial derivatives at the ith time step. (A_i is the same as $D^{(i)}$ in Chapter 7, Section 3.) If the explicit scheme in Chapter 7, Section 5, is used, then

$$B_i = (\Delta t)^{-1} I \text{ and } Q_i = (\Delta t)^{-1} I - A_i.$$

If, on the other hand, the implicit scheme in Chapter 7, Section 6, is used, then

$$B_i = (\Delta t)^{-1} I + A_i \text{ and } Q_i = (\Delta t)^{-1} I.$$

Finally, if the semi-implicit scheme in Chapter 7, Section 7, is used, then

$$B_i = (\Delta t)^{-1} I + A_i/2 \text{ and } Q_i = (\Delta t)^{-1} I - A_i/2.$$

Define the block-diagonal matrix B by

$$B = \begin{pmatrix} B_1 & & & \\ & B_2 & & \\ & & B_3 & \\ & & & \ddots \end{pmatrix}.$$

Define also the matrix Q, whose only nonzero blocks lie below the main block diagonal:

$$Q = \begin{pmatrix} 0 & & & \\ Q_1 B_1^{-1} & 0 & & \\ & Q_2 B_2^{-1} & 0 & \\ & & \ddots & \ddots \end{pmatrix}.$$

With these definitions, the coefficient matrix A that governs the time marching can be decomposed as the product

$$A = (I - Q)B.$$

This decomposition will be useful in the stability analysis below.

In what follows, we consider the semi-implicit scheme, which is the most complex one. The explicit and implicit schemes can be analyzed in a similar way. In order to have stability for the explicit scheme, however, one must make sure that Δt is not too large in comparison to the meshsize h. This condition indicates that the explicit scheme may require many more time levels than the implicit and semi-implicit schemes, which can use larger Δt.

8.3 Stability in Time Marching

Marching in time in the time-space grid is equivalent to inverting the block-bidiagonal matrix $A = (I - Q)B$. Of course, in practice A is never inverted explicitly; by "inverting" we mean solving the linear system $Au = \mathcal{F}$ for the unknown vector u. Although A^{-1} is never computed explicitly, its mathematical properties are important in the stability analysis.

Instability means that small errors in the initial data accumulate so rapidly that the numerical solution is useless because it no longer approximates the solution to the original PDE well. In our case, instability can result from applying either B^{-1} or $(I - Q)^{-1}$ to a vector. Let us examine these two processes separately and make sure that neither of them produces instability.

Let us first study the application of B^{-1} to a vector. As we have seen, the blocks B_i on the main block diagonal in B are just the sum of $(\Delta t)^{-1}I$ and a fraction of A_i (the discrete convection-diffusion spatial derivatives at the ith time step). Because A_i is diagonally dominant, its eigenvalues lie in the right half of the complex plane. Furthermore, it follows from [46] that A_i is also irreducible and, therefore, nonsingular. Thus, its eigenvalues lie strictly in the right half of the complex plane; that is, they have positive real parts. In fact, when the diffusion coefficient ε is small, it can be shown that the eigenvalues of A_i lie well away from the origin [38].

Thus, the contribution from the fraction of A_i can only increase the eigenvalues of B_i in magnitude, resulting in yet greater stability when B^{-1} is applied to a vector.

Let us make this reasoning more precise. Let

$$A_i = J_i^{-1} \Lambda_i J_i$$

be the Jordan form of A_i. Let J be the block diagonal matrix with J_i on its main block diagonal. Let Λ be the block-diagonal matrix with Λ_i on its main block diagonal. Then we have

$$\|JB^{-1}J^{-1}\| = \|(JBJ^{-1})^{-1}\| \le \Delta t.$$

This guarantees that the application of B^{-1} to a vector produces no instability.

Let us now verify that the application of $(I - Q)^{-1}$ to a vector also produces no instability. For this purpose, we assume that

$$\|J_{i+1}J_i^{-1}\| \le 1$$

for every i. This assumption indeed holds in some important model cases. For example, when the A_i's are normal ($A_i A_i^t = A_i^t A_i$), the J_i's are orthogonal ($\|J_i\| = \|J_i^{-1}\| = 1$), and, therefore,

$$\|J_{i+1}J_i^{-1}\| \le \|J_{i+1}\| \cdot \|J_i^{-1}\| = 1.$$

The assumption also holds whenever all the J_i's are the same, which happens when the convection coefficient $C(t, x)$ is actually independent of t (constant shock). In Chapter 9, Section 8, we'll see that the assumption also holds in other model cases of major interest.

Even when the above assumption doesn't hold, one can still proceed with the stability analysis below, provided that

$$\|J_i Q_i B_i^{-1} J_i^{-1}\| \le \|J_{i+1}J_i^{-1}\|^{-1}$$

for every i. This may happen when the eigenvalues of $\Delta t A_i$ lie well away from the origin, as indeed is the case when ε is small [38].

(When neither of the above assumptions holds, one should probably replace the above Jordan form by another form, which is based on the so-called pseudospectrum of A_i. In this form, the main-diagonal elements in Λ_i are not exactly the eigenvalues of A_i, but the

J_i's are nearly orthogonal, so $\|J_{i+1}J_i^{-1}\|$ is not much larger than 1. This is left to future research; here we stick to our original assumption $\|J_{i+1}J_i^{-1}\| \le 1$.)

Let us use our original assumption to bound the blocks in Q in the following sense:

$$\begin{aligned}\|J_{i+1}Q_iB_i^{-1}J_i^{-1}\| &= \|J_{i+1}J_i^{-1}(I - \Delta t\Lambda_i/2)(I + \Delta t\Lambda_i/2)^{-1}\| \\ &\le \|J_{i+1}J_i^{-1}\| \cdot \|(I - \Delta t\Lambda_i/2)(I + \Delta t\Lambda_i/2)^{-1}\| \\ &\le 1.\end{aligned}$$

(When the eigenvalues of $\Delta t A_i$ lie well away from the origin, this bound can actually be improved to a constant smaller than 1.)

As a result, we have

$$\|JQJ^{-1}\| \le 1.$$

Let M denote the number of time steps. Using the above results and the formula

$$(I - Q)^{-1} = \sum_{k=0}^{M-1} Q^k,$$

we have

$$\begin{aligned}\|J(I - Q)^{-1}J^{-1}\| &= \|\sum_{k=0}^{M-1}(JQJ^{-1})^k\| \\ &\le \sum_{k=0}^{M-1}\|JQJ^{-1}\|^k \\ &\le M.\end{aligned}$$

(When the eigenvalues of $\Delta t A_i$ lie well away from the origin, this upper bound can actually be improved to a constant independent of M.)

Thus, the application of $(I - Q)^{-1}$ to a vector also produces no instability. This guarantees stability in the time-marching process. In fact, we have

$$\begin{aligned}\|JA^{-1}J^{-1}\| &= \|JB^{-1}J^{-1}J(I - Q)^{-1}J^{-1}\| \\ &\le \|JB^{-1}J^{-1}\| \cdot \|J(I - Q)^{-1}J^{-1}\| \\ &\le \Delta t \cdot M = T\end{aligned}$$

(where T is the maximal time). This guarantees that small errors in $\mathcal{F}$ indeed produce only small errors in the numerical solution u.

8.4 Accuracy and Adequacy

Here, we use the above stability result to show that the numerical schemes are accurate and adequate in the sense described below.

Let $\tilde{u}$ be the numerical solution and u be the vector obtained from the solution $u(t, x)$ of the original PDE when confined to the discrete time-space grid. The discretization error

$$\tilde{u} - u$$

is the difference between the numerical solution $\tilde{u} = A^{-1}\mathcal{F}$ and the true values in u.

We say that a numerical scheme is accurate if the discretization error is sufficiently small. More precisely, the discretization error must approach zero as both h and Δt approach zero at the same time.

Of course, the discretization error is never available explicitly, because the solution of the original PDE is unknown. Fortunately, it can be estimated from the truncation error defined below.

The truncation error is the vector

$$\mathcal{F} - Au.$$

Although u is unknown, the truncation error can still be estimated when the solution of the original PDE is smooth and has a Taylor expansion. In fact, it can be shown that the truncation error at interior grid points (t, x) where $u(t, x)$ is smooth is as small as

$$\Delta t u_{tt}(t, x) + h u_{xx}(t, x)$$

for the explicit and implicit schemes and

$$(\Delta t)^2 u_{ttt}(t, x) + h u_{xx}(t, x)$$

for the semi-implicit scheme. Thus, the truncation error approaches zero whenever both h and Δt approach zero at the same time, as required.

The assumption that the solution of the PDE is smooth is fair. Indeed, even when there are discontinuities in the initial or boundary conditions, they are smoothed away by the diffusion term $-\varepsilon u_{xx}(t, x)$ (or $-\varepsilon(u_{xx}(t, x, y) + u_{yy}(t, x, y))$ in the two-dimensional case). Therefore, one can safely assume that the truncation error is indeed small.

In order to estimate the discretization error $\tilde{u} - u$, note that

$$\tilde{u} - u = A^{-1}(\mathcal{F} - Au).$$

In other words, the discretization-error vector is just A^{-1} times the truncation-error vector. Thus, in order to be accurate, the numerical scheme must be stable in the sense that applying A^{-1} to a vector can enlarge its norm only moderately. Using the result in Section 8.3 above, we indeed have

$$\begin{aligned}
\|J(\tilde{u} - u)\| &= \|JA^{-1}(\mathcal{F} - Au)\| \\
&= \|JA^{-1}J^{-1}J(\mathcal{F} - Au)\| \\
&\le \|JA^{-1}J^{-1}\| \cdot \|J(\mathcal{F} - Au)\| \\
&\le T\|J(\mathcal{F} - Au)\|.
\end{aligned}$$

This guarantees that the discretization error is indeed small, and the scheme is indeed accurate.

The accuracy property, however, is not always sufficient. In some important applications, the diffusion coefficient ε is very small and may in practice be even smaller than h and Δt. Assuming that ε is fixed as h and Δt approach zero is no longer realistic; it suits only problems with substantial diffusion and smooth solution but not problems with little diffusion and practically discontinuous solution. There is a need for a property that suits these cases as well: adequacy.

We say that the numerical scheme is adequate if the discretization error approaches zero as the three parameters h, Δt, and ε approach zero at the same time. (See [38] and Chapter 12, Section 6, in [39].)

The above discretization-error estimate can still be used to show not only accuracy but also adequacy. For adequacy, however, one should keep in mind that u, $\tilde{u}$, and A also depend on ε. Therefore, the derivatives used in the Taylor expansion may grow indefinitely as ε approaches zero. As a result, a truncation error as large as h^{-1} or $(\Delta t)^{-1}$ may appear next to lines of sharp variation in the solution.

Fortunately, the Jordan matrix J used in these estimates multiplies values at such grid points by very small numbers, whereas values at grid points away from sharp variation in the solution are multiplied by relatively very large numbers. We'll return to this subject in Chapter 9, Section 9.

8.5 Exercises

1. The heat equation is obtained from the convection-diffusion equation by setting the convection coefficient $C(t, x)$ to 0 and the diffusion coefficient ε to 1:

$$u_t(t, x) - u_{xx}(t, x) = F(t, x).$$

   Let A_i be the tridiagonal matrix representing the discrete second derivative with respect to the spatial variable x at the ith time step. Show that A_i is symmetric and diagonally dominant.

2. Use the above result to show that the stability condition in Section 8.3 holds, so the implicit and semi-implicit schemes are stable (and, hence, accurate) with respect to the usual norm.

3. Find a bound for $\Delta t/h^2$ for which the explicit scheme is also stable and, hence, accurate. The solution can be found in Section A.5 of the Appendix.

4. The time-dependent diffusion equation with variable coefficient is defined by

$$u_t(t, x) - (P(t, x)u_x(t, x))_x = F(t, x),$$

   where $P(t, x)$ is a given positive and differentiable function. The x-derivative at the midpoints is discretized by

$$Pu_x(i\Delta t, (j + 1/2)h) \doteq h^{-1}P(i\Delta t, (j + 1/2)h)\left[u(i\Delta t, (j + 1)h) - u(i\Delta t, jh)\right].$$

   The spatial term in the PDE is discretized by the divided difference of approximations of the above form at $x = (j + 1/2)h$ and $x = (j - 1/2)h$ [46]. Show that this discretization produces a tridiagonal matrix A_i that is symmetric and diagonally dominant.

5. Use the above result to show that the implicit and semi-implicit schemes are stable (and, hence, accurate) with respect to the usual norm.

6. Find a bound for $\Delta t/h^2$ for which the explicit scheme is also stable (and, hence, accurate). The solution can be found in Section A.5 of the Appendix.

7. Using the above guidelines, show that the above time-marching schemes are stable and accurate with respect to the usual norm for the heat equation in two spatial dimensions:

$$u_t(t, x, y) - u_{xx}(t, x, y) - u_{yy}(t, x, y) = F(t, x, y).$$

8. Similarly, show that the above time-marching schemes are stable and accurate with respect to the usual norm for the time-dependent diffusion equation in two spatial dimensions:

$$u_t(t, x, y) - (P(t, x, y)u_x(t, x, y))_x - (Q(t, x, y)u_y(t, x, y))_y = F(t, x, y),$$

where $P(t, x, y)$ and $Q(t, x, y)$ are given positive and differentiable functions and the spatial discretization is done as in the one-dimensional case above.

9. Assume that the convection coefficient $C()$ in the convection-diffusion equation is independent of t. (For example, $C = 1$ in the left half of the x-interval and $C = -1$ in the right half of it.) Show that the stability condition in Section 8.3 holds, and, hence, the implicit and semi-implicit schemes are stable, accurate, and adequate with respect to the norm used there.

10. For the general convection-diffusion equation, assume that the stability condition in Section 8.3 holds, so the implicit and semi-implicit schemes are stable (and, hence, accurate and adequate) with respect to the norm used there. Find the bound for Δt for which the explicit scheme is also stable (and, hence, accurate and adequate) with respect to that norm. The solution can be found in Section A.5 of the Appendix.

Chapter 9

Nonlinear Equations

In this chapter, we consider nonlinear PDEs such as the Riemann problem, a scalar conservation law with discontinuous initial conditions that may produce a shock wave. We describe the Godunov and random-choice schemes to solve this problem numerically with more general initial conditions. Furthermore, we extend the Godunov scheme to solve a Riemann problem that has been perturbed with a small amount of diffusion. We also describe Newton's iteration, in which the original nonlinear PDE is linearized around an initial approximation obtained from the numerical solution of the corresponding unperturbed problem and then relinearized subsequently around better and better approximate solutions. Finally, we show that the time-marching schemes for these linearized problems are stable, accurate, and adequate.

9.1 Nonlinear PDEs

So far, we have dealt only with linear PDEs, in which the coefficients are given functions that don't depend on the solution. This problem may be thought of as a fixed given frame into which one should fit the suitable solution.

In this chapter, we study quasi-linear PDEs in which the coefficients may well depend on the unknown solution. This problem can be thought of as a flexible frame into which one is supposed to fit a solution. The extra difficulty here is that the frame may change unexpectedly and reject a solution that seemed to fit. The task of finding a solution is thus much more challenging.

9.2 The Riemann Problem

A typical quasi-linear PDE is the Riemann problem (also known as the scalar conservation law): find a function $u(t, x)$ that satisfies the PDE

$$u_t(t,x) + u(t,x)u_x(t,x) = 0, \qquad -\infty < x < \infty,\ 0 < t < \infty,$$

and the initial conditions

$$u(0, x) = \begin{cases} a & \text{if } x \leq 0, \\ b & \text{if } x > 0, \end{cases}$$

where a and b are given real numbers.

When a is different from b, u is nondifferentiable and cannot satisfy the PDE in the usual (strong) sense. Instead, it satisfies it in the weak sense, which means that the integral of the left-hand side of the PDE over every subdomain of the original time-space domain is zero. In this integral, Green's formula can be used to avoid differentiation. We'll return to this subject in Section 9.4 and Chapter 11, Section 2.

Note that the Riemann problem is similar to the convection-diffusion equation in Chapter 7. The main differences are that here there is no diffusion ($\varepsilon = 0$), and the convection coefficient $C(t, x)$ is replaced by the unknown solution $u(t, x)$ itself. This nonlinearity is the source of all sorts of interesting phenomena.

9.3 Conflicting Characteristic Lines

Let us find lines $x(t)$ in the time-space domain along which the solution $u(t, x(t))$ is constant, or

$$\frac{d}{dt}u(t, x(t)) = 0.$$

These lines, called characteristic lines, are particularly important because they are the lines along which the initial data propagate.

Let us use the PDE to specify the characteristic lines. From the chain rule, we have

$$u_t(t, x(t)) + \frac{dx(t)}{dt} \cdot u_x(t, x(t)) = \frac{d}{dt}u(t, x(t)) = 0.$$

From the PDE, we have that the above equation is satisfied whenever

$$\frac{dx(t)}{dt} = u(t, x(t)).$$

Thus, the slope of characteristic lines issuing from points $x(0) \leq 0$ is $dx(t)/dt = a$, and the slope of characteristic lines issuing from points $x(0) > 0$ is $dx(t)/dt = b$. In other words, the state $u(0, x) = a$ to the left of 0 propagates to the right with speed a, and the state $u(0, x) = b$ to the right of 0 propagates to the right with speed b.

All this works well so long as $a \leq b$. The left state travels to the right more slowly than the right state, so they don't collide with each other. Between these states, there emerges a third, nonconstant state, which connects them linearly. This solution is called a rarefaction wave.

In Figure 9.1, the case $a < b < 0$ is displayed. Because both a and b are negative, the rarefaction wave actually travels to the left. In Figure 9.2, the case $a < 0 < b \leq -a$ is displayed. In this case, the state $u = a$ to the left of the origin travels to the left, the state $u = b$ to the right of the origin travels to the right, and these states are connected by a linear segment with slope that decreases gradually in time. (The cases $0 < a < b$ and $a < 0 < -a < b$ can be transformed to the above cases (respectively) by the transformation $x \to -x$ and $u \to -u$.)

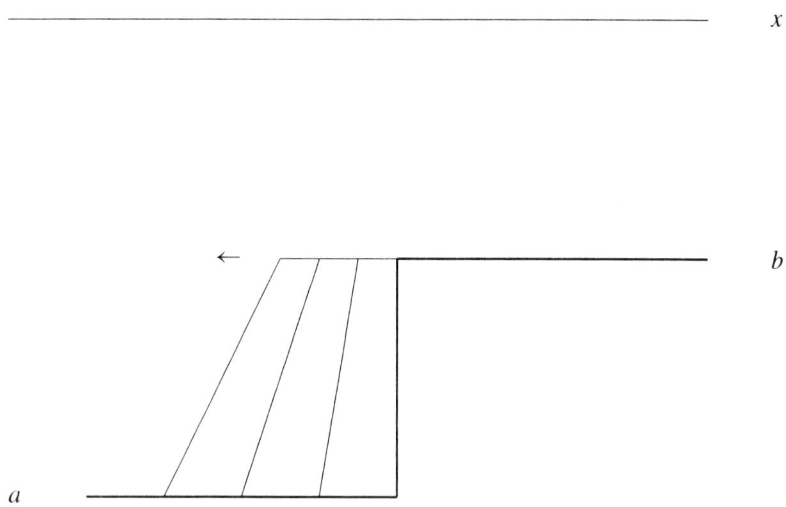

Figure 9.1. *The rarefaction wave that travels to the left ($a < b < 0$).*

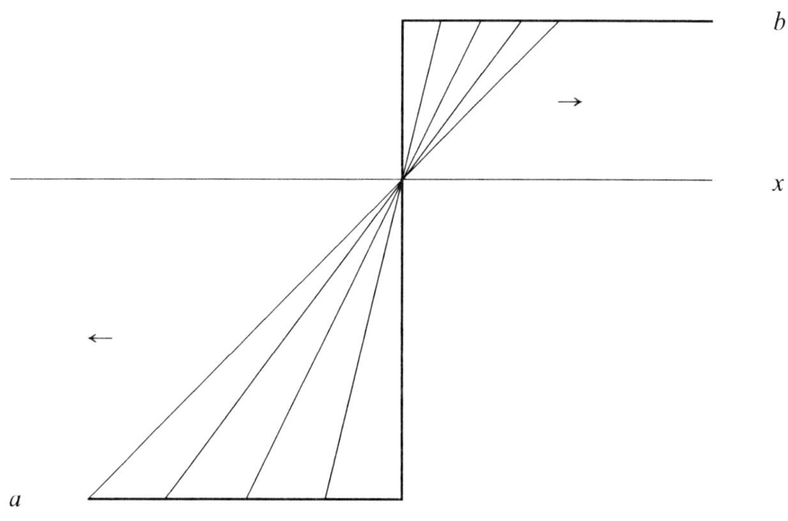

Figure 9.2. *The rarefaction wave whose right part travels to the right and left part travels to the left ($a < 0 < b \leq -a$).*

When $a > b$, on the other hand, the characteristic lines issuing from $x(0) < 0$ have a larger slope than those issuing from $x(0) > 0$. In fact, these characteristic lines collide with each other at $x = 0$ immediately at the beginning of the process. The rule for resolving this conflict is that the shock front at $x = 0$ travels to the right at the average speed $(a + b)/2$ [41]. This solution is called a shock wave.

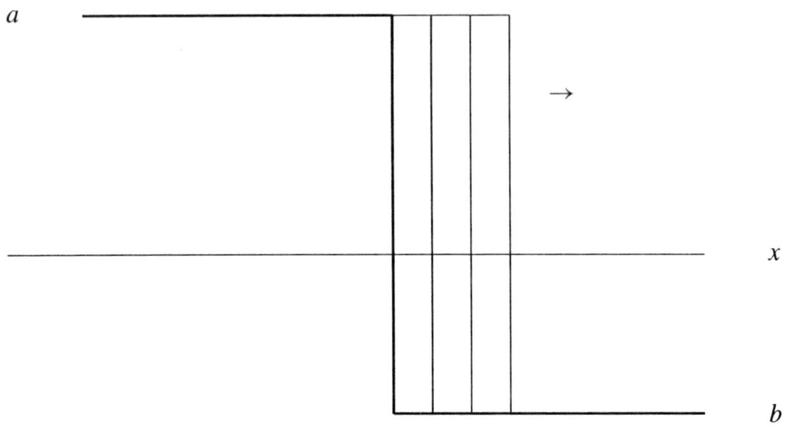

Figure 9.3. *The shock wave that travels to the right with speed* $(a+b)/2 > 0$.

In Figure 9.3, we display the case $(a+b)/2 > 0$, where the shock travels to the right. (The case $(a+b)/2 < 0$, where the shock travels to the left, can be transformed to the wave in Figure 9.3 by the transformation $x \to -x$ and $u \to -u$.)

The solution to the Riemann problem is, thus, either a shock or a rarefaction wave, depending on the parameters a and b. In what follows, we discuss further the properties of these solutions and use them in numerical schemes to solve more general versions of the Riemann problem.

9.4 The Godunov Scheme

As we have seen above, the Riemann problem has an analytic solution, so no numerical scheme is needed. Here, we consider a more general version of the Riemann problem, with general initial condition $u(0,x)$, which may assume not only two constant values a and b but also a whole range of real values.

The numerical time-marching scheme for this problem uses shock and rarefaction waves to form the so-called half time step at time $t = (i+1/2)\Delta t$, from which the numerical solution at the $(i+1)$th time step soon follows.

Let us rewrite the Riemann problem in a slightly more general way:

$$u_t(t,x)+(f(u(t,x)))_x = u_t(t,x)+f'(u(t,x))u_x(t,x) = 0, \quad -1 \le x \le 1,\ 0 < t < \infty,$$

where f is a given convex function of one variable. (In our case, for example, $f(q) = q^2/2$, so $f'(q) = q$ for every real number q.)

We also assume that the initial conditions are given by

$$u(0,x) = u_0(x), \quad -1 \le x \le 1,$$

where $u_0(x)$ is a given function. For simplicity, we assume that the boundary conditions are periodic:

$$u(t,-1) = u(t,1) \quad 0 \le t < \infty.$$

The Godunov scheme [41] is defined as follows. Let $u_{i,j}$ denote the value of the numerical solution at the point at row i and column j in the time-space grid. In other words,

$$u_{i,j} \doteq u(i\Delta t, jh).$$

Similarly, we also consider the numerical solution at the half time step $(i + 1/2)\Delta t$:

$$u_{i+1/2,j+1/2} \doteq u((i + 1/2)\Delta t, (j + 1/2)h).$$

The numerical solution at the half time step is helpful in calculating the numerical solution at the next time step, $(i + 1)\Delta t$. To this end, consider the small square s in Figure 9.4. In s, the midpoint of the bottom edge is (i, j), and the midpoint of the top edge is $(i + 1, j)$. The boundary of s, ∂s, consists of four edges: the top edge ∂s_1, the right edge ∂s_2, the bottom edge ∂s_3, and the left edge ∂s_4. Let $\vec{n}$ be the outer normal vector at ∂s:

$$\vec{n} = \begin{cases} (0, 1) & \text{in } \partial s_1, \\ (1, 0) & \text{in } \partial s_2, \\ (0, -1) & \text{in } \partial s_3, \\ (-1, 0) & \text{in } \partial s_4. \end{cases}$$

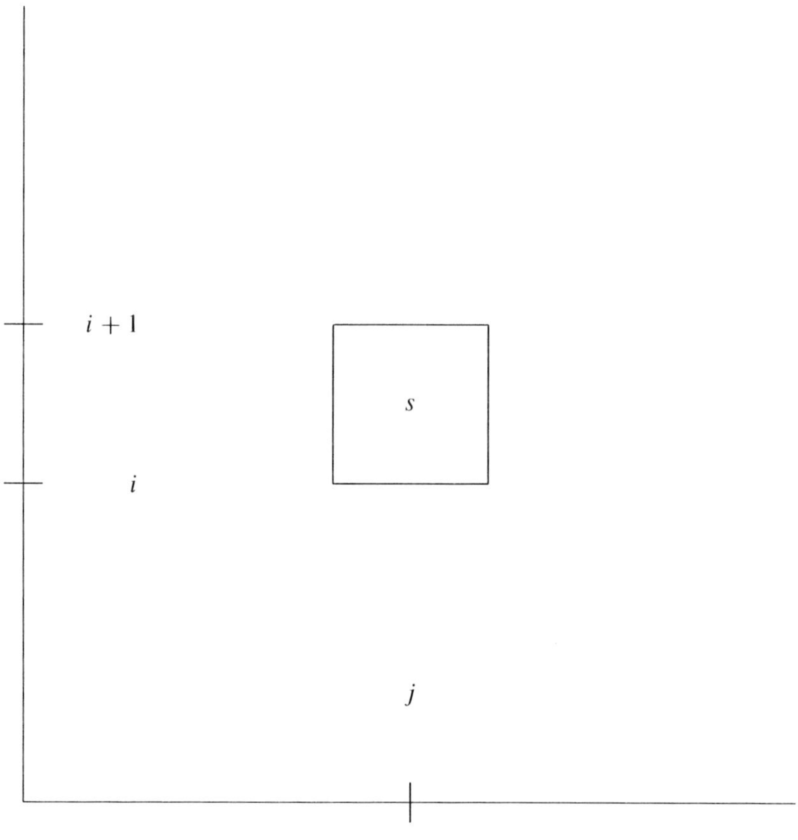

Figure 9.4. *The square s on which the conservation law is integrated to produce the Godunov scheme.*

Let ds be the length element along ∂s. Using Green's formula in the PDE and approximating the value of u at each edge of s by the value at its midpoint, we have

$$\begin{aligned}0 &= \int_s (u_t(t,x) + f(u(t,x))_x)dtdx \\ &= \int_{\partial s} (u(t,x), f(u(t,x))) \cdot \vec{n}\, ds \\ &= \int_{\partial s_1} u(t,x)\, dx - \int_{\partial s_3} u(t,x)\, dx \\ &+ \int_{\partial s_2} f(u(t,x))\, dt - \int_{\partial s_4} f(u(t,x))\, dt \\ &\doteq h\left(u_{i+1,j} - u_{i,j}\right) \\ &+ \Delta t\left(f(u_{i+1/2,j+1/2}) - f(u_{i+1/2,j-1/2})\right).\end{aligned}$$

The Godunov scheme is thus defined by

$$u_{i+1,j} \equiv u_{i,j} - (\Delta t/h)\left(f(u_{i+1/2,j+1/2}) - f(u_{i+1/2,j-1/2})\right),$$

where the values $u_{i+1/2,j+1/2}$ at the half time step are calculated by solving small local Riemann problems at every two adjacent cells, as in Figure 9.5.

To illustrate how the values $u_{i+1/2,j+1/2}$ at the half time step are calculated, let us return to our case, in which

$$f(u(t,x)) = \frac{1}{2}u^2(t,x).$$

In this case, we have

$$u_{i+1/2,j+1/2} \equiv \begin{cases} u_{i,j} & \text{if } u_{i,j} > u_{i,j+1} \text{ and } (u_{i,j} + u_{i,j+1})/2 \ge 0 \\ & \text{(local shock wave traveling to the right)}, \\ u_{i,j+1} & \text{if } u_{i,j} > u_{i,j+1} \text{ and } (u_{i,j} + u_{i,j+1})/2 < 0 \\ & \text{(local shock wave traveling to the left)}, \\ u_{i,j} & \text{if } 0 \le u_{i,j} \le u_{i,j+1} \\ & \text{(local rarefaction wave traveling to the right)}, \\ u_{i,j+1} & \text{if } u_{i,j} \le u_{i,j+1} \le 0 \\ & \text{(local rarefaction wave traveling to the left)}, \\ 0 & \text{if } u_{i,j} < 0 < u_{i,j+1} \\ & \text{(local rarefaction wave)}. \end{cases}$$

In order to use the half time step, we also need to define the values $u_{i+1/2,1/2}$ and $u_{i+1/2,N+1/2}$ at the endpoints. Using the periodic boundary conditions, this can be done either as before or by

$$u_{i+1/2,1/2} \equiv u_{i+1/2,N-1/2} \quad \text{and} \quad u_{i+1/2,N+1/2} \equiv u_{i+1/2,3/2}.$$

This completes the definition of the Godunov scheme.

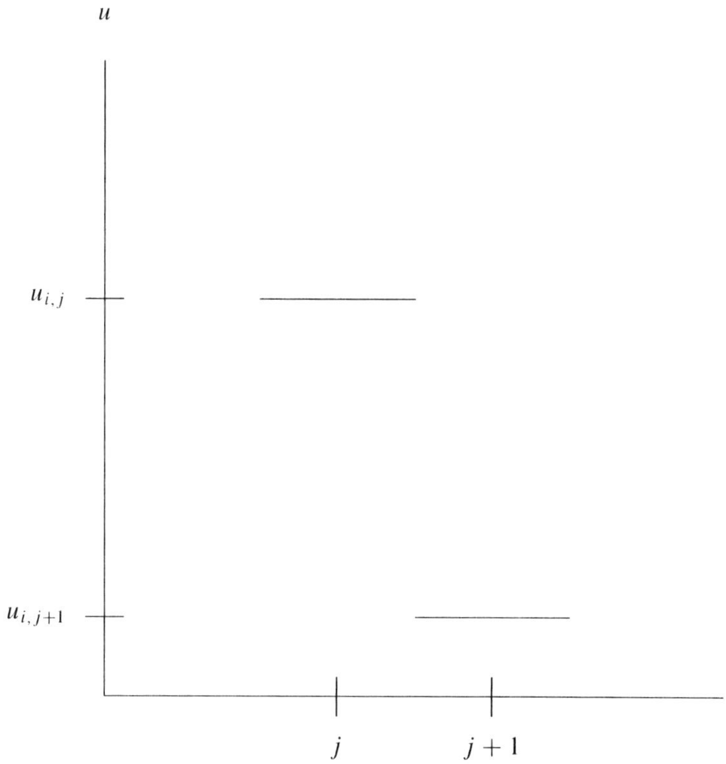

Figure 9.5. *The local Riemann problem in the two adjacent cells that contain the points j and $j+1$ in the ith time step. This local problem produces the half time step used in the Godunov scheme.*

9.5 The Random-Choice Scheme

Here, we describe the random-choice scheme [41] for the solution of the Riemann problem

$$u_t(t,x) + u(t,x)u_x(t,x) = 0, \qquad -1 \le x \le 1,\ 0 < t < \infty,$$

with general initial conditions

$$u(0,x) = u_0(x)$$

and periodic boundary conditions. In this scheme, the half time step is also calculated by solving small local Riemann problems as in Figure 9.5. This, however, is done in a slightly different way: a random number $-1 < \theta < 1$ is picked at each half time step, and $u_{i+1/2,j+1/2}$ takes the value of the solution to the local Riemann problem in Figure 9.5 at the point $(j+1/2+\theta/2)h$. More precisely, the random-choice scheme is defined by

$$u_{i+1/2,j+1/2} \equiv \begin{cases} u_{i,j} & \text{if } u_{i,j} > u_{i,j+1} \text{ and } \theta h/\Delta t \le (u_{i,j}+u_{i,j+1})/2 \\ & \text{(local shock wave)}, \\ u_{i,j+1} & \text{if } u_{i,j} > u_{i,j+1} \text{ and } \theta h/\Delta t \ge (u_{i,j}+u_{i,j+1})/2 \\ & \text{(local shock wave)}, \\ u_{i,j} & \text{if } \theta h/\Delta t \le u_{i,j} \le u_{i,j+1} \\ & \text{(local rarefaction wave)}, \\ u_{i,j+1} & \text{if } u_{i,j} \le u_{i,j+1} \le \theta h/\Delta t \\ & \text{(local rarefaction wave)}, \\ \theta h/\Delta t & \text{if } u_{i,j} < \theta h/\Delta t < u_{i,j+1} \\ & \text{(local rarefaction wave)}. \end{cases}$$

The $(i+1)$th time step is then produced from the $(i+1/2)$th time step in the same way that the $(i+1/2)$th time step was produced from the ith time step. Note that the random number θ may assume different values at different time steps. This completes the definition of the random-choice scheme.

9.6 The N-Wave

The N-wave is the solution $u(T,x)$ (at a sufficiently large time T) of the following version of the Riemann problem:

$$u_t(t,x)+u(t,x)u_x(t,x)=0, \qquad -1 \le x \le 1,\ 0<t<\infty,$$

with sinusoidal initial conditions of the form

$$u(0,x)=\sin(\pi x)$$

and periodic boundary conditions of the form

$$u(t,-1)=u(t,1), \qquad 0 \le t < \infty.$$

The reason for this name is that the solution has the shape of the letter N in English. Here, we solve for the N-wave numerically and indeed observe this shape.

We use the Godunov scheme with $\Delta t = h/2$ on a 200×200 time-space grid and obtain the results in Figure 9.6. The solution at the 200th time step indeed has a symmetric and straight N-shape, as required.

We have also applied the random-choice scheme and obtained the results in Figure 9.7. Here, the solution is slightly less straight and symmetric than before. The more accurate Godunov scheme is therefore used in the perturbed problem below.

9.7 Singular Perturbation

Here, we consider the singular perturbation of the Riemann problem, in which a small diffusion term is added to the PDE:

$$-\varepsilon u_{xx}(t,x)+u_t(t,x)+u(t,x)u_x(t,x)=0, \qquad -1<x<1,\ 0<t<\infty.$$

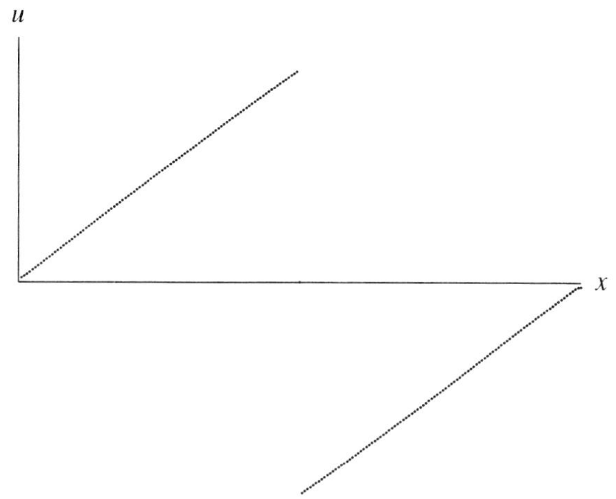

Figure 9.6. *The N-wave produced by the Godunov scheme at the 200th time step, with $\Delta t = h/2$ and a 200×200 x-t grid.*

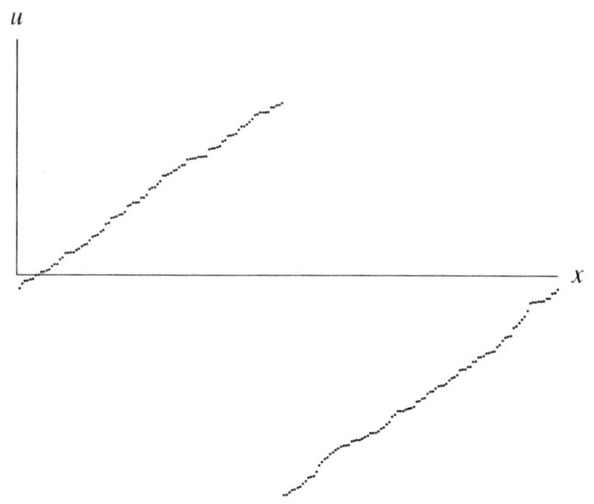

Figure 9.7. *The N-wave produced by the random-choice scheme at the 200th time step, with $\Delta t = h/2$ and a 200×200 x-t grid.*

The initial and boundary conditions are as before.

The original Godunov scheme should be modified slightly to apply to this problem. In fact, here the integration along the vertical edges ∂s_2 and ∂s_4 of the square s in Figure 9.4 is also done on the term $-\varepsilon u_x(t, x)$, so more terms must be added in the calculation of the $(i+1)$th time step in Section 9.4. These terms can be added in an explicit, implicit, or semi-implicit way. For simplicity, we consider here only the explicit way, in which the

Godunov scheme takes the modified form

$$u_{i+1,j} = u_{i,j} - \frac{(\Delta t)\varepsilon}{h^2}\left(2u_{i,j} - u_{i,j-1} - u_{i,j+1}\right) \\ - \frac{\Delta t}{h}\left(f(u_{i+1/2,j+1/2}) - f(u_{i+1/2,j-1/2})\right).$$

(The half time step is computed as before.) We refer to this scheme as the explicit Godunov scheme.

We have applied this scheme to the singular perturbation of the N-wave problem defined above. In order to maintain stability, we have used $\varepsilon = h/4$. (For larger values of ε, smaller values of Δt or the implicit or semi-implicit scheme must be used.) The results are displayed in Figure 9.8.

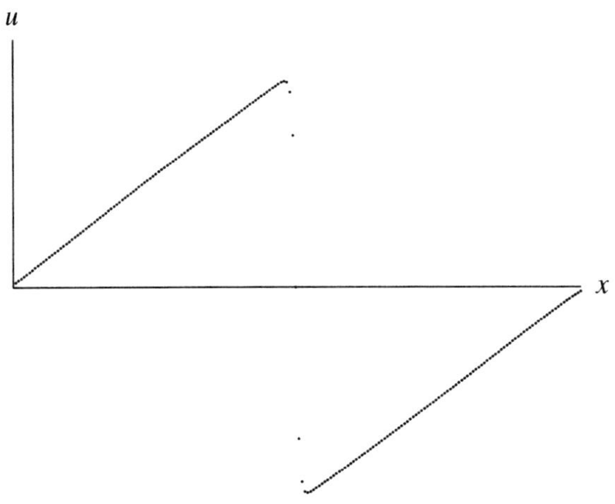

Figure 9.8. *The singularly perturbed N-wave produced by the explicit Godunov scheme at the* 200*th time step, with* $\Delta t = h/2$, $\varepsilon = h/4$, *and a* 200×200 x-t *grid.*

It can be seen that the solution has a symmetric and straight N-shape as before. It is only slightly smoothed at the middle point $x = 0$ due to the extra diffusion term, but the N-shape is still as clear as before.

9.8 Linearization

The only difference between the convection-diffusion equation in Chapter 7 and the singularly perturbed Riemann problem in Section 9.7 is that in the convection-diffusion equation the convection coefficient is a given function $C(t, x)$ that is known in advance, whereas in the singularly perturbed Riemann problem the convection coefficient is a function that is not known in advance and is actually the same as the unknown solution $u(t, x)$ itself. Therefore, in the convection-diffusion equation, the characteristic lines (or curves) are available as the solutions to the ODE

$$\frac{dx(t)}{dt} = C(t, x(t))$$

with initial condition

$$x(0) = x_0$$

(where $-1 < x_0 < 1$ is the fixed starting point of the characteristic curve). On this characteristic curve, the solution remains constant:

$$u(t, x(t)) = u(0, x(0)).$$

In other words, the data in the initial condition flow along the characteristic curves to form the solution there.

In the singularly perturbed Riemann problem, on the other hand, the characteristic curves solve an ODE that depends on the unknown solution $u(t, x)$ itself:

$$\frac{dx(t)}{dt} = u(t, x(t)),$$

with initial condition

$$x(0) = x_0$$

(where $-1 < x_0 < 1$ is the starting point of the characteristic curve). Because $u(t, x)$ is not yet available, the characteristic curves are also unavailable. Thus, the solution $u(t, x)$ determines the shape of characteristic curves that govern the data propagation to solve the original PDE. This circularity is the direct result of nonlinearity.

Can the linear PDE serve as a good approximation to the corresponding nonlinear problem? Probably not, since it can never predict the nonlinear phenomena (shock and rarefaction waves) in the original nonlinear problem. The only way in which it may form a fair approximation is when the convection coefficient $C(t, x)$ already incorporates the nonlinear waves, e.g., it is the solution of the unperturbed Riemann problem. We then say that the singularly perturbed Riemann problem has been linearized around $C(t, x)$. Because $C(t, x)$ already contains the required shock and rarefaction waves, it is sufficiently close to the solution, and the linearized problem indeed approximates the original nonlinear problem well.

The above discussion leads to the following algorithm for the solution of the singularly perturbed Riemann problem: let $u = (u_{i,j})$ be the numerical solution to the unperturbed Riemann problem, obtained from the Godunov scheme in Section 9.4 or the random-choice scheme in Section 9.5. Let $C(t, x)$ be a function that coincides with $u_{i,j}$ on the time-space grid. The algorithm requires solving the convection-diffusion equation with $C(t, x)$ as the convection coefficient. This solution is obtained numerically by using the explicit, implicit, or semi-implicit scheme in Chapter 7, Sections 5 to 7. Because this solution is a fair approximation to the required solution of the original nonlinear problem, it can be substituted for $C(t, x)$ to yield an even better linearization. This process can repeat iteratively, resulting in better and better linearizations, and better and better numerical solutions to the original nonlinear problem. This is the Newton (or Newton–Rabson, or Picard, or fixed-point) iteration.

In several model cases of major interest, we know that the explicit, implicit, and semi-implicit schemes used to solve the first linearized problem are indeed stable and, hence, also accurate and adequate. These model cases are the cases in which $C(t, x)$ is a shock or a rarefaction wave as in Figures 9.1 to 9.3. Indeed, in these cases, one can show that the stability condition in Chapter 8, Section 3,

$$\|J_{i+1}J_i^{-1}\| \leq 1,$$

holds, where i is the index of the current time step and J_i is the Jordan matrix for the matrix A_i that contains the discrete convection-diffusion terms at time $i\Delta t$ (also denoted by $D^{(i)}$ in Chapter 7, Section 3). The proof of this inequality is in Section A.6 of the Appendix.

For more general initial conditions, the solution to the unperturbed Riemann problem that is substituted initially for $C(t, x)$ may be much more complicated than the model waves in Figures 9.1 to 9.3. For sinusoidal initial conditions, for example, it converges with time to the N-wave in Figures 9.6 and 9.7. Although the present stability analysis no longer applies in these general cases, it can still be interpreted to indicate that time marching in the linearized problem should be stable and, hence, also accurate and adequate.

9.9 Adequacy in the Linearized Problem

This section is related to the theory in Chapter 8 and can be skipped by readers who are interested in the practical aspects only. Here, we study the adequacy of the time-marching schemes in Chapter 7, Sections 5 to 7, for the singularly perturbed Riemann problem linearized around a shock or a rarefaction wave. The result of this linearization is actually a convection-diffusion equation, with convection coefficient $C(t, x)$, as in Figures 9.1 to 9.3. This problem may serve as an interesting model for the first linearized problem in the above Newton iteration.

As discussed at the end of Chapter 8, Section 4, large truncation errors may appear at grid points where the solution $u(t, x)$ exhibits large variation. This problem is particularly relevant when adequacy is considered, because in this case the diffusion coefficient ε goes to zero together with the meshsize h and Δt, so there is no sufficient diffusion to smooth sharp variation. Therefore, the truncation error at these places can be as large as $(\Delta t)^{-1} + h^{-1}$.

Fortunately, when Dirichlet, Neumann, or mixed boundary conditions are used, the Jordan matrix J used in Chapter 8, Section 3, is practically diagonal, so applying it to a vector is equivalent to multiplying its components by the corresponding main-diagonal elements. Furthermore, these main-diagonal elements are particularly small at grid points where $u(t, x)$ exhibits large variation, such as shock fronts. Thus, one can show that the discretization error is small at least at grid points where $u(t, x)$ is rather smooth.

Indeed, it can be seen in Section A.6 of the Appendix that J can be decomposed as the product

$$J = RE$$

of the orthogonal (norm-preserving) matrix R and the diagonal matrix E. Assume first that the convection coefficient $C(t, x)$ is a shock wave, as in Figure 9.3. Then, the main-diagonal elements in E have the small value $e_{\min}$ at grid points at the shock front and grow exponentially as one gets farther away from the shock front. Moreover, when one gets farther and farther away from the shock front, the solution $u(t, x)$ approaches a constant value even more rapidly, so $u_{tt}(t, x)$ and $u_{xx}(t, x)$ (and, hence, also the truncation error) approach zero faster than the growth in the corresponding main-diagonal elements in E (see [38]).

Let E_i be the block in E that corresponds to the ith time level. This way, $J_i = R_i E_i$ is the Jordan matrix for A_i, the matrix that contains the discrete spatial derivatives at the ith time level. Assume also that

$$\varepsilon = O(h) = O(\Delta t)$$

(that is, ε approaches zero at least as rapidly as h, and h approaches zero at least as rapidly as Δt). As discussed at the end of Section A.6 of the Appendix, in this case $\Delta t E_i A_i E_i^{-1}$ is strictly diagonally dominant, so its eigenvalues are well away from the origin, and the stability estimate in Chapter 8, Section 3, improves to read

$$\|JA^{-1}J^{-1}\| = O(\Delta t)$$

(that is, $\|JA^{-1}J^{-1}\|$ approaches zero at least as rapidly as Δt). By denoting the numerical solution by $\tilde{u}$ and the solution of the linearized problem (in the grid) by u (as in Chapter 8, Section 4), we thus have

$$\begin{aligned} J(\tilde{u} - u) &= JA^{-1}J^{-1}J(\mathcal{F} - Au) = JB^{-1}(I - Q)^{-1}J^{-1}J(\mathcal{F} - Au) \\ &= \left(JBJ^{-1}\right)^{-1} \left(\sum_{n=0}^{M-1} \left(JQJ^{-1}\right)^n \right) J(\mathcal{F} - Au). \end{aligned}$$

Consider, in particular, the $(i+1)$st time level. The contribution to this time level from the previous one is done when $n = 1$ in the above sum through the block $J_{i+1} Q_i B_i^{-1} J_i^{-1}$ in the matrix JQJ^{-1}. In fact, the truncation error at the ith time level (multiplied by J_i) is first multiplied by $J_i Q_i B_i^{-1} J_i^{-1}$ (by which it can only decrease) then multiplied by $J_{i+1} J_i^{-1}$ to scale properly (that is, to scale by J_{i+1} rather than J_i). The contributions from former time levels (through $(JQJ^{-1})^n$) decrease exponentially with n. Thus, the error at the $(i+1)$st time level (multiplied by J_{i+1}) is governed by the truncation error at this time level (which is concentrated at the shock front), multiplied by J_{i+1} and by $\left(J_{i+1} B_{i+1} J_{i+1}^{-1}\right)^{-1} = O(\min(\Delta t, h))$. The conclusion is, thus, that the discretization error is of order 1 at most at a finite number of grid points around the shock front. At grid points farther away from the shock front, the discretization error must decrease exponentially.

Unfortunately, the above analysis proves adequacy only when the convection coefficient $C(t, x)$ is as in Figure 9.3 but not when it is a shock wave that is positive everywhere ($a > b > 0$). In the latter case, the main-diagonal elements in E decrease exponentially monotonically along each x-line, so large discretization errors ahead of the shock front (to the right of it) are not excluded in the above analysis. Still, it makes no sense to assume that instability should occur ahead of the shock front, and the present analysis still excludes discretization errors behind it.

Similarly, when the convection coefficient $C(t, x)$ is a rarefaction wave, as in Figure 9.1, the main-diagonal elements in E increase exponentially monotonically along each x-line, so discretization errors are excluded by the above analysis only at grid points behind the intermediate linear segment in the rarefaction wave (to the right of it).

Finally, when the convection coefficient $C(t, x)$ is a rarefaction wave, as in Figure 9.2, the situation is more complex. The main-diagonal elements in E decrease exponentially as one goes farther away from $x = 0$ in each x-line. Therefore, the above analysis excludes discretization errors in the intermediate linear segment but not in the constant states to its left and right. Still, since these constant states lie ahead of the wave front, it is unlikely that any instability should occur there.

9.10 The Inhomogeneous Case

Let us now consider the Riemann problem with a nonzero right-hand-side function $F(t, x)$:

$$u_t(t, x) + u(t, x)u_x(t, x) = F(t, x).$$

The Godunov scheme in Section 9.4 should be slightly modified to account for the nonzero right-hand side. In particular, at the $(i + 1)$th time step, one should also add the integral of $F(t, x)$ over the $\Delta t \times h$ cell s, which can be approximated by

$$\Delta t h F((i + 1/2)\Delta t, jh).$$

This defines the Godunov scheme at the $(i + 1)$th time step.

The $(i + 1)$th time step, however, depends on the numerical solution at the $(i + 1/2)$th half time step, where the Godunov scheme needs to be modified as well. This is done as follows. Recall that the value $u_{i+1/2,j+1/2}$ in Section 9.4 is equal to either $u_{i,j}$ or $u_{i,j+1}$, depending on the solution of the local Riemann problem in the interval $[jh, (j+1)h]$. This is because u must remain constant along the characteristic line $dx(t)/dt = u(t, x(t))$ issuing from the ith time step toward the point $((i+1/2)\Delta t, (j+1/2)h)$. In the inhomogeneous case, however, u is no longer constant along this line. Indeed, it follows from the chain rule that

$$\begin{aligned} du(t, x(t))/dt &= u_t(t, x(t)) + u_x(t, x(t))dx(t)/dt \\ &= u_t(t, x(t)) + u(t, x(t))u_x(t, x(t)) \\ &= F(t, x(t)). \end{aligned}$$

Thus, one should also add the contribution from the integral of $F(t, x(t))$ over the characteristic line $x(t)$ issuing from time $t = it\Delta t$ toward the point $((i + 1/2)\Delta t, (j + 1/2)h)$. (See Chapter 6, Section 6, and set $S \equiv 0$ there.) This contribution can be approximated by the term

$$(\Delta t/2)\sqrt{1 + u_{i,j}^2}F(i\Delta t, jh)$$

if the value $u_{i+1/2,j+1/2}$ in Section 9.4 is the same as $u_{i,j}$ or

$$(\Delta t/2)\sqrt{1 + u_{i,j+1}^2}F(i\Delta t, (j + 1)h)$$

if the value $u_{i+1/2,j+1/2}$ in Section 9.4 is the same as $u_{i,j+1}$. The above approximation to the integral of $F(t, x(t))$ along the characteristic line $x(t)$ that leads to $((i+1/2)\Delta t, (j+1/2)h)$ (see Figure 9.9) should then be added to the value of $u_{i+1/2,j+1/2}$ in Section 9.4 to produce the correct value in the inhomogeneous case. This completes the definition of the modified Godunov scheme in the inhomogeneous case.

In the next section, we use the above modification to extend the Godunov scheme to the vector conservation law.

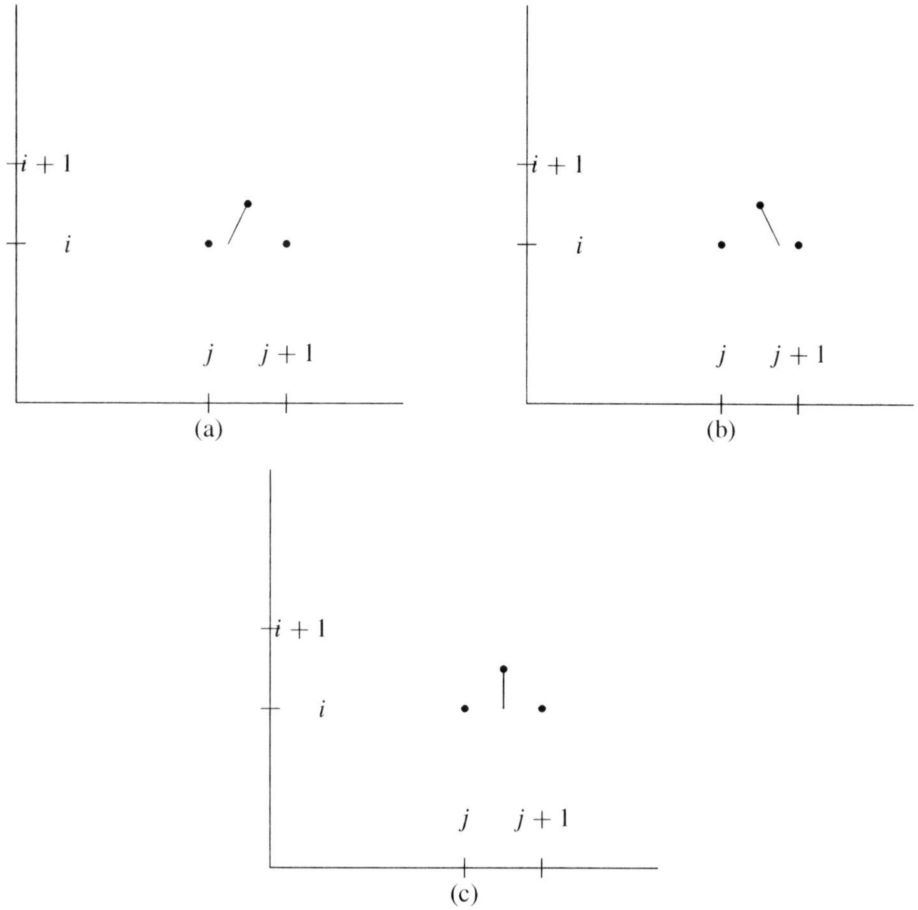

Figure 9.9. *The characteristic line (along which F is integrated to contribute to $u_{i+1/2,j+1/2}$ at the half time step) has slope $dx/dt = u_{i,j} > 0$ if the state $u_{i,j}$ on the left travels to the right* (a), *$dx/dt = u_{i,j+1} < 0$ if the state $u_{i,j+1}$ on the right travels to the left* (b), *or $dx/dt = 0$ if $u_{i,j} < 0 < u_{i,j+1}$* (c).

9.11 System of Nonlinear PDEs

Here, we consider the extension of the singularly perturbed Riemann problem in Section 9.7 to the case of two spatial dimensions x and y and two unknown functions $u(t, x, y)$ and $v(t, x, y)$, which are coupled in the following system of nonlinear PDEs:

$$
\begin{aligned}
&-\varepsilon \left(u_{xx}(t, x, y) + u_{yy}(t, x, y)\right) + u_t(t, x, y)\\
&+ u(t, x, y)u_x(t, x, y) + v(t, x, y)u_y(t, x, y)\\
&= F_1(t, x, y),\\
&-\varepsilon \left(v_{xx}(t, x, y) + v_{yy}(t, x, y)\right) + v_t(t, x, y)\\
&+ u(t, x, y)v_x(t, x, y) + v(t, x, y)v_y(t, x, y)\\
&= F_2(t, x, y),
\end{aligned}
$$

where $F_1(t,x,y)$ and $F_2(t,x,y)$ are given functions, and x and y are the independent variables in a two-dimensional domain, with suitable initial and boundary conditions. The above system of PDEs is also called a system of singularly perturbed conservation laws.

The characteristic lines for this system are defined in the three-dimensional (t,x,y) space. In fact, they are defined by the vector $(t, x(t), y(t))$ satisfying

$$dx(t)/dt = u(t,x(t),y(t)) \text{ and } dy(t)/dt = v(t,x(t),y(t)).$$

With this definition, we have from the chain rule that, when $\varepsilon = 0$,

$$\begin{aligned} du(t,x(t),y(t))/dt &= u_t(t,x(t),y(t)) + u(t,x(t),y(t))u_x(t,x(t),y(t)) \\ &\quad + v(t,x(t),y(t))u_y(t,x(t),y(t)) \\ &= F_1(t,x(t),y(t)), \\ dv(t,x(t),y(t))/dt &= v_t(t,x(t),y(t)) + u(t,x(t),y(t))v_x(t,x(t),y(t)) \\ &\quad + v(t,x(t),y(t))v_y(t,x(t),y(t)) \\ &= F_2(t,x(t),y(t)). \end{aligned}$$

Thus, the initial and boundary conditions about u and v travel along the characteristic lines (or curves) to form the solution in the entire time-space domain.

It is assumed that the right-hand-side functions F_1 and F_2 are independent of ε. This implies that the solution functions u and v must be smooth along the characteristic curves. Still, they could in theory oscillate rapidly (with frequency of up to $\varepsilon^{-1/2}$) in the directions perpendicular to the characteristic curves in the three-dimensional time-space domain. To prevent this, it is assumed that the initial and boundary conditions are also independent of ε. Because the initial and boundary data flow along the characteristic curves to the entire time-space domain, the solution functions u and v cannot oscillate frequently in any direction and must be smooth except possibly at fronts of shock or rarefaction waves (as in Figure 9.8). Therefore, the truncation error in the time-marching schemes for the linearized problem must be small, except possibly at these fronts.

The time-marching schemes can thus be used in the linearized systems in the Newton iteration. For this, however, we need good initial guesses for u and v to get the iteration started. These initial guesses should incorporate nonlinear phenomena such as shock and rarefaction waves. In fact, they can be obtained from the numerical solution of the unperturbed case. This numerical solution can be obtained by a vector version of the Godunov scheme. Better yet, the initial guess in Newton's iteration can be the numerical solution obtained from a vector version of the explicit Godunov scheme in Section 9.7.

The explicit Godunov scheme is used within a version of the alternating-direction implicit (ADI) approach [2]. For this purpose, let us rewrite the first equation in the form

$$\begin{aligned} &-\varepsilon u_{xx}(t,x,y) + u_t(t,x,y) + u(t,x,y)u_x(t,x,y) \\ &= \varepsilon u_{yy}(t,x,y) - v(t,x,y)u_y(t,x,y) + F_1(t,x,y). \end{aligned}$$

This equation is used to advance from the ith time step to the $(i+1)$th time step for the unknown function u. This is done as follows. The terms that depend on y are thrown to the right-hand side, where they are approximated by finite differences applied to the numerical solution at the ith time step. These terms are then treated as the inhomogeneous term in Section 9.10 above. The explicit Godunov scheme is then used to solve for u separately in each individual x-line. This produces the numerical solution u in the entire $(i + 1)$th time step.

Similarly, the explicit Godunov scheme is applied separately to each individual y-line to solve for v at the $(i + 1)$th time step. For this purpose, the second equation in the system is rewritten in the form

$$\begin{aligned} &-\varepsilon v_{yy}(t, x, y) + v_t(t, x, y) + v(t, x, y)v_y(t, x, y) \\ &= \varepsilon v_{xx}(t, x, y) - u(t, x, y)v_x(t, x, y) + F_2(t, x, y). \end{aligned}$$

Here, the terms that depend on u are thrown to the right-hand side, where they are approximated by finite differences at the ith time step and then treated just as an inhomogeneous term. The explicit Godunov scheme is then used independently in each individual y-line to solve for v and obtain the required numerical solution in the entire $(i + 1)$th time step. This completes the definition of the so-called alternating-direction Godunov scheme.

Because the above numerical solution contains the required information about shock and rarefaction waves, it can serve as a good initial guess in Newton's iteration. This iteration linearizes the original system of PDEs around the u and v calculated above; in other words, it uses them as known convection coefficients. Once the resulting linear system is solved numerically by some time-marching scheme, the numerical solution is used once again to linearize the original system and repeat the procedure. This is the Newton iteration, which converges rapidly to the required solution of the original system of PDEs.

9.12 Exercises

1. Implement the Godunov scheme using the "xtGrid" object in Chapter 7, Section 11.
2. Implement the random-choice scheme using the "xtGrid" object in Chapter 7, Section 11.
3. Use your code to solve the Riemann problem with sinusoidal initial conditions and periodic boundary conditions. Do you obtain the N-wave? Which scheme gives better results? Compare your results to those in Figures 9.6 and 9.7.
4. Rewrite your code using the short-memory approach, in which the numerical solution at a particular time step is dropped right after it is used to calculate the numerical solution at the next time step. Compare the numerical solution at the final time step to the result in the previous exercise. Are they the same? What are the advantages and disadvantages of the short-memory code?
5. Apply the explicit Godunov scheme to the singularly perturbed Riemann problem in Section 9.7 above. Are your results as close to the N-wave as those in Figure 9.8?
6. Develop implicit and semi-implicit versions of the Godunov scheme for the singularly perturbed Riemann problem in Section 9.7. In these versions, (half of) the diffusion term εu_{xx} is thrown to the left-hand side of the difference scheme in Section 9.7 and solved for implicitly. Can you use these versions to solve singularly perturbed Riemann problems with ε larger than that used in Figure 9.8?

7. Prove that the stability condition in Chapter 8, Section 3,

$$\|J_{i+1}J_i^{-1}\| \le 1$$

(where i is the index of the current time step and J_i is the Jordan matrix for the matrix A_i that contains the discrete spatial derivatives at the ith time step), indeed holds whenever the convection coefficient $C(t, x)$ is a shock wave, as in Figure 9.3, or a rarefaction wave, as in Figure 9.1 or 9.2. The solution can be found in Section A.6 of the Appendix.

8. Use your previous answer to show that the implicit and semi-implicit schemes in Chapter 7, Sections 6 and 7, are stable, accurate, and adequate with respect to the norm used in Chapter 8, Section 3, and, hence, may be suitable for the linearized problem in Section 9.8.

9. Show that the above is also true for explicit time marching, provided that Δt is sufficiently small. The solution follows from Section A.5 of the Appendix.

10. Use the "dynamicVector2" object in Section A.4 of the Appendix and the short-memory approach to implement the alternating-direction Godunov scheme in Section 9.11 to solve the system of nonlinear PDEs.

Chapter 10

Application in Image Processing

In this chapter, we present a nonlinear PDE that is useful for removing random noise from digital images. The initial condition in the initial-boundary-value problem is taken from the original noisy image. The solution to the PDE at the final time step produces the required denoised image. Finite-difference discretization is also introduced, and Newton's iteration is used to linearize the nonlinear term. A single multigrid iteration is then used to solve each individual linear system approximately. The algorithm is also extended to color images.

10.1 Digital Images

In this chapter, we use the objects introduced above to implement useful algorithms in image processing. For this purpose, we need to define some common terms.

A digital image is actually an $N_x \times N_y$ array of numbers (where N_x and N_y are positive integers). This array represents a realistic two-dimensional image as follows. Each entry (j, i) in the array is called a pixel; the number contained in the pixel represents the intensity of light in the corresponding point in the image. Usually, the numbers in the pixels are between 0 (zero light, or black) and 255 (full light, or white). The larger the number, the larger the amount of light at the corresponding point in the image and the closer it is to pure white. On the other hand, the smaller the number, the darker the corresponding point in the image and the closer it is to pure black. This is why the numbers in the pixels are called gray-levels or grayscales.

10.2 The Denoising Problem

The field of image processing deals with digital images and their storage, transfer, and other useful manipulations. An important problem in image processing is how to make a digital image clearer. The original image may suffer from two kinds of drawbacks: blur, which may be introduced when the photo is taken, and random noise, which may be introduced when the digital image is transferred or broadcast. Two major problems in image processing are, thus, deblurring (removing blur) and denoising (removing noise).

Another interesting problem in image processing is object segmentation, namely, detecting objects in the digital image. This problem is closely related to the problem of denoising and has important applications in medicine and other areas as well.

In what follows, we consider the problem of denoising, namely, removing random noise from a given noisy image, without spoiling its features. It turns out that this problem can be formulated as an initial-boundary-value problem. In this form, the problem can be solved by numerical methods.

10.3 The Nonlinear Diffusion Problem

Let us formulate the denoising problem as an initial-boundary-value problem. The domain is the rectangle $[0, L_x] \times [0, L_y]$ in which the original image is contained. At the initial time $t = 0$, the initial conditions are taken from the original, noisy image. As time progresses, a slight amount of diffusion is introduced to smooth out the random noise from the image. At time $t = T$, the final, denoised image is produced.

The diffusion introduced in the PDE is aimed to smooth the random noise out. It is essential, however, not to smooth out grayscale discontinuities that are an integral part of the original image. Thus, the diffusion must be small across discontinuity lines or in directions of significant change in grayscale in the digital image. If the grayscale at the pixels in the image is thought of as a function of the two spatial variables x and y, then the diffusion must be small along directions where the gradient of this function is large in magnitude, since the sharp variation in the function indicates true features of the image that must be preserved.

The diffusion coefficient must therefore be inversely proportional to the gradient of the solution. Since this solution is not yet known, the diffusion coefficient is not available either. This means that the problem is quasi-linear: it is nonlinear but may be linearized by setting the diffusion coefficient using some approximation to the solution.

The PDE is the time-dependent diffusion problem [31]

$$u_t(t, x, y) - (Pu_x(t, x, y))_x - (Pu_y(t, x, y))_y = 0$$

in the time-space domain

$$0 < t < T, \ 0 \le x \le L_x, \ 0 \le y \le L_y,$$

where P is the function given by

$$P \equiv P(u_x(t, x, y), u_y(t, x, y)) = \frac{1}{1 + \eta^{-1}(u_x^2(t, x, y) + u_y^2(t, x, y))},$$

where η is a parameter to be specified later. (Typically, η is the average of a discrete approximation to $u_x^2 + u_y^2$ over the pixels in the entire image.)

In order to make the problem well posed, one must also impose initial and boundary conditions. The boundary conditions are of homogeneous Neumann type everywhere in the boundary of the rectangular spatial domain. The initial condition $u(0, x, y)$ is a function that agrees with the original, noisy image in the rectangular grid in which the image is defined. In other words, $u(0, jh_x, ih_y)$ is the grayscale at the (j, i)th pixel in the original noisy digital image, where $h_x = L_x/N_x$ is the meshsize in the x spatial direction and $h_y = L_y/N_y$ is the meshsize in the y spatial direction. Thus, the grid

$$\{(jh_x, ih_y) \mid 0 \le j < N_x, \ 0 \le i < N_y\}$$

(Figure 7.4) coincides with the digital image, and the values of the solution u in it are as in the original, noisy image at the initial time $t = 0$ and as in the required denoised image at the final time $t = T$.

10.4 The Discretization

The terms in the above PDE that contain spatial derivatives are symmetric in the sense to be described in Chapter 11, Section 2. Thus, the discretization method must be defined so that the matrix that contains the discrete spatial derivatives is also symmetric. Furthermore, in Chapter 12, Section 4, we'll see that it is also desirable that this matrix be diagonally dominant. Thus, we already have rather clear guidelines toward the definition of a discretization method.

In fact, we already have a discretization method that enjoys the above properties. This discretization method is defined in the exercises at the end of Chapter 8 for linear diffusion problems. It is based on the observation that if the diffusion coefficient P is evaluated at midpoints of the form $(j+1/2, i)$ and $(j, i+1/2)$ in the grid, then the resulting matrices that contain the discrete spatial derivatives at a particular time step are symmetric and diagonally dominant [46]. In the exercises at the end of Chapter 8, it is shown that these properties are attractive because they guarantee that the implicit and semi-implicit schemes are stable and accurate with respect to the usual norm.

The finite-difference discretization is based on the approximation

$$(Pu_x)((j+1/2)h_x, ih_y) \doteq P_{j+1/2,i} h_x^{-1}(u((j+1)h_x, ihy) - u(jh_x, ih_y)),$$

where $P_{j+1/2,i}$ is an approximation of P at the corresponding midpoint:

$$P_{j+1/2,i} = \frac{1}{1 + \eta^{-1}(u_{x\ j+1/2,i}^2 + u_{y\ j+1/2,i}^2)},$$

where $u_{x\ j+1/2,i}$ and $u_{y\ j+1/2,i}$ are approximations to u_x and u_y, respectively, at the corresponding midpoint:

$$\begin{aligned} u_{x\ j+1/2,i} &= h_x^{-1}(u((j+1)h_x, ih_y) - u(jh_x, ih_y)), \\ u_{y\ j+1/2,i} &= (4h_y)^{-1}(u((j+1)h_x, (i+1)h_y) + u(jh_x, (i+1)h_y) \\ &\quad - u((j+1)h_x, (i-1)h_y) - u(jh_x, (i-1)h_y)). \end{aligned}$$

Note that the above definitions may use points that lie outside the grid and, hence, are not well defined. For example, when i indicates the top line in the grid, $i + 1$ is not well defined. Fortunately, using the homogeneous Neumann boundary conditions, $i + 1$ can be replaced by i in this case. Similarly, when i indicates the bottom line in the grid, $i - 1$ should be replaced by i in the above formulas. A similar approach is used when $j = N_x$ $(j + 1 \leftarrow j)$ and $j = 0\ (j - 1 \leftarrow j)$.

The above definition of P at the midpoints is now used to discretize the term $(Pu_x)_x$ at the grid points themselves, as in Chapter 7, Section 13. This completes the discretization of the first spatial term in the PDE.

The discretization of the second spatial term, $(Pu_y)_y$, is done in a similar way, by inverting the roles of x and y (and i and j) in the above formulas. The time derivative is discretized implicitly as in Chapter 7, Section 6. (The semi-implicit scheme in Chapter 7, Section 7, is probably more accurate, although not used here.) This completes the definition of the discretization method.

10.5 Linearization

The above implicit scheme requires the solution of a nonlinear system of equations at each time step. The nonlinearity is in the coefficients $P_{j+1/2,i}$ and $P_{j,i+1/2}$ defined above, which depend on the unknown discrete solution u. A linearization method is thus needed to solve the system of equations iteratively.

This linearization is provided in Newton's iteration. An initial approximation to the discrete solution u at the current time step is obtained from the previous time step. This approximation is used to define $P_{j+1/2,i}$ and $P_{j,i+1/2}$ at the midpoints, as above. These coefficients are used in the discretization to form the discrete linear system of equations. The solution of this system is an improved approximation to the solution at the current time step. This procedure is repeated iteratively, producing better and better approximate solutions at the current time step. When a sufficiently good approximation is reached, it is accepted as the numerical solution at the current time step, and the time marching can proceed to the next time step.

The Newton method requires the solution of a linear system of equations at each iteration to produce the improved approximation. It turns out that these linear systems don't have to be solved exactly. In fact, it is sufficient to solve them approximately, using one iteration of the multigrid linear-system solver in Chapter 17, Section 8. Indeed, because the multigrid method significantly reduces the residual, it provides the required improved approximation.

It also turns out that, with the implicit time marching described above, it is sufficient to use one time step only. The numerical solution at time $t = \Delta t$ is already denoised properly and can, hence, serve as the output of the denoising algorithm. This completes the definition of the denoising algorithm for grayscale images. In what follows, the algorithm is also extended to color images.

10.6 Color Images

So far, we have dealt with grayscale images, in which each pixel contains a single number to indicate the amount of light at the corresponding point in the actual image. Here, we consider color digital images, in which each pixel contains three numbers to indicate the intensities of the three basic colors at the corresponding point in the color image: red, green, and blue. The combination of these three colors with the intensities specified in the pixel uniquely produces the actual color at the corresponding point in the image. This digital representation is called the RGB code: R for red, G for green, and B for blue.

The color digital image can thus be defined mathematically as the set of three $N_x N_y$-dimensional vectors: r, containing the intensities of the red color in the pixels; g, containing the intensities of the green color in the pixels; and b, containing the intensities of the blue color in the pixels in the entire image. We refer to the vectors r, g, and b as color vectors.

In the above form, the color image is ready for mathematical manipulations to make it clearer. In particular, when the vectors r, g, and b are contaminated with random noise, a denoising algorithm is required to remove it. Fortunately, the above denoising algorithm can also be extended to color images. Furthermore, the coefficient matrices used in it preserve their good properties, such as symmetric positive definiteness and diagonal dominance.

10.7 Denoising Color Images

The denoising algorithm for color images is a natural extension of the above algorithm. Here, however, we have a system of three coupled, nonlinear PDEs, each of which solves for a different color. The solutions of the PDEs at the final time $t = T$, when confined to the discrete grid, produce the denoised color vectors. When these color vectors are combined, they produce the required denoised color image.

The system of coupled PDEs contains three nonlinear equations, one for each color:

$$\begin{aligned}
r_t(t,x,y) - (Pr_x(t,x,y))_x - (Pr_y(t,x,y))_y &= 0,\\
g_t(t,x,y) - (Pg_x(t,x,y))_x - (Pg_y(t,x,y))_y &= 0,\\
b_t(t,x,y) - (Pb_x(t,x,y))_x - (Pb_y(t,x,y))_y &= 0,
\end{aligned}$$

where $0 < t < T$ is the time variable; $0 \le x \le L_x$ and $0 \le y \le L_y$ are the spatial variables; $r(t,x,y)$, $g(t,x,y)$, and $b(t,x,y)$ are the unknown functions; and the diffusion coefficient P (which depends nonlinearly on $r_x(t,x,y)$, $r_y(t,x,y)$, $g_x(t,x,y)$, $g_y(t,x,y)$, $b_x(t,x,y)$, and $b_y(t,x,y)$) is defined by

$$P = \det(E)^{-1},$$

where E is the 2×2 matrix

$$E = \begin{pmatrix} 1 + (r_x^2 + g_x^2 + b_x^2)/\eta & (r_x r_y + g_x g_y + b_x b_y)/\eta \\ (r_x r_y + g_x g_y + b_x b_y)/\eta & 1 + (r_y^2 + g_y^2 + b_y^2)/\eta \end{pmatrix}.$$

(Compare this system with the system in [35, 42], where E^{-1} is used instead of $\det(E)^{-1}$.)

In order to complete the system of PDEs into a well-posed problem, one must also impose initial and boundary conditions for the functions r, g, and b. The boundary conditions are of homogeneous Neumann type:

$$r_n = g_n = b_n = 0,$$

where $\vec{n}$ is the outer normal vector at the boundary of the rectangular spatial domain. The initial conditions at $t = 0$ must agree with the original, noisy image in the grid: $r(0,x,y)$ agrees with the noisy red color vector in the grid, $g(0,x,y)$ agrees with the noisy green color vector in the grid, and $b(0,x,y)$ agrees with the noisy blue color vector in the grid. The solutions of the PDEs at the final time $t = T$, when confined to the grid, produce the denoised color vectors: $r(T,x,y)$ produces the denoised red color vector, $g(T,x,y)$ produces the denoised green color vector, and $b(T,x,y)$ produces the denoised blue color vector. These color vectors combine to produce the required denoised color image.

The discretization is the same as before. First, E is evaluated at the midpoints $(j + 1/2, i)$ and $(j, i + 1/2)$. Then, P is also evaluated there and used to construct the discrete spatial derivatives. The implicit scheme in Chapter 7, Section 6, is used to discretize the time derivatives. This scheme requires the solution of a large system of nonlinear difference equations at each time step.

This system is solved by Newton's iteration. The initial guess is obtained from the previous time step and used to calculate E and linearize the equations. This produces three independent linear systems for the color vectors r, g, and b at the current time step. Each of these linear systems is solved approximately by one multigrid iteration. Once Newton's iteration terminates, its result is accepted as the numerical solution at the current time step, and the process is ready to proceed to the next time step.

Actually, it turns out that one time step is sufficient; that is, $T = \Delta t$ provides the required denoised color vectors, which combine to the required denoised color image.

The parameter η may change from one Newton iteration to the next. Because it estimates the total variation in the image (which decreases as the image gets denoised), it should also decrease. A possible way to define η is

$$\eta = h_x h_y \left(\|r_x\|^2 + \|r_y\|^2 + \|g_x\|^2 + \|g_y\|^2 + \|b_x\|^2 + \|b_y\|^2 \right),$$

where the vectors r, g, and b are taken from the previous Newton iteration, their partial derivatives are approximated by divided forward differences, and the $\|\cdot\|^2$ function stands for the sum of the squares over the entire grid. This completes the definition of the denoising algorithm for RGB color images.

10.8 Numerical Examples

In this section, we apply the denoising algorithms to realistic images. For this purpose, we first take a 512×512 original grayscale image. We add to each pixel in the original image a uniformly distributed random noise with magnitude of at most 25% of the maximum intensity in the original image. Then, we apply the denoising algorithm, with $L_x = L_y = 1$ (unit square), $N_x = N_y = 512$, $h_x = h_y = 1/512$, and $\Delta t = 0.000025$. We use ten Newton iterations, with $\eta = 10000$ in the first two iterations (where the variation is particularly large due to the initial noise), $\eta = 1000$ in the next three iterations, and $\eta = 100$ in the final five iterations (where the noise is already reduced and the variation is significantly lower).

A similar test is also employed for the RGB color image. In this case, the random noise is added to every color in every pixel. Then, the denoising algorithm is applied. Again, 10 Newton iterations are used, with η as at the end of Section 10.7.

The denoising algorithms provide good denoising with little blur. The algorithms may yet improve by using smaller Δt and more Newton iterations with clever choices of η or by switching to the more accurate semi-implicit scheme.

For comparison, we also test the Wiener filter available in the MATLAB® library, applied separately to each color vector in the noisy image. This filter is based on the statistical characteristics of the input image. At each pixel, the new value is calculated as a weighted average of values in a subsquare around it of 10 pixels by 10 pixels. (This size is the minimal size to have reasonable denoising in this example.) However, this algorithm produces much more blur than the previous one. The grayscale and color images produced in the above experiments can be found on the Web page http://www.siam.org/books/cs01.

The present algorithms can also be extended to object-segmentation algorithms, which detect objects in grayscale and color digital images. This is left to future research.

10.9 Exercises

1. Modify the code in Section A.4 of the Appendix to implement the denoising algorithm for grayscale images. The solution can be found in Section A.7 of the Appendix.

2. Modify your code to implement the denoising algorithm for RGB color images. The solution can be found in Section A.7 of the Appendix.

Part IV

The Finite-Element Discretization Method

In every time step in the implicit and semi-implicit schemes used above, one actually needs to solve a time-independent subproblem. Thus, the original time-dependent problem has actually been reduced to a sequence of time-independent subproblems.

Similarly, Newton's iteration for the solution of a nonlinear problem also uses a sequence of linear subproblems. Thus, the effective solution of a single elliptic problem is the key to the successful solution of complex nonlinear time-dependent problems.

In this part, we thus focus on elliptic PDEs, with no time variable t at all. Instead, we assume that there are two independent spatial variables x and y. (The three-dimensional case, in which the spatial variable z is also used, is in principle the same.)

Our model problem is the pure-diffusion equation, which has no convection term in it at all. This equation is symmetric in a sense to be defined later and can be reformulated as a minimization problem. This is the basis for the finite-element discretization method.

Although the finite-difference schemes used above are suitable for rectangular domains that can be approximated by uniform grids, they are no longer applicable to complicated domains with curved boundaries. The more flexible finite-element method that uses triangles of variable size and shape is more suitable for this purpose. Furthermore, this method is also suitable for problems with variable and even discontinuous coefficients. Therefore, it deserves and indeed gets full attention here.

Although the finite-element method is originally defined for symmetric partial differential equations (PDEs), it is not limited to them. In fact, it is also applicable to nonsymmetric PDEs such as the convection-diffusion equation discussed above.

As discussed above, the most important advantage of the finite-element discretization method is the opportunity to use not only uniform (structured) grids but also unstructured triangle meshes to approximate complicated domains. Unfortunately, unstructured meshes are particularly tricky to implement, because they cannot be embedded naturally in standard data structures such as arrays. Furthermore, the topological information incorporated in the mesh must use sophisticated data-access patterns to be used effectively (Chapter 4, Section 7).

Fortunately, the high level of abstraction available in C++ is particularly useful for this purpose. Indeed, the finite-element and mesh objects implemented below are particularly transparent, effective, and user friendly. The object-oriented approach proves particularly suitable for implementing a general unstructured mesh and using it in the finite-element discretization method.

This part contains five chapters. The first one (Chapter 11) contains the mathematical background. The second one (Chapter 12) presents linear finite elements. The third one (Chapter 13) presents unstructured meshes and their object-oriented implementation. The fourth one (Chapter 14) describes an adaptive mesh-refinement algorithm and its implementation. The final chapter (Chapter 15) describes high-accuracy finite-element schemes.

Chapter 11

The Weak Formulation

In this chapter, we consider diffusion problems with variable (and possibly discontinuous and anisotropic) coefficients in complicated domains. For this kind of problem, the weak formulation is particularly suitable. Indeed, it is well posed in the sense that it has a unique solution. This is why it should be used to model physical phenomena and derive discretization methods.

11.1 The Diffusion Problem

This chapter contains some theoretical background relevant to the finite-element discretization method. Readers who are only interested in the practical aspect are advised to read at least the first two sections. More advanced readers may find the entire chapter useful to understand better the nature of finite-element schemes.

In Chapter 7, Section 13, we considered the convection-diffusion equation in one and two spatial dimensions. This equation contains two spatial-derivative terms: the diffusion term $-\varepsilon(u_{xx}+u_{yy})$ and the convection term $C_1u_x+C_2u_y$. The diffusion term is the principal term that determines the type of equation, i.e., second-order parabolic equation. Therefore, boundary conditions should be imposed at every point in the boundary of the spatial domain.

Nevertheless, for a very small diffusion coefficient ε, the convection term plays a major role in shaping the behavior of the solution. In fact, the information from the initial conditions progresses along the characteristic lines, as in the first-order hyperbolic equation. Furthermore, the boundary data flow along these characteristic lines from one side of the boundary to the other, where the boundary layer is introduced at which the solution must vary sharply to meet the other boundary condition.

In this chapter, we consider diffusion problems with no convection at all ($C_1 = C_2 = 0$). This makes the equation easier, because no boundary layers can form. Furthermore, the equation is symmetric in the sense to be discussed below. On the other hand, the present problem has a source of extra difficulty, which is missing in the convection-diffusion equation. Here, the diffusion coefficient may vary from point to point in the (x, y)-plane and may even be discontinuous across certain lines in the plane. In its general form, the diffusion equation reads

$$u_t - (Pu_x)_x - (Qu_y)_y + Ku = F,$$

where P, Q, K, and F are given functions of t, x, and y and $u \equiv u(t, x, y)$ is the unknown solution. It is assumed that this equation holds in a given time interval $[0, T]$ and a given domain Ω in the Cartesian (x, y)-plane. It is also assumed that initial conditions that specify u at the initial time $t = 0$ are given and that boundary conditions that specify u or u_n (the directional derivative of u in the direction that is normal to the boundary and points toward the outside of the domain) are also given at the boundary $\partial\Omega$.

In many realistic applications, the time-dependent process represented by $u(t, x, y)$ converges as $t \to \infty$ to a stable equilibrium state or stagnation state or steady state. Clearly, in the steady state, $u_t \equiv 0$, because u no longer changes in time. Thus, u is actually a function of the spatial variables x and y only: $u \equiv u(x, y)$. In this case, the given functions P, Q, K, and F take their limit values as $t \to \infty$ (if these limits indeed exist), so they are also functions of x and y only. Thus, the diffusion equation takes the form

$$-(Pu_x)_x - (Qu_y)_y + Ku = F, \quad (x, y) \in \Omega,$$

where P, Q, K, F, and the unknown u are functions in Ω. This PDE is accompanied by the boundary conditions

$$\begin{aligned} u &= G, \quad (x, y) \in \Gamma_0, \\ \alpha u + Pu_x n_1 + Qu_y n_2 &= G, \quad (x, y) \in \Gamma. \end{aligned}$$

Here,

- Γ_0 is the subset of the boundary $\partial\Omega$ where Dirichlet boundary conditions are imposed;
- $\Gamma = \partial\Omega \setminus \Gamma_0$ is the remainder of the boundary $\partial\Omega$, where mixed boundary conditions are imposed;
- G is a given function on $\partial\Omega$;
- α is a given function on Γ; and
- (n_1, n_2) is the outer normal vector to Ω; that is, it is a unit vector ($n_1^2 + n_2^2 = 1$), is normal to Γ (makes a right angle with the tangent to Γ), and points toward the outside of Ω.

We refer to this problem as the diffusion problem in Ω. In fact, this problem arises not only in the steady state of time-dependent problems but also in implicit and semi-implicit time-marching schemes for the numerical solution of time-dependent problems.

11.2 The Weak Formulation

In the convection-diffusion equation in Chapter 7, Section 13, the diffusion coefficient is the constant ε, so the finite-difference scheme is suitable. For problems with variable coefficients, on the other hand, this scheme may no longer be accurate. In particular, when the coefficients are discontinuous, the solution $u(x, y)$ may have no continuous derivatives, let alone Taylor expansion. In this case, the boundary-value problem in the above (strong) form may have no solution at all. The correct (well-posed) form should use integration rather than differentiation. This form is called the "weak" formulation, because it requires weaker assumptions than the original "strong" formulation. In particular, it no longer requires $u(x, y)$ to have continuous derivatives but only that the flux vector (Pu_x, Qu_y) be continuous. The weak formulation is the basis of the most useful finite-element discretization method.

In the original strong formulation of the diffusion problem as a PDE accompanied by boundary conditions, Pu_x and Qu_y must be differentiable. These assumptions may be too strong. For example, when P and Q are discontinuous, Pu_x and Qu_y may be continuous but not differentiable. In such cases, the strong formulation may have no mathematical solution at all.

Still, the physical phenomenon does exist and needs to be formulated mathematically. The mathematical model must be well posed in the sense that it has a unique solution, just like the original physical phenomenon. This model is provided by the weak formulation described below.

The weak formulation is based on integration rather than differentiation. Thus, it only requires that the flux (Pu_x, Qu_y), as well as K and F, be square-integrable over Ω with respect to the area element $dxdy$. (We say that a function w is square-integrable over Ω if its square w^2 is integrable over Ω; that is, $\int_\Omega w^2 dxdy < \infty$.) We also assume that α and G are square-integrable over Γ with respect to the length element ds.

Let us now define the weak formulation in detail. For every function v that is square-integrable over Ω with respect to the area element $dxdy$ (and also square-integrable over Γ with respect to the length element ds), define the linear functional $f(v)$ by

$$f(v) = \int_\Omega F v dxdy + \int_\Gamma G v ds.$$

For every two functions v and w with square-integrable derivatives in Ω, define the symmetric bilinear form $a(v, w)$ by

$$a(v, w) = \int_\Omega (Pv_x w_x + Qv_y w_y)\, dxdy + \int_\Omega K v w dxdy + \int_\Gamma \alpha v w ds.$$

The weak formulation is obtained from the strong one by using Green's formula in the PDE and boundary conditions as follows. Let v be a function in Ω that vanishes on Γ_0 and has square-integrable derivatives in Ω. Then, we have, from Green's formula and the boundary conditions,

$$\begin{aligned}
\int_\Gamma (Gv - \alpha uv)\, ds &= \int_\Gamma (Pu_x v n_1 + Qu_y v n_2)\, ds \\
&= \int_{\partial\Omega} (Pu_x v n_1 + Qu_y v n_2)\, ds \\
&= \int_\Omega \left((Pu_x v)_x + (Qu_y v)_y\right) dxdy \\
&= \int_\Omega \left((Pu_x)_x\, v + (Qu_y)_y\, v\right) dxdy \\
&+ \int_\Omega (Pu_x v_x + Qu_y v_y)\, dxdy.
\end{aligned}$$

Using the PDE itself, we have

$$\begin{aligned}
f(v) &= \int_\Omega F v dx dy + \int_\Gamma G v ds \\
&= -\int_\Omega \left((P u_x)_x + \left(Q u_y\right)_y\right) v dx dy \\
&+ \int_\Omega K u v dx dy + \int_\Gamma G v ds \\
&= \int_\Omega \left(P u_x v_x + Q u_y v_y\right) dx dy \\
&+ \int_\Omega K u v dx dy + \int_\Gamma \alpha u v ds \\
&= a(u, v).
\end{aligned}$$

The problem in its weak formulation is thus as follows: find a function u that agrees with G on Γ_0 and has square-integrable derivatives in Ω such that, for every function v that vanishes on Γ_0 and has square-integrable derivatives in Ω,

$$a(u, v) = f(v).$$

From the above discussion, it is clear that a solution to the strong formulation (if it exists) must also be a solution to the weak formulation. In Sections 11.3 to 11.6 below, it is shown that the weak formulation is more suitable than the strong one, because it is well posed in the sense that it has a unique solution, just like the original physical phenomenon. Indeed, a realistic physical phenomenon must have a solution, or it wouldn't exist. Furthermore, the solution must be unique, or the phenomenon would be unstable in the sense that every tiny perturbation in the initial or boundary data may produce enormous changes in the entire physical state, which is highly unlikely in realistic processes.

The mathematical model that approximates the original physical phenomenon must also be stable and well posed. Otherwise, every tiny inaccuracy in the model may result in an enormous error in the solution. In what follows, we show that the weak formulation is indeed well posed and is thus suitable for modeling real physical situations.

11.3 The Minimization Problem

In this section, we show that the above weak formulation is equivalent to the following minimization problem: consider the functional

$$g(v) \equiv \frac{1}{2} a(v, v) - f(v).$$

From the family of functions that agree with G on Γ_0 and have square-integrable derivatives in Ω, pick the function u for which $g(u)$ is minimal. This function is referred to as the solution to the minimization problem.

Later on, we will show that this minimization problem has a unique solution; this implies that the weak formulation (which is equivalent to it) is indeed well posed.

Let us assume that the functions P, Q, K, and α are bounded and nonnegative, so $a(v, v) \geq 0$ for every function v. In order to show the equivalence between the weak formulation and the minimization problem, we have to show that every solution to the one also solves the other. Assume first that u solves the minimization problem, and let us show that it also solves the weak formulation. Indeed, if u were not a solution to the weak formulation, then there would exist a function v that vanishes on Γ_0 and has square-integrable derivatives in Ω, and yet

$$a(u, v) \neq f(v).$$

Without loss of generality, we can assume that

$$a(u, v) < f(v).$$

(Otherwise, just replace v by $-v$.) Let ϵ be a positive parameter. For sufficiently small ϵ, we have

$$\begin{aligned}
g(u + \epsilon v) &= \frac{1}{2} a(u + \epsilon v, u + \epsilon v) - f(u + \epsilon v) \\
&= \frac{1}{2} a(u, u) + \epsilon a(u, v) + \frac{1}{2} \epsilon^2 a(v, v) - f(u) - \epsilon f(v) \\
&= g(u) + \epsilon (a(u, v) - f(v)) + \frac{1}{2} \epsilon^2 a(v, v) \\
&< g(u),
\end{aligned}$$

which contradicts the assumption that u minimizes g. Thus, u must also solve the weak formulation.

Conversely, assume now that u solves the weak formulation, and let us show that it must also solve the minimization problem. Indeed, let w be some function that agrees with G on Γ_0 and has square-integrable derivatives in Ω, and define $v = w - u$. Since v vanishes on Γ_0, we have

$$\begin{aligned}
g(w) &= \frac{1}{2} a(w, w) - f(w) \\
&= \frac{1}{2} a(u + v, u + v) - f(u + v) \\
&= \frac{1}{2} a(u, u) + a(u, v) + \frac{1}{2} a(v, v) - f(u) - f(v) \\
&= g(u) + \frac{1}{2} a(v, v) \\
&\geq g(u),
\end{aligned}$$

which implies that u also solves the minimization problem. This completes the proof that the weak formulation and the minimization problem are indeed equivalent to each other.

11.4 The Coercivity Property

In this section, we discuss the coercivity property of the quadratic form $a(w, w)$. This property is most useful for proving the existence and uniqueness of the solution to the minimization problem defined above.

For this purpose, however, we need some more assumptions:

1. The domain Ω is connected (every two points in it can be connected to each other by a curve that lies in it).

2. There is a constant $\eta > 0$ such that $P \geq \eta$ and $Q \geq \eta$ in Ω.

3. Either Γ_0 is nonempty, or $\int_\Gamma \alpha ds > 0$, or $\int_\Omega K dxdy > 0$.

These assumptions, in conjunction with the assumptions made in Section 11.3, guarantee that the quadratic form $a(w, w)$ is coercive in the sense that if w is a function with square-integrable derivatives in Ω such that $a(w, w) = 0$ and $w(r) = 0$ at some reference point $r \in \Omega$, then w must be the zero function $w \equiv 0$. Indeed, for every point $q \in \Omega$, $w(q)$ takes the form

$$w(q) = w(r) + \int_r^q (w_x s_1 + w_y s_2) ds,$$

where the integral is taken along some curve connecting r to q in Ω, and (s_1, s_2) is the unit vector tangent to this curve. But, from the assumptions, it is clear that both w_x and w_y must vanish on the curve, so the integral along it vanishes as well, which implies that $w(q) = 0$ as well.

So far, we have assumed the existence of the reference point r for which $w(r) = 0$. In some cases, however, this assumption can be relaxed. For example, consider the case in which Γ_0 is empty. According to the assumptions at the beginning of this section, we must then have $\int_\Gamma \alpha ds > 0$ or $\int_\Omega K dxdy > 0$. In either case, the required reference point r must exist: in the first case it lies in the support of α in Γ, whereas in the second case it lies in the support of K in Ω. (The support of a function is the set of points where the function is nonzero.)

When Γ_0 is nonempty, we must explicitly assume the existence of the reference point r. However, as we'll see below, we are mostly interested in functions w that vanish on Γ_0. For such functions, r could be any point in Γ_0.

11.5 Existence Theorem

Here, we show that the minimization problem indeed has a solution. In the next section, we'll show that this solution is also unique. Together, these sections show that the minimization problem (and, hence, also the weak formulation) is indeed well posed.

In order to show the existence of a solution to the minimization problem, let us reformulate it in a slightly different way. By the same reasoning as in the previous section, one can also show that $\int_\Omega (u^2 + u_x^2 + u_y^2) dxdy$ cannot be too large (where u is some solution to the minimization problem, if it exists). Indeed, if it is large, then $a(u, u)$ will be large as well. As a result, $g(u)$ will also be large, because $a(w, w)$ grows quadratically with w, whereas $f(w)$ grows only linearly. This contradicts the fact that u minimizes g. Therefore, there must exist a sufficiently large constant L_0 for which

$$\int_\Omega (u^2 + u_x^2 + u_y^2) dxdy \leq L_0$$

for any solution u to the minimization problem, if it exists.

The minimization problem can thus be reformulated as follows: in the family of functions v that agree with G on Γ_0 and satisfy

$$\int_{\Omega} (v^2 + v_x^2 + v_y^2)dxdy \leq 2L_0,$$

find a function u that minimizes g. Since this family is compact, the continuous functional g must indeed have a minimum in it. This proves the existence of a solution to the minimization problem.

11.6 Uniqueness Theorem

Let us now show the uniqueness of the solution to the minimization problem. To this end, note that the coercivity of the quadratic form $a(w, w)$ guarantees that the functional g is convex in the family of functions that agree with G on Γ_0 and have square-integrable derivatives in Ω. To see this, let v and w be distinct functions that agree with G on Γ_0 and have square-integrable derivatives in Ω. Let us first show that

$$a(v - w, v - w) > 0.$$

Indeed, if $a(v - w, v - w)$ is zero, then, since $v - w$ vanishes on Γ_0, the coercivity property implies that $v - w$ vanishes everywhere in Ω, which contradicts the assumption that v and w are distinct. The conclusion is, thus, that $a(v - w, v - w)$ must indeed be positive.

We are now ready to show that g is indeed convex in the family of functions that agree with G on Γ_0 and have square-integrable derivatives in Ω. To this end, let $0 < \beta < 1$ be some constant. Then we have

$$\begin{aligned}
&g(\beta v + (1 - \beta)w) \\
&= \frac{1}{2}a(\beta v + (1 - \beta)w, \beta v + (1 - \beta)w) - f(\beta v + (1 - \beta)w) \\
&= \frac{1}{2}\left(\beta^2 a(v, v) + 2\beta(1 - \beta)a(v, w) + (1 - \beta)^2 a(w, w)\right) \\
&- \beta f(v) - (1 - \beta)f(w) \\
&= \frac{1}{2}(\beta a(v, v) + (1 - \beta)a(w, w) - \beta(1 - \beta)a(v - w, v - w)) \\
&- \beta f(v) - (1 - \beta)f(w) \\
&< \frac{1}{2}(\beta a(v, v) + (1 - \beta)a(w, w)) - \beta f(v) - (1 - \beta)f(w) \\
&= \beta\left(\frac{1}{2}a(v, v) - f(v)\right) + (1 - \beta)\left(\frac{1}{2}a(w, w) - f(w)\right) \\
&= \beta g(v) + (1 - \beta)g(w).
\end{aligned}$$

This completes the proof that g is indeed convex in the family of functions that agree with G on Γ_0 and have square-integrable derivatives in Ω.

As a result, the convex functional g must have a unique minimum in the above family of functions. Indeed, if there are two distinct functions v and w for which g is minimized, then we have

$$g\left(\frac{v+w}{2}\right) < \frac{1}{2}(g(v)+g(w)) = g(v),$$

in contradiction to the assumption that v minimizes g. This completes the proof that the solution to the minimization problem is indeed unique.

Let us denote the unique solution to the minimization problem by u. From Section 11.3, it follows that u is also the unique solution to the weak formulation of the boundary-value problem.

As discussed in Section 11.2, a solution to the strong formulation must also solve the weak formulation. Therefore, the strong formulation must have at most one solution. If the strong formulation models the physical phenomenon well, then it surely has a solution. This is also a solution to the weak formulation, so both weak and strong formulations can serve as good models.

Unfortunately, this is not always the case, and the strong formulation may have no solution at all. For example, when P or Q is discontinuous, the flux may be nondifferentiable, and the strong formulation will have no solution whatsoever. The weak formulation, which has a unique solution, is thus more likely to model the original physical phenomenon well. Therefore, it is better suited to numerical modeling.

Because the solution to the weak formulation must have only square-integrable (rather than differentiable) derivatives, it can be well approximated by piecewise-linear continuous functions. This approach is used in the finite-element discretization method below.

11.7 Exercises

1. Show that the bilinear form $a(\cdot,\cdot)$ in Section 11.2 is symmetric in the sense that

$$a(v,w) = a(w,v)$$

for every two functions v and w with square-integrable derivatives.

2. Define the bilinear form $a(\cdot,\cdot)$ for the equation

$$-u_{xx}(x,y) - u_{yy}(x,y) + u(x,y) = F(x,y).$$

Assume that Dirichlet boundary conditions are imposed on part of the boundary and mixed boundary conditions are imposed on the other part, as in Section 11.1 above.

3. Show that the above bilinear form is symmetric and that the corresponding quadratic form is coercive. Conclude that the weak formulation is indeed well posed.

4. Define the bilinear form $a(\cdot,\cdot)$ for the equation

$$-u_{xx}(x,y) - u_{yy}(x,y) + u_x(x,y) + u_y(x,y) = F(x,y).$$

5. Show that the above bilinear form is nonsymmetric.

6. Define the bilinear form $a(\cdot, \cdot)$ for the diffusion equation in three spatial dimensions

$$-(Pu_x)_x - (Qu_y)_y - (Wu_z)_z = F,$$

where P, Q, W, F, and u are scalar functions of the three spatial variables x, y, and z. Show that the above bilinear form is symmetric and that the corresponding quadratic form is coercive under the assumptions in Section 11.4. Conclude that the weak formulation is indeed well posed.

Chapter 12

Linear Finite Elements

In this chapter, we describe the linear finite-element discretization method for the weak formulation of the diffusion equation. In particular, we show how the stiffness matrix should be assembled. We also show that, under certain conditions, the stiffness matrix is a diagonally dominant M-matrix. In the exercises at the end of the chapter, we also show how finite-element discretization can be used in time-dependent problems to produce stable and accurate time-marching schemes.

12.1 The Finite-Element Triangulation

The finite-difference discretization method in Chapter 7, Section 13, is particularly suitable for rectangular spatial domains such as the unit square $0 \leq x, y \leq 1$. Most realistic applications, however, use much more complicated domains that can no longer be approximated by uniform grids. Furthermore, they often use variable and even discontinuous coefficients, for which the strong formulation is no longer well posed. The finite-element discretization method [44] is particularly attractive in these cases, because it uses not only the well-posed weak formulation but also nonuniform, unstructured meshes that may approximate irregular domains well. Particularly suitable for this purpose is the mesh of triangles or triangulation.

Triangulation is defined in Chapter 4, Section 5, as a conformal mesh of triangles. Here, conformity means that adjacent triangles that share the same edge also share its endpoints as their joint vertices.

The vertices of the triangles in the triangulation are called nodes. Triangulation of a domain Ω in the Cartesian plane is triangulation with nodes in Ω and edges that are either shared by two triangles or have both endpoints on the boundary $\partial\Omega$. The smaller the edges that lie next to the boundary are, the better the approximation to curved parts in $\partial\Omega$ may be.

Triangulation is, thus, much more flexible than the structured uniform grid in Chapter 7, Section 13, and can thus approximate not only rectangular but also complicated domains. In particular, small triangles can be used near irregularities in the boundary to provide high resolution there, while larger triangles are used elsewhere.

12.2 The Discrete Weak Formulation

The triangulation of the domain Ω, denoted by T, induces a subspace of piecewise-linear continuous functions on which the original weak formulation is approximated. This subspace, denoted by V, contains the functions that are continuous in T and linear in each particular triangle in it.

With the notation in Chapter 11, Section 2, the discrete form of the weak formulation is as follows: find a function $\tilde{u}$ in V that agrees with G at every node in Γ_0 and also satisfies

$$a(\tilde{u}, v) = f(v)$$

for every function v in V that vanishes at every node in Γ_0. We refer to the above problem as the discrete weak formulation or discrete problem.

In fact, the discrete problem can be viewed as the restriction of the original weak formulation to V. As we'll see below, the number of degrees of freedom (or unknowns) in the discrete problem is the same as the number of nodes in T. Thus, it can be solved numerically on a digital computer.

12.3 The Stiffness System

Let us now rewrite the discrete weak formulation in an algebraic form that can be implemented on a computer. For every node j in the triangulation T, let ϕ_j be the function in V that has the value 1 at j and 0 at all the other nodes in T. In other words, ϕ_j vanishes on every triangle in T except those triangles that use j as a vertex, where ϕ_j decreases linearly from j toward the edge that lies across from it, where it vanishes. The function ϕ_j is also called the nodal basis function corresponding to j.

In what follows, $\tilde{u}$ denotes the solution to the discrete weak formulation. In fact, $\tilde{u}$ can be written as a sum over all the nodes in T:

$$\tilde{u} = \sum_{j \in T} x_j \phi_j,$$

where the unknown scalars x_j are the same as $\tilde{u}(j)$ (the value of $\tilde{u}$ at node j). Clearly, because $\tilde{u}$ agrees with G at nodes in Γ_0, we have $x_j = \tilde{u}(j) = G(j)$ for $j \in \Gamma_0$, so these x_j's are already known; the rest of the x_j's, which correspond to j's that don't lie on Γ_0, are yet to be solved for.

In order to find the unknown x_j's, one should substitute $\sum_j x_j \phi_j$ and ϕ_i for $\tilde{u}$ and v (respectively) in the discrete weak formulation:

$$\sum_{j \in T} x_j a(\phi_j, \phi_i) = a\left(\sum_{j \in T} x_j \phi_j, \phi_i\right) = a(\tilde{u}, \phi_i) = f(\phi_i).$$

This can be done for every $i \in T$ that doesn't lie on Γ_0. Now, let the nodes in T that don't lie on Γ_0 be numbered from 0 to $N - 1$. Define the vector of unknowns

$$x = (x_0, x_1, x_2, \ldots, x_{N-1}).$$

Define also the right-hand-side vector

$$f = (f_0, f_1, f_2, \ldots, f_{N-1}),$$

where

$$f_i = f(\phi_i) - \sum_{j \in T \cap \Gamma_0} G(j) a(\phi_j, \phi_i).$$

Finally, define the so-called stiffness matrix A as follows: for $0 \le i, j < N$, define

$$A_{i,j} = a(\phi_j, \phi_i).$$

With these definitions, the stiffness system is

$$Ax = f.$$

The solution x of this system is the vector whose components x_j are the same as $\tilde{u}(j)$. Clearly, $\tilde{u}$ is determined uniquely by these values. This completes the definition of the finite-element discretization method and its algebraic form.

12.4 Properties of the Stiffness Matrix

Here, we discuss some properties of the stiffness matrix. It turns out that it has some attractive properties, which indicate that the finite-element discretization is indeed appropriate.

The properties of the stiffness matrix A follow from the corresponding properties of the bilinear form $a(\cdot, \cdot)$ defined in Chapter 11, Section 2. First, the symmetry of $a(\cdot, \cdot)$ implies that A is also symmetric. Indeed,

$$A_{i,j} = a(\phi_j, \phi_i) = a(\phi_i, \phi_j) = A_{j,i}.$$

Furthermore, A is also positive semidefinite in the sense that, for every N-dimensional vector v,

$$v^t A v \ge 0.$$

Indeed,

$$\begin{aligned} v^t A v &= \sum_{0 \le i,j < N} v_i A_{i,j} v_j \\ &= \sum_{0 \le i,j < N} v_i a(\phi_j, \phi_i) v_j \\ &= a\left(\sum_{j=0}^{N-1} v_j \phi_j, \sum_{i=0}^{N-1} v_i \phi_i \right) \\ &\ge 0. \end{aligned}$$

Furthermore, the coercivity of the quadratic form $a(w, w)$ implies that A is also positive definite in the sense that, for every nonzero N-dimensional vector v,

$$v^t A v > 0.$$

Indeed, because v is nonzero, the function

$$\sum_{j=0}^{N-1} v_j \phi_j$$

is not identically zero. Still, it vanishes at nodes that lie in Γ_0, so it has a reference point on which it vanishes. Thanks to the coercivity property, we must have

$$a\left(\sum_{j=0}^{N-1} v_j \phi_j, \sum_{j=0}^{N-1} v_j \phi_j\right) > 0,$$

which implies that $v^t A v > 0$, as required.

In summary, the stiffness matrix A is symmetric and positive definite, or SPD for short.

In some special cases, the stiffness matrix A may have more attractive properties. Consider, for example, triangulation with angles that never exceed $\pi/2$ (90°). Assume also that the diffusion problem is isotropic ($P \equiv Q$). In this case, it is shown in Section 12.7 below that A is an L-matrix, which means that all of its off-diagonal elements are negative or 0:

$$A_{i,j} \leq 0, \qquad 0 \leq i, j < N, \; i \neq j.$$

Furthermore, in this case, A is also diagonally dominant in the sense described in Chapter 8, Section 1.

The combination of these two properties (L-matrix and diagonal dominance) implies that A is also an M-matrix in the sense that it has a nonnegative inverse [46]:

$$\left(A^{-1}\right)_{i,j} \geq 0, \qquad 0 \leq i, j < N.$$

This property is particularly attractive, because it indicates that the numerical solution is obtained from some weighted average of the conditions imposed at the boundary and the right-hand-side function F. In particular, in the Laplace equation, where $P \equiv Q \equiv 1$ and $F \equiv 0$, it implies that the maximum principle is observed not only in the original boundary-value problem but also in the discrete system. This indicates that the discrete system indeed approximates the original PDE well.

In the more general case where P and Q can be variable and F is not necessarily zero, the M-matrix property still indicates that the discretization is well done. Indeed, the inverse of the differential operator can be written as an integral operator with a nonnegative kernel, known as the Green function. The M-matrix property is just the discrete analogue of this property.

As discussed in Section 12.7 below, the above properties exist if the diffusion problem is isotropic ($P \equiv Q$) and the angles in the triangulation are either acute or right. In general, it is a good idea to keep the angles in the triangulation moderate and avoid degenerate triangles with too small or too large angles. Stretched triangles with very small angles should be used only when necessary, for example, at curved boundaries.

A mesh with moderate angles is called regular. Meshes with high regularity have a better chance of producing good numerical approximations. Increasing regularity in the mesh is thus an important guideline in mesh construction and is also used in the adaptive mesh refinement in Chapter 14, Section 2.

12.5 Calculating the Stiffness Matrix

In order to calculate the stiffness matrix A, we use a standard reference triangle on which the integration is carried out. The integral over a triangle in the mesh is calculated by mapping the triangle onto the reference triangle. The reference triangle, denoted by r, is displayed in Figure 12.2. The three typical (standard) nodal functions in r are

$$\begin{aligned}\phi_{1,0} &= x,\\ \phi_{0,1} &= y,\\ \phi_{0,0} &= 1 - x - y.\end{aligned}$$

Each of these linear functions has the value 1 at one of the vertices of r and 0 at the other two vertices. (This is why they are called "nodal.") The typical nodal functions are essential in the calculation of the stiffness matrix.

Let us now show how r is used to calculate the stiffness matrix. Let e denote a triangle in the mesh, as in Figure 12.1, with vertices denoted by i, j, and k. Let $M_e : r \to e$ be the

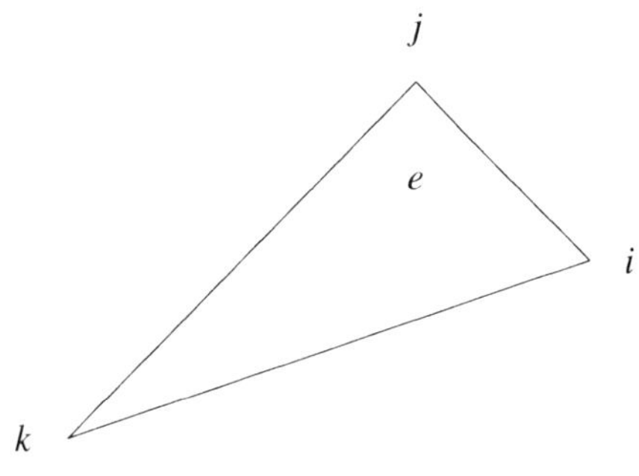

Figure 12.1. *The original finite element e in the mesh.*

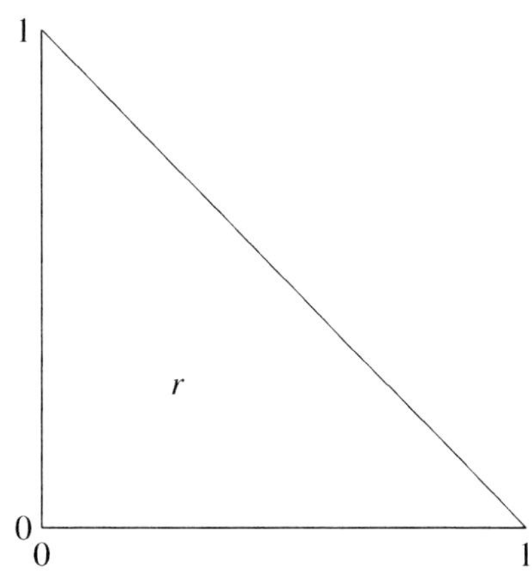

Figure 12.2. *The reference element r that is mapped to e by M_e.*

affine mapping that maps r onto e, namely, the mapping composed of a linear transformation followed by a shift by a constant vector:

$$M_e((x, y)) = S_e \begin{pmatrix} x \\ y \end{pmatrix} + s_e,$$

where (x, y) is any point in the Cartesian plane, S_e is a 2×2 matrix, and s_e is a two-dimensional constant vector. It is assumed that the vertices of r are mapped to the vertices of e as follows:

$$\begin{aligned} M_e((1, 0)) &= i, \\ M_e((0, 1)) &= j, \\ M_e((0, 0)) &= k. \end{aligned}$$

In what follows, we'll see that not the mapping M_e but rather the matrix S_e is the important factor in calculating the stiffness matrix. Fortunately, S_e can be calculated easily. Indeed, since

$$s_e = M_e((0, 0)) = k,$$

we have that

$$\begin{aligned} S_e \begin{pmatrix} 1 \\ 0 \end{pmatrix} &= M_e((1, 0)) - s_e = i - k, \\ S_e \begin{pmatrix} 0 \\ 1 \end{pmatrix} &= M_e((0, 1)) - s_e = j - k. \end{aligned}$$

Thus, the first column in the matrix S_e is the two-dimensional vector representing the difference of vertices $i - k$, and the second column in S_e is the two-dimensional vector $j - k$. Because these vertices are available for every triangle e in the mesh, the corresponding matrix S_e is also available.

The inverse mapping $M_e^{-1} : e \to r$ that maps e back onto r is also available:

$$M_e^{-1}((x, y)) = S_e^{-1} \begin{pmatrix} x \\ y \end{pmatrix} - S_e^{-1} s_e.$$

The Jacobians of M_e and M_e^{-1} (the 2×2 matrices that contain the linear parts of these mappings) are given by

$$\begin{aligned} \frac{\partial M_e}{\partial (x, y)} &= S_e, \\ \frac{\partial M_e^{-1}}{\partial (x, y)} &= S_e^{-1}. \end{aligned}$$

Let ϕ_i, ϕ_j, and ϕ_k be the nodal basis functions corresponding to the vertices i, j, and k in Figure 12.1. Then we have

$$\begin{aligned} \phi_i &= \phi_{1,0} \cdot M_e^{-1}, \\ \phi_j &= \phi_{0,1} \cdot M_e^{-1}, \\ \phi_k &= \phi_{0,0} \cdot M_e^{-1}, \end{aligned}$$

where '·' stands for composition of functions. By using the chain rule for calculating derivatives of composite functions, we have

$$\phi_{i\,x} = \left(\phi_{1,0} \cdot M_e^{-1}\right)_x = \phi_{1,0\,x}\left(S_e^{-1}\right)_{1,1} + \phi_{1,0\,y}\left(S_e^{-1}\right)_{2,1},$$
$$\phi_{i\,y} = \left(\phi_{1,0} \cdot M_e^{-1}\right)_y = \phi_{1,0\,x}\left(S_e^{-1}\right)_{1,2} + \phi_{1,0\,y}\left(S_e^{-1}\right)_{2,2},$$

where the x- and y-derivatives of ϕ_i are computed at a point $(x, y) \in e$, and the x- and y-derivatives of $\phi_{1,0}$ are computed at its inverse image $M_e^{-1}((x, y)) \in r$. In summary, the gradient of ϕ_i (the two-dimensional vector whose components are the x- and y-derivatives of ϕ_i) is given by

$$\nabla\phi_i = \nabla\left(\phi_{1,0} \cdot M_e^{-1}\right) = \nabla\left(\phi_{1,0}\right) S_e^{-1},$$

where the gradient of ϕ_i, $\nabla(\phi_i)$, is evaluated at a point $(x, y) \in e$, and the gradient of $\phi_{1,0}$ is evaluated at its inverse image $M_e^{-1}((x, y)) \in r$.

Further simplification is obtained by recalling that $\phi_{1,0} = x$, so its gradient is actually the constant vector $(1, 0)$. With this observation, the gradient of ϕ_i is just the first row in S_e^{-1}:

$$\nabla\phi_i = \left(\left(S_e^{-1}\right)_{1,1}, \left(S_e^{-1}\right)_{1,2}\right).$$

Similarly, for ϕ_j we have

$$\nabla\phi_j = \nabla\left(\phi_{0,1}\right) S_e^{-1},$$

where the gradient of ϕ_j is evaluated at a point $(x, y) \in e$, and the gradient of $\phi_{0,1}$ is evaluated at its inverse image $M_e^{-1}((x, y)) \in r$.

Again, one can simplify the above by recalling that $\phi_{0,1} = y$, so its gradient is just the constant vector $(0, 1)$. With this observation, the gradient of ϕ_j is just the second row in S_e^{-1}:

$$\nabla\phi_j = \left(\left(S_e^{-1}\right)_{2,1}, \left(S_e^{-1}\right)_{2,2}\right).$$

We are now ready to calculate the elements in the stiffness matrix A. For simplicity, we assume that $K \equiv 0$ in the PDE, so only the second-derivative terms remain in it. Let us use the notation

$$\text{diag}(P, Q) = \begin{pmatrix} P & 0 \\ 0 & Q \end{pmatrix}.$$

Now, $A_{i,j}$ in the stiffness matrix can be written as the sum of contributions from individual triangles:

$$A_{i,j} = \sum_e \int_e \nabla\phi_j \text{diag}(P, Q) \nabla\phi_i dxdy.$$

Here, the sum goes over all triangles in the mesh. Still, only those triangles that use both i and j as vertices (like the triangle in Figure 12.1) actually contribute to $A_{i,j}$. It is therefore common to say that $A_{i,j}$ (and, in fact, A) is "assembled" from the nonzero contributions from the individual finite elements.

Using the rule of the integral over a mapped domain, we have that the contribution from e to $A_{i,j}$ is

$$\begin{aligned}
&\int_e \nabla\phi_j \mathrm{diag}(P, Q)\nabla\phi_i dxdy \\
&= \int_r \nabla\phi_{0,1} S_e^{-1} diag(PM_e, QM_e) S_e^{-t} \nabla\phi_{1,0} dM_e(x, y) \\
&= \int_r (0, 1) S_e^{-1} diag(PM_e, QM_e) S_e^{-t} \begin{pmatrix} 1 \\ 0 \end{pmatrix} \left| \det\left(\frac{\partial M_e(x, y)}{\partial(x, y)}\right) \right| dxdy \\
&= \int_r (0, 1) S_e^{-1} diag(PM_e, QM_e) S_e^{-t} \begin{pmatrix} 1 \\ 0 \end{pmatrix} |\det(S_e)|\, dxdy.
\end{aligned}$$

In what follows, we'll see how these contributions are calculated explicitly in common examples.

12.6 Example: Rectangular Domain and Uniform Mesh

As mentioned above, finite elements are particularly suitable for the approximation of complicated domains. Here, however, we consider a simple example in which the domain is rectangular, as in Chapter 7, Section 13. The diffusion coefficients in this example can be variable and even discontinuous. The finite-element discretization method, which is based on the well-posed weak formulation, is thus most suitable.

In the present example, the domain Ω is the rectangle $[0, L_x] \times [0, L_y]$, $K \equiv 0$, and Γ_0 contains the upper and right edges and Γ contains the bottom and left edges in the rectangle.

For the finite-element mesh, we use a uniform triangle mesh as in Figure 12.3. We also assume that α is constant in each triangle in the mesh but may be discontinuous across the edges of the triangles.

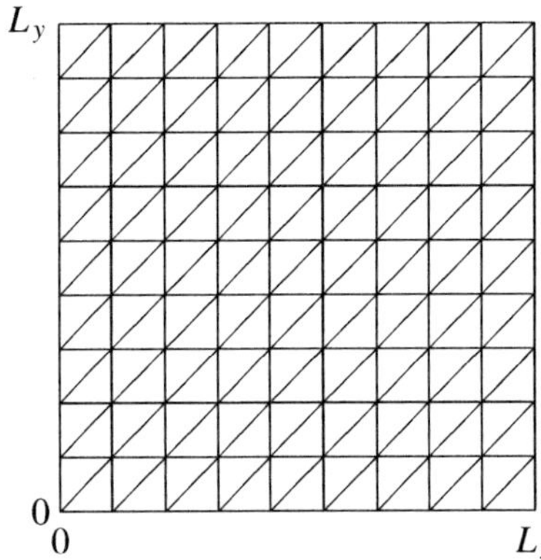

Figure 12.3. *The uniform finite-element mesh.*

Let e be a triangle in the mesh (as in Figure 12.4) with vertices i, j, and k. As in Section 12.5, let M_e map the reference triangle r onto e, and let S_e be its Jacobian. As shown in Section 12.5, the first column in S_e is the two-dimensional vector that represents the difference of vertices $i - k$, and the second column is $j - k$. For the right-angle triangle

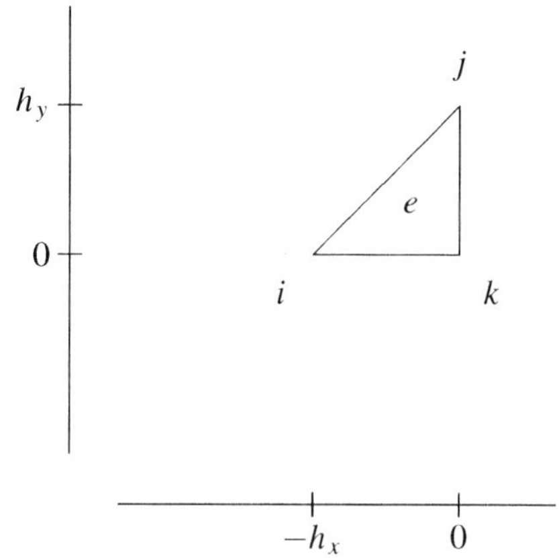

Figure 12.4. *The original finite element e in the uniform mesh. The reference element r is mapped to e by M_e.*

e of Figure 12.4, $i - k = (-h_x, 0)$ and $j - k = (0, h_y)$. Therefore, we have

$$S_e = \begin{pmatrix} -h_x & 0 \\ 0 & h_y \end{pmatrix}.$$

As a result, we have

$$\det(S_e) = -h_x h_y$$

and

$$S_e^{-1} = \begin{pmatrix} -h_x^{-1} & 0 \\ 0 & h_y^{-1} \end{pmatrix}.$$

The contribution from e to $A_{i,k}$ is, thus,

$$\begin{aligned}
&\int_e \nabla\phi_k \text{diag}(P, Q)\nabla\phi_i dxdy \\
&= \int_r \nabla\phi_{0,0} S_e^{-1} \text{diag}(PM_e, QM_e) S_e^{-t} \nabla\phi_{1,0} |\det(S_e)| dxdy \\
&= \int_r (-1, -1) S_e^{-1} \text{diag}(PM_e, QM_e) S_e^{-t} \begin{pmatrix} 1 \\ 0 \end{pmatrix} h_x h_y dxdy \\
&= -h_x^{-2} h_x h_y \int_r PM_e dxdy.
\end{aligned}$$

The other triangle that contributes to $A_{i,k}$ is the triangle t that lies below the edge connecting i to k (Figure 12.5). The contribution from t is calculated in a similar way. The rest of the triangles contribute nothing to the $A_{i,k}$-element, because either ϕ_i or ϕ_k vanishes on them. The final result is, thus,

$$A_{i,k} = -h_x^{-1} h_y \left(\int_r PM_e dxdy + \int_r PM_t dxdy \right).$$

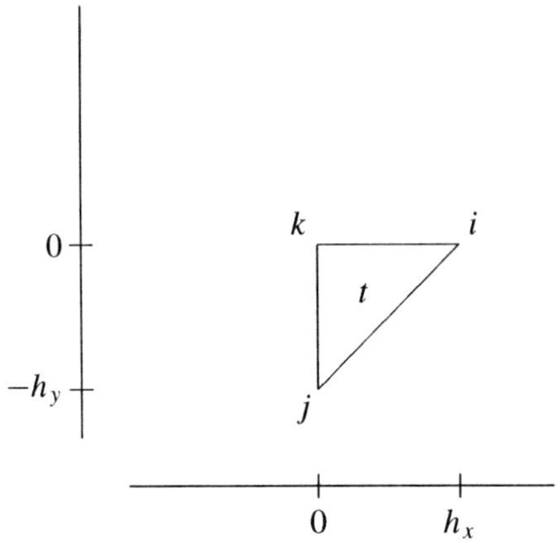

Figure 12.5. *The finite element t onto which the reference element r is mapped by* M_t.

A similar method is used to calculate $A_{i,k}$ for every two nodes i and k. Of course, when i and k don't lie in the same triangle, $A_{i,k}$ vanishes. Thus, A is sparse in the sense that most of its elements vanish. In an efficient implementation of sparse matrices, only nonzero matrix elements are stored (see Chapters 4 and 16).

The main-diagonal elements in A ($k = i$ in the above discussion) can be calculated using the same method. In this case, however, the six triangles that surround the node i contribute to $A_{i,i}$.

Let us now calculate the contribution to $A_{i,k}$ from the boundary conditions. Assume that the edge that connects i to k lies on the bottom edge of the rectangular domain, so there is no triangle t below it. In this case, the mixed boundary condition imposed on the bottom edge also contributes to $A_{i,k}$. In the following, we calculate this contribution.

Let $M_{i,k}$ denote the restriction of M_e to the interval $[0, 1]$, so that $M_{i,k}$ maps the interval $[0, 1]$ onto the interval $[k, i]$, with the Jacobian, or derivative, $-h_x$. For simplicity, assume also that α is constant at the line connecting i to k, and its value there is denoted by $\alpha_{i,k}$. The contribution from the mixed boundary conditions to $A_{i,k}$ is, thus,

$$
\begin{aligned}
&\int_{[i,k]} \alpha\phi_k\phi_i dx \\
&= \alpha_{i,k} \int_{M_{i,k}^{-1}([i,k])} \phi_{0,0}\phi_{1,0} \left| \frac{dM_{i,k}}{dx} \right| dx \\
&= \alpha_{i,k} \int_0^1 (1 - x - 0)x| - h_x|dx \\
&= \alpha_{i,k}h_x/6.
\end{aligned}
$$

The stiffness matrix for the present example has the same stencil (structure) as the difference operator in Chapter 7, Section 13. In fact, every node in the mesh is connected in A only to its nearest (immediate) neighbors, namely, the nodes to its right, left, top, and bottom. Thus, the stiffness matrix can actually be implemented most efficiently as a "difference2" object (Section A.4 of the Appendix).

In Chapter 13, we introduce a more general implementation, which is suitable not only for uniform meshes such as the present one but also for general unstructured meshes, which are used often in practical applications. With this implementation, the stiffness matrix has a far more complex structure and must be implemented as a more sophisticated sparse-matrix object (Chapter 16).

The remainder of this book deals with the construction and numerical solution of the stiffness system on general unstructured meshes. In realistic applications, where the domain is complicated and the solution of the weak formulation may have sharp variation, the finite-element triangulation must be highly unstructured. For example, the finite-element mesh can be rather coarse (low resolution) where the solution is expected to be rather smooth, and rather fine (high resolution) where the solution is expected to vary sharply. It is in this case that the advantage of C++ becomes most clear.

In what follows, we discuss unstructured meshes and give sufficient conditions with which the stiffness matrix has the attractive property of being an M-matrix. This property indicates that the discretization is indeed appropriate. The sufficient conditions provide guidelines for how unstructured meshes should be constructed.

12.7 M-Matrix Property in the Isotropic Case

This section gives motivation to why finite-element meshes should be regular in the sense that the triangles should have moderate angles. Readers who are mainly interested in the practical aspects can skip it and proceed to Section 12.8.

As discussed in Section 12.4 above, the M-matrix property is most attractive. Although the stiffness matrix is not always an M-matrix, it is possible to specify sufficient conditions that guarantee that it is. This indicates that the finite-element discretization method indeed makes sense.

Assume that the diffusion equation is isotropic; that is, the diffusion coefficients are the same in both the x and y spatial directions:

$$P(x, y) = Q(x, y), \quad (x, y) \in \Omega.$$

With this assumption, the contribution from e to $A_{i,j}$ at the end of Section 12.5 takes the form

$$(0, 1)S_e^{-1}S_e^{-t}\begin{pmatrix} 1 \\ 0 \end{pmatrix} |\det(S_e)| \int_r PM_e dxdy.$$

Note that the above integral won't be affected if S_e is multiplied by an orthogonal matrix on the left. Indeed, if S_e is replaced by OS_e for some orthogonal matrix O (orthogonality means that $O^t = O^{-1}$), then we will still have

$$(OS_e)^{-1}(OS_e)^{-t} = S_e^{-1}O^{-1}O^{-t}S_e^{-t} = S_e^{-1}S_e^{-t}$$

and

$$\det(OS_e) = \det(O)\det(S_e) = \det(S_e).$$

Therefore, we may assume without loss of generality that the edge connecting i to k in e is parallel to the x-axis, as in Figure 12.6, so that the y-coordinate of $i - k$ vanishes.

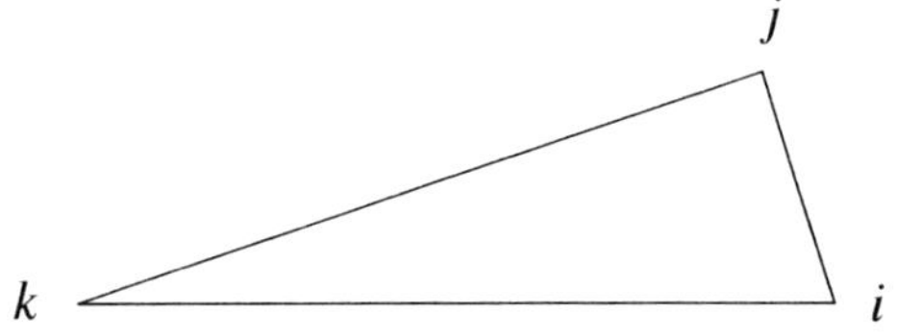

Figure 12.6. *The rotated finite element.*

For a finite element e as in Figure 12.6, S_e takes the upper triangular form

$$S_e = \begin{pmatrix} (S_e)_{1,1} & (S_e)_{1,2} \\ 0 & (S_e)_{2,2} \end{pmatrix},$$

with $(S_e)_{1,1} > 0$ and $(S_e)_{2,2} > 0$. Therefore, we have

$$\det(S_e) = (S_e)_{1,1}(S_e)_{2,2}.$$

Furthermore, S_e^{-1} takes the upper triangular form

$$S_e^{-1} = \begin{pmatrix} (S_e^{-1})_{1,1} & (S_e^{-1})_{1,2} \\ 0 & (S_e^{-1})_{2,2} \end{pmatrix}$$
$$= \begin{pmatrix} (S_e)_{1,1}^{-1} & -(S_e)_{1,1}^{-1}(S_e)_{1,2}(S_e)_{2,2}^{-1} \\ 0 & (S_e)_{2,2}^{-1} \end{pmatrix}.$$

As a result, the above integral that represents the contribution from e to $A_{i,j}$ takes the form

$$(S_e^{-1})_{2,2}(S_e^{-1})_{1,2}\,|\det(S_e)| \int_r PM_e dxdy$$
$$= -(S_e)_{1,2}(S_e)_{2,2}^{-1} \int_r PM_e dxdy$$
$$= -\cot(\theta) \int_r PM_e dxdy,$$

where θ is the (positive) angle between $i - k$ and $j - k$ in Figures 12.1 and 12.6, and cot stands for the cotangent function (cosine divided by sine).

We conclude from the above calculations that the contribution from e to $A_{i,j}$ depends not on the area of e but only on the cotangent of the angle θ that lies between $i - k$ and $j - k$. If θ is greater than $\pi/2$, then this cotangent is negative, so the contribution from e to $A_{i,j}$ is positive. In this case, A may not be an L-matrix unless there are negative contributions to $A_{i,j}$ from other triangles to balance this positive contribution. On the other hand, if triangles in the mesh have acute or right angles only, then all contributions to off-diagonal matrix elements in A are nonpositive, and A is an L-matrix.

Let us show that, for triangulation with acute or right angles only, A is also diagonally dominant. Since we have already seen that A is an L-matrix, it is sufficient to show that its row-sums are nonnegative. To see this, consider a node i in the mesh. The sum of the

ith row in A is the same as $a(v, \phi_i)$, where v is the sum of the nodal basis functions at i and the nodes surrounding it. But since these nodal basis functions are linear, v must be constant on the triangles surrounding i, so its gradient vanishes there. It is easy to see that the boundary conditions normally increase the row-sum. This shows that A is indeed diagonally dominant.

The combination of diagonal dominance and the L-matrix property implies that A is indeed also an M-matrix [46]. This property indicates that the finite-element discretization indeed produces a good approximation to the original problem (Section 12.4).

For less regular meshes with angles greater than $\pi/2$, the stiffness matrix may no longer be an M-matrix. Still, it is likely that the finite-element discretization will provide a good approximation on these meshes as well, provided that the irregularity is limited to a rather small area such as the neighborhood of irregular boundaries.

In the next section, we'll see that the M-matrix property may also apply to the more difficult case of highly anisotropic equations.

12.8 Highly Anisotropic Equations

Above, we studied the isotropic case, in which the diffusion coefficient in the x spatial direction is the same as the diffusion coefficient in the y spatial direction. We showed that, for triangulation with acute and right angles only, the stiffness matrix is an M-matrix, which indicates that the finite-element discretization is indeed appropriate. This led us to the guideline that the mesh should be kept as regular as possible.

Here, we consider the highly anisotropic case, in which the diffusion coefficients in the x and y spatial directions are totally different from each other. In this case, for some meshes, it is still possible to show that the stiffness matrix is an M-matrix by simply transforming the equation into an isotropic one.

Consider the following diffusion equation:

$$-(Pu_x)_x - (Qu_y)_y = F$$

(in a rectangle, with suitable boundary conditions), where $u \equiv u(x, y)$ is the unknown solution, and the diffusion coefficients P and Q are considerably different from each other. For a start, consider the case in which P and Q are constants:

$$P = 1 \quad \text{and} \quad Q = \varepsilon,$$

where ε is a small positive parameter.

Let us first consider the basic finite-difference scheme on a uniform grid as in Figure 7.4 above. Denote the discrete approximation by

$$u_{i,j} \doteq u(jh_x, ih_y),$$

where h_x and h_y are the meshsizes in the x- and y-directions, respectively. Then, as in Chapter 7, Section 13, the second spatial derivatives are approximated by

$$u_{xx}(jh_x, ih_y) \doteq h_x^{-2}(u_{i,j+1} + u_{i,j-1} - 2u_{i,j}),$$
$$u_{yy}(jh_x, ih_y) \doteq h_y^{-2}(u_{i+1,j} + u_{i-1,j} - 2u_{i,j}).$$

Let us study the adequacy of this scheme. For this purpose, we need to estimate the discretization error as the meshsizes and ε approach zero at the same time.

As explained in Chapter 8, Section 4, the discretization error is the inverse of the coefficient matrix times the truncation error. Let us first estimate the maximal possible enlargement of a vector by applying A^{-1} to it (the norm of A^{-1}). Fortunately, here A (and, hence, also A^{-1}) is SPD; therefore, the norm of A^{-1} is the same as its maximal eigenvalue. In order to estimate how large this eigenvalue could be, it is sufficient to estimate how small an eigenvalue of A could be. Well, A is just the sum of discrete xx- and yy-derivatives. Thanks to the fact that these discrete derivatives are in the divided form (they contain the factor h_x^{-2} or h_y^{-2}), even the smoothest discrete Fourier mode $\sin(j\pi h_x)\sin(i\pi h_y)$ (where j and i are the indices of grid point in the x- and y-directions, respectively) doesn't shrink under the application of A. In fact, it is just multiplied by a positive constant independent of h_x, h_y, and ε. This implies that $\|A^{-1}\|$ is indeed bounded independent of h_x, h_y, and ε.

All that is left to do, thus, is to estimate the truncation error. For the finite-difference scheme, it is just

$$h_x^2 u_{xxxx} + \varepsilon h_y^2 u_{yyyy},$$

where $u \equiv u(x, y)$ is the solution to the original PDE. Let us make the reasonable assumption that the right-hand side, F, is independent of ε. This implies that u must be smooth in the x-direction but can oscillate rapidly (with frequency up to $\varepsilon^{-1/2}$) in the y-direction. Thus, taking the y-derivative of u is the same as multiplying it by $\varepsilon^{-1/2}$, and the truncation error takes the form

$$h_x^2 + h_y^2/\varepsilon.$$

Thus, the finite-difference scheme is adequate so long as ε is not too small; i.e.,

$$\varepsilon \gg h_y^2$$

as h_x, h_y, and ε approach zero at the same time.

We can have adequacy because the grid uses the same x- and y-coordinates as the original PDE. In other words, the grid aligns with the strong and weak diffusion directions in the PDE. Indeed, if the grid were rotated by some angle, then the y-direction in which u oscillates rapidly would be oblique relative to the grid, and u would oscillate rapidly in both directions of the grid points, resulting in a truncation error as large as h_x^2/ε^2. In order to have adequacy, the grid must thus align with the directions in the original PDE, namely, the Cartesian x- and y-directions.

As we have seen above, the solution to the original PDE is rather smooth in the x-direction. This raises the thought that maybe h_x could be rather large, so a rather small number of grid points is required in the x spatial direction. Indeed, the above error estimate shows that h_x can be as large as

$$h_x = \varepsilon^{-1/2} h_y,$$

with truncation error practically the same as before:

$$h_x^2 u_{xxxx} + \varepsilon h_y^2 u_{yyyy} \doteq h_y^2/\varepsilon + h_y^2/\varepsilon.$$

In the following, we use this observation in the finite-element discretization.

Let us now consider the anisotropic diffusion equation with nonconstant diffusion coefficients P and Q, defined as in Figure 12.7. Here, the equation is highly anisotropic in the lower-left subsquare, where the diffusion in the y-direction is much stronger than in the x-direction, and in the upper-right subsquare, where the diffusion in the x-direction is much stronger than in the y-direction.

In order to benefit from the above observation that the meshsize in the strong-diffusion direction can be larger than that in the weak-diffusion direction, we must use finite elements rather than finite differences. Indeed, the discretization proposed in [10] (which is also supported by results from approximation theory) uses finite elements that are stretched in the strong-diffusion direction. Unfortunately, as discussed there, in order to preserve conformity across lines of discontinuity in the diffusion coefficients, one may need to compromise regularity, that is, use triangles with very small angles.

In order to overcome this problem, we propose in Figure 12.8 a mesh that is both conformal and reasonably regular. This mesh uses stretched $8h \times h$ and $h \times 8h$ finite elements, where h is the typical meshsize. In order to change gradually from stretched to unstretched finite elements across the discontinuity line in a diffusion coefficient, we use $\log_2 8 = 3$ layers of regular and conformal finite elements (Figure 12.8).

When mapped to the isotropic coordinates in the lower-left and upper-right subsquares (using the new coordinates $(x, \tilde{y})$ in the lower-left subsquare and $(\tilde{x}, y)$ in the upper-right subsquare, where $\tilde{x} = x/8$ and $\tilde{y} = y/8$), the conditions in Section 12.7 hold, implying that the coefficient matrix A is indeed a diagonally dominant M-matrix. This indicates that the discretization is indeed appropriate.

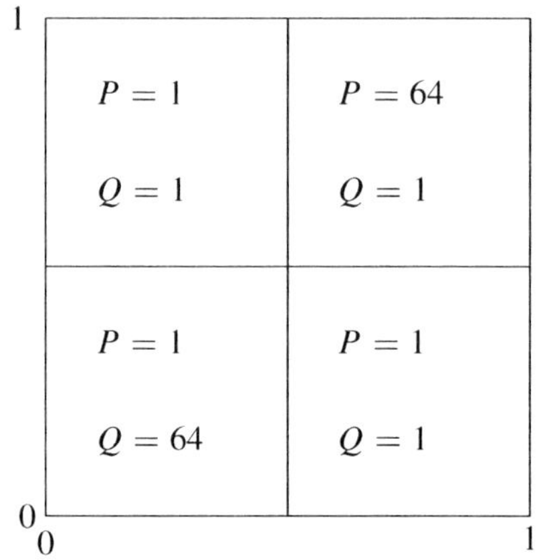

Figure 12.7. *The diffusion coefficients P and Q for the anisotropic diffusion equation.*

The present finite-element mesh depends on the coefficients in the original PDE. It would be advantageous to have a process that constructs the mesh more automatically, avoiding any special human observation. This automatic process could then be applied to any given problem as a black box, yielding the required locally stretched mesh independent of the special properties of the particular PDE.

A good candidate for this purpose is the process of adaptive refinement. In this process, the numerical solution is first computed on a coarse mesh. This mesh is then successively refined at places where the numerical solution exhibits large variation. For the above examples, this process produces low resolution in strong-diffusion directions, where the numerical solution is rather smooth, and high resolution in weak-diffusion directions, where it changes

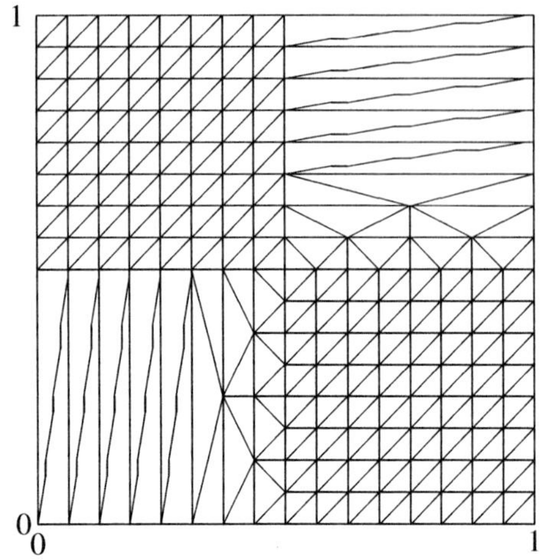

Figure 12.8. *The stretched finite-element mesh for the anisotropic diffusion equation.*

rapidly. This should automatically give a mesh similar to the mesh in Figure 12.8. We'll return to this algorithm and its detailed C++ implementation in Chapter 14.

12.9 Example: Circular Domain

The rectangular domain used above can be approximated by a uniform mesh. Domains with a curved boundary, such as the circle, are more difficult to approximate. The straight edges of the triangles can never coincide with a curved boundary; they can at best approximate it to acceptable accuracy, provided that sufficiently small triangles are used.

Consider, for example, the Poisson equation in the unit circle:

$$-u_{xx} - u_{yy} = \mathcal{F}, \quad x^2 + y^2 < 1,$$

with suitable boundary conditions.

Let us construct the finite-element mesh that approximates the circular boundary well. The construction is done successively: we start with a coarse mesh with only two triangles, which provides a poor approximation (Figure 12.9). This mesh is then improved by adding four more triangles that approximate the circular boundary better (Figure 12.10). The process repeats itself again and again: in the ith step, 2^i triangles are added near the boundary to approximate it better than before. The process stops when the approximation to the curved boundary is satisfactory, and the final mesh is used in the finite-element scheme.

To make things more concrete, let us construct the stiffness matrix for the mesh in Figure 12.10. (The construction of the stiffness matrix for the finest mesh is in principle the same.) Consider the big triangle $\triangle(i, k, l)$ with vertices at i, k, and l and the small triangle $\triangle(i, j, k)$ with vertices at i, j, and k in Figure 12.10. Both triangles contribute to the matrix element $A_{i,k}$ in the stiffness matrix A. However, it turns out that these contributions cancel

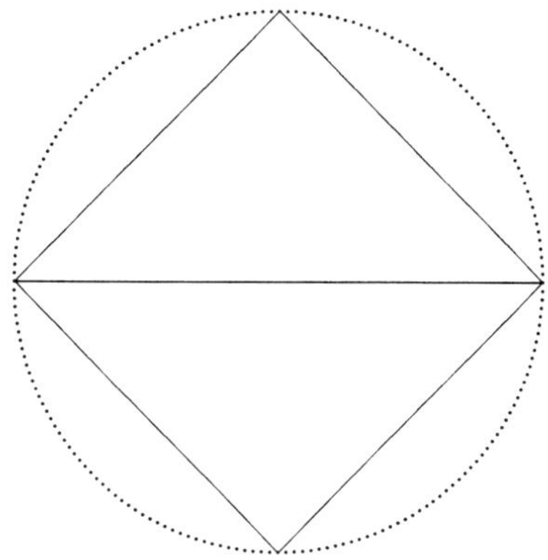

Figure 12.9. *The initial coarse mesh that provides a poor approximation to the circular boundary.*

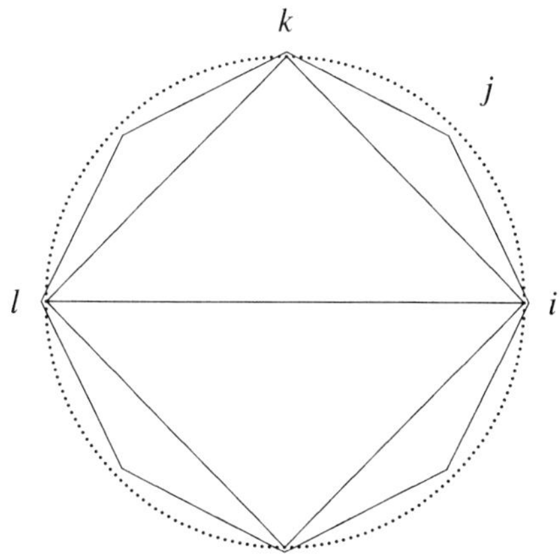

Figure 12.10. *The next, finer, mesh that better approximates the circular boundary.*

each other, so $A_{i,k}$ actually vanishes. Indeed, as calculated in Section 12.7, the contribution from $\triangle(i, k, l)$ is

$$\frac{\cot(\angle(k, l, i))}{2},$$

where $\angle(k, l, i)$ is the angle at the vertex l in $\triangle(i, k, l)$. Similarly, the contribution from $\triangle(i, j, k)$ is

$$\frac{\cot(\angle(i, j, k))}{2}.$$

Since these two angles lie on the same circle, it follows from Euclidean geometry that

$$\angle(k, l, i) + \angle(i, j, k) = \pi.$$

As a result,

$$\begin{aligned}\cot(\angle(k,l,i)) + \cot(\angle(i,j,k)) &= \cot(\angle(k,l,i)) + \cot(\pi - \angle(k,l,i)) \\ &= \cot(\angle(k,l,i)) - \cot(\angle(k,l,i)) \\ &= 0,\end{aligned}$$

implying that

$$A_{i,k} = 0.$$

On the other hand, the matrix element $A_{i,j}$ that couples the adjacent nodes i and j doesn't vanish. Indeed, since the only triangle that contributes to it is $\triangle(i, j, k)$, it follows from Section 12.7 that

$$A_{i,j} = -\frac{\cot(\angle(j,k,i))}{2} = -\frac{\cot(\pi/N)}{2},$$

where N is the number of nodes. (In the case in Figure 12.10, $N = 8$; in general, $N = 2^{k+1}$, where k is the number of refinement steps.) In summary, A takes the tridiagonal form

$$A = \text{tridiag}\left(-\frac{\cot(\pi/N)}{2}, \cot(\pi/N), -\frac{\cot(\pi/N)}{2}\right).$$

More precisely, here A is not exactly tridiagonal in the usual sense, because it also contains the nonzero elements

$$A_{0,N-1} = A_{N-1,0} = -\frac{\cot(\pi/N)}{2}$$

at its upper-right and bottom-left corners. It is more accurate to say that A is a circulant Toeplitz matrix with three nonzero diagonals.

The stiffness matrix constructed above may also be interpreted as the discretization of a one-dimensional Poisson equation (with periodic boundary conditions) along the circular boundary. Clearly, this is not a good scheme for the two-dimensional Poisson equation in the entire circle.

The reason for this failure is that we have concentrated too much on approximating the boundary well and ignored the interior of the domain. In practice, one must also refine the big triangles (e.g., $\triangle(i, k, l)$) to obtain a good approximation in the interior of the domain. This can be done adaptively as in Chapter 14, Section 2. The algorithm presented there also properly refines the interior of the domain.

The above mesh uses particularly small and narrow triangles near the boundary. These triangles are highly degenerate: they contain very small angles. This may lead to a large discretization error. In Chapter 14, Section 7, we propose a more balanced approach, in which the mesh is refined simultaneously both in the interior of the domain and at its boundary. This mesh is more regular and contains no degenerate triangles.

12.10 Exercises

1. Calculate the stiffness matrix for the PDE

$$-u_{xx}(x,y) - u_{yy}(x,y) + u(x,y) = F(x,y)$$

on a right-angle-triangle mesh as in Figure 12.3.

2. Calculate the stiffness matrix of the above PDE on a general triangle mesh. In particular, write the contribution from some triangle to some matrix element $A_{i,j}$ (where i and j are indices of vertices in this triangle) in terms of the mapping from the reference triangle to this triangle.

3. Show that the resulting stiffness matrix is SPD.

4. Calculate the stiffness matrix for the PDE

$$-u_{xx}(x, y) - u_{yy}(x, y) + u_x(x, y) + u_y(x, y) = F(x, y)$$

on a uniform triangle mesh with right-angle triangles as in Figure 12.3.

5. Show that the above matrix is nonsymmetric.

6. Calculate the stiffness matrix for the above PDE on a general triangle mesh. In particular, write the contribution from some triangle to some matrix element $A_{i,j}$ (where i and j are indices of vertices in this triangle) in terms of the mapping from the reference triangle to this triangle.

7. Show that the above matrix is nonsymmetric.

12.11 Advanced Exercises

1. Consider the equation

$$-(Pu_x)_x - (Qu_y)_y = F$$

in a domain Ω in the Cartesian plane, where P and Q are given uniformly positive functions and F is a given function. Write the algorithm that produces the stiffness matrix A on a given finite-element mesh. Consider Dirichlet, Neumann, and mixed boundary conditions of the form

$$Pu_x n_1 + Qu_y n_2 + \alpha u = g,$$

where $\vec{n} = (n_1, n_2)$ is the outer normal vector, α is a given nonnegative function, and g is a given function on $\partial\Omega$.

2. Show that the bilinear form for the above equation is symmetric. Conclude that A is symmetric as well.

3. Use the coercivity of the corresponding quadratic form to show that A is actually SPD.

4. Consider now the equation

$$-(Pu_x)_x - (Qu_y)_y + u = F.$$

Write the algorithm that produces the stiffness matrix $A + K$, where A is as before and K results from the free term u in the PDE.

5. Show that K is SPD.

6. Consider the time-dependent diffusion equation

$$u_t - (Pu_x)_x - (Qu_y)_y = F.$$

Write the algorithms that use the above finite-element discretization in explicit, implicit, and semi-implicit time discretization. The solution follows from Chapter 19, Section 4.

7. Write the algebraic representations of your algorithms by replacing the identity matrix I used in Chapter 8, Section 2, by K.

8. Define

$$B \equiv (\Delta t)^{-1} K$$

for the explicit scheme,

$$B \equiv (\Delta t)^{-1} K + A$$

for the implicit scheme, and

$$B \equiv (\Delta t)^{-1} K + A/2$$

for the semi-implicit scheme. Show that B is inverted in each time step in the time marching.

9. Show that B above is SPD.

10. Define the energy inner product by

$$(u, v)_B \equiv (u, Bv) = u^t Bv$$

(where u and v are any two vectors).
A matrix M is symmetric with respect to the energy inner product if

$$(Mu, v)_B = (u, Mv)_B$$

for every two vectors u and v. Show that $B^{-1}A$ and $B^{-1}K$ are symmetric with respect to the energy inner product.

11. Define the energy norm of a matrix M by

$$\|M\|_B \equiv \max_{v \neq 0} \left(\frac{(M^t v, M^t v)_B}{(v, v)_B} \right)^{1/2}.$$

It is well known that, if M^t is symmetric with respect to the energy inner product, then the above quantity is also equal to the modulus of the largest eigenvalue of M^t. In other words,

$$\|M\|_B = \max_{v \neq 0} \left| \frac{(v, M^t v)_B}{(v, v)_B} \right|.$$

12. Use the SPD property of A and K to show that, for every nonzero vector v,

$$\frac{\left(v, (\triangle t)^{-1} K v\right)}{\left(v, \left((\triangle t)^{-1} K + A\right) v\right)} \quad \text{and} \quad \frac{\left(v, \left((\triangle t)^{-1} K - A/2\right) v\right)}{\left(v, \left((\triangle t)^{-1} K + A/2\right) v\right)}$$

are smaller than 1 in magnitude, and, for sufficiently small $\triangle t$,

$$\frac{\left(v, \left((\triangle t)^{-1} K - A\right) v\right)}{\left(v, (\triangle t)^{-1} K v\right)}$$

is also smaller than 1 in magnitude.

13. Use the above results to show that the block submatrices $Q_i B_i^{-1}$ in Chapter 8, Section 3, are smaller than 1 in terms of the energy norm induced by B_i. Use this result to show stability and accuracy in the time marching.

14. Assume that the function α in the above mixed boundary conditions is constant in each particular boundary edge, e.g., an edge leading from node i to node j that both lie on the boundary segment where the mixed boundary conditions are imposed. Show that the contribution to $A_{i,i}$ from this edge is positive and twice as large as the contribution to $A_{i,j}$.

15. Consider a particular triangle in the mesh, with vertices at nodes i, j, and k. Show that the contribution to $K_{i,i}$ from this triangle is positive and the same as the sum of the contributions to $K_{i,j}$ and $K_{i,k}$ (which are positive as well). (You can use the reference triangle to calculate these contributions.)

16. As in Section 12.7, assume that the diffusion is isotropic ($P \equiv Q$) and the angles in the triangles in the mesh never exceed $\pi/2$. Show that A and $A + K$ are diagonally dominant, even when mixed boundary conditions are imposed.

17. Use your answers to the exercises at the end of Chapter 11 and the guidelines in this chapter to calculate the stiffness matrix for the diffusion problem in three spatial dimensions:

$$-(Pu_x)_x - (Qu_y)_y - (Wu_z)_z = F,$$

where P, Q, W, F, and u are scalar functions of the three spatial variables x, y, and z. For finite elements, use tetrahedra. A tetrahedron is a three-dimensional shape with four vertices (connected to each other by four straight lines) and four flat triangular sides. A tetrahedron is denoted by a set of four points in the three-dimensional Cartesian space, which denote its vertices. For example, the unit cube $[0, 1]^3$ can be written as the union of the following five disjoint tetrahedra:

$$\begin{gathered}\{(0,0,0), (1,0,0), (0,1,0), (0,0,1)\},\\ \{(1,1,0), (1,0,0), (0,1,0), (1,1,1)\},\\ \{(1,0,1), (0,0,1), (1,1,1), (1,0,0)\},\\ \{(0,1,1), (1,1,1), (0,0,1), (0,1,0)\},\\ \{(1,0,0), (0,1,0), (0,0,1), (1,1,1)\}.\end{gathered}$$

Note that, in this case, the matrix S_e is of order 3. The reference element now has four vertices, $(1, 0, 0)$, $(0, 1, 0)$, $(0, 0, 1)$, and $(0, 0, 0)$, with the corresponding standard nodal functions x, y, z, and $1 - x - y - z$, respectively. (In general, the finite-element mesh is not necessarily structured, so the angles in the tetrahedra in it can take any value. In this exercise, however, we deal only with a uniform three-dimensional mesh in which every cubic cell can be divided into five tetrahedra as above.)

18. Assume that the above three-dimensional diffusion equation is isotropic ($P \equiv Q \equiv W$) and the angles in each side of each tetrahedron never exceed $\pi/2$. Extend the result in Section 12.7, and show that the stiffness matrix is diagonally dominant.

Chapter 13

Unstructured Finite-Element Meshes

In this chapter, we introduce the C++ implementation of general, (unstructured) finite-element meshes. The complex data structures used here are best implemented in C++. The hierarchy of objects goes from the "mesh" object at the high level, through the "finiteElement" object at the intermediate level, to the "node" and "point" objects at the low level. The "mesh" object is then used to assemble the stiffness matrix required in the finite-element scheme.

13.1 Concrete and Abstract Objects

In practice, the finite-element mesh can be highly unstructured and nonuniform. The approximation of complicated domains may require a large number of small and irregular finite elements to fit to the boundary. Furthermore, for problems with discontinuous coefficients, the solution may be irregular across the discontinuity lines, particularly where they have corners [44]. In such locations, large numbers of small finite elements are required to provide sufficiently high resolution.

It is thus clear that general finite-element meshes have no uniform structure. Therefore, their implementation on the computer cannot use standard arrays. More sophisticated data structures are necessary.

Let us illustrate how inefficient a naive implementation with arrays would be. Because it is not assumed that the mesh is uniform, a two-dimensional array can no longer imitate its geometry. In fact, the storage method must be independent of any geometrical property. The nodes must be stored in an N-dimensional array V (where N is the number of nodes). The mesh, however, contains not only nodes but also edges to connect them to each other. How are these edges implemented?

One can argue that the edges don't have to actually be implemented, because they are stored in the stiffness matrix, where $A_{i,j} \neq 0$ only if i is connected to j in the mesh. However, we are concerned here with the situation before the stiffness matrix is available. It is necessary to store the information in the mesh beforehand to help construct the stiffness matrix.

One possible solution is to add another N-dimensional array W of integers. This array stores the index of the node that is connected to the corresponding node in V. For example, if i is connected to j, then $W[i] = j$.

Still, the W array can store at most one edge per node, while in general a node can be connected to many other nodes. One must thus use not only one but also many more W-arrays to store all these other edges as well. In fact, the number of W-arrays must be as large as the maximal number of edges that can use a particular node. This involves considerable overhead in terms of computer memory, because most nodes may be connected to only a few nodes and never use most of the W-arrays.

The above attempt to implement the mesh suffers from another drawback. The maximal number of nodes and edges per node must be determined in advance and cannot change dynamically. The implementation is thus not only expensive but also limited. A more natural and flexible approach is required.

A much better idea is to use connected lists. Each node should have a connected list of integers that indicate the nodes to which it is connected in the mesh. The entire mesh is then implemented as a list of connected lists of integers.

Unfortunately, because a list is actually an array of addresses, it is not sufficiently flexible in terms of adding and dropping items. In other words, the above implementation doesn't allow adding new nodes and removing old ones. An even more flexible approach is to use a connected list rather than a list. The entire mesh is then implemented as a connected list of nodes, each of which is by itself a connected list of integers. One can clearly see how the implementation gets more and more complicated, obscure, and hard to write and use.

Furthermore, our main job is to construct the stiffness matrix. With the above implementation, this task is practically impossible, because the finite-element object is missing.

In mathematical terms, a mesh is actually a nonoriented graph embedded in the plane with no crossing edges. This graph is actually defined in terms of nodes and edges only. Here, however, we are dealing with a finite-element mesh, for example, a mesh of triangles as in Chapter 4, Section 5. We have not yet used this important property. In fact, we declined to implement the finite-element object, the missing link between the "node" object and the "mesh" object.

The object-oriented approach tells us to think about the objects that are actually used in the mathematical framework. Here, these objects are the finite elements. So far, we have been busy with more elementary objects such as nodes and edges. These are not the objects we should be concerned with. We should rather concentrate on the finite-element object at the high level of the implementation. This approach is called "downward implementation" in Chapter 4, Section 7.

One may rightly ask why the concept of finite element is necessary in the actual implementation. After all, the mesh is well defined in terms of nodes and edges only, with no mention of finite elements whatsoever. Furthermore, although we humans can see triangles in the mesh in Chapter 4, Section 5, how can the computer be expected to "see" them? Surely, the computer can never understand what "triangle" means. At best, it can store its vertices, but this returns to our previous (naive) implementation, where nodes are stored and triangles are ignored. Moreover, implementing a triangle as a triplet of nodes may do for a single, dangling triangle, but not for a triangle that takes part in a mesh, where adjacent triangles share the same nodes. But if every triangle is defined in terms of three nodes, how does the computer know that a node in one triangle may coincide with a node in another triangle, and that any change to the one should affect the other as well?

The answer to these questions is one and the same. The conditions imposed on the mesh in Chapter 4, Section 5, actually imply the existence of abstract triangles that must be implemented. Triangles exist only in the minds of us humans to help us imagine the mathematical structure obtained by these conditions. They are never stored physically in the computer memory. The mesh is defined and indeed stored in terms of nodes alone, but the triangle is still a most useful conceptual tool to refer to a triplet of nodes as a whole. The triangle never contains its vertices physically, but merely refers to them. In this sense, the triangle object is far more abstract than the concrete nodes stored in the computer memory. Therefore, the triangle object should contain not actual nodes but rather three addresses of nodes. Triangles that share the same node should both have its address.

Thus, we use here two levels of abstraction. In the more concrete (low) level, the nodes in the mesh are stored in the computer memory as pairs of numbers to indicate their x- and y-coordinates in the Cartesian plane. In the more abstract (high) level, triangles are stored as triplets of pointers to their vertices (see Figure 13.1).

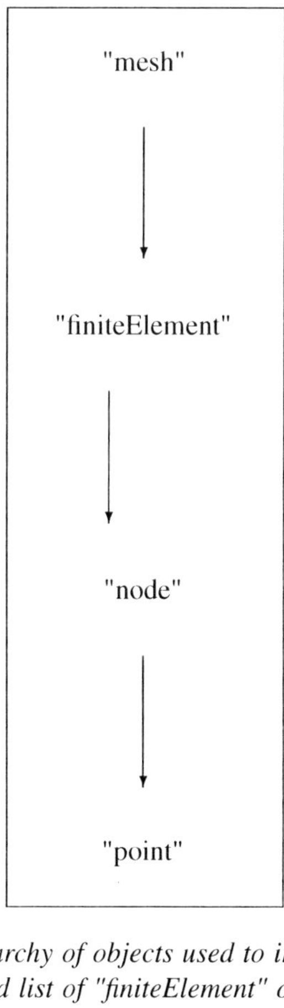

Figure 13.1. *The hierarchy of objects used to implement an unstructured mesh: the "mesh" object is a connected list of "finiteElement" objects, each of which is a list of (pointers to) "node" objects, each of which contains a "point" object to indicate its location in the Cartesian plane.*

Our complete plan to implement the finite-element mesh is thus as follows. The mesh is a connected list of triangles, so it is flexible in the sense that it is easy to drop unnecessary triangles from it and insert new ones. The triangle object has three pointers to its three vertices. This object is particularly useful in the calculation of the stiffness matrix. These vertices are implemented as node objects in the low level of implementation.

Note that because no arrays are used, no natural indexing of nodes is available. A node object should thus contain not only data about its geometric location but also its index among the nodes in the mesh. This index is eventually used to access the numerical solution of the stiffness system at this particular node. For example, the numerical solution at the node indexed by i is x_i, where x is the solution to the stiffness equation $Ax = f$.

13.2 The Node Object

We start with the most basic object in the implementation: the node. This elementary object will be used later to construct finite elements and the mesh itself.

As discussed in Section 13.1 above, a node may be shared by two or more finite elements. Therefore, the "node" object must contain information not only about its location in the Cartesian plane but also about the number of finite elements that share it. Thus, the "node" object contains three fields: the first to specify the point in the Cartesian plane where it is located, the second to specify its index among the nodes in the mesh, and the third to specify how many finite elements share it.

The type of the "location" field is the template 'T', to be specified in compilation time. In our case, 'T' is chosen to be the two-dimensional "point" object of Chapter 2, Section 18. Still, the use of the 'T' template also gives us the opportunity to substitute for 'T' the three-dimensional "point3d" object of Chapter 2, Section 18, whenever three-dimensional problems are encountered.

The location of the node in the Cartesian plane can be read by the public "operator()" function. Thus, if a node 'n' is defined, then "n()" is its location in the Cartesian plane. The "operator()" used here takes no arguments whatsoever. This is also why it cannot be replaced by "operator[]", which must take exactly one argument.

Initially, the integer data fields are filled with trivial values. In particular, the "index" field containing the index of the node is set to a meaningless negative value. A meaningful nonnegative value is assigned to it once the mesh is ready.

The integer data field "sharingElements" that indicates how many finite elements share the node is initially set to zero, indicating that this is a dangling node that belongs to no finite element as yet. Later on, when finite elements are defined, this information will prove most useful. The value increases by 1 whenever a new finite element that shares it is created and decreases by 1 whenever such a finite element is destroyed. Therefore, we have to define functions that increase, decrease, and read the value of the "sharingElements" field:

```
template<class T> class node{
  T location;
  int index;
  int sharingElements;
public:
  node(const T&loc=0., int ind=-1, int sharing=0)
```

```
    : location(loc),index(ind),sharingElements(sharing){
    } // constructor

    node(const node&n):location(n.location),index(n.index),
      sharingElements(n.sharingElements){
    } // copy constructor

    const node& operator=(const node&);
    ~node(){} // destructor
```

So far, we have implemented the constructors and destructor and declared the assignment operator, to be defined later on. Next, we define some more member functions that allow reading and accessing the private data in an individual "node" object:

```
    const T& operator()() const{
      return location;
    } // read the location

    int getIndex() const{
      return index;
    } // read index

    void setIndex(int i){
      index=i;
    } // set index
```

The following member functions read and manipulate the number of finite elements that share the current node:

```
    int getSharingElements() const{
      return sharingElements;
    } // read number of elements that share this node

    void moreSharingElements(){
      sharingElements++;
    } // increase number of elements that share this node

    int lessSharingElements(){
      return !(--sharingElements);
    } // decrease number of elements that share this node

    int noSharingElement() const{
      return !sharingElements;
    } // indicate a dangling node
  };
```

This concludes the block of the "node" class. In the following, we implement the assignment operator declared above:

```
template<class T>
const node<T>&
node<T>::operator=(const node<T>&n){
  if(this != &n){
    location = n.location;
    index = n.index;
    sharingElements = n.sharingElements;
  }
  return *this;
} // assignment operator
```

Finally, we define a function that prints the data in the "node" object to the screen:

```
template<class T>
void print(const node<T>&n){
  print(n());
  printf("index=%d; %d sharing elements\n",
      n.getIndex(),n.getSharingElements());
} // print a node
```

13.3 The Finite-Element Object

Here, we define the finite-element object. As discussed in Section 13.1, this object contains not nodes but rather pointers to nodes. This way, two adjacent finite elements that share the same node each contain a copy of its address. A change to the node can thus be made from either one of the sharing finite elements.

In fact, the finite-element object contains N pointers to nodes, where the template N is to be specified in compilation time. In our two-dimensional case, where triangles are used, N is set to 3. The template N may take values greater than 3 in the implementation of high-order finite elements (see Chapter 15) and tetrahedra ($N = 4$) in three-dimensional applications.

The type of nodes pointed at in the finite-element object is "node<T>", where the template 'T' is to be filled later in compilation time. In our application, 'T' is "point", so the vertices are of type "node<point>", which implements the node in a two-dimensional finite-element mesh. The template 'T' may also take more complicated types such as "point3d" to implement three-dimensional meshes.

As discussed in Section 13.1 above, a pointer-to-node contained in a finite element may point to a node that is also pointed at in another finite element. Therefore, the constructors, destructor, and assignment operator do not always create or destroy nodes; most often, they only change the number of finite elements that share them. One should keep in mind, though, that this implementation breaks an important rule of thumb.

As mentioned in Chapter 2, Section 10, the C++ programmer is advised to write the constructors, destructor, and assignment operator and not rely on the default functions supplied by the compiler, which may do the wrong thing. Consider, for example, the following dummy class:

```
template<class T> class container{
      T* pointer;
    public:
     container(const container&c) :
        pointer(c.pointer ? new T(*c.pointer) : 0){
    } // copy constructor

    ~container(){
      delete pointer;
    } // destructor
};
```

This is considered good programming. The "container" object contains a pointer-to-'T' field named "pointer". The copy constructor is defined as follows. If the copied object has a meaningful value in the content of its "pointer" field, then the constructed object should also have the same value in the content of its own "pointer" field. This is done by the "new" command that allocates memory for a new 'T' object and returns its address.

The destructor is also defined properly. The "pointer" field is destroyed by the "delete" command, which automatically invokes the destructor of the 'T' class to destroy its content and free the memory occupied by it.

What would happen if the above functions weren't written? The default functions of the C++ compiler would be used instead. The situation would be as if the following functions were written:

```
    container(container&c):pointer(c.pointer){
    } // nothing is copied
    ~container(){} // *pointer is not destroyed
```

This is definitely not what we want. The constructor constructs nothing but another address for the same 'T' object. This is why the argument in the copy constructor cannot be declared "constant" as before: the 'T' object contained in it can now change through the newly constructed "container" object that also has its address.

The default destructor is also inappropriate. When a "container" object is destroyed, its "pointer" field is automatically destroyed by the compiler. This means that the variable that contains the address of the 'T' object is indeed destroyed. However, the 'T' object itself, although no longer accessible through this pointer, still occupies valuable memory.

This example shows clearly why one should write explicit constructors and destructor, making sure that they do the right thing. The same is true for the assignment operator.

After understanding the rules clearly, it is now time to break them. In fact, we do want the constructors to construct nothing but more references to existing nodes. Similarly, we want the destructor to destroy nothing but only reduce the number of these references. Therefore, the present constructors, destructor, and assignment operator should be written in an unusual way that violates the rule of thumb. The constructors and assignment operator should create no new nodes but merely new variables that contain the addresses of existing ones, thus providing more ways to access them. Similarly, the destructor should destroy no node that may still be used by some other finite element; it should only destroy the variable that contains its address, so it is no longer accessible through the destroyed finite element. Instead of actually removing the vertices, the destructor of the finite-element object only decreases their "sharingElements" fields to indicate that they are now being shared by fewer finite elements. Only those nodes whose "sharingElements" field is reduced to 0 are physically destroyed, because they belong to no finite element.

From the above discussion, it is clear that there is no point in deriving the finite element from the list of nodes. Indeed, in this case, most of the member functions in the "list" class in Chapter 3, Section 4 (including constructors, destructor, and assignment operator), would have to be overridden anyway along the above guidelines. Furthermore, in a list, the number of items is determined only in run time. This is unnecessary for a finite element, which always has a fixed number of vertices that never changes during the run. It makes more sense to define the finite element explicitly with no inheritance, with number of vertices 'N' that is fixed in compilation time:

```
template<class T, int N> class finiteElement{
    node<T>* vertex[N];
  public:
    finiteElement(){
      for(int i=0; i<N; i++)
        vertex[i] = new node<T>;
    } // default constructor

    finiteElement(node<T>&,node<T>&,node<T>&);
    finiteElement(finiteElement<T,N>&);
    const finiteElement<T,N>&
        operator=(finiteElement<T,N>&);
    ~finiteElement();

    node<T>& operator()(int i){
      return *(vertex[i]);
    } // read/write ith vertex
    const node<T>&
    operator[](int i)const{
      return *(vertex[i]);
    } // read only ith vertex
```

So far, we have declared the constructors, destructor, and assignment operator and defined a function that reads a particular vertex in the finite element. Next, we define functions that manipulate the "index" field in the individual vertices in the finite element:

```
    void resetIndices(){
      for(int i=0; i<N; i++)
        vertex[i]->setIndex(-1);
    } // reset indices to -1
```

The "resetIndices" function in the above code resets all the indices of the finite-element vertices back to their initial value -1. However, since the vertices are of class "node" whose "index" field is private, this change must be done through the "setIndex" function, which is declared public in the "node" class. The same is true for the "indexing" function that renumbers the indices of vertices in increasing order:

```
      void indexing(int&count){
        for(int i=0; i<N; i++)
          if(vertex[i]->getIndex()<0)
            vertex[i]->setIndex(count++);
      }  //  indexing the vertices
  };
```

This concludes the block of the "finiteElement" class. Next, we define the constructors declared above:

```
  template<class T, int N>
  finiteElement<T,N>::finiteElement(
      node<T>&a, node<T>&b, node<T>&c){
    vertex[0]=a.noSharingElement() ? new node<T>(a)  : &a;
    vertex[1]=b.noSharingElement() ? new node<T>(b)  : &b;
    vertex[2]=c.noSharingElement() ? new node<T>(c)  : &c;
    for(int i=0; i<N; i++)
      vertex[i]->moreSharingElements();
  }  //  constructor

  template<class T, int N>
  finiteElement<T,N>::finiteElement(finiteElement<T,N>&e){
    for(int i=0; i<N; i++){
      vertex[i] = e.vertex[i];
      vertex[i]->moreSharingElements();
    }
  }  //  copy constructor
```

Next, we define the assignment operator declared above:

```
  template<class T, int N>
  const finiteElement<T,N>&
  finiteElement<T,N>::operator=(finiteElement<T,N>&e){
    if(this != &e){
      for(int i=0; i<N; i++)
        if(vertex[i]->lessSharingElements())
          delete vertex[i];
      for(int i=0; i<N; i++){
        vertex[i] = e.vertex[i];
        vertex[i]->moreSharingElements();
      }
    }
    return *this;
  }  //  assignment operator
```

Note that the arguments in the constructors and assignment operator cannot be declared constant because the addresses of nodes in them are assigned to nonconstant pointers. Recall that pointer-to-constant cannot be assigned to pointer-to-nonconstant, because its constant content could then be changed through the latter.

Next, we define the destructor declared above:

```
template<class T, int N>
finiteElement<T,N>::~finiteElement(){
  for(int i=0; i<N; i++)
    if(vertex[i]->lessSharingElements())delete vertex[i];
} //  destructor
```

Next, we define the binary "operator<" that checks whether a given node is a vertex in a given finite element. If the node 'n' is a vertex in the finite element 'e', then "n<e" returns the index of 'n' in the "vertex" field in 'e' plus 1. Otherwise, it returns 0.

The "operator<" is implemented by comparing the address of 'n' to the addresses of the vertices in 'e'. Only if the address is the same, that is, 'n' actually coincides with one of the vertices in 'e', does "n<e" return a nonzero value:

```
template<class T, int N>
int
operator<(const node<T>&n, const finiteElement<T,N>&e){
  for(int i=0; i<N; i++)
    if(&n == &(e[i]))return i+1;
  return 0;
} // check whether a node n is in a finite element e
```

Note that the '<' symbol takes here a totally different meaning from its usual meaning.

Finally, we define the function that prints the vertices in the finite element:

```
template<class T, int N>
void print(const finiteElement<T,N>&e){
  for(int i=0; i<N; i++)
    print(e[i]);
} // printing a finiteElement
```

The "typedef" command is used below for short and convenient notation in the code. This command makes two different terms identical, so they mean exactly the same thing in every code:

```
typedef finiteElement<point,3> triangle;
typedef finiteElement<point3d,4> tetrahedron;
```

Here, "triangle" is short for a "finiteElement" object with three vertices in the Cartesian plane, and "tetrahedron" is short for a "finiteElement" object with four vertices in the three-dimensional Cartesian space. These notations will be useful in what follows.

13.4 The Mesh Object

As discussed in Chapter 4, Section 9, the unstructured mesh must be implemented in a flexible way that allows triangles to be inserted and removed easily and efficiently. The suitable data structure is the connected list in Chapter 3, Section 5.

It is more natural to think of the mesh as a collection of finite elements rather than a container that has finite elements in it. The "is a" approach in Chapter 2, Section 19, thus seems more appropriate than the "has a" approach.

In light of the above points, the "mesh" template class is derived from a connected list of objects of type 'T' that will be specified later (in compilation time) as some kind of finite element. In most of the present applications, where the finite elements are triangles, the "mesh" object is actually a connected list of triangles.

The hierarchy of objects used to implement the mesh is displayed in Figure 13.1. The "mesh" object at the highest level is a connected list of "finiteElement" objects. The "finiteElement" object in the intermediate level is a list of (pointers to) "node" objects. Finally, the "node" object at the lowest level contains a "point" object to store the technical information about its geometrical location in the Cartesian plane.

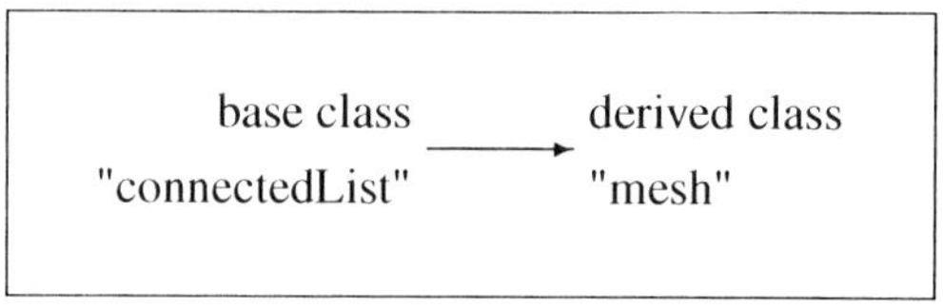

Figure 13.2. *Schematic representation of inheritance from the base class "connectedList" to the derived class "mesh".*

In Figure 13.2, we show schematically how the "mesh" class is derived from the base "connectedList" class. As discussed above, this approach allows us to add more finite elements to the mesh and remove existing ones from it by simply using the "append", "dropNextItem", "dropFirstItem", "insertNextItem", and "insertFirstItem" functions available in the base "connectedList" class.

The constructor in the "mesh" class uses an argument of type "finiteElement", which is copied to the first item in the underlying connected list of finite elements. As usual, this argument is passed by reference to save extra calls to the copy constructor of the "finiteElement" class. However, this argument cannot be declared constant because it is copied to another "finiteElement" object, namely, the first finite element in the mesh. As explained in Section 13.3, this copying increases the "sharingElements" fields in the vertices of the copied object, so it must be nonconstant.

Here is the block of the "mesh" template class:

```
template<class T>
class mesh : public connectedList<T>{
public:
  mesh(){
  } // default constructor
```

```
    mesh(T&e){
      item = e;
    } //  constructor
```

The following member functions are only declared here and defined later on. The function "indexing()" that assigns indices to nodes will be defined soon. The "refine" and "refineNeighbor" functions will be defined in Chapter 14, Sections 5 and 6:

```
    int indexing();
    void refineNeighbor(node<point>&,node<point>&,node<point>&);
    void refine(const dynamicVector<double>&,double);

};
```

This concludes the block of the "mesh" class. Next, we define the member functions that were only declared above but not defined. We start with the "indexing()" function.

The "indexing" function assigns indices to the nodes in the mesh in increasing integer order. For this purpose, it uses the "indexing" member function of the "finiteElement" class, which assigns indices in increasing order to the vertices of an individual finite element:

```
template<class T>
int mesh<T>::indexing(){
  for(mesh<T>* runner = this;
      runner; runner=(mesh<T>*)runner->next)
    runner->item.resetIndices();
```

In this short loop, the triangles in the mesh are scanned by the pointer "runner", and the indices of their vertices are set to -1 using the "resetIndices()" member function of the "finiteElement" class. In this loop, "runner" is advanced to the address of the next item in the underlying connected list. This address is stored in the field "next" inherited from the base "connectedList" class. However, this field is of type pointer-to-connectedList, so it must be converted to type pointer-to-mesh before it can be assigned to "runner". This is done by adding the prefix "(mesh*)".

A similar loop is used to index the vertices of the triangles in the mesh. This is done by the "indexing()" member function of the "finiteElement" class, which indexes vertices whose "index" field has not yet been set and is still -1:

```
  int count=0;
  for(mesh<T>* runner = this;
      runner; runner=(mesh<T>*)runner->next)
    runner->item.indexing(count);
  return count;
} //  indexing the nodes in the mesh
```

The above function also has an output: it returns the number of nodes in the mesh. This information can be quite helpful. In addition, the indices assigned to the nodes in the above function are used to refer to the corresponding unknowns in the stiffness system.

Here is a simple application that constructs and prints a mesh of three triangles. Note that when the "print" function is called, the mesh is interpreted as a connected list, and the "print" function of Chapter 3, Section 5, is invoked:

```
int main(){
  node<point> a(point(1,1));
  node<point> b(point(2,2));
  node<point> c(point(2,0));
  node<point> d(point(3,1));
  node<point> e(point(3,3));
  triangle t1(a,b,c);
  triangle t2(t1(1),t1(2),d);
  triangle t3(t2(0),t2(2),e);
```

Now, we have constructed the three required triangles. Note that, once a node is placed in a triangle, it is referred to as a vertex of this triangle rather than by its original name. This way, when an existing vertex of some triangle is also used as a vertex in a new triangle, its "sharingElements" field increases to reflect the fact that it is now shared by more triangles.

Next, we use the above triangles to form the required mesh:

```
  mesh<triangle> m(t1);
  m.append(t2);
  m.append(t3);
  t1.~triangle();
  t2.~triangle();
  t3.~triangle();
  m.indexing();
  print(m);
  return 0;
}
```

13.5 Assembling the Stiffness Matrix

We are now ready to construct the stiffness matrix, as in Chapter 12, Section 5. Thanks to the above classes, the implementation is straightforward and transparent. Constructing the stiffness matrix is also called assembling, because the contributions from the different triangles in the mesh are assembled to form the required matrix elements.

The assembling is done by scanning the triangles in the mesh. When a triangle e with vertices i, j, and k as in Figure 12.1 is encountered, the corresponding elements $A_{i,j}$, $A_{j,k}$, and $A_{i,k}$ in the stiffness matrix are incremented by the corresponding contribution from the integral over e, calculated as in Chapter 12, Section 5.

The loop that scans the triangles in the mesh uses a pointer-to-mesh variable named "runner". Since the "mesh" class is derived from the "connectedList" class, it would seem that "runner" can jump from item to item in it simply by writing "runner = runner->next". However, since the "next" field in the base "connectedList" class is of type pointer-to-connectedList rather than pointer-to-mesh, it must be converted to pointer-to-mesh before its value can be assigned to "runner". This conversion is done explicitly by adding the prefix "(mesh*)" just before the "next" variable. This conversion returns the address in "next" interpreted as the address of the "mesh" object rather than merely the address of the "connectedList" object.

Usually, this conversion is considered risky, because in theory "next" can point to an object of class "connectedList" or any other class derived from it, with completely different member functions that could produce the wrong results. Fortunately, here the "next" field belongs to a "mesh" object and, therefore, must also point to a "mesh" object, so no risk is taken.

For each triangle scanned in the above loop, those elements in the stiffness matrix that couple vertices in it are incremented. In order to increment the correct matrix elements, the "index" field in these vertices must be used. This field contains the index of the vertex in the vector of unknowns x in the stiffness system.

Although the "index" field is private in the "node" class in Section 13.2, it can still be read as follows. Assume that we are at the middle of the loop, when "runner" points to the partial mesh consisting of the remaining triangles that have not yet been scanned. This partial mesh is then accessed simply by writing "*runner". Then, the "operator()" of the base "connectedList" class is invoked to access the first item in it simply by writing "(*runner)()". Then, the "operator[]" of the "finiteElement" class is invoked to read the 'i'th vertex in the triangle by writing "(*runner)()[i]". Finally, the "getIndex" function of the "node" class is invoked to get the required index of the 'i'th vertex. This index is denoted by the capital letter 'I' in the code below.

The assembling code is of a rather high level, as it implements the mathematical definition of the stiffness matrix. Therefore, we may assume that some helpful low-level objects are available, although they have not yet been actually implemented. In particular, we assume that a class "stiffnessMatrix" is available, with an "operator()" that returns a nonconstant reference to the specified matrix element. In other words, if 'A' is a "stiffnessMatrix" object, then "A(I,J)" refers to the "(I,J)"th element in it. These assumptions are sufficient for now; later on, the imaginary object "stiffnessMatrix" will be replaced by a more suitable object to store the stiffness matrix (Chapter 16, Section 5).

For simplicity, we consider only the case with constant diffusion coefficients $P \equiv Q \equiv 1$ (the Poisson equation). As discussed in Chapter 12, Section 5, the contribution to the stiffness matrix from the triangle e is based on integration over the reference triangle r in Figure 12.2. The integrand is the gradient of a nodal function in r times S_e^{-1} times S_e^{-t} times the gradient of a nodal function in r. The gradients of these nodal functions ($\phi_{0,0}$, $\phi_{1,0}$, and $\phi_{0,1}$) are stored in the three "point" objects "gradient[0]", "gradient[1]", and "gradient[2]", respectively:

```
point gradient[3];
gradient[0] = point(-1,-1);
gradient[1] = point(1,0);
gradient[2] = point(0,1);
```

The 2×2 matrix S_e in Chapter 12, Section 5, is stored in a "matrix2" object named 'S'. The first column in 'S' should be $i-k$ in Figure 12.1, and the second column should be $j-k$. In order to define these columns correctly, one must have access to the geometric location of the vertices in the Cartesian plane. This information is obtained as follows: "runner" points initially to the entire mesh and is then advanced gradually to point to subsequent submeshes that contain fewer and fewer triangles. In other words, "*runner" is the connected list whose first item is the current triangle under consideration. Using the "operator()" of the base "connectedList" class, this triangle is obtained as "(*runner)()". Now, by applying "operator[]" in the "finiteElement" class, the 'i'th vertex in this triangle is obtained as "(*runner)()[i]". Finally, by applying the "operator()" of the "node" class, the required location of this vertex in the Cartesian plane is obtained as "(*runner)()[i]()". The matrix 'S' is thus defined as follows:

```
for(const mesh<triangle>* runner = &m; runner;
    runner=(const mesh<triangle>*)runner->readNext()){
  matrix2 S((*runner)()[1]() - (*runner)()[0](),
            (*runner)()[2]() - (*runner)()[0]());
```

We are now ready to calculate the integrand as the gradient of a nodal function in r times S_e^{-1} times S_e^{-t} times the gradient of a nodal function in r. For this purpose, we use the "det()", "inverse()", and "transpose()" functions implemented at the end of Section A.2 of the Appendix:

```
  matrix2 Sinverse = inverse(S);
  matrix2 weight =
      abs(det(S)/2) * Sinverse * transpose(Sinverse);
```

The contributions from the current triangle in the loop are calculated in a nested loop over its vertices. In particular, the vertices denoted by 'i' and 'j' contribute to the "(I,J)"th matrix element, where 'I' and 'J' are the corresponding row indices in the stiffness matrix:

```
  for(int i=0; i<3; i++)
    for(int j=i; j<3; j++){
      int I = (*runner)()[i].getIndex();
      int J = (*runner)()[j].getIndex();
      A(I,J) += gradient[j]*weight*gradient[i];
    }
} // assembling the stiffness matrix
```

Actually, in the nested loop over the vertices of a triangle, only pairs 'i' and 'j' for which 'i' is less than or equal to 'j' are used; contributions from pairs for which 'i' is greater than 'j' will be added to the stiffness matrix A later by substituting

$$A \leftarrow A + A^t - \text{diag}(A).$$

Indeed, thanks to the symmetry in the original bilinear form $a(\cdot,\cdot)$, the contribution from e to the matrix element $A_{I,J}$ is the same as the contribution to $A_{J,I}$. It is thus unnecessary to calculate it twice; it would be more efficient to add it only to $A_{I,J}$ now, and postpone its addition to $A_{J,I}$ to the next part of the code, where the transpose A^t will be added as above.

13.6 Exercises

1. Assume that the numerical solution is contained in a dynamic vector v, with dimension equal to the number of nodes in the mesh. (The "index" field in a "node" object is the same as the index of the component in v that contains the value at that node.) Assume also that the solution has a singularity at the origin, so one is interested in the maximum solution away from it. Write a function "maxNorm" that calculates the maximum modulus of the components in v that correspond to nodes of distance 0.1 or more from the origin. (You may assume that the function is a member of the "mesh" class.) The solution can be found in Section A.8 of the Appendix.

2. Write a function "refineBoundary" that successively refines a coarse mesh as in Figure 12.9 until it approximates the circular boundary well (Figure 12.10). You may assume that the function is a member of the "mesh" class. The solution can be found in Section A.8 of the Appendix.

3. Assume that the right-hand side of the PDE is $F(x, y) \equiv 1$. Write a function that takes a "mesh" object as an argument and returns the right-hand-side vector in the stiffness system. (You may use the polynomials of two variables in Chapter 5, Sections 13 and 14.)

4. Modify the code in Section 13.5 so that it also calculates and assembles the contribution to the stiffness matrix from the mixed boundary conditions (see Chapter 12, Section 6).

5. Apply your code to the uniform mesh in Figure 12.3 and verify that the stiffness matrix is indeed symmetric and diagonally dominant.

6. Apply your code to the mesh that approximates a circle as in Figure 12.10. Verify that the stiffness matrix is indeed tridiagonal as in Chapter 12, Section 9.

7. Modify the above code to assemble the stiffness matrix for the PDE

$$-u_{xx}(x, y) - u_{yy}(x, y) + u(x, y) = F(x, y).$$

   (You may use the polynomials of two variables in Chapter 5, Sections 13 and 14 to assemble the contribution from the free term in the PDE.) Verify that the stiffness matrix is indeed symmetric.

8. Modify the above code to assemble the stiffness matrix for the PDE

$$-u_{xx}(x, y) - u_{yy}(x, y) + u_x(x, y) + u_y(x, y) = F(x, y).$$

   Verify that the stiffness matrix is indeed nonsymmetric.

9. Write the analogous code that assembles the stiffness matrix for a three-dimensional finite-element mesh. Use the "matrix3" and "mesh<tetrahedron>" objects and your answers to the exercises at the end of Chapter 12.

Chapter 14
Adaptive Mesh Refinement

In this chapter, we present an adaptive approach toward the construction of the finite-element mesh. This approach uses an initial coarse mesh and refines it until sufficient resolution is obtained wherever needed. In each refinement step (or level), some triangles in the mesh are split (refined), yielding extra accuracy there. The decision of where to refine is made adaptively, using the numerical solution that has just been computed on the current mesh. The algorithm is well implemented in C++ using the above objects.

14.1 Local Refinement

Constructing a finite-element mesh may be particularly difficult. The domain is often complicated and irregular, and the mesh must be highly nonuniform to capture different kinds of phenomena in it. Constructing it manually by writing a special code line for each node is completely impractical. An automatic process that will do the job iteratively or recursively is clearly necessary.

The automatic process uses a sequence of refinement steps or levels. The process starts from a coarse mesh, with coarse (big) triangles. At a particular refinement step, each coarse triangle is divided into two smaller triangles. This produces the next (finer) refinement level. The process continues iteratively, producing finer and finer meshes. At the final (finest) refinement level, the triangles are so small that every subtle variation in the solution is well captured. Therefore, the finest mesh is used to calculate the required numerical solution for the original boundary-value problem.

The above process actually uses global refinement. Every big triangle in the mesh in a particular refinement level is divided, regardless of its place in the domain. Unfortunately, with this approach, the number of nodes and triangles may grow exponentially with the number of refinement levels, producing prohibitively large meshes in terms of both storage and computation time. A more economic approach is clearly necessary.

In local refinement, not every triangle in a particular refinement level is divided. In fact, only those triangles that satisfy some refinement criterion are divided. This criterion may specify their location in the domain; in this case, only triangles in some subdomain are subsequently refined, whereas all the others remain coarse. This subdomain may, for example,

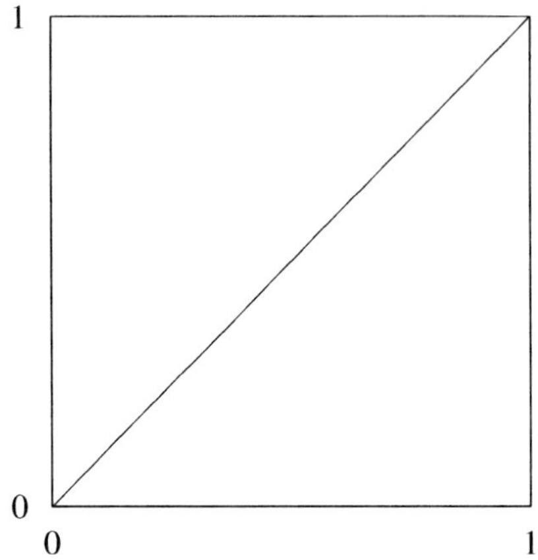

Figure 14.1. *The initial coarse mesh that approximates the square poorly.*

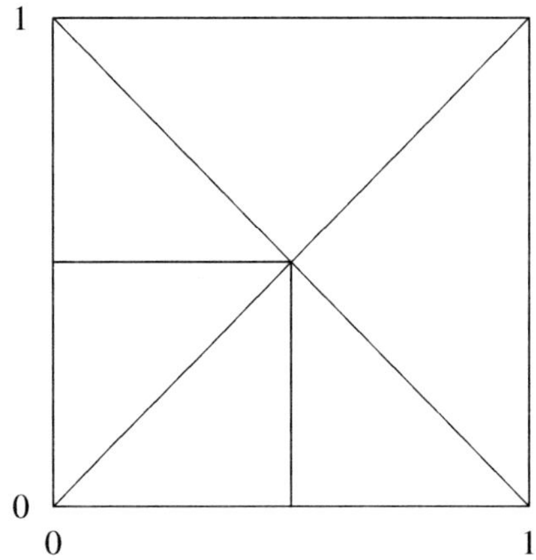

Figure 14.2. *The second, finer, mesh resulting from one step of local refinement. It is assumed here that extra accuracy is needed only at the origin.*

contain the neighborhood of an irregular boundary, as well as other places where the solution is expected to have sharp variation. Elsewhere, where the solution is probably smooth, relatively coarse triangles can be used, which saves a lot of storage and computation time.

Thus, in local refinement, only those triangles where extra accuracy is absolutely necessary are refined, while the others remain coarse. More specifically, a triangle should be refined only if it is located near an irregularity in the boundary or solution. In Figures 14.1 to 14.3, for example, it is assumed that the solution is irregular (has unusually sharp variation) only at the origin, so extra refinement is needed only there. With local refinement, the number of nodes (and triangles) grows only moderately with the number of refinement levels yet produces sufficient accuracy wherever needed.

The above local-refinement criterion is still manual: it depends on the properties of the original boundary-value problem as they are interpreted by us humans. Unfortunately,

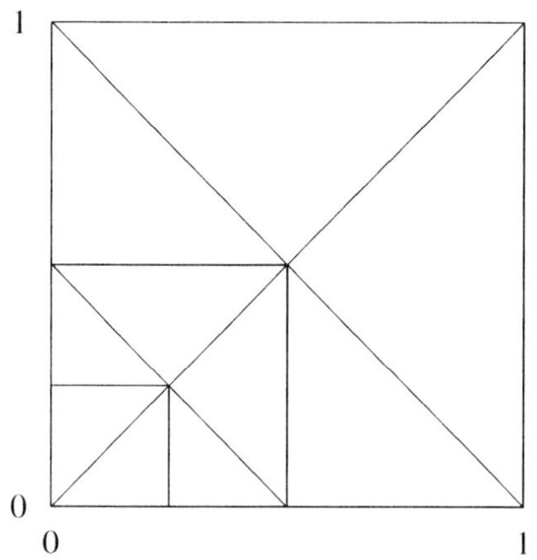

Figure 14.3. *The third, yet finer, mesh resulting from the second local-refinement step. It is assumed here that extra accuracy is needed only at the origin.*

these properties, let alone their interpretation, are often unclear, which could lead to errors of judgment in defining the proper refinement criterion. A more automatic refinement criterion is also necessary, as discussed below.

14.2 Adaptive Refinement

Local refinement requires a criterion to tell us where extra refinement is needed. If we had known in advance where the solution to the PDE had had particularly large variation and possible irregularities, then we would refine only there. Unfortunately, this information is in general unavailable. One should thus use the best approximate solution available: the coarse numerical solution.

In order to make a clever decision on where to refine, one should form and solve the stiffness system on the coarse (unrefined) mesh. The resulting (coarse) numerical solution can then be used to help one decide where to refine. More specifically, further refinement is required only where the variation in the coarse solution is large. In this area, more nodes are needed to capture the behavior of the solution of the original PDE. In the rest of the domain, where only a small variation is observed in the coarse solution, the solution to the PDE is probably smooth, so no extra nodes are needed, and the triangles can remain coarse.

Using the coarse solution to decide where to refine is called adaptive refinement. Indeed, in each refinement level, the mesh adapts itself to the nature of the numerical solution computed on it and refines locally according to its special properties. In what follows, we present the detailed algorithm, including the precise criterion for refinement.

14.3 The Adaptive-Refinement Algorithm

The adaptive-refinement algorithm is displayed schematically in Figure 14.4. The precise definition of this algorithm is as follows. (The algorithm uses some small predetermined threshold, say 0.01.)

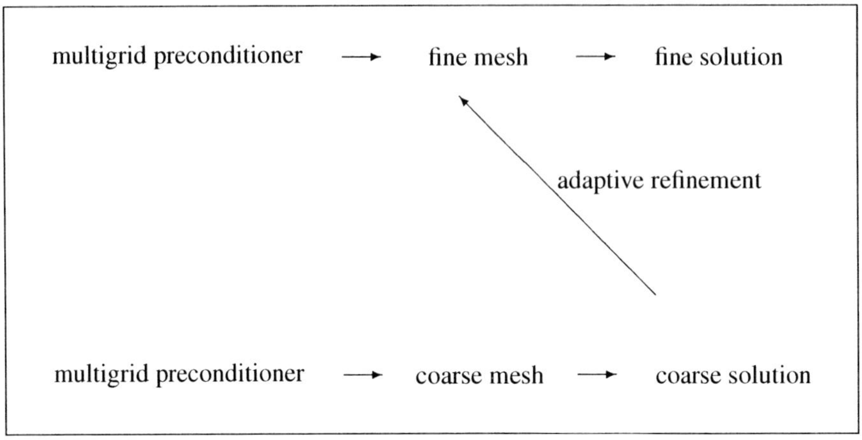

Figure 14.4. *The adaptive-refinement algorithm: the numerical solution obtained by the multigrid preconditioner at a particular mesh is used to refine it further and produce the next level of refinement.*

Algorithm 14.1.

1. *Let T be the initial coarse finite-element triangulation (the set of triangles in the mesh).*

2. *Construct the stiffness matrix A and the right-hand side f corresponding to T.*

3. *Solve the stiffness system*

$$Ax = f$$

for the vector of unknowns x.

4. *Let E be the set of edges in T.*

5. *Scan the edges in E one by one in some order. For every edge e encountered in this scanning, do the following:*

   - *Denote the endpoints of e by i and j (see Figure* 14.5*).*
   - *If e lies in the interior of the domain, then there are two triangles that share it. If e lies next to the boundary, then there is only one triangle that uses it. Let*

$$t = \Delta(i, j, k)$$

   be the triangle with vertices i, j, and k that uses e (see Figure 14.5*).*
   - *If*

$$|x_i - x_j| > threshold,$$

   then divide t into the two triangles

$$t_1 = \Delta(i, (i + j)/2, k),$$
$$t_2 = \Delta(j, (i + j)/2, k),$$

where $(i + j)/2$ is the midpoint in e (see Figure 14.6). In other words, include t_1 and t_2 in T instead of t. Do the same to the other triangle that shares e, if it exists.

6. *If the resolution of the mesh is not yet sufficiently high, then go back to step 2.*

7. *Use x as the numerical solution of the boundary-value problem.*

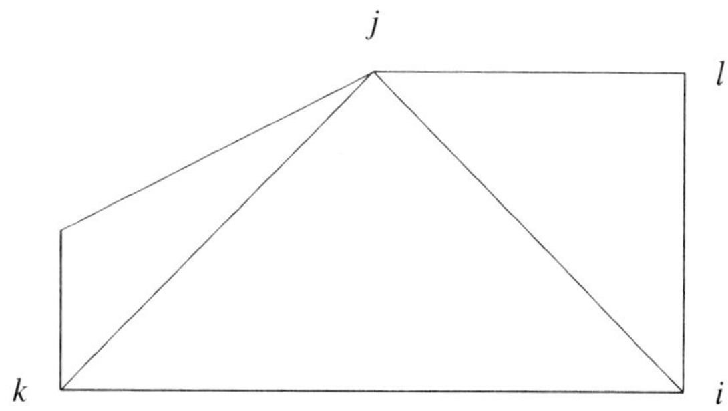

Figure 14.5. *The original coarse finite-element mesh.*

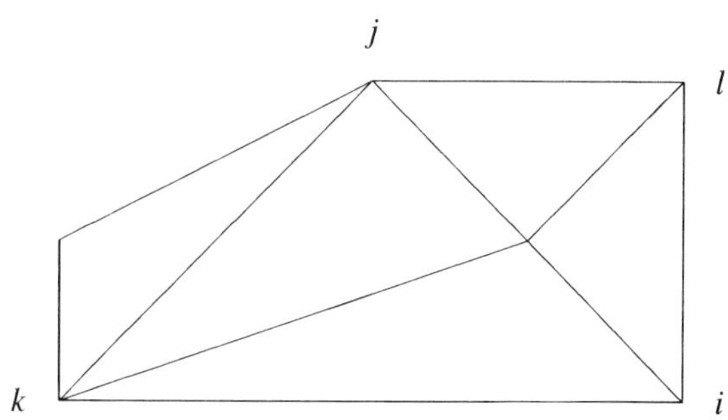

Figure 14.6. *The adaptively refined finite-element mesh.*

14.4 Preserving Conformity

The finite-element meshes considered in this book are conformal in the sense that if a node lies on an edge of some triangle, then it must also be a vertex of this triangle. In other words, a node cannot lie on an edge unless it is an endpoint of this edge. Here, we show how this important property is preserved in mesh refinement.

The algorithm in Section 14.3 has the important property of preserving conformity: if the initial (coarse) mesh is conformal, then all the subsequent finer meshes are conformal as well, including the final (finest) mesh that is eventually used in the actual numerical modeling. For this reason, once the node $(i + j)/2$ in Figure 14.6 is added to the mesh in the refinement step, both triangles that share the edge leading from i to j must be divided, as is indeed done in Figure 14.6. In fact, if one of these triangles (say, $\triangle(i, j, l)$) were not divided, then it would contain a node $(i + j)/2$ that is not a vertex of it, in violation of the conformity rule. By insisting that this triangle also be divided as in Figure 14.6, we guarantee that conformity is indeed preserved.

Let us explain why conformity is so important. The finite-element space in Chapter 12, Section 2, contains all the functions that are continuous in the entire mesh and linear in each particular triangle in it. If, for instance, $\triangle(i, j, l)$ had not been divided in the refinement step, then the nodal basis function $\phi_{(i+j)/2}$ would be either discontinuous (across the edge leading from i to j) or nonlinear in it. Preserving conformity by dividing $\triangle(i, j, l)$ as well guarantees that $\phi_{(i+j)/2}$ is indeed a continuous and piecewise-linear function, as a proper nodal basis function should be. This way, the function space on the coarse mesh is contained in the function space on the fine mesh, which implies that the fine mesh indeed provides a better numerical approximation to the original PDE.

14.5 Preserving Conformity in Practice

Here, we show how conformity is preserved in actual practice. Recall that the finite-element mesh is implemented as a connected list of triangles (Chapter 13, Section 4). In order to scan the edges in the mesh as in the algorithm in Section 14.3, one must scan the triangles themselves (or the items in the connected list) and consider for refinement every edge (pair of vertices) in each of them. When an edge that is an appropriate candidate for refinement is found in some triangle, its midpoint is considered as a new node and used as a vertex in the two new triangles that are added to the mesh instead of this triangle. This midpoint ($(i + j)/2$ in Figure 14.6) must also be used to divide the adjacent triangle that shares the same edge that has just been divided, or the resulting mesh would not be conformal. This process is displayed in Figure 14.7.

The task of dividing the adjacent triangle cannot wait; it must be completed immediately, before the midpoint $(i + j)/2$ becomes unavailable. Indeed, because a triangle is a triplet of pointers-to-nodes rather than nodes, the very object $(i + j)/2$ must also be used to divide the adjacent triangle. It would be wrong to defer the division of the adjacent triangle until it is scanned in the loop, because then a new "node" object would have to be constructed to represent the midpoint, resulting in two different "node" objects representing the same node $(i + j)/2$.

This is the motivation for the "refineNeighbor" function implemented below. This function takes three "node" arguments that represent i, j, and $(i + j)/2$ in Figure 14.6 and uses them to search, find, and divide the adjacent triangle.

Because the "node" object that represents the midpoint $(i+j)/2$ already exists, it must also be used in the division of the adjacent triangle found in the "refineNeighbor" function. Therefore, it must be passed to it by reference, along with the nodes i and j. In fact, these three nodes must be passed by reference-to-nonconstant-node, because they change when the adjacent triangle is divided. Indeed, their "sharingElements" fields change when the coarse adjacent triangle is replaced by two fine triangles.

The adjacent triangle is found by using the "operator<" in Chapter 13, Section 3, which checks whether or not a node is a vertex in a triangle. If the node is indeed a vertex in the triangle, then "operator<" returns its index in the list of vertices of that triangle plus one. Otherwise, it returns zero.

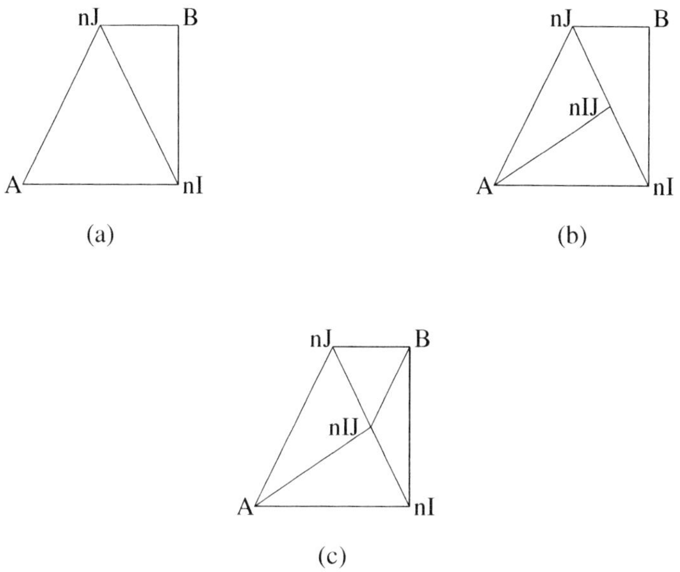

Figure 14.7. *The coarse triangle with vertices A, nI, and nJ* (a) *is divided into two smaller triangles by the new line leading from A to nIJ* (b). *In order to preserve conformity, its neighbor on the upper right is also divided by a new line leading from nIJ to B* (c) *in the "refineNeighbor()" function.*

The "operator<" function is called twice for each triangle: if both nodes i and j are vertices in some triangle, then it must be the required adjacent triangle. The third vertex is then located by straightforward elimination, because it must be the vertex that is neither i nor j. The adjacent triangle is then replaced by two smaller triangles, denoted by "t1" and "t2".

The search for the adjacent triangle uses the recursive structure of the mesh (which is actually a connected list of triangles). In fact, if the first item in the connected list is proved not to be the adjacent triangle, then the "refineNeighbor" function is called recursively to check the rest of the triangles in the mesh. For this purpose, it is applied to the contents of the "next" field in the "mesh" object.

The "next" field, however, is inherited from the base "connectedList" class as pointer-to-connectedList rather than pointer-to-mesh. Therefore, it must be converted explicitly into pointer-to-mesh before the "refineNeighbor" function can be applied recursively to it. Usually, this is a risky practice, because in theory "next" can point to a "connectedList" object or any other object derived from it, with a completely different "refineNeighbor" function that can do completely different things. Fortunately, here "next" must point to a "mesh" object, so the recursive call is safe.

Here is the actual implementation of the "refineNeighbor" function:

```
void mesh<triangle>::refineNeighbor(node<point>&nI,
          node<point>&nJ, node<point>&nIJ){
  int ni = nI < item;
  int nj = nJ < item;
```

Here, the arguments "nI" and "nJ" represent the nodes i and j in Figure 14.6, respectively. If they are both vertices in the first triangle in the mesh, "item", then the integers "ni" and "nj" take their indices in the list of vertices in this triangle plus 1. These integers are now used to identify the third vertex in this triangle:

```
    if(ni&&nj){
```

Here, we enter the "if" block that checks whether "item" is indeed an adjacent triangle that uses "nI" and "nJ" as its own vertices. All that is left to do is to identify the third vertex in it and divide it into two smaller triangles. This is done as follows. First, we identify the integer "nk", the index of the third vertex in the list of vertices in "item":

```
        ni--;
        nj--;
        int nk = 0;
        while((nk==ni)||(nk==nj))
          nk++;
```

Next, the integer "nk" is used to form two small triangles "t1" and "t2" to replace the adjacent triangle "item":

```
        triangle t1(nI,nIJ,item(nk));
        triangle t2(nJ,nIJ,item(nk));
        insertNextItem(t2);
        insertNextItem(t1);
        dropFirstItem();
    }
    else
```

Finally, we consider the possibility that "item" is not the adjacent triangle. In this case, we apply the "refineNeighbor" function recursively to the next triangle in the mesh to check whether or not it is the required adjacent triangle. However, the "next" field must be converted from pointer-to-connectedList into pointer-to-mesh before the recursive call can be made:

```
        if(next)
          ((mesh<triangle>*)next)->refineNeighbor(nI,nJ,nIJ);
  }  //  refine the neighbor of a refined triangle
```

14.6 Mesh Refinement in Practice

In this section, we implement the refinement step in the adaptive-refinement algorithm using the "mesh" class of Chapter 13, Section 4. Because the "mesh" object is actually a connected list of triangles, it is only natural to use recursion to complete this task.

The "refine" member function of the "mesh" class completes a single refinement step on the current "mesh" object. This function is outlined as follows. First, the first triangle in the mesh (the first item in the underlying connected list) is considered for refinement; that is, it is checked if there is any significant jump in the numerical solution at its vertices. If such a jump from vertex i to vertex j is indeed found, then the triangle is divided into two smaller triangles: "t1", with vertices i, $(i+j)/2$, and k, and "t2", with vertices j, $(i+j)/2$, and k (see Figure 14.6). The "refineNeighbor" function is then used to search the rest of the mesh and find and divide the adjacent triangle, if it exists. The first triangle "item" is then replaced by the two "fine" triangles "t1" and "t2". The "refine" function is then called recursively to consider for refinement the rest of the triangles in the mesh. In particular, it also checks whether the edge leading from i to k in "t1" or the edge leading from j to k in "t2" should also be divided. The new edge that emerges from $(i+j)/2$ is not divided in this refinement step any more, because the "index" field in this new node is still -1, so it is excluded from any further refinement in this step.

The key factor in the function is the construction of the node "itemij" representing the midpoint $(i+j)/2$. This is done by writing

```
node<point> itemij = (item[i]()+item[j]())/2.;
```

Here "item[i]" and "item[j]" are two vertices in the first triangle "item". Using the "operator()" of the "node" class in Chapter 13, Section 2, we have that "item[i]()" and "item[j]()" are the "point" objects representing the points i and j in Figure 14.6, respectively. The "point" object that represents the midpoint between i and j is then converted implicitly into the required node object "itemij".

Each triangle 't' in the mesh contains three vertices, which can be accessed by "t(0)", "t(1)", and "t(2)". This access is made by the "operator()" of the "finiteElement" class in Chapter 13, Section 3, that returns a nonconstant reference-to-node, because the "sharingElements" fields in the nodes may change in the refinement step. It is assumed that the largest edge in the triangle is the edge leading from "t(0)" to "t(2)". In order to have high regularity, this edge should be considered for division before the other edges, so that the fine triangles produced by this division will have moderate angles. This approach is indeed used in the present implementation. The vertices in the fine triangles "t1" and "t2" are again ordered in the same way, so that the next refinement step also preserves regularity. In other words, the above assumption holds inductively for the recursive call to the "refine" function, which guarantees high regularity.

Here is the detailed implementation of the "refine" function:

```
void mesh<triangle>::refine(const dynamicVector<double>&v,
          double threshold){
  for(int i=0; i<3; i++)
    for(int j=2; j>i; j--)
      if((item[i].getIndex() >= 0)&&
       (item[j].getIndex() >= 0)&&
       (abs(v[item[i].getIndex()] -
       v[item[j].getIndex()])>threshold)){
```

We are now in the middle of a nested loop over the vertices in the first triangle in the mesh, "item". By now, we have found a pair 'i' and 'j' that represents an edge that should be divided according to the refinement criterion and has not been divided in the present refinement step. We proceed to define the midpoint in this edge:

```
            node<point> itemij = (item[i]()+item[j]())/2.;
```

and the third vertex in the triangle, numbered by 'k':

```
            int k=0;
            while((k==i)||(k==j))
              k++;
```

These points are then used to construct the two halves of the triangle "item":

```
            triangle t1(item(i),itemij,item(k));
            triangle t2(item(j),t1(1),item(k));
```

The smaller triangles "t1" and "t2" are first used to find the triangle adjacent to "item" and divide it:

```
            if(next)
              ((mesh<triangle>*)next)->
                refineNeighbor(item(i),item(j),t1(1));
```

Then, they are placed in the mesh instead of the original triangle "item":

```
            insertNextItem(t2);
            insertNextItem(t1);
            dropFirstItem();
```

By now, we have divided the first triangle in the mesh and its neighbor, provided that the refinement criterion holds. The mesh has therefore changed, and new triangles have been introduced, which need to be considered for refinement as well. Therefore, we have to call the "refine()" function recursively here. This call can only divide edges that do not use the new node "itemij", whose "index" field is -1:

```
            refine(v, threshold);
            return;
          }
```

Finally, if the first triangle in the mesh does not satisfy the refinement criterion and remains unrefined, then the "refine()" function is applied recursively to the rest of the mesh contained in the "next" variable, after this variable is converted explicitly from pointer-to-connectedList to pointer-to-mesh:

```
      if(next)
        ((mesh<triangle>*)next)->refine(v, threshold);
  } // refinement step
```

14.7 Automatic Boundary Refinement

In the above discussion, it is assumed that the initial coarse mesh approximates the boundary sufficiently well, so extra refinement is needed only in the interior of the domain. Unfortunately, this is a highly unrealistic assumption. In most practical cases, the boundary is curved and irregular and is approximated rather poorly by the coarse mesh (see, e.g., Figure 12.9). It is thus desirable that the refinement process refine not only in the interior of the domain but also near its curved boundary.

Here we modify the adaptive-refinement algorithm to do just this. Let us first illustrate how it works in a circle. In Figure 14.8, we display the coarse mesh that is passed to the adaptive-refinement algorithm as input. This mesh contains only four triangles, which provide a rather poor approximation to both the interior and the boundary of the domain. Thus, it has to be refined not only in the sense of dividing the existing triangles but also in the sense of adding more triangles next to the circular boundary.

This is indeed done in the finer mesh in Figure 14.9. For simplicity, this mesh is produced under the assumption that the coarse numerical solution changes significantly only between the points nI and nJ in Figure 14.8, so a midpoint nIJ must be added between them. This implies that the upper-right triangle in Figure 14.8 should be divided into two smaller triangles, as is indeed done in Figure 14.9. However, this is not the end of the story: two extra triangles are also added between the edge leading from nI to nJ and the circular boundary. This way, the approximation improves not only in the interior of the domain but also at its boundary.

14.8 Implementation of Automatic Boundary Refinement

Let us show how easy it is to implement automatic boundary refinement in the present framework. Indeed, all that has to be done is to modify the "refineNeighbor()" function in Section 14.5. This function refines the neighbor (edge-sharing) triangle of a refined triangle. But what if there is no neighbor triangle? This implies that the edge under consideration must be a boundary edge, namely, an edge that lies next to the boundary. Automatic boundary refinement requires that two extra triangles should then be added between it and the curved boundary.

Thus, one only needs to detect the boundary edges and add two small triangles between them and the curved boundary. Fortunately, this is easy enough. Recall that the "refineNeighbor()" function in Section 14.5 contains a recursive call in its final "else" question. This recursive call is used only if no neighbor triangle has yet been found, and the search must therefore continue among the rest of the triangles in the mesh. If the entire connected list of triangles has been scanned and no neighbor has been found, then the edge must be a boundary edge.

The end of the connected list of triangles is reached when the "next" field is equal to 0. This indicates that the edge leading from nI to nJ is indeed a boundary edge, so two extra triangles should be added between it and the boundary.

Automatic boundary refinement is thus implemented simply by modifying the final "else" question in the "refineNeighbor()" function to read as follows:

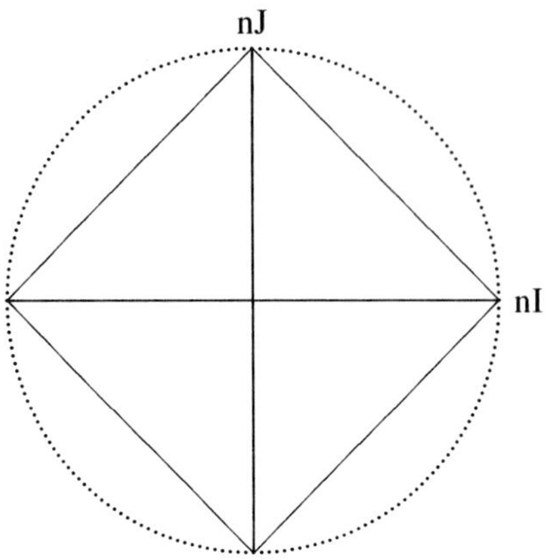

Figure 14.8. *The coarse mesh that serves as input for the adaptive-refinement algorithm with automatic boundary refinement.*

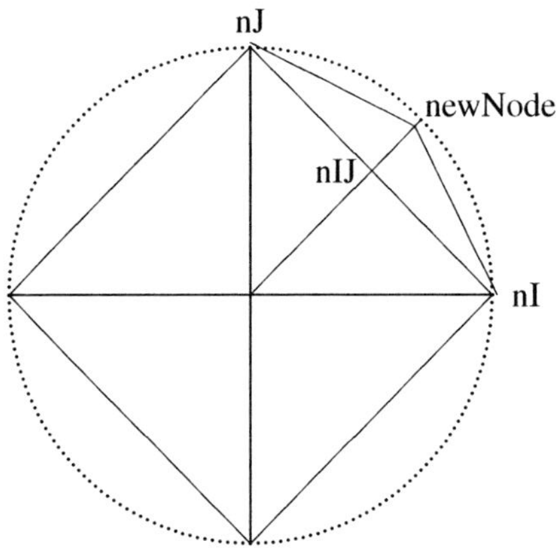

Figure 14.9. *The finer mesh, in which the upper-right triangle is refined and two extra triangles are also added to better approximate the upper-right part of the circular boundary.*

```
else{
  if(next)
    ((mesh<triangle>*)next)->refineNeighbor(nI,nJ,nIJ);
  else{
    node<point>
     newNode((1./sqrt(squaredNorm(nIJ()))) * nIJ());
    triangle t1(nI,nIJ,newNode);
    triangle t2(nJ,nIJ,t1(2));
    insertNextItem(t2);
    insertNextItem(t1);
  }
}
```

The interior "else" block in this code considers the case in which no neighbor triangle exists in the entire mesh, which implies that the edge leading from nI to nJ is indeed a boundary edge. In this case, two extra triangles, named "t1" and "t2", are added between it and the circular boundary.

In the above, we assume that the domain is circular, so the above procedure should take place in every boundary edge. The extension to other convex domains is straightforward. In more general domains, a boundary edge does not necessarily lie next to a convex boundary segment. One should thus check whether the midpoint between nI and nJ indeed lies within the domain before applying the above procedure. This is discussed and illustrated next.

14.9 Nonconvex Domains

As we have seen above, levels of refinement are useful in refining further not only in the interior of the domain but also next to a curved boundary like that in Chapter 12, Section 9. This is done by adding smaller and smaller triangles to the inside of the circle.

This procedure can be employed not only in a circle but also in more complicated domains with curved boundaries, provided that they are convex. For nonconvex domains, the curved boundary must be approached from the outside in.

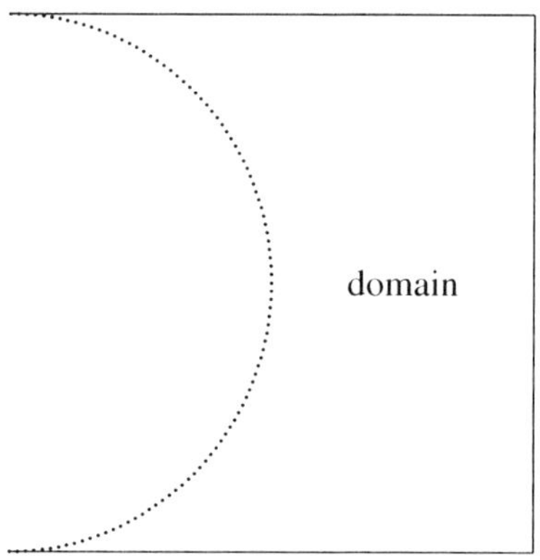

Figure 14.10. *The nonconvex domain in which the PDE is defined.*

Consider, for example, the domain in Figure 14.10. The left edge in it is concave, so it cannot be approximated from the inside out, as before. Instead, it is approached from the outside in, as follows.

Assume that the initial (coarse) mesh is as in Figure 14.11. This mesh contains only three triangles, and the left, curved edge is approximated rather poorly. In order to improve the approximation, we proceed in the same spirit as in adaptive refinement.

In adaptive refinement, the midpoint between node 1 and node 2 in Figure 14.11 is connected to node 4 to divide the upper triangle into two smaller triangles. Here, however, this midpoint lies outside the domain; therefore, it is replaced by the nearest point across from it on the curved boundary. This point is then connected to nodes 1, 2, and 4 to form the required two smaller triangles instead of the original upper triangle. The same is done

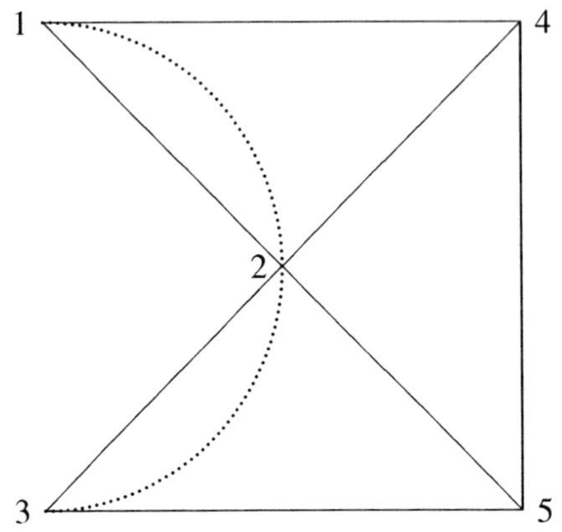

Figure 14.11. *The original coarse mesh that gives a poor approximation to the nonconvex domain.*

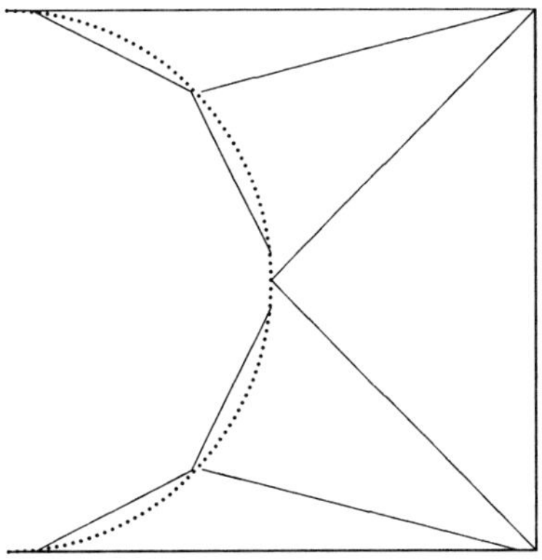

Figure 14.12. *The refined mesh that gives a better approximation to the nonconvex domain.*

for the lower coarse triangle in Figure 14.11. The resulting fine mesh is displayed in Figure 14.12.

The above procedure may be repeated, yielding better and better approximations to the curved boundary on the left. Better yet, it can be combined with adaptive refinement to produce an improved algorithm that refines simultaneously in both the interior of the domain and on its boundary:

Algorithm 14.2.

1. *Let T be the initial coarse finite-element triangulation, that is, the set of triangles in the mesh.*

2. *Construct the stiffness matrix A and the right-hand side f corresponding to T.*
3. *Solve the stiffness system*

$$Ax = f$$

for the vector of unknowns x.

4. *Let E be the set of edges in T.*
5. *Scan the edges in E one by one in some order. For every edge e encountered in this scanning, do the following:*
   (a) *Denote the endpoints of e by i and j (see Figure 14.5).*

   (b) *If e lies in the interior of the domain, then there are two triangles that share it. If e lies next to the boundary (e is "boundary edge"), then there is only one triangle that uses it. Let*

   $$t = \Delta(i, j, k)$$

   be a triangle with vertices i, j, and k that uses e (Figure 14.5).

   (c) *Let $m = (i + j)/2$ be the midpoint between i and j.*

   (d) *If e is a boundary edge, then m can lie outside the domain and must be replaced by a boundary point that is nearest to it in some sense. This is done as follows: let l be the point on the line leading from k to m that also lies on the boundary, and substitute*

   $$m \leftarrow l.$$

   This guarantees that m lies either on the boundary or in the interior of the domain.

   (e) *If*

   $$|x_i - x_j| > threshold,$$

   then divide t into the two triangles

   $$t_1 = \Delta(i, m, k),$$
   $$t_2 = \Delta(j, m, k)$$

   (Figure 14.6). In other words, include t_1 and t_2 in T instead of t.

   (f) *Do the same to the other triangle that shares e, if it exists. If e is a boundary edge, then no such triangle exists. Still, if m lies in the interior of the domain, then we need to add two more triangles next to the boundary. This is done as follows: continue the line leading from k to m until it meets the boundary at a point denoted by l. Add to T the triangles*

   $$t_3 = \Delta(i, m, l),$$
   $$t_4 = \Delta(j, m, l).$$

6. *If the mesh is not yet sufficiently fine, then go back to step 2.*
7. *Use x as the numerical solution of the boundary-value problem on the final mesh T.*

14.10 Exercises

1. Implement the adaptive-refinement algorithm in Section 14.3 using the "refine" function. (You may assume that a function "solve" that solves the stiffness system at the current mesh is available.) The solution can be found in Section A.15 of the Appendix.

2. Modify the "refine" function so that every edge is refined, regardless of the numerical solution (global refinement). This can actually be done by choosing a zero threshold in the adaptive-refinement algorithm.

3. Assume that the domain is circular. Modify the "refine" function so that the boundary edges are always refined.

4. Use graphic software (or LaTeX) to print the meshes resulting from the adaptive-refinement algorithm. Verify that the meshes are indeed conformal.

5. Currently, the "refine" function uses an internal loop with index 'j' that decreases from 2 to 'i'+1, where 'i' = 0, 1, 2 scans the list of vertices in the current triangle. Modify this loop so that 'j' increases from 'i'+1 to 2. What is the effect of this change on the refinement?

Chapter 15
High-Order Finite Elements

In this chapter, we describe quadratic and cubic finite elements, which may be used to improve the accuracy of a discretization. We show that the present object-oriented framework indeed provides the required tools to implement these methods. In particular, the "polynomial" object, along with its multiplication and integration functions, is most helpful in assembling the required stiffness matrix.

15.1 High-Order vs. Linear Finite Elements

So far, we have used only linear finite elements. This means not only that the elements are triangles with straight sides but also that the function that approximates the solution of the original PDE is continuous in the entire mesh and linear in each particular triangle in it. Indeed, this function can be written as a linear combination of the nodal basis functions, which are continuous in the entire mesh and linear in each triangle. More specifically, the nodal basis function ϕ_i assumes the value 1 only at the ith node and vanishes at all the other nodes. Thus, ϕ_i vanishes in all the triangles that don't use node i as a vertex. In triangles that do use it as a vertex, on the other hand, ϕ_i decreases linearly away from node i until it hits the edge that lies across from it, where it vanishes. As a result, ϕ_i is indeed continuous and piecewise linear throughout the mesh, as required. The numerical solution obtained from the discrete stiffness system is the best continuous piecewise-linear function in terms of minimizing the quadratic functional associated with the original boundary-value problem (Chapter 11, Section 3).

In many applications, however, this approximation is insufficient. In fact, the solution of the original PDE may have sharp variation inside individual triangles as well. This behavior cannot be approximated well by linear functions. Although one can use a large number of small triangles to approximate the solution better in such areas, this can lead to a considerable increase in the number of nodes and, hence, the number of unknowns in the discrete stiffness system. It may be better to improve the quality of the approximation within each particular triangle.

For this purpose, one may switch to finite elements of higher order. The elements still have the same triangular shape as before. However, the functions that form the approximate solution can now be not only linear but also quadratic or even cubic (or, in general, polynomials of degree $k > 1$ in x and y) in each particular triangle. The approximate solution is now the best continuous piecewise-quadratic (cubic) function in terms of minimizing the quadratic functional associated with the original boundary-value problem.

Clearly, piecewise-quadratic or cubic functions have a better chance of approximating a solution with large variation well. This extra accuracy, however, comes at the price of increasing the number of degrees of freedom in each particular triangle. In fact, a quadratic polynomial in x and y is characterized by six coefficients and a cubic polynomial by ten coefficients, whereas a linear function is characterized by three coefficients only. Because the number of degrees of freedom is larger than before, the order of the stiffness matrix is also larger and more difficult to solve. Still, the extra cost involved in solving the stiffness system may well be worth it for the sake of better accuracy.

In what follows, we describe the discretization method in more detail.

15.2 Quadratic Finite Elements

The quadratic finite elements considered here have the same triangular shape as before. They are quadratic only in the sense that the function that is used to approximate the solution of the original PDE is quadratic in each finite element and continuous in the entire mesh. Indeed, it is a linear combination of the quadratic nodal basis functions described below.

In linear finite elements, the nodal basis function ϕ_i assumes the value 1 at node i and the value 0 at all other nodes. Furthermore, it decreases linearly away from node i and vanishes at the edges that lie across from it in triangles that use it as a vertex. Actually, the nodal basis functions at a particular finite element can be defined in terms of three typical (standard) nodal functions, defined in the reference triangle in Figure 12.2, which is mapped to the finite element under consideration. Each typical nodal function is linear in the reference triangle and assumes the value 1 at one vertex and 0 at the other two vertices. The typical nodal functions are useful in assembling the stiffness system.

In order to have an even better approximation, the quadratic finite-element method uses continuous functions that are not only linear but also quadratic in each particular finite element. In other words, the solution to the original PDE is now approximated by a linear combination of not only linear but also quadratic nodal basis functions. In order to define the quadratic nodal basis functions, it is sufficient to define typical nodal functions in the reference triangle in Figure 15.1 and use the mapping of this triangle to the finite element under consideration.

In the reference triangle, a quadratic function (polynomial of degree 2 in the spatial variables x and y) is determined by the six coefficients of the terms $1, x, y, x^2, xy$, and y^2 in it. This means that the function has six degrees of freedom in its definition and hence can also be defined uniquely by its values at six distinct points in the triangle, e.g., the three vertices (numbered 1, 2, 3) and three midpoints of edges (numbered 4, 5, 6) in Figure 15.1. In fact, a typical nodal function in the reference triangle assumes the value 1 at one of these points and vanishes at the other five.

Although the quadratic finite-element method uses twice as many points as the linear finite-element method, it may well be worth it for the sake of extra accuracy.

Let us now define explicitly the typical nodal functions in the reference triangle in

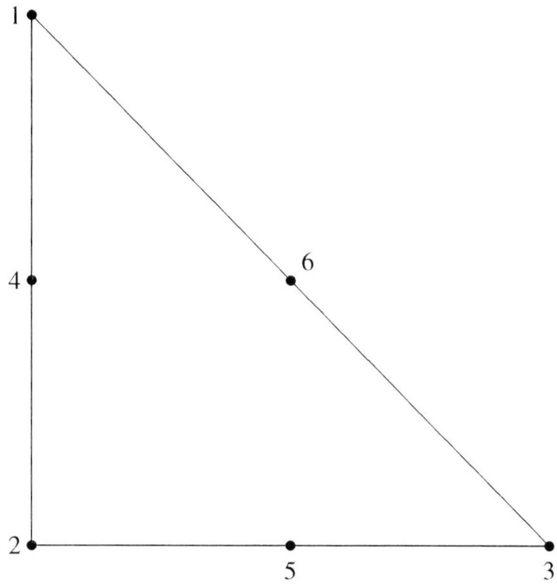

Figure 15.1. *The reference triangle in the quadratic finite-element method. The nodes are numbered* 1, 2, 3, 4, 5, 6. *A typical quadratic nodal function assumes the value* 1 *at one of these nodes and* 0 *at the others.*

Figure 15.1:

$$\begin{aligned}
\phi^{(1)} &\equiv 2y(y - 1/2), \\
\phi^{(2)} &\equiv 2(1 - x - y)(1/2 - x - y), \\
\phi^{(3)} &\equiv 2x(x - 1/2), \\
\phi^{(4)} &\equiv 4y(1 - x - y), \\
\phi^{(5)} &\equiv 4x(1 - x - y), \\
\phi^{(6)} &\equiv 4xy.
\end{aligned}$$

Note that the typical nodal functions indeed satisfy

$$\phi^{(i)}(j) = \begin{cases} 1 & \text{if } j = i, \\ 0 & \text{if } j \neq i, \end{cases}$$

where i and j are integers between 1 and 6. The typical nodal functions can now be used to assemble the stiffness matrix, as in Chapter 12, Section 5. The solution to the discrete stiffness system $Ax = f$ contains the components x_i that are the coefficients of the nodal basis functions ϕ_i in the expansion of the required approximate solution. This completes the definition of the quadratic finite-element discretization method.

The implementation of quadratic finite elements is similar to that of linear ones in Chapter 13. The only difference is that the "triangle" object used there should be replaced by the "quadraticTriangle" object defined by

```
typedef finiteElement<point,6> quadraticTriangle;
```

which contains the six required nodes in Figure 15.1. The entire mesh is then implemented as a "mesh<quadraticTriangle>" object.

Assembling the stiffness matrix is done in a similar way as that in Chapter 13, Section 5. The only difference is that here the gradient of a typical quadratic nodal function is no longer constant but rather a polynomial of degree 1 in the spatial variables x and y. The implementation of such polynomials, including their multiplication and integration, is available in Chapter 5, Sections 13 and 14.

15.3 Cubic Finite Elements

The cubic finite elements considered here have the same triangular shape as before. They are cubic in the sense that the function that approximates the solution to the original PDE can be not only quadratic but also cubic within each particular finite element. In other words, it can be written in each finite element as a polynomial of degree 3 in the spatial variables x and y. Thus, it has 10 degrees of freedom in it: the coefficients of the terms 1, x, y, x^2, xy, y^2, x^3, x^2y, xy^2, and y^3. Therefore, it is determined uniquely in each particular finite element in terms of its values at 10 distinct points in it.

This is also true for the typical cubic nodal functions defined in the reference triangle in Figure 15.2, which is mapped onto the finite element under consideration. Each typical nodal function assumes the value 1 at one of the ten nodes in Figure 15.2 and 0 at the other

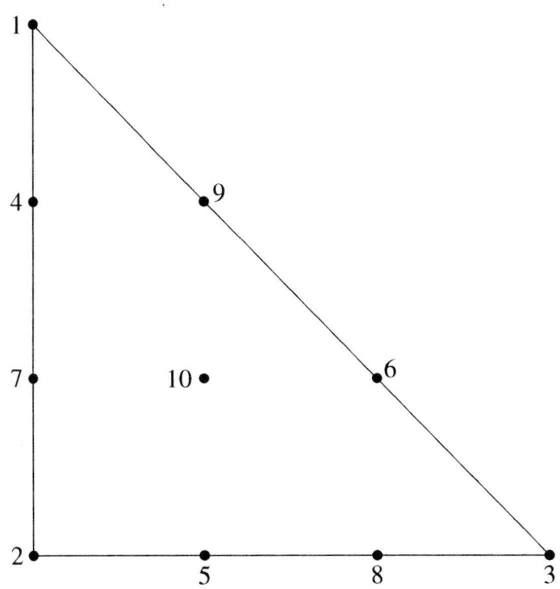

Figure 15.2. *The reference triangle in the cubic finite-element method. The nodes are numbered* 1, 2, 3, 4, 5, 6, 7, 8, 9, 10. *A typical cubic nodal function assumes the value* 1 *at one of these nodes and* 0 *at the others.*

nine. More explicitly, the typical cubic nodal functions are defined as follows:

$$\begin{aligned}
\phi^{(1)} &\equiv (9/2)y(y-1/3)(y-2/3),\\
\phi^{(2)} &\equiv (9/2)(1-x-y)(2/3-x-y)(1/3-x-y),\\
\phi^{(3)} &\equiv (9/2)x(x-1/3)(x-2/3),\\
\phi^{(4)} &\equiv (27/2)y(y-1/3)(1-x-y),\\
\phi^{(5)} &\equiv (27/2)x(1-x-y)(2/3-x-y),\\
\phi^{(6)} &\equiv (27/2)xy(x-1/3),\\
\phi^{(7)} &\equiv (27/2)y(1-x-y)(2/3-x-y),\\
\phi^{(8)} &\equiv (27/2)x(x-1/3)(1-x-y),\\
\phi^{(9)} &\equiv (27/2)xy(y-1/3),\\
\phi^{(10)} &\equiv 27xy(1-x-y).
\end{aligned}$$

Note that the typical cubic nodal functions indeed satisfy

$$\phi^{(i)}(j) = \begin{cases} 1 & \text{if } j = i, \\ 0 & \text{if } j \neq i, \end{cases}$$

where i and j are integers between 1 and 10. The typical cubic nodal functions are useful in assembling the stiffness matrix, as in Chapter 12, Section 5. The solution of the resulting stiffness system produces the required approximate solution to the original PDE. This completes the definition of the cubic finite-element discretization method.

The implementation of cubic finite elements is similar to that of linear ones in Chapter 13. The only difference is that the "triangle" object used there is replaced by the "cubicTriangle" object defined by

```
typedef finiteElement<point,10> cubicTriangle;
```

which contains the 10 required nodes in Figure 15.2. The entire mesh is then implemented as a "mesh<cubicTriangle>" object.

Assembling the stiffness matrix is done in a similar way as that in Chapter 13, Section 5. The only difference is that here the gradient of a typical cubic nodal function is no longer constant but rather a polynomial of degree 2 in the spatial variables x and y. The implementation of such polynomials, including their multiplication and integration, is available in Chapter 5, Sections 13 and 14.

15.4 Pros and Cons

In Table 15.1, we give a brief summary of the pros and cons of linear and high-order finite elements. The comparison assumes that the same number of degrees of freedom (unknowns) is used in both methods.

High-order finite elements provide extra accuracy, particularly inside individual finite elements and near the boundary. However, this comes at the price of a more complicated assembling procedure, which must use the "polynomial" object in Chapter 5, Section 13.

Table 15.1. *Pros and cons of linear and high-order finite elements. It is assumed that the same number of degrees of freedom (unknowns) is used in both methods.*

	high-order finite elements	linear finite elements
pros	higher accuracy	straightforward assembling
		sparser stiffness matrix
cons	more complex assembling	lower accuracy
	denser stiffness matrix	

Furthermore, the resulting stiffness matrix is denser than the one obtained from linear finite elements. There is, thus, a tradeoff between extra accuracy and extra computational cost.

Actually, one could also have extra accuracy with linear finite elements by just increasing the number of nodes and triangles in the mesh. Although this approach also increases the order of the stiffness matrix, its sparsity remains the same, so efficient linear-system solvers can still be used. Furthermore, increasing the number of triangles in the mesh allows a good numerical approximation of possible discontinuities across the edges. Therefore, it provides a better chance of approximating solutions with large variation and possible shocks well.

In the applications in this book, we use linear finite elements only. The present codes can easily be modified to use quadratic and cubic finite elements using the "polynomial" object in Chapter 5, Section 13. This is left as an exercise.

15.5 Exercises

1. Verify that the typical quadratic nodal functions in the reference triangle in Figure 15.1 indeed satisfy
$$\phi^{(i)}(j) = \begin{cases} 1 & \text{if } j = i, \\ 0 & \text{if } j \neq i, \end{cases}$$
where i and j are integers between 1 and 6.

2. Prove that a quadratic nodal basis function ϕ_i is indeed continuous in the entire mesh. Use the fact that a polynomial of degree 2 is defined uniquely in an edge in Figure 15.1 by its value at three distinct points along it.

3. Modify the code in Chapter 13, Section 5, to assemble the stiffness matrix for the quadratic finite-element discretization method. Use the polynomials of two variables in Chapter 5, Sections 13 and 14, to implement the required gradients.

4. Verify that the typical cubic nodal functions in the reference triangle in Figure 15.2 indeed satisfy
$$\phi^{(i)}(j) = \begin{cases} 1 & \text{if } j = i, \\ 0 & \text{if } j \neq i, \end{cases}$$
where i and j are integers between 1 and 10.

5. Prove that a cubic nodal basis function ϕ_i is indeed continuous in the entire mesh. Use the fact that a polynomial of degree 3 is defined uniquely in an edge in Figure 15.2 by its value at four distinct points along it.

6. Modify the code in Chapter 13, Section 5, to assemble the stiffness matrix for the cubic finite-element discretization method. Use the polynomials of two variables in Chapter 5, Sections 13 and 14, to implement the required gradients.

Part V

The Numerical Solution of Large Sparse Linear Systems of Equations

The finite-element discretization method produces a large system of algebraic equations called the stiffness system. The number of equations in this system, N, is as large as the number of nodes in the mesh. Actually, the system can be written algebraically as

$$Ax = f,$$

where A is the coefficient (stiffness) matrix of order N, f is the given N-dimensional right-hand-side vector, and x is the N-dimensional vector of unknowns. The matrix A is sparse in the sense that most of the elements in it are zero. Once this system is solved numerically, the ith component in x, x_i is the value of the numerical solution of the partial differential equation (PDE) at the node of index i in the list of nodes in the mesh. The approximate solution at points that are not in the grid can also be obtained by some interpolation.

The numerical solution of a linear system of the form $Ax = f$ is, thus, an essential part of the numerical solution of PDEs. In the past, before computers became available, this task had to be done manually using Gaussian elimination (Chapter 2, Section 21). Of course, this approach is impractical unless the order of the system is very small.

Fortunately, the digital computer freed the human mind from the exhausting work of such calculations. All that is left for us humans to do is to write a correct and efficient computer program (code) and feed it to the computer.

Since the machine is much more powerful than the human mind in terms of memory and computational ability, it can solve a system with a much higher order. The number of unknowns (or grid points) can be much larger than before and the numerical model much more accurate and realistic. Thus, the digital computer has started a new era in experimental and applied science and engineering.

The opportunity to solve realistic numerical models increased the appetite of scientists and engineers to make further progress and solve even more advanced and complicated applications. Unfortunately, it turns out that the power of the digital computer is not unlimited. When the order of a system is too large, traditional Gaussian elimination is impractical because it requires prohibitively large time and memory resources. In fact, in some cases the computer can work for hours, days, and even weeks before it completes the calculations.

It is therefore essential to use the computational power offered by the computer wisely and develop more efficient algorithms to solve large sparse linear systems of equations. These algorithms must be much more efficient than traditional Gaussian elimination in terms of both time and storage. They must preserve the sparsity of the coefficient matrix A and avoid extensive fill-in. The implementation must take advantage of this sparsity, store only nonzero matrix elements, and disregard zero ones. This is the subject of this part.

This part contains three chapters. In the first one (Chapter 16), the present object-oriented approach is used to implement sparse matrices efficiently along with useful functions such as arithmetic operations. In the second one (Chapter 17), efficient iterative methods for solving large sparse linear systems are presented and implemented. In the third one (Chapter 18), parallel implementation is discussed.

Chapter 16

Sparse Matrices and Their Implementation

In this chapter, we implement a sparse matrix as a list of rows. Each individual row is implemented as a connected list of nonzero matrix elements. These objects allow efficient implementation of arithmetic operations such as matrix times vector and matrix times matrix.

16.1 Sparse vs. Dense Matrices

As we've already seen in Chapter 12, Section 5, most elements in the stiffness matrix A are 0. In fact, $A_{i,j}$ is nonzero only if i and j are vertices in the same triangle. In other words, the stiffness matrix A is sparse.

The naive implementation of matrices as in Chapter 2, Section 20, is suitable for dense matrices, of which most elements are nonzero. For sparse matrices, on the other hand, a much more efficient approach is to store only the nonzero elements and ignore all the zero elements. This approach makes arithmetic operations much more efficient and saves a lot of time and storage (see Chapter 4, Section 3).

For example, consider the "difference" object in Chapter 7, Section 12, which implements a tridiagonal matrix of order N by storing only its main diagonal and the diagonal above and below it. The total memory required for this object is only $3N$ numbers. Of course, this is much more efficient than storing all N^2 numbers in the full matrix, most of which are 0 anyway.

Sparse matrices are thus a most useful tool in practical implementations. The data structures used in their implementation, however, are no longer simple arrays, as in Chapter 2, Section 20, but rather more complicated structures such as connected lists (Chapter 4, Section 4). The items in the connected list should be objects that contain two fields: one to store the actual value of the element and the other to store the index of the column in which it lies. The sparse matrix should thus be implemented as a list of such connected lists.

In summary, we need three new objects: the "rowElement" object, which contains data about a particular element; the "row" object, which is a connected list of "rowElement" objects; and the "sparseMatrix" object, which is a list of "row" objects. These objects are implemented below.

16.2 The Matrix-Element Object

We start by implementing the element in some row in the sparse matrix. The "rowElement" class contains two data fields. The first field is of type 'T', to be specified later in compilation time. This field contains the value of the element. The second field is of type "integer" and contains the index of the column in which the element is located in the matrix.

Here is the block of the "rowElement" class:

```
template<class T> class rowElement{
    T value;
    int column;
  public:
    rowElement(const T& val=0, int col=-1)
        : value(val),column(col){
    } //  constructor

    rowElement(const rowElement&e)
        : value(e.value),column(e.column){
    } //  copy constructor

    const rowElement& operator=(const rowElement&e){
      if(this != &e){
        value = e.value;
        column = e.column;
      }
      return *this;
    } //  assignment operator

    ~rowElement(){} //  destructor
```

Although the data fields are private, they can still be read by the public "getValue" and "getIndex" member functions:

```
    const T& getValue() const{
      return value;
    } //  read the value

    int getColumn() const{
      return column;
    } //  return the column
```

Next, we define some useful arithmetic operations. For example, the "+=" operator can take either a 'T' or a "rowElement" argument. In either case, its value is incremented by the corresponding value of the argument:

```
    const rowElement&
    operator+=(const T&t){
```

```
      value += t;
      return *this;
    } // adding a T

    const rowElement&
    operator+=(const rowElement<T>&e){
      value += e.value;
      return *this;
    } // adding a rowElement

    const rowElement& operator-=(const T&t){
      value -= t;
      return *this;
    } // subtracting a T

    const rowElement&
    operator-=(const rowElement<T>&e){
      value -= e.value;
      return *this;
    } // subtracting a rowElement

    const rowElement&
    operator*=(const T&t){
      value *= t;
      return *this;
    } // multiplying by a T

    const rowElement& operator/=(const T&t){
      value /= t;
      return *this;
    } // dividing by a T
};
```

This concludes the block of the "rowElement" class.

Next, we define the binary '<', '>', and "==" nonmember operators. These operators take two "rowElement" arguments and compare their "column" fields. For example, "operator<" returns 1 if the column index of the first argument is smaller than that of the second one, and 0 otherwise. This priority order will be used later to preserve increasing column order in matrix rows:

```
template<class T>
int
operator<(const rowElement<T>&e, const rowElement<T>&f){
  return e.getColumn() < f.getColumn();
} //  smaller column index
```

```
template<class T>
int
operator>(const rowElement<T>&e, const rowElement<T>&f){
  return e.getColumn() > f.getColumn();
} //   greater column index

template<class T>
int
operator==(const rowElement<T>&e, const rowElement<T>&f){
  return e.getColumn() == f.getColumn();
} //   same column
```

We also define some useful nonmember binary arithmetic operators that involve a "rowElement" object and a scalar:

```
template<class T>
const rowElement<T>
operator+(const rowElement<T>&e, const T&t){
  return rowElement<T>(e) += t;
} //   rowElement plus a T

template<class T>
const rowElement<T>
operator+(const T&t, const rowElement<T>&e){
  return rowElement<T>(e) += t;
} //   T plus rowElement

template<class T>
const rowElement<T>
operator-(const rowElement<T>&e, const T&t){
  return rowElement<T>(e) -= t;
} //   rowElement minus T

template<class T>
const rowElement<T>
operator*(const rowElement<T>&e, const T&t){
  return rowElement<T>(e) *= t;
} //   rowElement times a T

template<class T>
const rowElement<T>
operator*(const T&t, const rowElement<T>&e){
  return rowElement<T>(e) *= t;
} //   T times rowElement

template<class T>
```

```
const rowElement<T>
operator/(const rowElement<T>&e, const T&t){
  return rowElement<T>(e) /= e;
} //   rowElement divided by a T
```

Finally, we define a function that prints a "rowElement" to the screen:

```
template<class T>
void print(const rowElement<T>&e){
  print(e.getValue());
  printf("column=%d\n",e.getColumn());
} // print a rowElement object
```

16.3 The Row Object

In this section, we introduce the "row" class that implements a row in a sparse matrix. The "row" object is actually a connected list of row elements. This inheritance is displayed schematically in Figure 16.1.

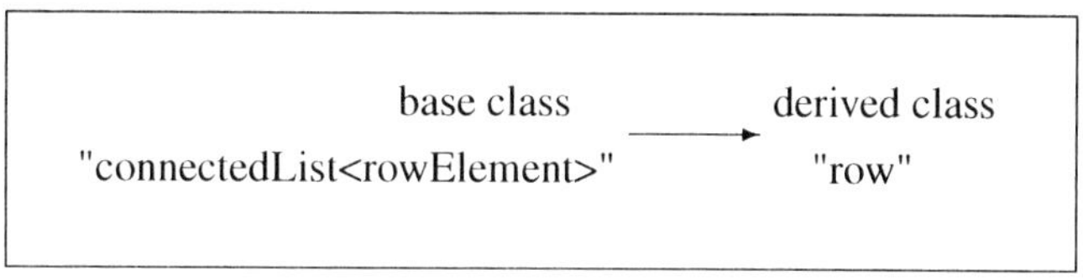

Figure 16.1. *Schematic representation of inheritance from the base class "connectedList<rowElement>" to the derived class "row".*

The "row" class is also a template class. The template 'T' that indicates the type of value of the element is to be specified later in compilation time.

It is assumed that the elements in the "row" object are ordered in increasing order. The priority order of "rowElement" objects used for this purpose is induced from the priority order of their corresponding columns.

Here is the actual implementation of the "row" class and required member functions:

```
template<class T>
class row : public connectedList<rowElement<T> >{
  public:
     row(const T&val=0,int col=-1){
       item=rowElement<T>(val,col);
     } // constructor
```

So far, we have implemented only the constructor. Next, we define functions that read the data in the first element in the row:

```
const rowElement<T>& operator()() const{
  return item;
} // read only first item

const T& getValue() const{
  return item.getValue();
} // read only first-item value

int getColumn() const{
  return item.getColumn();
} // read only first-item column
```

Since the "row" class is derived from the "connectedList" class, it can use its public and protected members. In particular, it can use the functions that insert or drop items. Still, the "row" class contains its own local versions of the "insertNextItem", "insertFirstItem", and "append" functions. These versions are different from the original versions in Chapter 3, Section 5, since they take two arguments to specify the value and column of the inserted element. The definitions of the present versions call the corresponding original version. In these calls, the prefix "connectedList::" precedes the function name to indicate that an original version is called:

```
void insertNextItem(const T&val, int col){
  rowElement<T> e(val,col);
  connectedList<rowElement<T> >::insertNextItem(e);
} // insert a rowElement as second item

void insertFirstItem(const T&val, int col){
  rowElement<T> e(val,col);
  connectedList<rowElement<T> >::insertFirstItem(e);
} // insert a rowElement at the beginning

void append(const T&val, int col){
  rowElement<T> e(val,col);
  connectedList<rowElement<T> >::append(e);
} // append a rowElement at the end of row
```

The recursive structure of the base "connectedList" class is particularly useful. The definitions of many member functions use recursive calls applied to the "next" field that points to the rest of the row. However, the "next" field inherited from the base "connectedList" class is of type pointer-to-connectedList rather than pointer-to-row. Therefore, it must be converted explicitly to pointer-to-row before the recursive call can take place. This is done by adding the prefix "(row*)".

Usually, this conversion is considered risky because in theory "next" can point to a "connectedList" object or any object derived from it, with a completely different interpretation of the recursively called function. Fortunately, here "next" must point to a "row" object, so the conversion is safe.

The following function uses recursion to compute the sum of the elements in the row:

```
const T rowSum() const{
  return next ? getValue()
      + (*(row<T>*)next).rowSum() : getValue();
} // row-sum
```

Recursion is also used in the following "operator[]" function. This function takes an integer argument 'i' and returns a copy of the value of the element in column 'i', if it exists. If, on the other hand, there is no such element in the current row, then it returns 0.

This is carried out using the recursive pattern of the "row" object. First, the "column" field in the first element is examined. If it is equal to 'i', then the required element has been found, and its value is returned as output. If, on the other hand, it is greater than 'i', then there is no hope of finding the required element because the elements are ordered in increasing column order, so 0 is returned. Finally, if it is smaller than 'i', then the "operator[]" is applied recursively to the rest of the row.

As before, the "next" field must be converted explicitly from pointer-to-connectedList to pointer-to-row before recursion can be applied to it. This is done by adding the prefix "(row*)".

The value returned by the "operator[]" function is of type constant-T rather than reference-to-constant-T. (Indeed, the words that precede the function name in the code are "const T" rather than "const T&".) This is because, as discussed above, the function may also return the zero value. This local constant cannot be referred to and must be stored in the temporary unnamed variable returned by the function:

```
const T operator[](int i) const{
  return (getColumn() == i) ? getValue() :
    next&&(getColumn() < i) ? (*(row*)next)[i] : 0.;
} // read only the value at column i
```

Recursion is also used in the rest of the member functions, such as the following arithmetic operators that involve a row and a scalar:

```
const row& operator*=(const T&t){
  item *= t;
  if(next) *(row*)next *= t;
  return *this;
} // multiply by a T

const row& operator/=(const T&t){
  item /= t;
  if(next) *(row*)next /= t;
  return *this;
} // divide by a T
```

Fortunately, the "+=" operator that adds a row to the current row (while preserving increasing column order) is already available from the base "connectedList" class, so there is no need to rewrite it.

The following binary operator computes the inner product of a row and a vector. This operation will be used later to compute the product of a sparse matrix and a vector:

```
const T
operator*(const dynamicVector<T>&v) const{
  return
   next ? getValue() * v[getColumn()]
     + *(row*)next * v
     : getValue() * v[getColumn()];
} // row times vector (inner product)
```

The following function renumbers the columns with new numbers contained in a vector of integers named "renumber". To increase efficiency, this vector is passed to the function by reference. As before, recursion is applied to the "next" field after its type is converted from pointer-to-connectedList to pointer-to-row:

```
void
renumberColumns(const dynamicVector<int>&renumber){
  item =
   rowElement<T>(getValue(),renumber[getColumn()]-1);
  if(next)
    (*(row<T>*)next).renumberColumns(renumber);
} // renumber columns
```

We also declare here two more member functions, to be defined later:

```
  void dropItems(const dynamicVector<int>&);
  void dropPositiveItems(int, const T&, double);
};
```

This concludes the block of the "row" class. Next, we define the "dropItems()" member function declared above. (The definition of the "dropPositiveItems" is different only in the criterion that is used to drop items.)

This "dropItems" function takes as argument a vector of integers named "mask". The zeroes in this vector indicate that the row elements in the corresponding columns should be dropped. As in Chapter 3, Section 5, this is done by looking ahead to the next element and dropping it if appropriate. For this purpose, the "row" object that contains the rest of the elements in the row is first accessed as "*next". Then, the first item in this "row" object (which is actually the second element in the current row) is accessed as "(*next)()", using the "operator()" in the base "connectedList" class. Now, the column of this element can be read, and if the corresponding component in "mask" vanishes, then this element is dropped by the "dropNextItem" function of the base "connectedList" class:

```
template<class T>
void
row<T>::dropItems(const dynamicVector<int>&mask){
  if(next){
    if(!mask[(*next)().getColumn()]){
      dropNextItem();
```

We are now in the "if" block that assumes that we have indeed dropped the second element from the row, so the row is now shorter. Therefore, we can apply the "dropItems" function recursively to it:

```
        dropItems(mask);
      }
```

If, on the other hand, the second element in the original row has not been dropped, then the "dropItems" function is applied recursively to the "next" field after its type is converted from pointer-to-connectedList to pointer-to-row:

```
      else
        (*(row<T>*)next).dropItems(mask);
```

Finally, the first row element is also dropped, provided that the dropping criterion is satisfied and it is not the only element left in the row:

```
      if(!mask[getColumn()])dropFirstItem();
    }
  } //  "masking" the row by a vector of integers
```

Actually, this code segment can be removed from this function and placed in another function that drops only the first element, if appropriate. This might increase the efficiency, because as it stands there are some unnecessary repetitive checks.

Functions that use recursion may call themselves many times. Therefore, one should be careful to avoid expensive operations in them, such as construction of big objects like dynamic vectors. This is why the "mask" vector in the above "dropItems" function is passed by reference. This way, unnecessary calls to the copy constructor of the "dynamicVector" class are avoided.

Finally, we define some nonmember binary arithmetic operators that involve a row and a scalar:

```
template<class T>
const row<T>
operator*(const row<T>&r, const T&t){
  return row<T>(r) *= t;
} //  row times T

template<class T>
const row<T>
operator*(const T&t, const row<T>&r){
  return row<T>(r) *= t;
} //  T times row

template<class T>
const row<T>
operator/(const row<T>&r, const T&t){
  return row<T>(r) /= t;
} //  row divided by a T
```

16.4 The Sparse-Matrix Object

As discussed in Chapter 4, Section 9, the most efficient way to implement a sparse matrix is as a list of connected lists or a list of "row" objects. This way, only the nonzero matrix elements are stored and participate in calculations, whereas the zero matrix elements are ignored. Although connected lists have their own drawbacks in terms of efficiency, because they use indirect indexing that may slow down the performance due to more expensive data access, this drawback is far exceeded by the advantage of avoiding trivial calculations. Furthermore, in some cases it is possible to map the connected list to a more continuous data structure and make the required computations in it.

The hierarchy of objects used to implement the sparse matrix is displayed in Figure 16.2. The "sparseMatrix" object in the highest level is implemented as a list of "row" objects, where the "row" object is by itself implemented as a connected list of "rowElement" objects. The "rowElement" object at the lowest level contains a template parameter 'T' to store the value of the element.

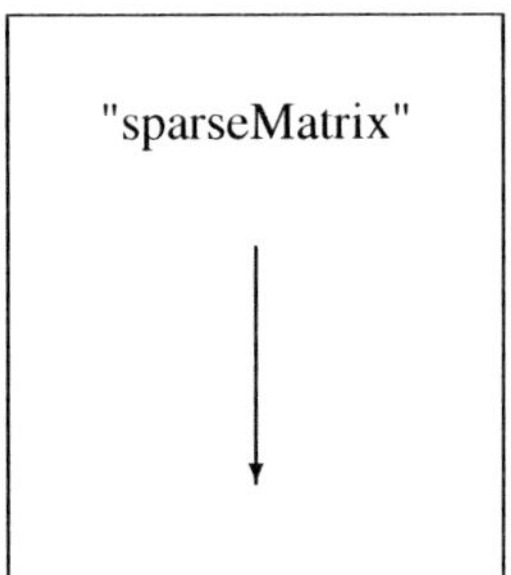

Figure 16.2. *The hierarchy of objects used to implement the sparse matrix: the "sparseMatrix" object is a list of "row" objects, each of which is a connected list of "rowElement" objects, which use the template 'T' to refer to the type of value of the matrix elements.*

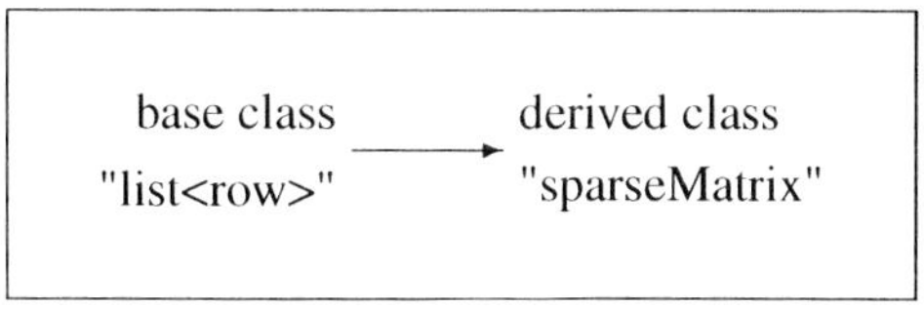

Figure 16.3. *Schematic representation of inheritance from the base class "list<row>" to the derived class "sparseMatrix".*

We now implement the sparse matrix as a list of rows. The "sparseMatrix" class is derived from a list of "row" objects (see Figure 16.3). Therefore, it enjoys access to the public and protected members of the "list" class in Chapter 3, Section 4. The additional member functions defined in the "sparseMatrix" class often loop over all the items in the underlying list of rows. In this loop, member functions of the "row" class in Section 16.3 are often used.

For example, the product of a matrix and a vector is implemented as follows. Let A be a sparse matrix with N columns, and let v be an N-dimensional vector. Let $a^{(i)}$ be the ith row in A. Then the ith component in Av is calculated by

$$(Av)_i = (a^{(i)}, v)$$

(inner product of row and vector, available in the "row" class). Thus, the implementation uses a loop over the rows, repeating the above calculation for each row $a^{(i)}$.

Another interesting example is the product of a matrix and a matrix. The algorithm described in Chapter 2, Section 20, is not very useful here, because it uses mainly column operations. A more suitable algorithm is the following one, which uses row operations only. Let A be a matrix with N rows and B be a matrix with N columns. Let $b^{(i)}$ be the ith row in B. Then the ith row in BA can be written as

$$(BA)^{(i)} = b^{(i)}A.$$

The calculation of each row in BA requires, therefore, the linear combination of rows in A with coefficients in $b^{(i)}$. This linear combination can be calculated using only operations with "row" objects: the '*' operator of the "row" class and the "+=" operator of the base "connectedList" class:

```
template<class T>
class sparseMatrix : public list<row<T> >{
  public:
    sparseMatrix(int n=0){
      number = n;
      item = n ? new row<T>*[n] : 0;
      for(int i=0; i<n; i++)
        item[i] = 0;
    } // constructor
```

```
sparseMatrix(int n, const T&a){
  number = n;
  item = n ? new row<T>*[n] : 0;
  for(int i=0; i<n; i++)
    item[i] = new row<T>(a,i);
} // constructor with T argument

sparseMatrix(mesh<triangle>&);
~sparseMatrix(){} // destructor
```

So far, we have defined the constructor and the destructor and also declared a constructor with a "mesh" argument, to be defined later on. We now define a function that takes two integer arguments, say 'i' and 'j', and reads the "(i,j)"th matrix element:

```
const T operator()(int i,int j) const{
  return (*item[i])[j];
} // (i,j)th element (read only)
```

Next, we define functions that return the number of rows in the matrix, the number of columns, and the order of a square matrix:

```
int rowNumber() const{
  return number;
} // number of rows

int columnNumber() const;
int order() const{
  return max(rowNumber(), columnNumber());
} // matrix order
```

Finally, we declare some member and friend functions. Some of these functions are defined in Section A.9 of the Appendix, whereas others are defined and used in the iterative methods in Chapter 17:

```
const sparseMatrix& operator+=(const sparseMatrix<T>&);
const sparseMatrix& operator-=(const sparseMatrix<T>&);
const sparseMatrix<T>& operator*=(const T&);
friend const sparseMatrix<T>
    operator*<T>(const sparseMatrix<T>&,
        const sparseMatrix<T>&);
friend const sparseMatrix<T>
    diagonal<T>(const sparseMatrix<T>&);
friend const sparseMatrix<T>
    transpose<T>(const sparseMatrix<T>&);
friend void
    GaussSeidel<T>(const sparseMatrix<T>&,
    const dynamicVector<T>&, dynamicVector<T>&);
```

```
    friend void
        symmetricGaussSeidel<T>(const sparseMatrix<T>&,
        const dynamicVector<T>&, dynamicVector<T>&);
    const sparseMatrix factorize(double);
    const dynamicVector<T>
        forwardElimination(const dynamicVector<T>&)const;
    const dynamicVector<T>
        backSubstitution(const dynamicVector<T>&)const;
    const dynamicVector<int> coarsen() const;
    const sparseMatrix<T> createTransfer();
};
```

This concludes the block of the "sparseMatrix" class. In the next section, we define the constructor that actually assembles the sparse stiffness matrix.

16.5 Assembling the Sparse Stiffness Matrix

In this section, we provide the full implementation of the calculation of the stiffness matrix according to the guidelines in Chapter 12, Section 5. The code has already been given in part in Chapter 13, Section 5. However, that code is incomplete because the required "sparseMatrix" object was not yet available there. Here, we have every required object and are therefore ready to write the complete code.

The function that assembles the stiffness matrix is actually a constructor that constructs a new sparse matrix. This approach is particularly convenient, because the required sparse matrix can be initialized with the correct number of rows, and the rows in it are also initialized with the correct row elements.

The sparse stiffness matrix is constructed in a constructor function declared in the block of the "sparseMatrix" class in Section 16.4. According to the rules of C++, a constructor must be a member function. The constructor takes an argument of type "mesh" and applies to it the "indexing" function of Chapter 13, Section 4, which assigns indices to the nodes and also returns their number, which is also the number of rows in the constructed "sparseMatrix" object:

```
template<class T>
sparseMatrix<T>::sparseMatrix(mesh<triangle>&m){
  item = new row<T>*[number = m.indexing()];
  for(int i=0; i<number; i++)
    item[i] = 0;
```

The assembling is done as in Chapter 13, Section 5. The triangles in the mesh are scanned, and the indices of vertices in each triangle, denoted by 'I' and 'J', are used to calculate the contribution from this triangle to the "(I,J)"th element in the stiffness matrix:

```
  point gradient[3];
  gradient[0] = point(-1,-1);
  gradient[1] = point(1,0);
  gradient[2] = point(0,1);
```

```
    for(const mesh<triangle>* runner = &m;
        runner;
        runner=(const mesh<triangle>*)runner->readNext()){
      matrix2 S((*runner)()[1]() - (*runner)()[0](),
                (*runner)()[2]() - (*runner)()[0]());
      matrix2 Sinverse = inverse(S);
      matrix2 weight =
          abs(det(S)/2) * Sinverse * transpose(Sinverse);
      for(int i=0; i<3; i++)
        for(int j=i; j<3; j++){
          int I = (*runner)()[i].getIndex();
          int J = (*runner)()[j].getIndex();
```

So far, the code is the same as in Chapter 13, Section 5. Here, it starts to differ from it and assemble directly into the constructed "sparseMatrix" object. More specifically, the contribution to the "(I,J)"th element should be added to the element with column index 'J' in the 'I'th row. For this purpose, it is particularly convenient to construct a new "row" object, named 'r', which contains only this contribution. The row 'r' is then added to the 'I'th row (if it exists), using the "+=" operator of the base "connectedList" class, or initializes it (if it does not yet exist):

```
          if(item[I]){
            row<T> r(gradient[j]*weight*gradient[i],J);
            *item[I] += r;
          }
          else
            item[I] =
              new row<T>(gradient[j]*weight*gradient[i],J);
        }
    }
}  //  assembling into the sparse stiffness matrix
```

As discussed at the end of Chapter 13, Section 5, the above matrix is not yet the required stiffness matrix until its transpose is added to it. Thus, the required stiffness matrix for a mesh 'm' is constructed as follows:

```
sparseMatrix<double> A(m);
A += transpose(A) - diagonal(A);
```

The sparse matrix 'A' is now the required stiffness matrix.

16.6 Exercises

1. Implement arithmetic operations with sparse matrices, such as addition, subtraction, multiplication by scalar, matrix times dynamic vector, and matrix times matrix. The solution can be found in Section A.9 of the Appendix.

2. Implement the "columnNumber" member function that returns the number of columns (the maximum column index in the elements in the rows). The solution can be found in Section A.9 of the Appendix.

3. Implement the "diagonal" function that returns the main diagonal of a sparse matrix. The solution can be found in Section A.9 of the Appendix. Does this function have to be a member of the "sparseMatrix" class? Why?

4. Write the "transpose" function that takes a sparse matrix A as a constant argument and returns its transpose A^t. The solution can be found in Section A.9 of the Appendix. Does this function have to be a friend of the "sparseMatrix" class? Why?

5. Write the "HermitAdjoint" function that takes a complex sparse matrix A (a "sparseMatrix<complex>" object) as a constant argument and returns its Hermitian adjoint (the complex conjugate of the transpose)

$$A^* \equiv \bar{A}^t.$$

   Is it possible to write these functions as one template function, with the template 'T' being either "double" or "complex"?

6. Write the "truncate()" member function that drops off-diagonal elements that are too small in magnitude. For example, the call "A.truncate(η)" drops elements $A_{i,j}$ for which

$$|A_{i,j}| < \eta |A_{i,i}|$$

   (where η is a small parameter). You may use the "truncateItems" function in Chapter 3, Section 5, to drop elements from the individual rows.

7. Rewrite your code from the exercises at the end of Chapter 13, only this time assemble the stiffness matrices as sparse matrices.

Chapter 17

Iterative Methods for Large Sparse Linear Systems

In this chapter, we consider large sparse linear systems of equations such as those obtained from the discretization of PDEs. The sparse coefficient matrix is implemented in efficient data structures that avoid storing trivial matrix elements. Iterative linear-system solvers, which preserve the sparsity of the original coefficient matrix, should thus be used. We describe and implement basic and advanced iterative methods such as relaxation, incomplete factorization, multigrid, and acceleration techniques. The present object-oriented framework is particularly useful in the implementation.

17.1 Iterative vs. Direct Methods

As we've seen in Chapters 7 and 12, the numerical discretization of PDEs often leads to a large sparse linear system of equations of the form

$$Ax = f,$$

where A is the coefficient matrix of order N, f is a given N-dimensional vector, and x is the N-dimensional vector of unknowns. Here, the ith component in x, x_i, is the numerical approximation to the solution of the PDE at the ith point in the grid or node in the mesh (in some predetermined order).

Most often, the coefficient matrix A is sparse; that is, most of the elements in it are 0. For example, when A is the stiffness matrix arising from the finite-element discretization of the diffusion equation, $A_{i,j} \neq 0$ only if i and j correspond to nodes in the same triangle in the mesh.

The naive implementation of the matrix object in Chapter 2, Section 20, is obviously highly inefficient here, because it stores all the elements in the matrix, including the zeroes. A much better implementation that avoids storing the zeroes is the implementation in Chapter 16. In order to benefit from this implementation, one should make sure that the coefficient matrix also remains sparse in the solution process.

The most common and general linear-system solver, Gaussian elimination, fails to preserve sparsity. In fact, this method may lead to a large number of zero elements in A being replaced by nonzero elements (fill-in) during the elimination process. The fill-in phenomenon is particularly expensive in terms of both time and storage, because each new nonzero element requires memory allocation using the expensive "new" command.

Due to the above problem, although Gaussian elimination is robust and reliable, it is prohibitively slow and expensive for most practical applications. In this chapter, we consider iterative linear-system solvers, which avoid fill-in and preserve the sparsity of the matrix.

17.2 Iterative Methods

In iterative methods, one picks some initial N-dimensional vector $x^{(0)}$ (initial guess) to approximate the numerical solution x. Of course, the initial error $x^{(0)} - x$ may be quite large; still, one can improve the approximation and reduce the error iteratively as described below.

Since the numerical solution x is unknown, the error $x^{(0)} - x$ is unknown as well. In fact, if it were known, then it could be subtracted from $x^{(0)}$, yielding immediately the required numerical solution:

$$x^{(0)} - \left(x^{(0)} - x\right) = x.$$

In practice, although the error is unavailable, it can still be approximated, yielding subsequent improvements to the initial approximation $x^{(0)}$.

The only quantity that is available to approximate the error is the residual:

$$f - Ax^{(0)} = -A\left(x^{(0)} - x\right).$$

Thus, the residual is related to the error through the matrix A. In other words, the error can be obtained from the residual by inverting A. Since we don't know how to do this, we'll have to do it approximately.

If A^{-1} were available, then we could multiply the above equation by it, obtain the error, and have the solution x immediately. Of course, A^{-1} is not available, so we must approximate it to the best of our ability. We hope that the residual, multiplied by the approximate inverse of A, will provide a sufficiently good approximation to the error. The approximate error will then be subtracted from $x^{(0)}$, giving a better approximation to x. This procedure is repeated iteratively, yielding better and better approximations to x. When a sufficiently good approximation to x is reached, the iteration terminates. This is the idea behind iterative methods.

Let's now define the iterative method in detail. Let $\mathcal{P}$ be an easily invertible matrix of order N that approximates A in some sense (to be discussed later). By "easily invertible," we mean that $\mathcal{P}$ is constructed in such a way that the linear system

$$\mathcal{P}e = r$$

(where r is a given N-dimensional vector and e is the N-dimensional vector of unknowns) is easily solved. Of course, $\mathcal{P}$ is never inverted explicitly, because this may require a prohibitively large number of computations.

The iterative method is defined as follows. For $i = 0, 1, 2, \ldots,$ define

$$x^{(i+1)} = x^{(i)} + \mathcal{P}^{-1}\left(f - Ax^{(i)}\right).$$

In other words, in every iteration, the approximation $x^{(i)}$ is improved by adding to it the residual $f - Ax^{(i)}$ multiplied on the left by the inverse of $\mathcal{P}$. As mentioned above, this inverse is never computed explicitly.

In practice, the iteration consists of two steps. First, the equation

$$\mathcal{P}e = f - Ax^{(i)}$$

is solved. As mentioned above, it is assumed that this is an easy task. Then, the improved approximation is computed in the second step:

$$x^{(i+1)} = x^{(i)} + e.$$

Let's see what happens to the error during the iteration. The new error at the $(i+1)$th iteration can be written as

$$x^{(i+1)} - x = x^{(i)} - x + \mathcal{P}^{-1}\left(f - Ax^{(i)}\right) = \left(I - \mathcal{P}^{-1}A\right)\left(x^{(i)} - x\right).$$

Thus, if $I - \mathcal{P}^{-1}A$ is small in some norm, then the error becomes smaller and smaller during the iteration, until a sufficiently small error is achieved and the iteration terminates. Therefore, $\mathcal{P}$ should approximate A in spectral terms; that is, it should have, if possible, approximately the same eigenvalues and eigenvectors as A. It is particularly important that $\mathcal{P}$ imitate the eigenvectors of A that correspond to small eigenvalues (in magnitude), because these eigenvectors are prominent in the error and hardest to annihilate.

The effect of multiplying the residual by $\mathcal{P}^{-1}$ on the left is actually the same as the effect of multiplying the original system by $\mathcal{P}^{-1}$ on the left:

$$\mathcal{P}^{-1}Ax = \mathcal{P}^{-1}f.$$

This is why the matrix $\mathcal{P}$ is also called a preconditioner: it improves the condition of the original system by practically replacing the original coefficient matrix A by $\mathcal{P}^{-1}A$, whose eigenvalues are much better bounded away from 0.

Of course, it is impossible to have a preconditioner $\mathcal{P}$ that is both a good spectral approximation to A and easily invertible. The challenge is to find a good compromise between these two desirable properties and come up with a preconditioner that is not too expensive to invert and yet approximates A sufficiently well to guarantee rapid convergence to the numerical solution x.

The multiplication of the original system by $\mathcal{P}^{-1}$ on the left is also attractive in terms of the error estimate. Indeed, the only available quantity that is reduced during the iteration is the preconditioned residual

$$r = \mathcal{P}^{-1}Ax^{(i)} - \mathcal{P}^{-1}f.$$

Now, the error is related to r by

$$x^{(i)} - x = \left(\mathcal{P}^{-1}A\right)^{-1}r.$$

If $\mathcal{P}$ is indeed a good spectral approximation to A, then the eigenvalues of $\mathcal{P}^{-1}A$ are well away from 0, and, hence, $\mathcal{P}^{-1}A$ has a well-bounded inverse. Therefore, a small r also means a rather small error, as required.

17.3 Gauss–Seidel Relaxation

The most basic family of iterative methods is the family of relaxation methods. In relaxation, the components of the approximate solution $x^{(k)}$ at the kth iteration are scanned and improved (updated, relaxed) one by one. Once all the components $x_i^{(k)}$ $(0 \le i < N)$ are relaxed, then the relaxation sweep is over, the $(k+1)$th iteration is complete, and the improved approximation, $x^{(k+1)}$, is formed.

Let us now specify how the components are updated. In the Gauss–Seidel relaxation method, the ith component, $x_i^{(k)}$, is updated in such a way that the ith equation in the linear system is satisfied. Of course, this equation is satisfied only temporarily and is violated again as soon as the next component, $x_{i+1}^{(k)}$, is changed. Still, the entire relaxation sweep may provide an improved approximation $x^{(k+1)}$.

The entire relaxation sweep can be written as follows. First, initialize $x^{(k+1)}$ by $x^{(k)}$:

$$x^{(k+1)} \equiv x^{(k)}.$$

Then, update the components of $x^{(k+1)}$ one by one as follows: for $i = 0, 1, 2, \ldots, N-1$, do

$$x_i^{(k+1)} \leftarrow x_i^{(k+1)} + \left(A_{i,i}\right)^{-1} \left(f - Ax^{(k+1)}\right)_i$$

(where $\leftarrow$ stands for substitution). Once the relaxation sweep is complete, an improved new approximation $x^{(k+1)}$ is formed.

Using the product of row and vector defined in Chapter 16, Section 3, we can implement the Gauss–Seidel iteration most elegantly as follows. This implementation is also efficient in terms of storage, because $x^{(k+1)}$ occupies the same storage as $x^{(k)}$:

```
template<class T>
void GaussSeidel(const sparseMatrix<T>&A,
        const dynamicVector<T>&f, dynamicVector<T>&x){
  for(int i=0; i<f.dim(); i++)
    x(i) += (f[i] - *A.item[i] * x) / A(i,i);
} //  Gauss-Seidel relaxation
```

Note that, in order to have access to the "item" field of the sparse matrix 'A', the "Gauss-Seidel" function must be a friend of the "sparseMatrix" class in Chapter 16, Section 4, as is indeed declared there.

17.4 Jacobi Relaxation

In the Gauss–Seidel relaxation, the components are relaxed one by one. The relaxation of the ith component, $x_i^{(k)}$, uses the already updated components $x_0^{(k+1)}, x_1^{(k+1)}, \ldots, x_{i-1}^{(k+1)}$. The order in which the relaxation takes place is, thus, important. In the Jacobi relaxation method, on the other hand, all the components are relaxed independently, using only components from $x^{(k)}$, not from $x^{(k+1)}$. This approach, although inferior to the previous one in terms of convergence rate, is more suitable for parallel implementation (see Chapter 18, Section 15).

The Jacobi relaxation method can be written as follows: for every $0 \le i < N$,

$$x_i^{(k+1)} \equiv x_i^{(k)} + \left(A_{i,i}\right)^{-1} \left(f - Ax^{(k)}\right)_i .$$

In fact, the Jacobi iteration can also be written in vector form as follows:

$$x^{(k+1)} \equiv x^{(k)} + diag(A)^{-1} \left(f - Ax^{(k)}\right).$$

Actually, here $diag(A)$ (the main diagonal of A) serves as a preconditioner.

If we use the arithmetic operations in the "dynamicVector" and "sparseMatrix" classes, the implementation of the Jacobi iteration is particularly straightforward. First, however, we need to define an "operator/" function that takes a vector v and a matrix A as arguments and returns the vector $diag(A)^{-1}v$:

```
template<class T>
const dynamicVector<T>
operator/(const dynamicVector<T>&v,
            const sparseMatrix<T>&A){
  dynamicVector<T> result(v);
  for(int i=0; i<v.dim(); i++)
    result(i) /= A(i,i);
  return result;
}  //  vector divided by the main diagonal of matrix
```

This operator is now used in the Jacobi iteration that computes $x^{(k+1)}$ and places it in the same storage that was occupied by $x^{(k)}$:

```
template<class T>
void Jacobi(const sparseMatrix<T>&A,
        const dynamicVector<T>&f, dynamicVector<T>&x){
  x += (f - A * x) / A;
}  //  Jacobi relaxation
```

Note that, since the priority rules of user-defined operators are the same as the priority rules built in C for arithmetic operations, one must use parentheses to form the residual before dividing it by the main diagonal of A. Using the above function, one can now use a loop of 100 Jacobi iterations as follows:

```
  for(int i=0; i<100; i++)
    Jacobi(A,f,x);
  print(x);
```

In general, it is recommended to avoid constructing big objects such as dynamic vectors in long loops, because this construction may be particularly time-consuming. Here, however, the "Jacobi()" function called in the loop constructs no extra objects, because the arguments are passed to it by reference and it returns no value.

The Jacobi iteration uses the old approximation $x^{(k)}$ to form the residual and compute the new iteration $x^{(k+1)}$. In this respect, it is analogous to the explicit scheme in Chapter 7, Section 5, which uses the solution at a particular time step to advance to the next time step.

The Jacobi iteration converges to the numerical solution x whenever A is diagonally dominant [46]. However, the convergence may be extremely slow. For example, for the Poisson equation with Dirichlet boundary conditions discretized by finite differences on a uniform grid, more than N^2 Jacobi iterations are required to converge with reasonable accuracy.

17.5 Symmetric Gauss–Seidel

The Gauss–Seidel iteration converges not only whenever A is diagonally dominant but also whenever A is symmetric and positive definite (SPD) [46]. Although the convergence rate is better than that of the Jacobi iteration, it is still prohibitively slow for many problems. For example, the Gauss–Seidel iteration converges only twice as fast as the Jacobi iteration for the Poisson equation with Dirichlet boundary conditions discretized by finite differences on a uniform grid. The search for a more efficient iterative method is, thus, still on.

The Gauss–Seidel iteration can also be written in terms of a preconditioning method, as in Section 17.2, with $\mathcal{P}$ being the lower triangular part of A. However, this preconditioner is nonsymmetric even when A is SPD, which is a considerable drawback (see Section 17.11). In order to have a symmetric preconditioner, one should perform another relaxation sweep in the reverse order. The two complete sweeps form a so-called symmetric Gauss–Seidel relaxation, which is implemented as follows:

```
template<class T>
void symmetricGaussSeidel(const sparseMatrix<T>&A,
        const dynamicVector<T>&f, dynamicVector<T>&x){
  for(int i=0; i<f.dim(); i++)
    x(i) += (f[i] - *A.item[i] * x) / A(i,i);
  for(int i=f.dim()-2; i>=0; i--)
    x(i) += (f[i] - *A.item[i] * x) / A(i,i);
} // symmetric Gauss-Seidel relaxation
```

When written in terms of preconditioning, as in Section 17.2, the symmetric Gauss–Seidel iteration has an SPD preconditioner $\mathcal{P}$ whenever A is SPD (Chapter 18, Section 16).

The Gauss–Seidel relaxation depends on the particular order of the unknowns. Different orders may produce slightly different convergence rates.

17.6 The Normal Equation

When A is nonsymmetric or indefinite, the Gauss–Seidel iteration does not necessarily converge. Still, convergence can be guaranteed when the original equation is multiplied in advance by the transpose of A:

$$A^t A x = A^t f.$$

This system of equations is called the normal equation. Because $A^t A$ is always SPD, the Gauss–Seidel iteration applied to the normal equation always converges to x.

The Gauss–Seidel iteration applied to the normal equation is known as the Kacmarz iteration [45]. Although this iteration always converges to the numerical solution x, the convergence may be extremely slow. The Kacmarz iteration may be useful only when no other iterative method converges, such as for highly indefinite systems. For better-posed systems such as those arising from diffusion equations, more efficient iterative methods should be used.

17.7 Incomplete Factorization

In this section, we describe the incomplete LU (ILU) factorization of the sparse matrix A. This factorization is then used in the ILU iterative method for the numerical solution of the linear system $Ax = f$. The ILU iteration is usually considered much more efficient than the above relaxation methods in terms of convergence rate (that is, it usually requires fewer iterations to converge to x), particularly when it is accelerated by one of the acceleration techniques in Section 17.11 below. However, as we shall see, this is not always the case, and ILU may well be inferior to symmetric Gauss–Seidel for SPD examples.

As discussed in Chapter 2, Section 21, the decomposition (or factorization) of the matrix A as the product

$$A = LU,$$

where L is a lower triangular matrix with main-diagonal elements that are all equal to 1 and U is an upper triangular matrix, is computed in Gaussian elimination. This decomposition is then used to give the solution of the linear system as

$$x = A^{-1}f = U^{-1}L^{-1}f.$$

Of course, U^{-1} and L^{-1} are never calculated explicitly. Instead, x can be calculated in two steps. First, the lower triangular system

$$Ly = f$$

is solved for the unknown vector y (forward elimination in L). Then, the upper triangular system

$$Ux = y$$

is solved for the unknown vector x (back substitution in U). This direct method gives the desired numerical solution x.

As discussed in Section 17.1, the direct method for finding x is impractical for large sparse linear systems because of the large amount of fill-in produced in Gaussian elimination. In other words, the L and U factors are no longer sparse and have to be stored as expensive dense matrices. The ILU factorization attempts to avoid this problem by producing approximate L and U matrices (incomplete factors), in which elements that are too small in some sense are dropped [26, 15]. These inexact factors are now used to form the preconditioner

$$\mathcal{P} = LU.$$

The iteration that uses this preconditioner is called the ILU iteration.

If the approximate factors L and U produced by the incomplete factorization are close enough to the original L and U of Gaussian elimination, then the ILU iteration should converge rapidly to the numerical solution x.

Here is one of the many possible algorithms to construct the approximate L and U factors. The algorithm uses some small predetermined threshold, say 0.1, to detect small matrix elements that should be dropped. It is also assumed that the pivots $U_{i,i}$ are not too small in magnitude, so the algorithm doesn't fail due to division by (almost) 0. (This is guaranteed, e.g., when the original matrix A is an M-matrix [26].)

Algorithm 17.1.

1. *Initialize* $L = (L_{i,j})_{0 \le i,j < N}$ *to be the identity matrix* I.
2. *Initialize* $U = (U_{i,j})_{0 \le i,j < N}$ *to be the same matrix as* A.
3. *For* $i = 0, 1, 2, 3, \ldots, N - 1$, *do the following:*
   - *For* $j = 0, 1, 2, \ldots, i - 1$, *do the following:*
     (a) *Define*
     $$factor = U_{i,j}/U_{j,j}.$$
     (b) *Set*
     $$U_{i,j} \leftarrow 0.$$
     (c) *If* $|factor| \ge threshold$, *then do the following:*
       - *For* $k = j + 1, j + 2, \ldots, N - 1$, *set*
       $$U_{i,k} \leftarrow U_{i,k} - factor \cdot U_{j,k}.$$
       - *Set*
       $$L_{i,j} \leftarrow factor.$$
   - *For* $j = i + 1, i + 2, \ldots, N - 1$, *if*
   $$|U_{i,j}| \le threshold \cdot |U_{i,i}|,$$
   then set
   $$U_{i,j} \leftarrow 0.$$

Next, we provide the detailed implementation of the ILU factorization. The function "factorize" is a member of the "sparseMatrix" class, so it can access its "item" field, which contains the addresses of the rows. Using this access privilege, the "factorize" function modifies the current "sparseMatrix" object and converts it from the original matrix A into the upper triangular factor U. In the process, it also creates the lower triangular factor L and returns it as output. Of course, a copy of the original matrix A must also be stored elsewhere for safekeeping before the function is called, because the current "sparseMatrix" object is changed.

Here is the actual code:

```
template<class T>
const sparseMatrix<T>
sparseMatrix<T>::factorize(double threshold){
  sparseMatrix<T> L(rowNumber());
```

Here, we have defined the "sparseMatrix" object 'L' that will eventually be returned by the function. This object is slightly different from the incomplete factor L, because its main-diagonal elements are missing. Because all these elements are equal to 1, there is no need to store them explicitly. Only the first row in 'L' contains a zero main-diagonal element as its only element:

```
    L.item[0] = new row<T>(0.,0);
```

Like the above incomplete-factorization algorithm, the "factorize" function uses three nested loops. The outer "for" loop scans the rows in the original matrix:

```
    for(int i=0; i<rowNumber(); i++){
```

The inner "while" loop creates the elements in the 'i'th row in 'L' and eliminates the corresponding elements in the current "sparseMatrix" object. At the end of this loop, the 'i'th row in the current "sparseMatrix" object will start from column 'i', so it will be the required 'i'th row in U, and the 'i'th row in 'L' will also be complete:

```
      while(item[i]&&(item[i]->getColumn() < i)){
```

The innermost nested loop is executed in the "+=" operator, in which a fraction of a previous U-row is subtracted from the current row, provided that this fraction is not too small:

```
        T factor = item[i]->getValue() /
              item[item[i]->getColumn()]->getValue();
        if(abs(factor) >= threshold){
          row<T> r =
              (-factor) * *item[item[i]->getColumn()];
          *item[i] += r;
```

Elements encountered in the "while" loop in the 'i'th row of the current "sparseMatrix" object are appended at the end of the 'i'th row in 'L' (provided that they are not too small). To do this, one must make sure that the 'i'th row in 'L' already exists. Otherwise, it should be created using the "new" command:

```
          if(L.item[i])
            L.item[i]->append(factor, item[i]->getColumn());
          else
            L.item[i]=new row<T>(factor,item[i]->getColumn());
        }
```

The element that has been appended at the end of the 'i'th row in 'L' must be dropped from the beginning of the 'i'th row in U:

```
        item[i]->dropFirstItem();
      }
```

This completes the "while" loop. If, however, this loop has terminated and no element in the 'i'th row in the original matrix is sufficiently large to be placed in 'l', then the 'i'th row in 'L' is not yet constructed. In this case, it is constructed here as the zero row:

```
      if(!L.item[i])L.item[i] = new row<T>(0.,0);
```

So far, we have made sure that the elements in the 'i'th row in 'L' are not too small in magnitude, which guarantees sparsity. But what about U, which is contained in the current "sparseMatrix" object? After all, the 'i'th row in it may have filled in with the fractions of previous rows subtracted from it!

Here we also drop small elements from the 'i'th row in U. By "small" we mean smaller (in magnitude) than the threshold times the corresponding main-diagonal element. (Usually, the threshold is 0.1 or 0.05. Zero threshold leads to complete factorization, which is actually Gaussian elimination.)

```
      item[i]->
        truncateItems(threshold * abs(item[i]->getValue()));
    }
```

This completes the "for" loop over the rows. The incomplete lower triangular factor 'L' is now returned:

```
    return L;
  }  //  incomplete LU factorization
```

This completes the "factorize" function. Note that the incomplete factor 'L' cannot be returned by reference, because the local variable 'L' disappears at the end of the function block, so a reference to it is a reference to nothing. It must be returned by value, that is, copied to a temporary unnamed "sparseMatrix" object that stores it until it has been placed or used. This is indicated by the word that appears before the function name at the beginning of the function block: "sparseMatrix" rather than "sparseMatrix&".

The implementation of the function "forwardElimination" (forward elimination in L) and "backSubstitution" (back substitution in U) is given in Section A.11 of the Appendix. In order to have access to the individual rows, these functions must be declared as members of the "sparseMatrix" class, as is indeed done in Chapter 16, Section 4. Assuming that these functions are available, the ILU iteration

$$x^{(k+1)} = x^{(k)} + U^{-1}L^{-1}\left(f - Ax^{(k)}\right)$$

is implemented as follows:

```
  template<class T>
  void ILU(const sparseMatrix<T>&A,
      const sparseMatrix<T>&L,
      const sparseMatrix<T>&U,
      const dynamicVector<T>&f, dynamicVector<T>&x){
    x +=
      U.backSubstitution(L.forwardElimination(f - A * x));
  }  //  ILU iteration
```

It is assumed that the incomplete factors L and U have been calculated once and for all in an early call to the "factorize" function, so they can be passed to the "ILU" function by reference for further use. It is, of course, highly inefficient to recalculate them in each ILU iteration. The new iteration $x^{(k+1)}$ is stored in the same vector in which $x^{(k)}$ was stored to save memory.

It turns out that the ILU iteration is particularly suitable for nonsymmetric linear systems such as those in Chapter 7, Sections 6 and 7. For SPD systems like those arising from the discretization of the diffusion equation, however, the symmetric Gauss–Seidel relaxation may be more suitable. Many more ILU versions [18] can be implemented by modifying the above code.

17.8 The Multigrid Method

The ILU iteration described above is still not sufficiently efficient for large-scale problems. This is because, by dropping certain matrix elements, the incomplete factorization fails to approximate well the nearly singular eigenvectors of A, which are usually prominent in the error.

The multigrid iterative method attempts to approximate these important error modes on a coarser grid. Although this grid contains fewer points, the global error modes are still well approximated and can be solved for.

The coarse grid is actually a subset of the original set of indices of unknowns:

$$c \subset \{0, 1, 2, \ldots, N-1\}.$$

One possible algorithm to define the subset c is as follows. The algorithm uses some small predetermined threshold, say 0.05. It is assumed that the main-diagonal elements in A are not too small in magnitude.

Algorithm 17.2.

1. *Initialize c to contain all the unknowns:*

$$c = \{0, 1, 2, \ldots, N-1\}.$$

2. *For $i = 0, 1, 2, \ldots, N-1$, do the following:*
   - *if $i \in c$, then, for every $0 \le j < N$ for which $j \ne i$ and*

$$|A_{i,j}| \ge threshold \cdot |A_{i,i}|,$$

   drop j from c.

3. *For $i = 0, 1, 2, \ldots, N-1$, do the following:*
   - *if $i \notin c$ and for every $j \in c$*

$$\frac{A_{i,j}}{A_{i,i}} \ge -threshold,$$

   then add i back to c.

(For other versions, see [9], [33], and [39].)

Once the coarse grid c has been defined, we denote the number of elements in it by $|c|$. The coarse grid c can be stored in the computer as an N-dimensional vector v with integer components. If $i \notin c$, then $v_i = 0$, and if $i \in c$, then $v_i > 0$. It is a good idea to let the nonzero v_i's indicate the suitable order in c, that is, use $v_i = 1, 2, 3, \ldots, |c|$ to number the indices $i \in c$.

We also denote by f the set of indices that are excluded from c:

$$f \equiv \{0, 1, 2, \ldots, N-1\} \setminus c.$$

The original set of indices of unknowns

$$\{0, 1, 2, \ldots, N-1\} = c \cup f$$

is also called the fine grid.

In order to approximate the original problem on the coarse grid, we must have a method of transferring information between the coarse and fine grids. In other words, we must have a prolongation operator P to transform a function defined on the coarse grid into a function defined on the fine grid and a restriction operator R to transform a fine-grid function into a coarse-grid function. In fact, P is a rectangular $N \times |c|$ matrix, and R is a rectangular $|c| \times N$ matrix.

Let us introduce a straightforward algorithm to define P. (The definition of R will follow soon.) The algorithm uses a small predetermined threshold to drop from P elements that are too small in magnitude. In fact, P is constructed from the elements in A that agree with the L-matrix property; for example, if A has positive main-diagonal elements, then only its negative off-diagonal elements are used in P. This method of construction is necessary to guarantee that the multigrid method is efficient for matrices that are neither L-matrices nor diagonally dominant.

The above definition of c guarantees that, for every $i \in f$, the ith row in A contains at least one negative off-diagonal element that lies in a column $j \in c$ and whose magnitude is not too small. (Otherwise, i would have been dropped from f and added to c.) These negative off-diagonal elements are scaled in P so that they become positive. Therefore, P contains nonnegative elements only and actually represents an interpolation operator that uses values at points in c to define values at points in f as well.

It is assumed that the main-diagonal elements in A are not too small in magnitude. Here is the algorithm to define P.

Algorithm 17.3.

1. *Initialize P by $P = A$.*

2. *For every $i \in f$ and $j \in c$, if*

$$\frac{P_{i,j}}{P_{i,i}} > -threshold,$$

then drop $P_{i,j}$ from P and replace it by 0:

$$P_{i,j} \leftarrow 0.$$

3. *For every index $i \in c$, replace the ith row*

$$(P_{i,0}, P_{i,1}, \ldots, P_{i,N-1})$$

in P by the standard unit row $e^{(i)}$, of which all components vanish except the ith one, which is equal to 1.

4. *For every $j \in f$, drop the corresponding column from P. (In this step, P becomes rectangular.)*

5. *For every $i \in f$, divide the ith row in P by its row-sum. (After this step, the row-sums in P are all equal to 1.)*

Then, define

$$R \equiv P^t.$$

Finally, the matrix Q (of order $|c|$) that approximates A on the coarse grid is defined by

$$Q \equiv RAP.$$

The inverse of the preconditioner $\mathcal{P}$ in Section 17.2 can now be defined by

$$\mathcal{P}^{-1} \equiv PQ^{-1}R.$$

The iteration that uses this preconditioner is called the two-level correction. Note that the iterative methods in Section 17.2 require only the application of $\mathcal{P}^{-1}$, not $\mathcal{P}$; therefore, the two-level correction is well defined.

When the two-level correction is preceded and followed by several iterations of some relaxation method (say, ILU or symmetric Gauss–Seidel), the entire procedure is called two-level iteration.

Of course, Q^{-1} is never computed explicitly. Instead, only a low-order linear system of the form

$$Qe = r$$

is solved, where r is a given $|c|$-dimensional vector and e is the $|c|$-dimensional vector of unknowns. In fact, this low-order system can be solved approximately by a single iteration of the same method, called recursively. This approach produces the multigrid iteration or V-cycle (Figure 17.1).

The main computational work in the multigrid iteration is done in the relaxations carried out in each level before and after the coarse-grid correction. Choosing a suitable relaxation method within multigrid is thus most important. For nonsymmetric systems (Section 17.12), ILU with no fill-in seems to be the optimal choice. For symmetric systems, the symmetric Gauss–Seidel relaxation may be preferable. Both of these methods, however, are efficient only on the usual sequential computer. On parallel computers, the so-called damped Jacobi relaxation method seems more attractive (see Chapter 18, Section 15). This method differs from the Jacobi method in Section 17.4 in that only a fraction of the residual is added:

```
x += 0.5 * (f - A * x) / A;
```

In the present applications, however, we stick for simplicity to the symmetric Gauss–Seidel relaxation within multigrid.

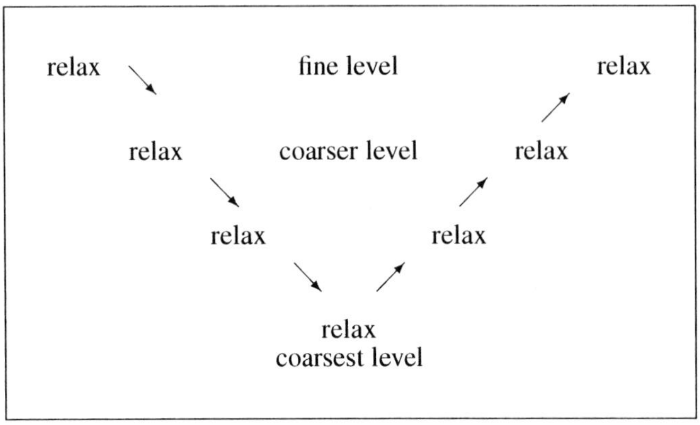

Figure 17.1. *The multigrid iteration has the shape of the letter V: first, relaxation is used at the fine grid; then, the residual is transferred to the next, coarser, grid, where a smaller V-cycle is used recursively to produce a correction term, which is transferred back to the fine grid; finally, relaxation is used again at the fine grid.*

17.9 Algebraic Multigrid (AMG)

The prolongation matrix P of Section 17.8 is particularly suitable for diagonally dominant matrices, for which the theory in Chapters 10 to 12 in [39] applies. For more general systems, the multigrid iteration may still converge well, even though no theory applies.

Other multigrid versions may differ from the above method in the way they define the prolongation operator P. The AMG method in [9, 33] differs from the above version in one detail only: before columns are dropped from P, as in Algorithm 17.3 in Section 17.8, the elements in them are distributed among the remaining elements in the same row. (This is in the spirit of [18].)

The algorithm to construct P is as follows. (The algorithm uses some small predetermined threshold, say 0.05.)

Algorithm 17.4.

1. *Initialize P by $P = A$.*
2. *For every $i \in f$ and $0 \le j < N$, if $i \ne j$ and*

$$\frac{P_{i,j}}{P_{i,i}} > -threshold,$$

then drop $P_{i,j}$ from P and replace it by 0:

$$P_{i,j} \leftarrow 0.$$

3. *Define the matrix B by $B = P$.*
4. *For every index $i \in c$, replace the ith row*

$$(P_{i,0}, P_{i,1}, \ldots, P_{i,N-1})$$

in P by the standard unit row $e^{(i)}$, of which all components vanish except the ith one, which is equal to 1.

5. *For every i and j that are both in f and satisfy $i \neq j$ and $P_{i,j} \neq 0$, define*

$$W_{i,j} \equiv \sum_{k \in c,\ P_{i,k} \neq 0} B_{j,k}.$$

6. *For every i and j that are both in f and satisfy $i \neq j$ and $P_{i,j} \neq 0$, add a fraction of $P_{i,j}$ to every nonzero element $P_{i,k}$ that lies in the same row and in column $k \in c$ as follows:*

$$P_{i,k} \leftarrow P_{i,k} + \frac{B_{j,k}}{W_{i,j}} P_{i,j}.$$

7. *For every $j \in f$, drop the corresponding column from P. (In this step, P becomes rectangular.)*

8. *For every $i \in f$, divide the ith row in P by its row-sum. (After this step, the row-sums in P are all equal to 1.)*

The rest of the details of the definition are as in Section 17.8; that is, $R = P^t$ and $Q = RAP$. The coarse-grid equation $Qe = r$ is solved approximately by one AMG iteration, called recursively.

Because AMG has no theoretical background, it is not clear whether it works for special diagonally dominant examples such as the one in Figure 7.1 in [39]. The present numerical examples show practically no difference between the multigrid versions.

17.10 Implementation of Multigrid

As we've seen in Section 17.8, the multigrid iteration relies strongly on recursion. The coarse-grid system of the form $Qe = r$ is solved approximately by one multigrid iteration (V-cycle), which is applied to it recursively. This recursive call requires an even coarser grid, and the recursive process continues until the coarsest grid is reached, where a small system is solved (see Figure 17.1). Naturally, these matrices should thus be stored in a recursively defined object.

The "multigrid" class is indeed defined recursively. It contains not only the "sparseMatrix" objects required to store the fine-grid coefficient matrix and transfer operators to and from the coarse grid, but also a pointer-to-multigrid field that contains the address of the next (coarser) "multigrid" object.

This recursive structure is useful not only in the multigrid V-cycle but also in basic member functions of the "multigrid" class, such as constructors and assignment operators.

The structure of the "multigrid" object is displayed schematically in Figure 17.2. The "multigrid" object contains five "sparseMatrix" objects: 'A', 'U', 'L', 'P', and 'R'. The function "createTransfer" constructs the rectangular matrices 'P' and 'R', which transfer information between the fine and coarse grids. More precisely, 'R' transforms fine-grid vectors into coarse-grid vectors, and 'P' transforms coarse-grid vectors into fine-grid vectors.

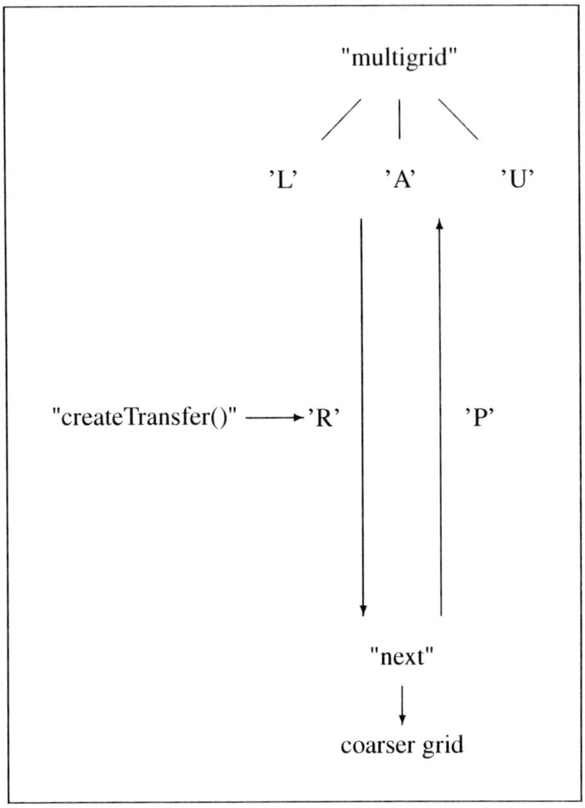

Figure 17.2. *The "multigrid" object contains the square matrices A, U, and L and the rectangular matrices R and P, which are used to transfer information to and from the coarse grid.*

The information about the coarse grid, which is the next "multigrid" object in the multigrid hierarchy, is pointed at by the pointer-to-multigrid "next". This recursive pattern is used later in the implementation of the V-cycle.

The following code uses some predetermined parameters. "useILU" is an integer parameter that determines whether ILU or symmetric Gauss–Seidel is used as the relaxation method within the V-cycle. (In most of the present applications, "useILU" is zero.) "gridRatio" determines whether the next-grid matrix is sufficiently small and deserves to be constructed or the grid hierarchy should terminate. (In most of the present applications, "gridRatio" is 0.95. We also use a negative "gridRatio" when we want to use a trivial grid hierarchy that contains only the original grid.)

Here is the block of the "multigrid" class:

```
const double gridRatio=0.95;
template<class T> class multigrid{
      sparseMatrix<T> A;
      sparseMatrix<T> U;
```

```
        sparseMatrix<T> L;
        sparseMatrix<T> P;
        sparseMatrix<T> R;
        multigrid* next;
    public:
        multigrid():next(0){
        }  //  default constructor
```

The copy constructor copies the sparse matrices in the "multigrid" argument and then calls itself recursively to copy the coarse-grid matrices:

```
        multigrid(const multigrid&mg)
            : A(mg.A),U(mg.U),L(mg.L),P(mg.P),R(mg.R),
             next(mg.next ? new multigrid(*mg.next) : 0){
        }  //  copy constructor
```

The most important function in the "multigrid" class is the constructor that uses a "sparseMatrix" argument to construct all the required sparse matrices in the entire grid hierarchy. This is done most elegantly in the initialization list as follows. Recall that the fields in the object are not necessarily constructed in their order in the initialization list but rather in the order in which they are declared in the class block. Here, the fields in the "multigrid" class are 'A', 'U', 'L', 'P', 'R', and "next" (in that order). First, 'A' is initialized to be the same as the matrix that is passed as an argument. Now, 'U' is initialized to be the same as 'A' and then changes and takes its final form when 'L' is initialized using the "factorize" function of Section 17.7, which modifies 'U' and returns 'L'. Similarly, 'P' is initialized to be the same as 'A' and then changes and takes its final form when 'R' is constructed using the "createTransfer" function, which modifies 'P' and returns 'R'. Finally, the "next" field is filled with the address of a new "multigrid" object, which is created using a recursive call to the same constructor itself, with the low-order matrix Q ('R' times 'A' times 'P') as argument. A necessary condition, however, to construct this new "multigrid" object is that it indeed contains much smaller matrices. In fact, the order of the matrices in it must be smaller than "gridRatio" times the order of 'A'. Otherwise, the hierarchy of the grids terminates, and "next" takes the zero value:

```
        multigrid(const sparseMatrix<T>&m)
          : A(m), U(useILU ? A : 0),
          L(useILU ? U.factorize(0.05) : 0),P(A),
           R(P.createTransfer()),
           next(R.rowNumber()<=gridRatio*A.rowNumber() ?
           new multigrid(R*A*P) : 0){
       }  //  constructor with matrix argument
```

Here is the definition of the destructor:

```
        const multigrid<T>& operator=(const multigrid<T>&);
        ~multigrid(){
          delete next;
          next = 0;
        }  //  destructor
```

The "Vcycle" and "print" functions are only declared here and defined later.

```
        const dynamicVector<T>&
          Vcycle(const dynamicVector<T>&, dynamicVector<T>&);
        friend void print<T>(const multigrid<T>&);
};
```

This completes the block of the "multigrid" class.

The V-cycle uses some more predetermined integer parameters:

1. "Nu1" is the number of prerelaxations that are used before the coarse-grid correction. ("Nu1" = 1 is used here.)

2. "Nu2" is the number of post-relaxations that are used after the coarse-grid correction. ("Nu2" = 1 is used here.)

3. "cycleIndex" is the number of coarse-grid corrections. ("cycleIndex" = 1 is used here.)

4. "NuCoarse" is the number of relaxations in the coarsest grid. ("NuCoarse" = 1 is used here.)

Here is the actual implementation of the multigrid V-cycle:

```
const int Nu1=1,Nu2=1,cycleIndex=1,NuCoarse=1,useILU=0;
template<class T>
const dynamicVector<T>&
multigrid<T>::Vcycle(
          const dynamicVector<T>&f, dynamicVector<T>&x){
   if(next){
     for(int i=0; i<Nu1; i++){
       if(useILU)
         ILU(A,L,U,f,x);
       else
         symmetricGaussSeidel(A,L,U,f,x);
     }
```

So far, we have completed the first "Nu1" relaxations, whatever relaxation method is used. We are now ready to do the coarse-grid correction:

```
     dynamicVector<T> residual = f - A * x;
     dynamicVector<T> correction(R.rowNumber(), 0.);
     for(int i=0; i<cycleIndex; i++)
       next->Vcycle(R * residual, correction);
     x += P * correction;
```

Now, we are ready to proceed with another "Nu2" relaxation:

```
      for(int i=0; i<Nu2; i++){
        if(useILU)
          ILU(A,L,U,f,x);
        else
          symmetricGaussSeidel(A,L,U,f,x);
      }
    }
```

If, however, we have already arrived at the coarsest level, then no coarse-grid correction is needed, and only relaxations are used:

```
    else
      for(int i=0; i<NuCoarse; i++){
        if(useILU)
          ILU(A,L,U,f,x);
        else
          symmetricGaussSeidel(A,L,U,f,x);
      }
    return x;
  } //  multigrid V-cycle
```

This completes the implementation of the multigrid V-cycle.

17.11 Preconditioned Conjugate Gradients (PCGs)

The iterative methods discussed so far have no memory in the sense that, once the ith iteration $x^{(i)}$ is computed, all previous iterations $x^{(0)}, x^{(1)}, \ldots, x^{(i-1)}$ are dropped and never used again. This property is attractive in terms of computer storage, because $x^{(i)}$ can occupy the same storage occupied by the previous iteration $x^{(i-1)}$. However, it is inefficient in terms of computer time, because the previous iterations contain valuable information that could be used to accelerate the convergence to the numerical solution x.

The PCG method combines the previous iterations with the current one to create an optimal approximation to x in the sense that it minimizes the sum of the squares of the algebraic error in $x^{(0)}$ plus the subspace spanned by the residuals of the previous iterations. This subspace, known as the Krylov subspace, can be defined as follows. Let

$$\tilde{r} = \mathcal{P}^{-1}\left(f - Ax^{(0)}\right) = x^{(1)} - x^{(0)}$$

be the preconditioned residual at the initial iteration $x^{(0)}$. The Krylov subspace of dimension i is

$$K_i(\tilde{r}) = \text{span}\left\{\tilde{r}, \mathcal{P}^{-1}A\tilde{r}, \left(\mathcal{P}^{-1}A\right)^2\tilde{r}, \ldots, \left(\mathcal{P}^{-1}A\right)^{i-1}\tilde{r}\right\}.$$

In order to construct the Krylov subspace, one must know how to apply the matrix $\mathcal{P}^{-1}A$ to a vector. This task may be tricky, because $\mathcal{P}^{-1}$ is not always available explicitly. Still, the application of $\mathcal{P}^{-1}A$ to a vector can easily be obtained from the iterative method as follows.

Let $\tilde{x}$ be the vector to which $\mathcal{P}^{-1}A$ should be applied, and let $\acute{x}$ be the vector obtained by applying the iterative method to $\tilde{x}$, with the zero right-hand-side vector $\vec{0}$. Then we have

$$\tilde{x} - \acute{x} = \tilde{x} - \left(\tilde{x} + \mathcal{P}^{-1}\left(\vec{0} - A\tilde{x}\right)\right) = \mathcal{P}^{-1}A\tilde{x}.$$

The PCG algorithm produces the optimal linear combination of vectors in $x^{(0)} + K_i(\tilde{r})$ in the sense of minimizing the algebraic error in terms of the energy norm induced by A. As i increases, the results produced by PCG converge to x. Of course, when $i = N$, the Krylov subspace is actually the entire N-dimensional vector space, so the result produced by PCG is mathematically exactly the same as x. However, the convergence of PCG is usually so rapid that even for i much smaller than N the result produced by PCG is sufficiently close to x and can serve as a suitable numerical solution.

In order to have the above minimum property, it is necessary that both A and $\mathcal{P}$ be SPD. When A is SPD, it can be shown that, for iterative methods such as Jacobi, symmetric Gauss–Seidel, ILU, and multigrid, $\mathcal{P}$ is SPD as well (see Chapter 18, Section 16 in this book and Chapter 10 in [39]). Therefore, the PCG method can be applied to these iterative methods to accelerate their convergence to the numerical solution x. Furthermore, the PCG method provides a stopping criterion to determine when the approximation to x is sufficiently accurate and the iteration may terminate.

Let us now describe the PCG algorithm in detail. In the following algorithm, the inner product of two vectors v and w is defined by

$$(v, w) \equiv v^t w.$$

If the vectors are complex, then it is defined by

$$(v, w) \equiv v^t \bar{w}.$$

The iteration terminates when the initial preconditioned residual has been reduced by six orders of magnitude (in terms of the energy norm of the preconditioned system).

Algorithm 17.5.

1. *Let $\tilde{x}$ be the initial-guess vector.*

2. *Let $\acute{x}$ be the result of applying the iterative method to $\tilde{x}$:*

$$\acute{x} = \tilde{x} + \mathcal{P}^{-1}(f - A\tilde{x}).$$

3. *Define the residual vector*

$$r = f - A\tilde{x}.$$

4. *Define the preconditioned residual vector*

$$\tilde{r} = \acute{x} - \tilde{x} = \mathcal{P}^{-1}(f - A\tilde{x}).$$

5. *Initialize the vectors p and $\tilde{p}$ by $p = r$ and $\tilde{p} = \tilde{r}$. ($\tilde{p}$ will serve as a direction vector to improve the approximate solution $\tilde{x}$.)*

6. *Define the scalars* $\gamma_0 = \gamma = (r, \tilde{r})$.

7. *Apply the iterative method to* $\tilde{p}$ *with zero right-hand side:*

$$\acute{p} = \tilde{p} - \mathcal{P}^{-1} A \tilde{p}.$$

8. *Compute the vector*

$$\tilde{w} = \tilde{p} - \acute{p} = \mathcal{P}^{-1} A \tilde{p}.$$

9. *Compute the vector*

$$w = A\tilde{p}.$$

10. *Compute the scalar*

$$\alpha = \gamma / (\tilde{p}, w).$$

11. *Update the approximate solution* $\tilde{x}$ *by*

$$\tilde{x} \leftarrow \tilde{x} + \alpha \tilde{p}.$$

12. *Update the residual by*

$$r \leftarrow r - \alpha w.$$

13. *Update the preconditioned residual by*

$$\tilde{r} \leftarrow \tilde{r} - \alpha \tilde{w}.$$

14. *Compute the scalar*

$$\beta = (r, \tilde{r}) / \gamma.$$

15. *Update* γ *by*

$$\gamma \leftarrow \beta \gamma.$$

16. *Update the direction vector* $\tilde{p}$ *by*

$$\tilde{p} \leftarrow \tilde{r} + \beta \tilde{p}.$$

17. *If* $\gamma / \gamma_0 > 10^{-12}$, *then go to step* 7.

The above algorithm converges to x whenever A and $\mathcal{P}$ are SPD. The convergence is particularly rapid when the basic iterative method is good (e.g., multigrid), so the preconditioner $\mathcal{P}$ is a good spectral approximation to A. However, when A or $\mathcal{P}$ is not SPD, the PCG method does not necessarily converge to x. Other Krylov-space methods that are not limited to the SPD case are required.

One such acceleration method is the conjugate gradient squared (CGS) method in [43]. When the CGS method is applied to the preconditioned system

$$\mathcal{P}^{-1} A x = \mathcal{P}^{-1} f,$$

one obtains a sequence of vectors that converges to x much more rapidly than the original sequence $x^{(i)}$, regardless of whether or not A or $\mathcal{P}$ is SPD.

Other general Krylov-space acceleration methods that can be applied to the preconditioned system are the transpose-free quasi-minimal residual (TFQMR) method in [16], which is actually a stable version of CGS, and the general minimal residual (GMRES) method in [34]. The implementation of these methods uses "sparseMatrix" and "dynamicVector" objects only. The detailed implementation can be found in Section A.13 of the Appendix.

17.12 Multigrid for Nonsymmetric Linear Systems

So far, we have considered symmetric systems with symmetric coefficient matrix $A = A^t$. But what if the coefficient matrix A is nonsymmetric; that is, $A \neq A^t$? In this case, the prolongation matrix P should be slightly different from that defined in the multigrid algorithm in Section 17.8. The algorithm to construct R and P should read as follows. First, P is defined by applying Algorithm 17.3 to A^t rather than A. Then, R is defined by $R = P^t$. Then, only P is redefined by applying Algorithm 17.3 to $(A + A^t)/2$ rather than A. Finally, Q is defined as usual by $Q = RAP$. In summary, R is defined from A^t, then P is redefined from $(A + A^t)/2$, and then Q is defined by $Q = RAP$. (This approach was introduced in [11] and extended in Chapter 12 of [39] to unstructured problems.)

The construction of the grid hierarchy then proceeds recursively at the coarser levels as well, with the same redefinition of the prolongation matrix.

It also turns out that, for nonsymmetric systems, ILU is better than Gauss–Seidel versions not only as a preconditioner but also as a relaxation method within the multigrid V-cycle, provided that it is used with no fill-in at all (see Chapter 12 in [39]).

For nonsymmetric systems, the PCG acceleration method is no longer applicable. More general acceleration methods, which are not limited to SPD matrices and preconditioners, are required. These acceleration techniques (e.g., CGS, TFQMR, and GMRES) use the preconditioned system

$$\mathcal{P}^{-1}Ax = \mathcal{P}^{-1}f$$

(where $\mathcal{P}$ is the preconditioner) rather than the original system $Ax = f$. In this system, the application of the coefficient matrix $\mathcal{P}^{-1}A$ to a vector is done as in Section 17.11.

17.13 Domain Decomposition and Multigrid

The key factor in the multigrid algorithm is the transfer of information from the fine to the coarse grid and vice versa. In particular, the prolongation operator P should transform a vector v (defined on the coarse grid c) into an extended vector Pv (defined on the original mesh) with energy norm as small as possible. In other words, Pv must be close to a nearly singular eigenvector of the stiffness matrix A.

So far, we have defined P in purely algebraic terms using only the elements in A. Here, we consider another approach, which also uses the geometry of the domain to construct the coarse grid c and the prolongation operator P. This approach can also be used to understand the original approach better and to develop further improvements.

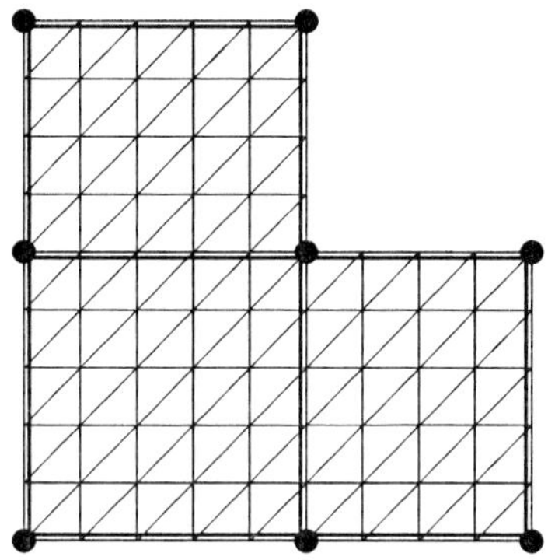

Figure 17.3. *The domain decomposition. The bullets denote nodes in the coarse grid c.*

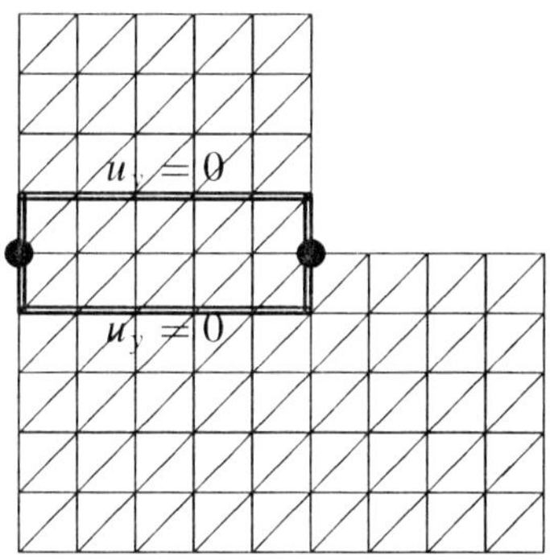

Figure 17.4. *The first prolongation step, in which the known values at the bullets are prolonged to the line connecting them by solving a homogeneous subproblem in the strip that contains this line.*

In the domain-decomposition approach, the original domain is viewed as the union of disjoint subdomains. The vertices of these subdomains form the coarse grid c (Figure 17.3). The prolongation operator P consists of two steps: the first step extends a vector defined on c to the nodes that lie on the edges of subdomains, and the second extends it further to the interiors of the subdomains.

For a given coarse-grid vector v, the first prolongation step produces the values of the extended vector Pv at nodes that lie on the edges of the subdomains (between coarse-grid nodes). The second prolongation step uses these values as Dirichlet data to solve a homogeneous PDE in each individual subdomain (Figure 17.5). The numerical solution of these subproblems produces Pv in the interiors of the subdomains as well. Because these subproblems are independent of each other, they can be solved simultaneously in parallel.

The first prolongation step, which extends v to the edges in Figure 17.3, is trickier than the second one. Clearly, it should use the original values of v at the endpoints of each edge as Dirichlet data for a one-dimensional equation in this edge. However, it is not clear how this one-dimensional equation should be defined.

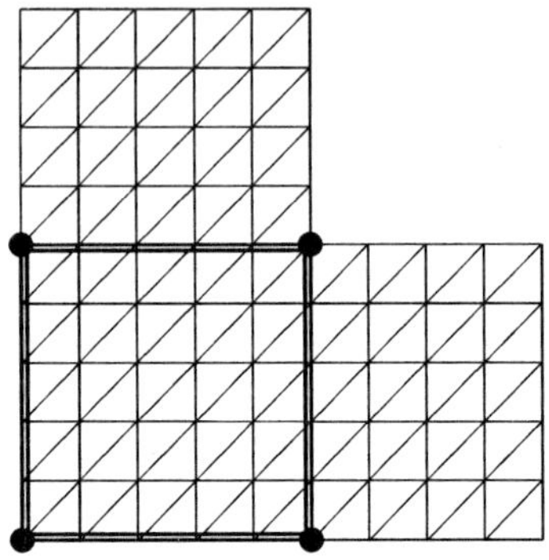

Figure 17.5. *The second prolongation step, in which the known values at the bullets and edges are prolonged to the interior of the subdomain by solving a homogeneous Dirichlet subproblem.*

In [4], it is assumed that the diffusion coefficients are constant in each individual subdomain. A one-dimensional diffusion equation is then formed in each particular edge, with the diffusion coefficient being the average of the original diffusion coefficients on both sides of this edge. This idea can also be extended to the Helmholtz equation:

$$-u_{xx} - u_{yy} - K^2 u = F.$$

In this case, the reduced one-dimensional equation along the edges of the subdomains is of the form

$$-u_{qq} - K^2 u/2 = 0,$$

where $\vec{q}$ is the unit vector tangent to the edge. This idea is formulated algebraically in Chapter 8 of [39].

A more general definition of the first prolongation step is illustrated in Figure 17.4. First, a homogeneous PDE is solved in a thin strip containing the edge under consideration. The boundary conditions for this subproblem are of Dirichlet type at the endpoints of the edge, where the values of the original vector v are available, and of homogeneous Neumann type elsewhere. The numerical solution of this subproblem provides the required values of Pv in the edge.

Actually, the subproblem in the strip can be reduced further to a problem on the edge only. Consider, for example, the strip in Figure 17.4. The discrete homogeneous Neumann conditions on the top and bottom edges of this strip can be used to assume that the numerical solution is constant on each vertical line in it. The unknowns on the top and bottom edges of the strip can thus be eliminated, resulting in a reduced subproblem on the edge alone. In other words, the linear system on the nodes in the strip is "lumped" into a tridiagonal system on the nodes in the edge only. The solution to this tridiagonal system produces the required values of Pv in the edge (Chapter 11 of [39]). Actually, the black box multigrid method [11] can also be obtained as a special case of this approach by assuming that each edge contains only one node between its endpoints.

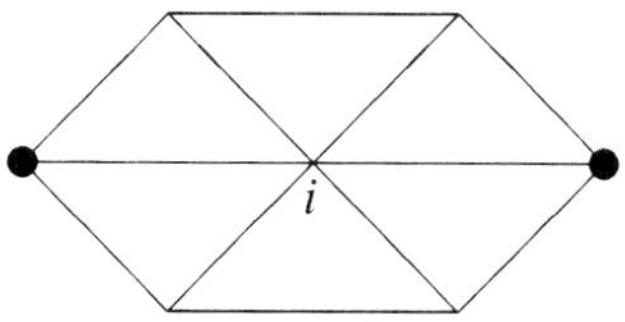

Figure 17.6. *Prolonging to node i by solving a homogeneous Dirichlet–Neumann subproblem in the "molecule" of finite elements that surround it.*

The idea of solving a Dirichlet–Neumann homogeneous subproblem in the strip and then using the numerical solution to produce the required values of Pv in the edge contained in it is formulated in a more general way in the AMGe method [8]. No subdomains or edges are used. The coarse grid c is constructed algebraically using the coefficient matrix A only. The prolongation operator P is defined as follows. At each node $i \in f$, $(Pv)_i$ is defined by solving the homogeneous PDE numerically in the "molecule" of finite elements that surround the node i (Figure 17.6). The boundary conditions for this subproblem are of Dirichlet type at nodes in c (where v is available) and of homogeneous Neumann type elsewhere. The numerical solution of this subproblem at i is then accepted as the prolonged value $(Pv)_i$. This defines $(Pv)_i$ at every $i \in f$, so no second prolongation step is needed.

In the AMGM method [21], the above approach is reformulated in pure algebraic terms. The molecules are defined algebraically and can also be used in the recursive calls in the multigrid V-cycle.

We conclude this section by showing that the domain-decomposition approach also leads to the present definition of P in Section 17.8 above. Instead of solving a subproblem on a molecule, we assume that the prolonged vector Pv should satisfy the ith equation in the homogeneous linear system

$$\sum_{j\in c} A_{i,j}v_j + \sum_{j\in f} A_{i,j}(Pv)_j = 0.$$

Our algebraic molecule is, thus, the set of unknowns j for which $A_{i,j} \neq 0$. Of these unknowns, v_j is available for each $j \in c$, which is the algebraic analogue to the Dirichlet conditions. Furthermore, we use the algebraic analogue of the discrete homogeneous Neumann conditions to assume that $(Pv)_j$ is the same for every $j \in f$ in the molecule, which leads to the definition

$$(Pv)_i \equiv \frac{-\sum_{j\in c} A_{i,j}v_j}{\sum_{j\in f} A_{i,j}},$$

which is practically the same as the definition used in Section 17.8.

The above definition applies to every $i \in f$, so it actually completes the definition of P. Still, one could use the resulting values of Pv to improve the prolongation with a second step. In this step, the algebraic Dirichlet conditions are used at the unknowns $j \neq i$, using the values $(Pv)_j$ from the first prolongation step:

$$(Pv)_i \equiv \frac{-\sum_{j\in c} A_{i,j}v_j - \sum_{j\in f,\ j\neq i} A_{i,j}(Pv)_j}{A_{i,i}}.$$

This amounts to an extra Jacobi relaxation on the values $(Pv)_i$ at each $i \in f$. The second prolongation step can thus be formulated more compactly as follows. Define the diagonal matrix D by

$$D_{i,i} \equiv \begin{cases} 1 & \text{if } i \in f, \\ 0 & \text{if } i \in c. \end{cases}$$

Modify the prolongation matrix P in Section 17.8 by the substitution

$$P \leftarrow (I - diag(A)^{-1}DA)P$$

(where I is the identity matrix of the same order as A).

The rest of the multigrid algorithm is as before ($R = P^t$, $Q = RAP$, and recursion). When this modified preconditioner is used in PCG in the applications in Part VI below, the number of PCG iterations is reduced by up to 50%. Unfortunately, the amount of setup time required to construct the "multigrid" object increases substantially due to the extra matrix-times-matrix operations. This is why we stick to the original definition of P and don't use the above modification here. This modification may be worthwhile when a powerful parallel computer is available to carry out matrix-times-matrix operations efficiently. This is the subject of the next chapter.

17.14 Exercises

1. Does the "GaussSeidel" function have to be a friend of the "sparseMatrix" class? Why?

2. Implement the Kacmarz iteration for the solution of real or complex sparse linear systems. The Kacmarz iteration is equivalent to the Gauss–Seidel iteration applied to the normal equation

$$A*Ax = A*f,$$

where $A* \equiv \bar{A}^t$ is the Hermitian adjoint of A. Use the "sparseMatrix<complex>" class of Chapter 16, Section 4. The solution can be found in Section A.10 of the Appendix.

3. Rewrite the "factorize()" function in Section 17.7 above in such a way that no fill-in is ever used; that is, an element in L or U can be nonzero only if the corresponding element in the original matrix A is nonzero. The solution can be found in Section A.11 of the Appendix.

4. Implement the "forwardElimination" and "backSubstitution" functions declared in Chapter 16, Section 4. The solution can be found in Section A.11 of the Appendix.

5. Follow Algorithm 17.2 in Section 17.8 and implement the "coarsen" member function of the "sparseMatrix" class. This function returns a vector of integers v, with $v_i = 0$ if and only if i is excluded from the coarse grid ($i \in f$). The nonzero components in v should have the monotonically increasing values $v_i = 1, 2, 3, \ldots, |c|$ for $i \in c$. The solution can be found in Section A.12 of the Appendix.

6. Follow Algorithm 17.3 in Section 17.8 and implement the "createTransfer" member function of the "sparseMatrix" class. This function transforms the current matrix from A to P. (It is assumed that the coefficient matrix A is also stored elsewhere, so the current matrix can be safely changed.) The columns with index j for which $v_j = 0$ in the previous exercise should be dropped using the "dropItems(v)" function of Chapter 16, Section 3 (applied to each row). Then, the column indices should be renumbered according to their number in c, using the "renumberColumns(v)" function in Chapter 16, Section 3. The "createTransfer" function also returns the matrix $R = P^t$ as output. The solution can be found in Section A.12 of the Appendix.

7. Modify the "createTransfer" function of the previous exercise so that it constructs the matrix P of the AMG algorithm in Section 17.9 and returns its transpose $R = P^t$. The solution can be found in Section A.12 of the Appendix.

8. Implement the PCG algorithm to accelerate the basic multigrid iteration. Use the "sparseMatrix" and "dynamicVector" classes. The solution can be found in Section A.13 of the Appendix.

9. Show that, for every real matrix A, A^tA is SPD. Conclude that PCG is applicable to the normal equations provided that an SPD preconditioner is used.

10. Modify your PCG code to apply to complex systems as well by making sure that the inner product of two complex vectors v and w is defined by

$$(v, w) \equiv v^t \bar{w}$$

in the '*' operator in the "dynamicVector" class.

11. Rewrite your PCG code as a template function that is also suitable for systems with a complex coefficient matrix of type "sparseMatrix<complex>". The solution can be found in Section A.13 of the Appendix.

12. Show that, for a complex matrix A, A^*A is the Hermitian adjoint of itself and is also positive definite. Conclude that PCG is applicable to the normal equation

$$A^*Ax = A^*f,$$

provided that the preconditioner is also the Hermitian adjoint of itself and positive definite.

Chapter 18
Parallelism

In this chapter, we describe parallel computers and parallelizable algorithms. In particular, we consider parallel architectures with distributed memory and a hypercube communication network. We also introduce a fair method of comparing parallel and sequential computers for a particular computational problem. We illustrate how this comparison method works for the numerical solution of SPD large sparse linear systems. Finally, we explain how the parallel implementation should take place in the low-level C++ code.

18.1 Parallel vs. Sequential Computers

When the digital computer was first invented in the 1950s, there seemed to be no computational task it couldn't do: from data management, through arithmetic calculations, to numerical modeling of real physical phenomena. The computational problems in those days, however complicated for the human mind, were easy enough for the machine. The scale of a problem (number of degrees of freedom or unknowns) used to be a few thousand at most. Such small-scale problems were no challenge even for the early computer, which could perform thousands of arithmetic operations per second.

As time went by, the computational power of the digital computer increased. Modern computers can perform millions of arithmetic operations per second, and supercomputers can perform even billions of arithmetic operations per second. However, the scale of computational problems has increased as well. Modern problems in numerical modeling, from weather forecasting to nuclear simulations, may contain millions of degrees of freedom. However efficient the algorithms that solve these problems may be, their complexity (operation count) still grows superlinearly with the scale of the problem. This means that even modern computers may not be sufficiently powerful to solve them in acceptable time.

The traditional computer is sequential in the sense that it can perform only one operation at a time. This is why its computational power, however large, may still be insufficient for solving large-scale problems. This drawback motivated the introduction of the parallel computer, which can perform more than one operation at a time.

The main difference between sequential and parallel computers is in terms of the number of processors available to carry out computations. The sequential computer contains three basic elements: memory to store data, a single processor to carry out calculations, and input/output devices to communicate with other sources of data. One processor can only do one thing at a time, be it data fetching, arithmetic operation, or data update. The parallel computer, on the other hand, contains more than one processor, so it may perform several operations at a time. Each processor must have access to the memory and input/output (I/O) devices. Thus, the parallel computer has a better chance of solving the original problem in acceptable time.

18.2 Parallelizable Algorithms

Computational problems are solved by an algorithm: a list of instructions that should be completed one by one. This list of instructions is often so long that it could never be completed by a human; a digital computer must be used.

Thus, the algorithm must be written in context-free formal language, which can be understood by the computer with no ambiguity. For this purpose, it must be written in a high-level programming language. This is called implementation. On parallel computers, however, the implementation may be different from that used on sequential ones, to exploit better the extra computational power they offer.

In order to have good parallel implementation, the nature of the algorithm must first be studied and well understood. In particular, some instructions in the algorithm are independent of each other and can be carried out concurrently. We then say that the algorithm is parallelizable: it can benefit from parallel implementation, in which different processors work on different instructions at the same time [30]. We say that an algorithm is well parallelizable if the most time-consuming instructions in it are indeed independent of each other and can be completed in parallel.

18.3 Inherently Sequential Algorithms

Some algorithms, however, are inherently sequential in the sense that the most time-consuming instructions cannot be carried out at the same time. For example, if each instruction uses data computed in the previous one, then the original, sequential order of the instructions must be kept. Such algorithms cannot benefit from parallel computers. In fact, the traditional sequential computer may complete them as rapidly as the parallel computer, thanks to its straightforward and efficient architecture.

18.4 Architecture

The architecture of a computer is the way basic elements are organized in it and connected to each other. Obviously, the three basic elements in the computer must be connected to each other: the processor must be connected to the memory to fetch and update data and, most importantly, obtain instructions from the file containing the executable program. Furthermore, the processor must also be connected to the I/O devices to exchange data with the user. The memory, however, doesn't have to be connected directly to the I/O devices, because all required communication can be carried out through the processor.

Actually, the memory is split into two parts: primary memory and secondary memory. The primary memory is small but easily accessed, so data that are used often are better stored in it. The secondary memory, on the other hand, is big but not easily accessed, so it should contain files and other data that are not frequently used. When a process is executed, data are stored temporarily in the primary memory, where they are available for further use. Once the process terminates, its output is placed in its permanent address in the secondary memory.

It is thus advisable to access the secondary memory only when necessary. Data that have already been placed in the primary memory should be exploited as much as possible before being returned to the secondary memory.

The primary memory is sometimes called the cache, because data are stored in it as in an easily accessed cache. Actually, the cache may be divided into several levels, from the most easily accessed cache in the highest level to the least easily accessed cache at the lowest level. It is advisable to store data in the cache according to the number of times they are expected to be used: often-used data should be stored in the highest level, whereas rarely used data should be stored in the lowest level.

The following algorithms are designed with cache in mind; that is, they exploit the cache fully and avoid accessing the secondary memory, if possible.

18.5 Cache-Oriented Relaxation

A cache-oriented algorithm is an algorithm that is specifically designed to exploit the cache in full and minimize access to the secondary memory. The principle is to carry out as many useful calculations as possible with data that are already available in the cache before returning them to their permanent location in the secondary memory. In what follows, we describe a cache-oriented version of the Gauss–Seidel relaxation method in [13].

The standard Gauss–Seidel relaxation is not cache oriented in the sense that it cannot be implemented efficiently in terms of cache use. Indeed, the unknowns are relaxed one by one in the order in which they are stored in the vector of unknowns. The relaxation of each particular unknown uses the values calculated before in the relaxation of previous unknowns. The value calculated at the relaxation of a particular unknown cannot be stored in the primary memory until the end of the relaxation sweep; it must be transferred to the secondary memory to make room for the unknowns to be relaxed next. The result is too many accesses to the secondary memory, without fully exploiting values that are calculated in the current relaxation sweep and are available in the cache.

The cache-oriented version of the Gauss–Seidel relaxation, which fully uses data that are already in the cache, is as follows. First, divide the vector of unknowns into chunks that fit in the cache. For example, if the number of unknowns is 1000 and the cache can contain the data required to relax 256 unknowns, then divide the vector of unknowns into four chunks of 250 unknowns each. Then, transfer from the secondary memory to the cache the data required to relax the unknowns in the first chunk. Then, do as many relaxations as possible on the unknowns in this chunk (or at least most of them), to make full use of their updated values, which are already in the cache. In each of these relaxations, use the newest values available from the previous relaxation in this chunk. Then, place the newest values calculated for the unknowns in this chunk back in the secondary memory, and repeat

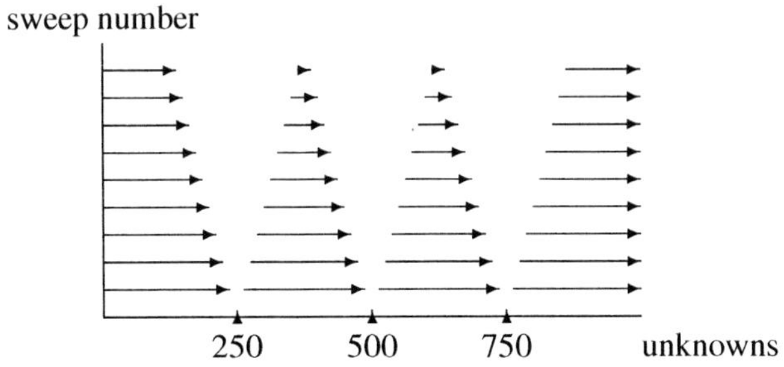

Figure 18.1. *The cache-oriented relaxation for a tridiagonal system: first part, in which the unknowns in each chunk are relaxed over and over again using data from the previous relaxation.*

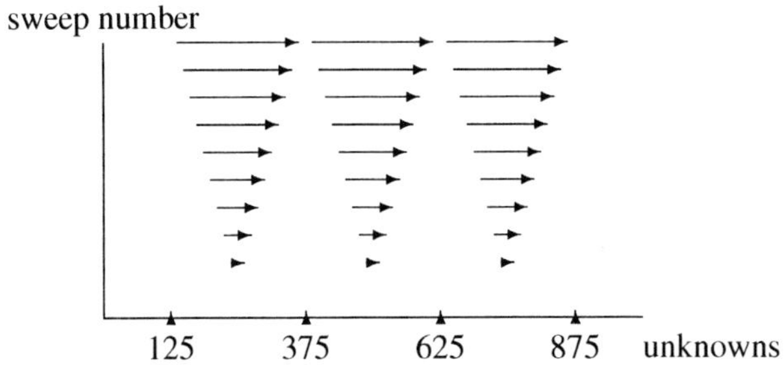

Figure 18.2. *The cache-oriented relaxation for a tridiagonal system: second part, in which the unknowns in each intermediate (interface) chunk are relaxed over and over again using data from the previous relaxation and the first part described above.*

the same procedure in the next chunk (containing the next 250 unknowns). Then, because there are probably unknowns at the interface between chunks that have not yet been relaxed enough times, repeat the same procedure in the intermediate chunk that contains the 250 unknowns from the 126th unknown to the 375th unknown. The procedure is repeated in the other chunks and intermediate chunks between them, until the final intermediate chunk (containing the 626th to 875th unknowns) has also been relaxed enough times.

The data required to relax a particular chunk may be transferred to the cache in one delivery as a whole. Thus, for the price of only 14 accesses to the secondary memory (to fetch and store the values in the four chunks and three intermediate chunks), we have all the unknowns relaxed many times.

In Figures 18.1 and 18.2, this procedure is illustrated for a tridiagonal system in which each unknown is coupled only with the unknowns immediately before and after it. In this case, the above algorithm is equivalent to 125 consecutive Gauss–Seidel iterations. Indeed, the arrows in Figure 18.2 complete the arrows in Figure 18.1 into 125 full arrows, which stand for 125 full (slightly reordered) relaxation sweeps.

18.6 Schwarz Block Relaxation

The main principle in cache-oriented algorithms is to make as many useful calculations as possible with data that are already available in the cache. This principle leads to the following improvement of the above algorithm. Since the data required to relax the unknowns in a particular chunk are already in the cache, why not keep relaxing there until convergence? In other words, why not relax all the unknowns in the chunk together, or solve the subsystem of equations corresponding to them, with all other unknowns kept fixed? Actually, this subsystem can be solved not by standard relaxation but rather by a more effective method such as multigrid or PCG. Once this is done, the residual in the chunk becomes 0, which is analogous to the zero residual at a particular unknown after being relaxed in standard Gauss–Seidel relaxation. This procedure is then repeated in the other chunks as well. This completes the block Gauss–Seidel or alternating Schwarz relaxation.

Actually, one could repeat the above procedure in the intermediate chunks used above. This would yield an alternating Schwarz relaxation with overlapping subdomains. The basic iteration may also be accelerated by PCG or CGS.

The principle of cache-oriented algorithms tells us to carry out only useful calculations in each individual chunk. It may therefore be more efficient not to solve each subproblem exactly but rather to apply to it only one multigrid iteration, to reduce the residual in the chunk substantially.

Although the number of calculations in each chunk is much larger than in standard relaxation, no extra access to the secondary memory is needed, so the cost of block relaxation should be comparable to that of standard relaxation. Block relaxation, though, is usually much more economic in terms of the number of iterations required to converge to the numerical solution of the original system.

The above algorithms are designed to suit the architecture of traditional sequential computers. In what follows, we also describe parallel architectures and algorithms that are suitable for them.

18.7 Parallel Architectures

So far, we have considered sequential computers and algorithms that are suitable for sequential architecture. Here, we move on to the subject of parallel architectures.

In a parallel computer, several processors are available to carry out operations concurrently. Each processor may have its own cache to help in the calculations. It must also be connected to the secondary memory to access data. This connection may be either direct or indirect, as discussed below.

18.8 Shared Memory

In shared-memory architecture, the memory is shared by all the processors, so each processor has access to every piece of data in it (Figure 18.3). This architecture is convenient to use, because there is no need to worry about the availability of data. However, the hardware might be rather expensive, because it must contain extensive wiring or a general mechanism to transfer data from everywhere in the memory to every processor. Because it must be capable of delivering data from everywhere to everywhere, this mechanism might also be rather slow. Furthermore, delivery of data could be delayed due to traffic jams or conflicts when several processors attempt to access the same area in the memory. Thus, the capability to deliver data from everywhere to everywhere requires greater cost but is rarely used in practical algorithms, where each processor uses data only from very few specific places in the memory.

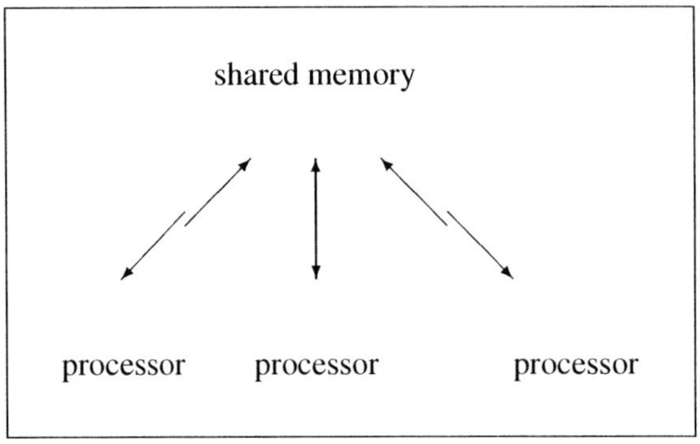

Figure 18.3. *Shared-memory architecture: each processor has direct access to the entire memory.*

18.9 Distributed Memory

In common architectures, the memory is distributed among the processors. Each processor has access only to the portion of memory assigned to it; other processors can obtain data from this portion only by requesting them explicitly from this processor (Figure 18.4). This request must be sent through the communication network that connects the processors to each other.

Unlike in shared-memory architecture, here the burden to transfer data lies mostly on the software rather than the hardware. Indeed, the programmer is responsible for writing explicit commands to request and deliver data. Fortunately, this is done only when necessary; it is a price worth paying for the sake of avoiding the slow and expensive hardware required to share memory.

Code designed for distributed-memory architecture is often of type multiple instruction and multiple data or MIMD (as opposed to SIMD in Chapter 1, Section 10). In this pattern, each individual processor uses not only different data from its private memory but also different instructions from a private file of commands.

The main advantage in distributed memory is that extra communication is used only for specific tasks to deliver specific data to the processors that require them. This strategy seems more efficient than sharing the entire memory by overburdening the hardware with a slow and expensive general-transfer mechanism. Still, a distributed-memory architecture must also contain a communication network through which messages are sent from processor to processor. This network must support efficient communication without imposing too much on the hardware.

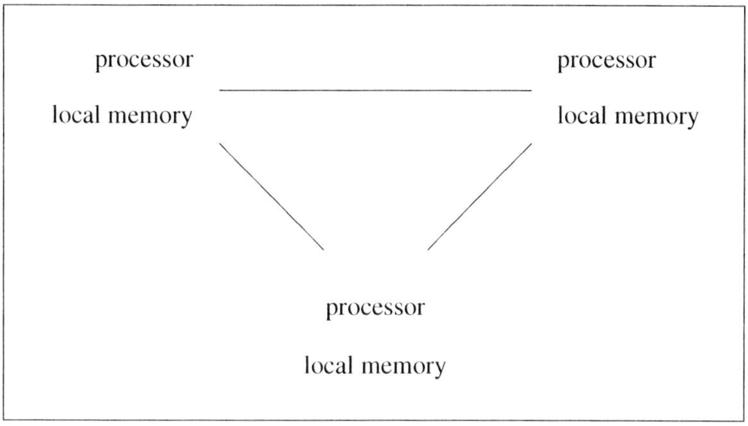

Figure 18.4. *Distributed-memory architecture: each processor has access only to its own local memory but not to the memories of other processors. If a processor needs data from the memory of another processor, then it must send a request through the communication wires, indicated by straight lines.*

18.10 Communication Network

The distributed-memory architecture may be thought of as a network of independent computers. Each processor may actually be viewed as an individual computer, with its own private memory. The communication network between the processors is, thus, the key feature in the parallel architecture.

Usually, there is one processor (the main processor) that is responsible for reading the executable program and assigning particular tasks from it to individual processors. This processor may be connected directly by a special wire to each individual processor to allow a continuous flow of instructions. This wire must be used for this purpose alone; the other processors can't use it to send messages to the main processor. Furthermore, the processors can't communicate with each other through the main processor, due to possible traffic jams. An efficient communication network between the processors that actually execute the program is thus necessary.

In networks, there is usually a tradeoff between hardware and software efficiency. Networks with many connections support straightforward communication at the price of heavy, expensive, and inefficient hardware. In fact, in a particular application, many connection lines may never be used and yet overburden the hardware of the system. In the extreme example of the so-called connection machine, every two processors are connected to each other by a wire, so they can communicate directly with each other (Figure 18.5). In this architecture, the number of connection lines is actually a quadratic function of K, the number of processors:

$$\binom{k}{2} = K(K-1)/2.$$

This is far too many; the number of connection lines that are actually used in an application is usually only a linear function of K.

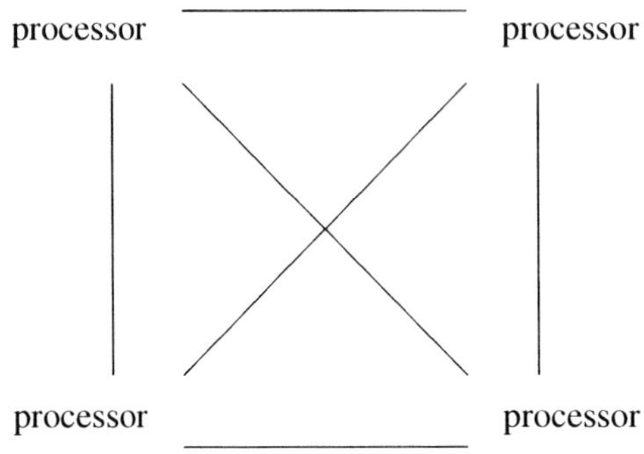

Figure 18.5. *A connection machine with four processors: every processor is connected to every other processor by a wire.*

On the other hand, simple networks with a small number of connection lines may have efficient hardware yet require extra effort and resources in terms of software and communication time. In the extreme example of the line architecture (Figure 18.6), the processors are ordered in a straight line. Each processor is connected directly only to the processors that lie immediately before and after it. For K processors, this network contains only $K - 1$ connection lines. However, the indirect data transfer often required in it is not very efficient: sending a message from one end of the line to the other end requires passing it through the entire line of processors, wasting their time and needlessly blocking the network. Furthermore, the data transferred may be crucial for the algorithm to proceed, so all the processors must wait idle until the message is delivered.

processor —— processor —— processor —— processor

Figure 18.6. *Four processors connected in a line: each processor is connected physically only to its immediate neighbors.*

There must be a compromise between these two extreme architectures, with a reasonably efficient data-transfer scheme that uses a moderate number of connection lines. This architecture is described next.

18.11 Hypercube

The desired compromise between the above extreme architectures is the hypercube (Figure 18.7). In a hypercube, each processor is assigned a binary index and is connected directly only to those processors with binary indices that differ from its own binary index in one digit only [17]. In fact, if the processors are numbered from 0 to $K - 1$, then the binary index of the ith processor is just the binary representation of i (Chapter 1, Section 18).

Let's describe this structure in some more detail. Assume for simplicity that K is a power of 2, say $K = 2^n$. Let the binary index of a processor be the binary representation of

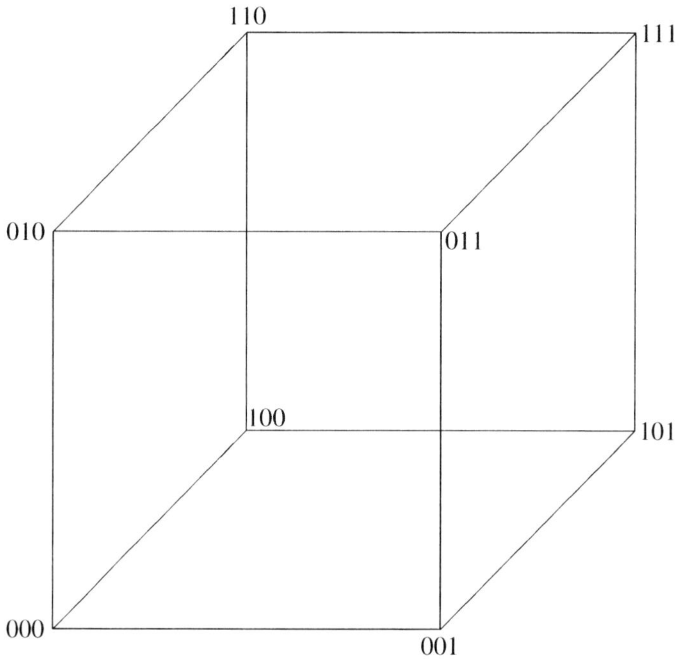

Figure 18.7. *A three-dimensional hypercube (cube) with eight processors in its nodes, numbered by the binary integers* 000 *to* 111*. Each node is connected only to the three nodes that differ from it by one binary digit only.*

its number, which contains n binary digits (0 or 1), including leading zeroes. For example, when $n = 4$, the zeroth processor is indexed by 0000, the seventh processor is indexed by 0111, and so on. The communication network is then defined by connecting each processor to the processors with indices that differ from its own index in one digit only. In the above example, the seventh processor is connected to the processors with binary indices

$$1111, \ 0011, \ 0101, \ \text{and} \ 0110$$

only. The number of processors to which a processor is connected is, thus, n ($n = 4$ in the above example). Because the network is symmetric (processor i is connected to processor j by the same wire by which processor j is connected to processor i), the total number of connections is

$$K(\log_2 K)/2.$$

This number is rather small, yet it supports data transfer from any processor to any other processor in at most $\log_2 K$ steps. Indeed, the binary indices of any two processors may differ in at most $n = \log_2 K$ digits, so messages can be exchanged between them indirectly through the processors with indices that are obtained by changing one digit at a time in the binary index of the sending processor until the binary index of the receiving processor is

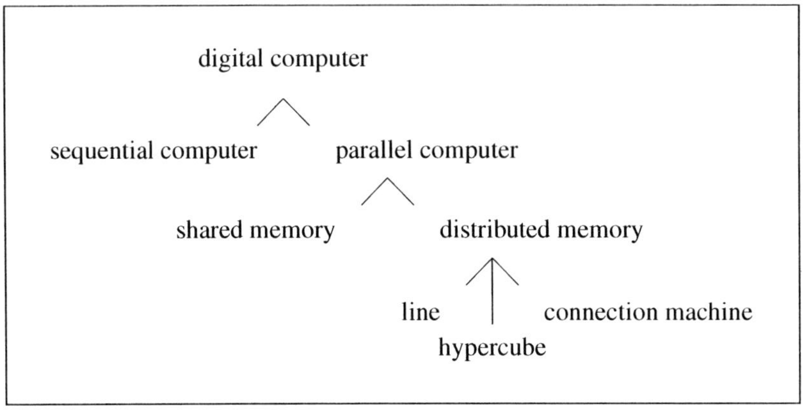

Figure 18.8. *Different kinds of digital computers and architectures.*

formed. For example, the zeroth processor in the above example can send a message to the seventh processor through the path

$$0000 \to 0001 \to 0011 \to 0111.$$

Actually, it follows from Chapter 1, Section 18, that there are six different paths that lead from one processor to the other.

The different kinds of architectures are summarized in Figure 18.8. In what follows, we show the advantage of parallel computers in general and the hypercube architecture in particular in numerical algorithms.

18.12 Example: Multiplication of Sparse Matrices

Here, we consider the parallel implementation of the problem of multiplying two sparse matrices. This problem is particularly important in the numerical solution of large sparse linear systems, as discussed below.

Some robust iterative linear-system solvers use the normal equation rather than the original one. This equation is obtained from the original linear system

$$Ax = f$$

by multiplying it by A^t:

$$A^t Ax = A^t f.$$

For example, the Kacmarz iteration in Chapter 17, Section 6, is just the Gauss–Seidel iteration applied to the normal equations. This method is robust in the sense that it always converges to the algebraic solution x. The convergence, however, may be very slow. It is thus advisable to use the symmetric Kacmarz iteration, which is just the symmetric Gauss–Seidel iteration applied to the normal equation. This iteration can then be further accelerated by PCG, as is shown in Section 18.16.

The calculation of $A^t A$ may be particularly time-consuming, because it may require a large number of operations with matrix rows, each of which invokes calls to the constructor of the "row" class. This constructor is particularly expensive, because it requires memory allocation by the "new" command. An efficient parallel implementation is clearly necessary.

Fortunately, the algorithm that multiplies two sparse matrices is inherently parallelizable. Indeed, if A and B are two sparse matrices, then a row in AB is calculated as a linear combination of rows in B, with coefficients from the corresponding row in A. More specifically, if $A^{(i)}$, $B^{(i)}$, and $(AB)^{(i)}$ denote the ith rows in A, B, and AB, respectively, then

$$(AB)^{(i)} = \sum_j A_{i,j} B^{(j)},$$

where the sum goes over the columns j in which nonzero elements $A_{i,j}$ lie in $A^{(i)}$. Thus, $(AB)^{(i)}$ is a function of $A^{(i)}$ and several rows in B and can be calculated independently of any other row in AB. In the extreme case, in which the number of processors, K, is as large as the number of rows in A, the task of calculating each row in AB can be assigned to a different processor, so all the rows can be calculated concurrently in parallel.

The algorithm to do this is particularly well implemented on the hypercube. Indeed, assume that the ith processor $(0 \leq i < K)$ holds in its own memory the ith rows in A and B. In order to calculate the ith row in AB, this processor must also have the rows $B^{(j)}$ for which $A_{i,j} \neq 0$. Because A is sparse, the number of these rows is rather small. These rows are now requested by the ith processor from the processors that hold them and are delivered to it in at most $\log_2 K$ steps. This completes the algorithm for calculating AB on a hypercube in parallel.

In general, some rows in A may contain relatively many nonzero elements. A processor that holds such a row has more work to do, which may delay the completion of the algorithm. In this case, a more balanced distribution of the total work among the individual processors is needed in advance. Long rows should be broken into pieces by the main processor and assigned to different processors. In practice, however, K is much smaller than the number of rows in A, so each processor is responsible for a chunk of rows rather than a single row. Good load balancing can then be achieved without breaking individual rows.

18.13 Low-Level C++ Implementation

The advantage of C++ in the context of parallelism is the opportunity to parallelize only the low-level code, while the high-level code remains unchanged. Because the low-level implementation is completely hidden from and never accessed by high-level programmers (Figure 18.9), it can be changed with no need to notify them, provided that the interface functions still take and return the same arguments as before.

In particular, the matrix-times-matrix and matrix-times-vector operations can be parallelized as in Section 18.12 inside the "sparseMatrix" class in Chapter 16, Section 4, without the users of this class ever knowing about it.

Similarly, arithmetic operations with vectors can be parallelized inside the "dynamicVector" class in Chapter 3, Section 3, with no need to change any code that uses it. High-level programmers can keep using these operations in the same way as before, particularly in the implementation of well-parallelizable algorithms.

The low-level programmer can thus work independently on the optimal parallel implementation of basic operations with vectors and matrices. For this purpose, the programmer

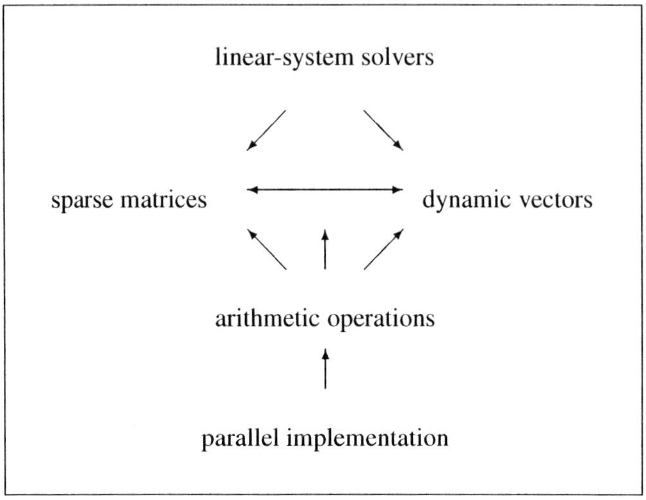

Figure 18.9. *The implementation of the "sparseMatrix" and "dynamicVector" objects in the low-level code is completely hidden from the algorithms that use them, so it can be parallelized with no need to change the high-level code, provided that the low-level interface functions still take and return the same types of arguments as before.*

can assume the existence of an efficient data-transfer scheme in the communication network. Using this assumption, the programmer can freely use functions such as "send()" and "receive()" to send messages from processor to processor and receive them. The requirements about what arguments these functions should take and what value they should return should then be passed to yet another low-level programmer, who is particularly experienced with communication networks and hardware.

The "P++" library in [25] implements vectors in parallel. Each vector is split into chunks, which are assigned to the individual processors. Each processor is responsible for carrying out the arithmetic operation in the chunk assigned to it. For operations that also involve matrices, such as matrix-times-vector, however, one should resort to algorithms such as that in Section 18.12 above, with no need to break individual vectors.

The low-level code, however, must be modified whenever one turns from one parallel computer to another. Each parallel computer may have its own schemes to assign memory to processors and communicate between them. These schemes should be taken into account in the low-level code to come up with optimal load-balancing and communication strategies. Again, the high-level code should not be affected by this change.

In the next section, we introduce a method of comparing different architectures.

18.14 Comparing Architectures

Clearly, a good algorithm should solve the computational problem efficiently. There are, however, several possible ways to interpret the concept of efficiency. The traditional way, used in complexity theory, is to count the numbers of elementary operations and data elements used in the algorithm and express them as functions of the number of degrees of freedom in the original problem. This methodology provides a standard way to compare different algorithms for a particular problem.

Algorithms that seem promising in terms of standard operation count may still disappoint if they are not well implemented. For example, accessing the secondary memory too often may deteriorate the performance considerably. On the other hand, algorithms that look unattractive in terms of complexity theory may still be practical if implemented properly. Suitable implementation is thus most important in practical complexity estimates.

Actually, complexity theory disregards implementation issues and ignores the possibility that certain algorithms may have no efficient implementation in terms of cache access or parallelism. In fact, standard complexity estimates don't tell to what extent algorithms are cache oriented and well parallelizable.

More practical complexity estimates take into account implementation issues such as cache access and parallelism. They assume that a particular architecture is used and compare algorithms in terms of time and storage requirements when implemented as efficiently as possible on this particular architecture. This approach is particularly relevant to owners of such architecture.

Thus, this practical approach provides a way of comparing different algorithms not in general, theoretical terms but rather in more concrete, practical terms relevant to a particular architecture. With this approach, the owner of the architecture can choose the most efficient algorithm for the problem under consideration.

Besides comparing different algorithms, one may also want to compare different computers and architectures. This question is particularly relevant to people who plan to purchase a new computer and wish to have a method of comparing different computers. For this purpose, one should consider a relevant computational problem and a common algorithm to solve it and estimate its time and storage requirements when implemented as efficiently as possible on each computer. The computer on which the algorithm performs best is the one that should probably be purchased.

Usually, the above approach can't really distinguish between computers that belong to the same family, that is, use the same architecture. Still, the practical complexity estimates are informative enough to distinguish between different architectures and advise the customer what architecture to buy.

The above method of comparing different architectures is, however, not entirely fair. After all, the test algorithm may have a particularly good implementation on one of them, but not on the others. It is more fair to let each architecture use its own algorithm, which is best implemented on it. Thus, the improved approach uses different algorithms on different architectures. More specifically, the relevant computational problem is first agreed upon. Then, for each architecture, the algorithm that is best for it is designed. (This task uses the method of comparing different algorithms on the same architecture, as discussed above.) The practical complexity is then estimated for each algorithm implemented on the corresponding architecture. The architecture with the algorithm with the least time and storage requirements is the one that should eventually be purchased.

In what follows, we consider the problem of solving large sparse linear systems. In order to compare sequential and parallel architectures as in Figure 18.10, we design a suitable algorithm for each of these architectures and estimate their performance.

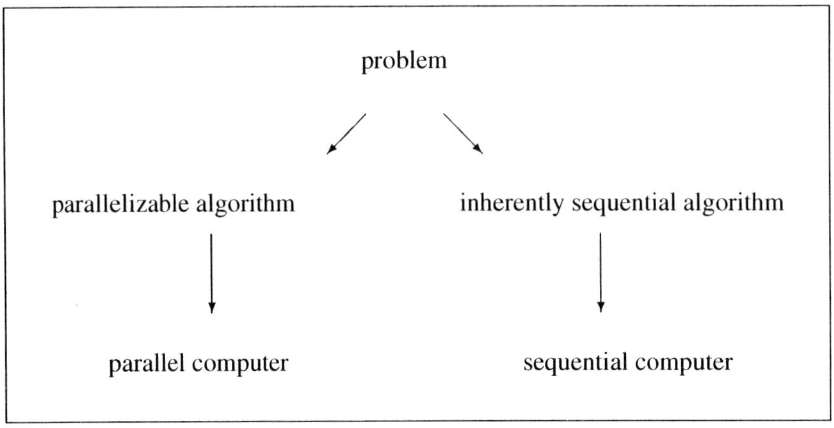

Figure 18.10. *Fair comparison between parallel and sequential computers for a particular problem: the performance of each computer is estimated with the algorithm that is most suitable for it in terms of efficient implementation.*

18.15 Jacobi Relaxation

As discussed in Section 18.5 above, the Gauss–Seidel relaxation method is inherently sequential. Indeed, each unknown is relaxed using the new values of previously relaxed unknowns, so it must wait patiently until these new values are available. The relaxation of a particular unknown cannot be concurrent with the relaxation of other unknowns: because the relaxations depend on each other, they must take place one at a time. As a result, only one processor is working at a time, while all the others remain idle. This is not a good use of computational resources.

Does this mean that parallel computers cannot help to solve large sparse linear systems? Definitely not. One should only switch to a more parallelizable algorithm and consider its implementation on the parallel computer under consideration. Such an algorithm is the Jacobi relaxation in Chapter 17, Section 4.

Unlike the Gauss–Seidel relaxation, the Jacobi relaxation is inherently parallelizable. Indeed, the relaxation at a particular unknown uses the old values of the other unknowns, as they were before the relaxation started. Therefore, the relaxation of a particular unknown can be carried out independently of the relaxation of the other unknowns. Thus, all the unknowns can be relaxed concurrently in parallel, using only data about their old values before the relaxation started.

The parallel implementation is thus as follows. The vector of unknowns is split into K chunks, where K is the number of processors. Each processor is assigned one chunk and is responsible for the Jacobi relaxation in it. Before the relaxation starts, the processors exchange information about the current values of the unknowns assigned to them. (Actually, only processors with chunks that are coupled in the original system should exchange this information.) These transfers can be carried out on a hypercube in at most $\log_2 K$ communication steps. Then, the Jacobi relaxation is carried out in all the processors simultaneously. This completes one Jacobi relaxation sweep.

With the above approach, the Jacobi relaxation can actually still be implemented as before (Chapter 17, Section 4). The only change is in the low-level implementation of the basic arithmetic operations with matrices and vectors, which is now done in parallel as in Section 18.13 above. The advantage of this approach is that the parallel implementation is completely hidden in the low-level code that implements the required matrix and vector objects.

The inner products used in PCG can also be calculated independently in the individual chunks assigned to the different processors. In the end, the contributions from these chunks should be assembled to form the required inner product. This requires another $\log_2 K$ communication steps on a hypercube. Fortunately, all these details are again hidden in the low-level code that implements the vector-times-vector operator. This completes the definition of the parallel implementation of PCG with a Jacobi preconditioner.

We are now ready to make a fair comparison of parallel and sequential architectures for solving large sparse SPD linear systems. On the one hand stands the standard sequential digital computer, which uses PCG with a symmetric Gauss–Seidel preconditioner. On the other hand stands the distributed-memory parallel architecture with hypercube communication network, which uses PCG with a Jacobi preconditioner, implemented as above. The result of this comparison may assist the user in deciding whether the expected gain from this particular parallel architecture is indeed worth investing in.

For the sake of a fair comparison, we must first show that both algorithms indeed converge to the solution. This is done below.

18.16 Convergence Analysis

The following analysis gives sufficient conditions for the convergence of PCG applied to basic relaxation methods. The results guarantee that the present preconditioners are indeed robust and suitable for comparing the power of different architectures. (Readers who are not particularly interested in this analysis may skip this section.)

It is assumed that the coefficient matrix A is SPD, so PCG converges if the preconditioner is also SPD. Thus, all that is left to do is to find out whether the preconditioner of the relaxation method is SPD or not.

In what follows, we use the fact from Chapter 17, Section 2, that, for a preconditioner $\mathcal{P}$, the iteration matrix is

$$I - \mathcal{P}^{-1}A.$$

(The iteration matrix is the matrix by which the error is multiplied in each iteration.)

Conversely, if the iteration matrix is available, then the inverse of the preconditioner can be obtained from it by

$$\mathcal{P}^{-1} = \left(I - \left(I - \mathcal{P}^{-1}A\right)\right) A^{-1}.$$

These facts are used below to calculate the preconditioner of a double relaxation, in which the original relaxation sweep is followed by a slightly different relaxation sweep.

Let us assume that two relaxation methods are used one after the other. Let us denote the preconditioner in the first relaxation method by L and the preconditioner in the second one by U. The iteration matrix for the double relaxation is given by

$$\left(I - U^{-1}A\right)\left(I - L^{-1}A\right).$$

The inverse of the preconditioner of the double relaxation is, thus,

$$\left(I - \left(I - U^{-1}A\right)\left(I - L^{-1}A\right)\right)A^{-1} = U^{-1} + L^{-1} - U^{-1}AL^{-1}$$
$$= U^{-1}(L + U - A)L^{-1}.$$

In other words, the preconditioner of the double relaxation is given by

$$L(L + U - A)^{-1}U.$$

Assume further that the second relaxation method is the adjoint of the first one, namely,

$$U = L^t.$$

Obviously, the preconditioner of the double relaxation is then symmetric. Therefore, we only have to verify that it is also positive definite to guarantee the convergence of PCG applied to it.

This is indeed the case for the symmetric Gauss–Seidel relaxation, for which L is just the lower triangular part of A (including the main diagonal), and U is the upper triangular part of A (including the main diagonal). (These definitions are displayed schematically in Figure 18.11.) In this case, we have

$$L + U - A = diag(A)$$

(the main diagonal of A). Because A is positive definite, the main-diagonal elements in it are positive, and the convergence of PCG applied to the symmetric Gauss–Seidel relaxation follows.

Let us now consider the double Jacobi relaxation method, composed of two consecutive Jacobi relaxations:

$$L = U = diag(A).$$

In this case,

$$L + U - A = 2diag(A) - A.$$

When A is diagonally dominant, this matrix is also diagonally dominant and, hence, positive definite. Thus, the diagonal dominance of A is a sufficient condition for the convergence of PCG applied to the double Jacobi relaxation.

Figure 18.11. *Symmetric Gauss–Seidel relaxation: schematic description of the triangular preconditioners L and U as windows through which elements of A are seen.*

Note that although PCG converges when applied to the Jacobi relaxation, it may diverge when applied to the double Jacobi relaxation. Indeed, when $2diag(A) - A$ is indefinite, PCG may diverge when applied to the double Jacobi relaxation. Nevertheless, it still converges when applied to the Jacobi relaxation, as shown above. This is why Jacobi rather than double Jacobi relaxation is used in the numerical experiments below.

The above analysis also applies to block relaxation (Section 18.6). The only difference is that matrix elements should be replaced by square blocks. In this case, $L + U - A$ is the block-diagonal part of A, which is indeed SPD whenever A is. Thus, PCG with the symmetric block Gauss–Seidel preconditioner is guaranteed to converge. In fact, the convergence is often faster than with the standard symmetric Gauss–Seidel preconditioner, because more work is invested in relaxing all the unknowns in a block together.

In our comparison of sequential and parallel architectures, we stick to standard (rather than block) symmetric Gauss–Seidel relaxation (for the sequential computer) and Jacobi relaxation (for the parallel computer). Each method requires a different number of iterations (within PCG) to converge to the required solution. This difference must be incorporated in the comparison as well. In the next section, we apply both methods to several test problems and compare the number of iterations required for convergence.

18.17 Examples from Harwell–Boeing

The Harwell–Boeing collection of sparse matrices available on the Internet[1] contains sparse stiffness matrices arising from practical applications in numerical modeling, which are particularly suitable to serve as test problems for our purpose: comparing sequential and parallel architectures. Here, we focus on SPD matrices arising from problems in structural mechanics. These examples are stored in the file "bcsstruc2.data" in the Harwell–Boeing collection.

In Chapter 4, Section 3, we already described briefly how the matrices are stored in the file. Here, we describe this efficient storage scheme in some more detail and introduce a short code that places it in a "sparseMatrix" object, ready to be used in iterative methods.

The matrices are stored in three sequences of numbers. It is easier to describe these sequences from the last one to the first one. The third sequence contains the nonzero elements in the matrix, ordered column by column. The second sequence contains the row indices of these elements, in the same order. Thus, the length of the second sequence is the same as that of the third one, which is the same as the number of nonzero elements in the matrix. (Compare with the array of edges used in Chapter 4, Section 3.)

These sequences don't contain all the information about the matrix. Indeed, they don't tell us where each column starts and where it ends. Therefore, we can't break them back into the original columns and obtain the original matrix. For this, we need to specify where exactly the sequences should be broken. The place in the third sequence where a particular column starts is just the index in it of the first nonzero element that belongs to this column. In the terminology in Chapter 4, Section 3, the column starts at the virtual address of its first nonzero element. This index (or virtual address) is called here the column pointer, because it indicates the start of the column in the third sequence.

The column pointers are listed in the first sequence in the file. This sequence ends with an extra 0 to mark its end, so the number of numbers in it is the same as the number of columns plus 1.

[1] http://math.nist.gov/MatrixMarket/data/Harwell-Boeing/

With this additional information, one can now reconstruct the original matrix. Indeed, in order to get back a particular column, one should just approach the third sequence at the index indicated in the column pointer and proceed in this sequence until (and not including) the index indicated in the column pointer of the next column. (If the last column is required, then one should proceed until the end of the sequence.) This way, one has every nonzero element in this column. By doing the same in the second sequence, one also has the row indices of these nonzero elements. This gives the column exactly as it is in the original matrix. By doing the same for each column, one gets back the original matrix. This shows that this storage scheme is indeed appropriate.

The "readSparseMatrix" function below reads the matrix elements in the above sequences. It is assumed that these sequences are stored in a file named "boeingMatrix" in the directory in which the program runs.

The number of columns in the original matrix and the number of nonzero elements in it are available in the heading in the Harwell–Boeing collection. For simplicity, it is assumed here that this heading has been deleted from the "boeingMatrix" file.

The "readSparseMatrix" function is a member of the "sparseMatrix" class, so it has access to the individual rows in it. Indeed, because the items in the "list" class are declared as "protected" rather than "private", the "sparseMatrix" class derived from it has access to them.

It is assumed that the current "sparseMatrix" object on which the "readSparseMatrix" function operates has the same number of "row" objects as the number of columns in the original sparse matrix. It is also assumed that these "row" objects are not yet constructed, so they should be constructed in the body of the "readSparseMatrix" function.

The function uses an integer argument to indicate the number of nonzero elements in the original sparse matrix. The three sequences in the file are first placed in three dynamic vectors for easier access. The first is a dynamic vector of integers whose length is the same as the matrix order plus 1. This vector contains the column pointers described above. The second is a dynamic vector of integers containing the row indices in the second sequence. The length of this vector is the same as the number of nonzeroes in the matrix. The third is a dynamic vector of "float" numbers, containing the actual values of the nonzero elements in the matrix. The length of this vector is the same as the number of nonzeroes in the matrix:

```
template<class T>
void sparseMatrix<T>::readSparseMatrix(int nonZeroes){
  dynamicVector<int> columnPtr(rowNumber()+1);
  dynamicVector<int> rowIndex(nonZeroes);
  dynamicVector<float> values(nonZeroes);
  FILE* fp = fopen("boeingMatrix","r");
  for(int i=0;i<=rowNumber();i++)
    fscanf(fp,"%d",&columnPtr(i));
  for(int i=0;i<nonZeroes;i++)
    fscanf(fp,"%d",&rowIndex(i));
  for(int i=0;i<nonZeroes;i++)
    fscanf(fp,"%f",&values(i));
```

These three dynamic vectors are now used to construct the required sparse matrix. Actually, what is constructed is the transpose of the required matrix, because the columns in the original matrix are read into "row" objects in the current "sparseMatrix" object. This drawback will be fixed soon after the "readSparseMatrix" is called by calling the "transpose" function.

The outer loop below scans the column pointers in the first dynamic vector. Each column pointer indicates the index at which the column starts. An inner loop is then used to copy the nonzero elements in this column into the corresponding "row" object in the current "sparseMatrix" object:

```
    for(int i=0;i<rowNumber();i++)
      for(int j=columnPtr[i];j<columnPtr[i+1];j++){
        if(item[i])
          item[i]->append((T)values[j-1],rowIndex[j-1]-1);
        else
          item[i] =
            new row<T>((T)values[j-1],rowIndex[j-1]-1);
      }
}  //  read sparse matrix from Harwell-Boeing collection
```

Nonsymmetric matrices are indeed stored fully in the file. In symmetric matrices, on the other hand, only the lower triangular part is stored in the file. To have the full matrix, one should add the transpose minus the diagonal part. For example, the fifth sparse matrix in the "bcsstruc2.data" file is read as follows:

```
    sparseMatrix<double> A(11948);
    A.readSparseMatrix(80519);
    A += transpose(A) - diagonal(A);
```

In Table 18.1, we report the number of symmetric Gauss–Seidel (or Jacobi) relaxations used within PCG to reduce the energy norm of the preconditioned residual by six orders of magnitude. The order of the matrix and the number of nonzero elements in the entire matrix are also reported in the table. Note that this number of nonzeroes is different from the number used as argument in the "readSparseMatrix" function, which refers to the lower triangular part only.

Table 18.1. *Number of symmetric Gauss–Seidel (or Jacobi) relaxations used within PCG applied to four SPD stiffness matrices (examples 2 to 5 in the file "bcsstruc2.data" in the Harwell–Boeing collection), where "order" is the matrix order and "nonzeroes" is the number of nonzero elements in the entire matrix.*

example	order	nonzeroes	PCG-SGS	PCG-Jacobi
(2)	3948	117816	162	464
(3)	4884	290378	51	150
(4)	10974	428650	743	2220
(5)	11948	149090	220	558

It can be seen from the table that the Jacobi method requires about three times as many iterations as the symmetric Gauss–Seidel method. Still, one should bear in mind that the Jacobi method is twice as cheap, because it uses only one relaxation sweep, whereas the symmetric Gauss–Seidel method uses both forward and backward sweeps. Thus, PCG with Jacobi uses about 50% more relaxation sweeps and 200% more inner products than PCG with a symmetric Gauss–Seidel preconditioner.

The ILU preconditioner in Chapter 17, Section 7, is another inherently sequential preconditioner that can also be used to test the performance of sequential computers. Although ILU is superior to both Gauss–Seidel and Jacobi preconditioners in terms of the number of CGS iterations required to solve nonsymmetric problems, it seems inferior to symmetric Gauss–Seidel in the present SPD cases. This is why it is not used in the present tests.

The multigrid preconditioner requires considerably fewer iterations to converge. However, because of its extra cost, it seems not to be more efficient than symmetric Gauss–Seidel for the present examples.

18.18 Block Jacobi Relaxation

Because data transfer is the most time-consuming task, avoiding it is most important. This principle motivated the block Gauss–Seidel relaxation in Section 18.6. In this relaxation method, the vector of unknowns is split into chunks, which are then relaxed one by one.

The relaxation of a chunk means that all the unknowns in it are relaxed together, so the residual in it vanishes. This is done by solving the subsystem corresponding to it, with all other unknowns kept fixed. This subsystem can by itself be solved iteratively by PCG, using data that are already available to the processor.

Still, the block Gauss–Seidel relaxation is inherently sequential. Relaxing a chunk requires the new values obtained from relaxing the previous chunks, and, hence, must wait until they are calculated. Thus, the blocks cannot be relaxed simultaneously, but only sequentially, one by one.

In the block Jacobi relaxation, on the other hand, the relaxation in a chunk uses only old values that are available from the previous iteration. Therefore, the relaxation in a particular chunk doesn't have to wait for the result of the relaxation in any other chunk. Thus, all the chunks can be relaxed simultaneously in parallel.

Each individual processor is assigned a chunk and is responsible for relaxing it. Before the relaxation starts, the processors exchange data about the old values from the previous iteration. Once the relaxation starts, each processor works independently on its own chunk, using no extra data transfer.

Unlike in the standard Jacobi relaxation, here the individual processor uses its data intensively to carry out many calculations and actually solves the subsystem corresponding to its chunk. Is this extra work worthwhile? Although it requires no extra data transfer, it still requires extra computation time. It would be worthwhile only if it contributed to reducing the total number of outer PCG iterations that use the block Jacobi relaxation as a preconditioner to solve the original system. Like PCG with a standard Jacobi preconditioner, this iteration is guaranteed to converge to the algebraic solution (Section 18.16). But does it require fewer iterations to converge?

Well, it usually does. Indeed, because a chunk is relaxed as a whole, it can be viewed as a single degree of freedom. The block Jacobi relaxation may thus be viewed as a standard Jacobi relaxation applied to a smaller system, in which the number of unknowns is the same as the number of chunks. Because the number of iterations in PCG grows superlinearly with the number of unknowns, the number of iterations with the block Jacobi preconditioner should be much smaller than with the standard Jacobi preconditioner.

Nevertheless, we usually don't need to solve the individual subsystems exactly. After all, our principle says that available data should be exploited to do useful calculations only. Here, "useful" means contributing to reducing the number of outer PCG iterations. Using inexact solvers for the subproblems in the individual chunks may be more efficient in terms of computation time, while still not using too many outer PCG iterations to solve the entire system.

18.19 Exercises

1. How would you implement the inner products required in PCG on a hypercube with K processors? What is the number of communication steps required? What is the arithmetic operator that should be changed to do this? Do functions that call it need to be changed as well? Why?

2. Assume that an arithmetic operation requires α seconds to complete, where α is a small positive parameter depending on the computer. What is the computation time required in the above parallel implementation of the inner product?

3. Assume that a communication step between two processors that are connected directly in the hypercube requires $\beta+\gamma B$ seconds, where β and γ are small positive parameters depending on the architecture, and B is the number of bytes contained in the message. What is the communication time (or communication overhead) required in the above parallel implementation of the inner product?

4. The speedup is the time required to solve the problem on a sequential computer divided by the time required to solve it on the parallel computer under consideration. Write the estimated speedup of the hypercube for the solution of Example (4) in Table 18.1 as a function of α, β, γ, and K. (Remember that the Jacobi preconditioner used on the hypercube requires more iterations than the symmetric Gauss–Seidel preconditioner used on the sequential computer.)

5. In the above speedup estimate, use the typical values $\alpha = 10^{-6}$, $\beta = 10^{-4}$, and $\gamma = 10^{-7}$. Show that, in the above example, the computation time dominates the communication time when $K = 10$, the communication time dominates the computation time when $K = 1000$, and they are comparable when $K = 100$.

6. Use your answer to the previous exercise to show that the speedup increases when the number of processors grows from 10 to 100 but decreases when it grows further from 100 to 1000.

7. Use the "dropItems" function in Chapter 16, Section 3, to construct the block submatrices along the main diagonal of the coefficient matrix, and use them to define functions that implement the block Jacobi and block Gauss–Seidel relaxation methods.

8. Use the above block relaxation methods as preconditioners in PCG, and solve example (4) in Table 18.1.

9. Repeat the above speedup estimates, but this time with your iteration count with block Jacobi and symmetric block Gauss–Seidel preconditioners on the parallel and sequential architectures, respectively. Is there any change in comparison with your previous speedup estimates? Do you think that block relaxation methods provide a fair way to compare parallel and sequential architectures?

10. Rewrite the functions in the "dynamicVector" class so that they are ready for parallel implementation on K processors. In this implementation, each loop is broken into K independent subloops. (To increase efficiency, each subloop should scan a subset of vector components that are stored continuously in the computer memory.) The solution can be found in Section A.14 of the Appendix.

11. Rewrite the functions in the "sparseMatrix" class so that they are ready for parallel implementation on K processors. In this implementation, each loop on the rows in the matrix is broken into K independent subloops. The solution can be found in Section A.14 of the Appendix.

12. Write the low-level code required to communicate between processors in a hypercube. The answer follows from Chapter 1, Section 18.

13. In the previous exercise, what is the best way to choose paths to communicate between processors in terms of minimizing the probability of traffic jams? Is the "path()" function in Chapter 1, Section 18, optimal in that sense? Run your (block) Jacobi code to compare different path-constructing schemes in terms of communication overhead.

Part VI

Applications

In this part, we use the present algorithms and their object-oriented implementation in several partial differential equations (PDEs) that arise often in practice. In particular, we consider complicated nonrectangular domains for which highly unstructured meshes are required.

The PDEs are discretized using linear finite elements in a hierarchy of adaptively refined meshes. The initial mesh is too coarse and approximates poorly both the interior and the boundary of the domain. Therefore, the initial mesh is refined successively in the interior of the domain and at its boundary, producing the required accuracy wherever needed. In this process, the number of nodes increases moderately, resulting in an efficient solution technique.

The stiffness system is solved in each particular refinement level to provide the information required for local refinement. For this purpose, the preconditioned conjugate gradient (PCG) method with the multigrid preconditioner is used. The resulting (coarse) numerical solution is then used adaptively to refine the mesh and advance to the next refinement level. The numerical solution at the final and finest mesh is accepted as the numerical solution of the original PDE.

The above algorithm is referred to as the adaptive-refinement algorithm with multigrid preconditioner. It works well for scalar diffusion problems with discontinuous and anisotropic coefficients, as illustrated in the first chapter in this part (Chapter 19). It needs some modification to work for systems of PDEs such as the linear elasticity equations, as shown in the second and third chapters in this part (Chapters 20 and 21). The final chapter (Chapter 22) indicates how it could also be applied to the Helmholtz and Maxwell equations in the field of electromagnetics.

Chapter 19

Diffusion Equations

In this chapter, we apply the adaptive-refinement algorithm to a diffusion equation with discontinuous and anisotropic coefficients in a complicated nonrectangular domain. We show that this algorithm gives good accuracy with a moderate number of nodes. We also show that the multigrid preconditioner is more efficient than other preconditioners in solving the individual stiffness systems.

19.1 The Boundary-Value Problem

Here, we introduce the diffusion equation used in the present numerical tests. The PDE is

$$-(Pu_x)_x - (Qu_y)_y = 0$$

in the domain in Figure 19.1, where the diffusion coefficients P and Q are equal to 1 in most of the domain, except its lower-left quarter, where they are very large:

$$P(x, y) = \begin{cases} 100 & \text{if } x < 0 \text{ and } y < 0, \\ 1 & \text{otherwise;} \end{cases}$$

$$Q(x, y) = \begin{cases} 1000 & \text{if } x < 0 \text{ and } y < 0, \\ 1 & \text{otherwise.} \end{cases}$$

The boundary conditions are also indicated in Figure 19.1. On the circular part of the boundary, mixed boundary conditions of the form

$$Pu_x n_1 + Qu_y n_2 + u = 0$$

are given, where $\vec{n} = (n_1, n_2)$ is the outer normal vector. The letter D in the figure indicates the part of the boundary where Dirichlet boundary conditions of the form

$$u(x, y) = \sin(x + y)$$

are given. Finally, homogeneous Neumann boundary conditions of the form

$$u_n = 0$$

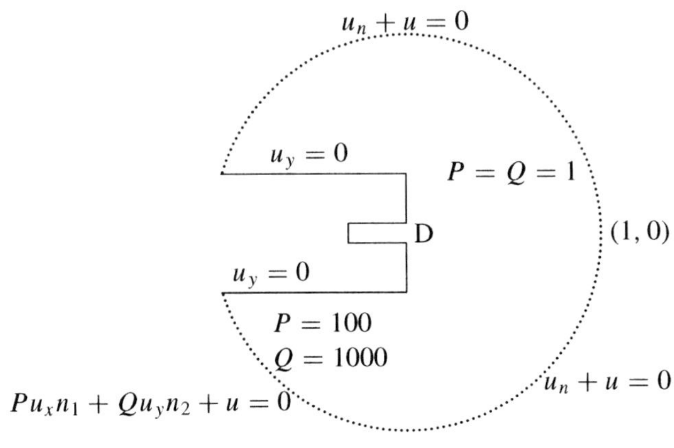

Figure 19.1. *The domain, diffusion coefficients, and boundary conditions for the present diffusion example.*

are given on the rest of the boundary. This completes the introduction of the boundary-value problem. In the next section, we discuss the finite-element discretization.

19.2 The Finite-Element Mesh

The finite-element mesh is constructed by the adaptive-refinement algorithm with automatic boundary refinement (Chapter 14, Section 8). The initial mesh contains only 11 nodes and 9 triangles, providing a rather poor approximation of the domain and its boundary (Figure 19.2). Still, this mesh will improve during the refinement process and eventually have sufficient resolution both in the interior of the domain and at the boundary.

In order to test the accuracy of the algorithm, we first use global refinement by setting the threshold in Chapter 14, Section 3, to 0. Of course, this is not a practical approach, because the number of nodes increases rapidly and has already reached 107000 at the eighth refinement level. Still, it gives a good reference point for testing more practical approaches.

We view the numerical solution produced by global refinement as practically exact. Numerical solutions obtained with local refinement are then compared to it to assess their accuracy. For this purpose, we consider the rightmost point in the domain, $(1, 0)$. With global refinement, the numerical solution at the eighth level of refinement produces the value

$$u((1, 0)) = 0.2016.$$

To this value, we compare the value obtained in several kinds of adaptive refinement. When the threshold in Chapter 14, Section 3, is 0.05, the mesh refines until it contains about 2000 nodes. In further refinement levels, practically no nodes are added. This is a small and attractive number; however, it produces the value $u((1, 0)) = 0.25$, which is not sufficiently accurate. When the threshold decreases to 0.025, no more than 7000 nodes are used in the mesh. Still, the accuracy is insufficient: $u((1, 0)) = 0.16$. Therefore, we use the even

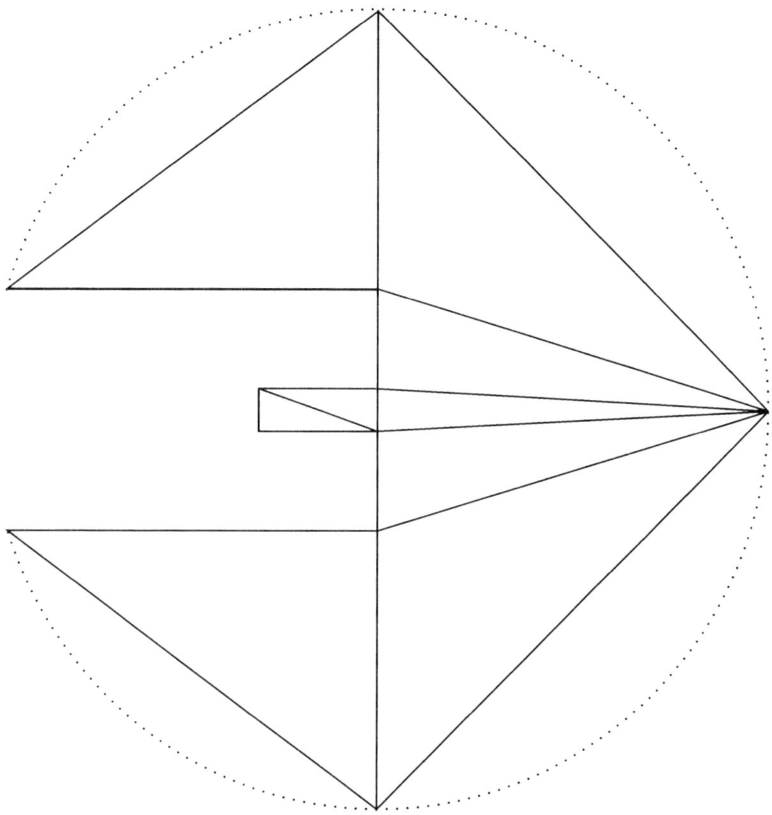

Figure 19.2. *The initial (coarse) mesh that contains only* 11 *nodes and* 9 *triangles and is refined further in the adaptive-refinement algorithm both in the interior of the domain and at the circular boundary.*

smaller threshold of 0.01, which produces the more accurate value of

$$u((1,0)) = 0.196.$$

The number of nodes required for this is still moderate: it is six times as small as with global refinement (see Table 19.1).

In Figure 19.3, one can see how the mesh refines in the first five levels of refinement. Thanks to automatic boundary refinement, the mesh refines not only in the interior of the domain but also near the circular boundary. The nodes are particularly clustered around the irregular part of the boundary (where Dirichlet boundary conditions are given). In the lower-left quarter of the domain, on the other hand, the strong diffusion prevents any considerable variation, so only a few nodes are needed. In the next section, we discuss the linear-system solver used to produce the numerical solutions on these meshes.

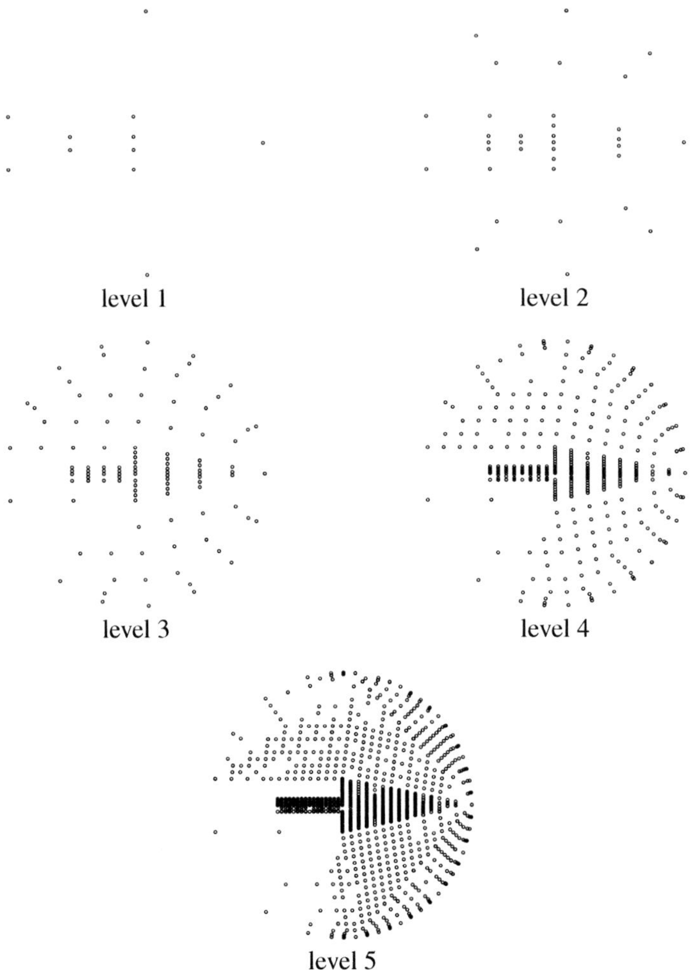

Figure 19.3. *Distribution of nodes in the first five levels of refinement in the diffusion example. The adaptive refinement uses a threshold of* 0.01 *and automatic boundary refinement. The strong diffusion in the lower-left quarter of the domain prevents large variation, so no extra refinement is needed.*

19.3 The Linear-System Solver

In each particular refinement level, the stiffness system is solved by the PCG iteration in Chapter 17, Section 11. We test four different preconditioners: symmetric Gauss–Seidel (Chapter 17, Section 5), ILU (Chapter 17, Section 7), multigrid (Chapter 17, Section 8), and AMG (Chapter 17, Section 9). The multigrid preconditioners are implemented in a V-cycle with one prerelaxation, one post-relaxation, and one relaxation to solve approximately the coarsest-grid system ("Nu1", "Nu2", and "NuCoarse" in Chapter 17, Section 10, are set to 1). The relaxation method used in this V-cycle is symmetric Gauss–Seidel ("useILU" in

Table 19.1. *Number of PCG iterations used in each refinement level in the adaptive-refinement algorithm (with threshold of* 0.01 *and automatic boundary refinement) applied to the diffusion example. The computation time of a multigrid iteration is the same as that of five symmetric Gauss–Seidel iterations. (The setup time is negligible.)*

level	nodes	$u(1,0)$	PCG-MG	PCG-AMG	PCG-ILU	PCG-SGS
1	11	0.000	5	5	3	5
2	34	0.343	5	5	6	11
3	106	0.247	8	9	9	23
4	340	0.233	12	12	17	41
5	1069	0.215	18	18	33	71
6	3050	0.203	22	22	63	121
7	8242	0.201	36	35	121	217
8	18337	0.196	54	51	340	382

Chapter 17, Section 10, is set to 0). (ILU is unsuitable for this purpose unless no fill-in is used.) We also use the parameter "gridRatio" = 0.95 in Chapter 17, Section 10. With these choices, the total computation time of the multigrid V-cycle is about five times as large as that of a symmetric Gauss–Seidel iteration.

All four preconditioners are symmetric and positive definite (SPD), so PCG is guaranteed to converge. Still, it can be seen in Table 19.1 that they differ in terms of convergence rate. There is practically no difference between the multigrid preconditioners in Chapter 17, Sections 8 and 9; they are both superior to both ILU and symmetric Gauss–Seidel. In fact, in some cases ILU and symmetric Gauss–Seidel converge prohibitively slowly, whereas the multigrid preconditioners converge in acceptable time.

19.4 Implicit Time Marching

In time-dependent diffusion equations, the solution may depend on the time variable t as well, and the term u_t is also added to the PDE (see Chapter 11, Section 1). This term contributes

$$\int_\Omega u_t v dx dy$$

to the bilinear form $a(u, v)$ used in the weak formulation in Chapter 11, Section 2, where Ω is the spatial (x, y)-domain. When time marching is used to discretize this term, it takes the form

$$(\Delta t)^{-1}\left(\int_\Omega u^{(i)} v dx dy - \int_\Omega u^{(i-1)} v dx dy\right),$$

where (i) stands for the current time level and $(i-1)$ stands for the previous one. When the time marching is (semi-) implicit, the diffusion terms are also evaluated at the current (ith) time level. Thus, implicit time marching requires in each time level the solution of a PDE, as in Section 19.1 above, with the extra free term $(\Delta t)^{-1}u$:

$$-(Pu_x)_x - (Qu_y)_y + (\Delta t)^{-1}u = F,$$

Table 19.2. *Number of PCG iterations used in each refinement level in the adaptive-refinement algorithm (with threshold of* 0.01 *and automatic boundary refinement) applied to an implicit time-marching step (with* $\Delta t = 1$*) in the time-dependent diffusion example.*

level	nodes	$u(1,0)$	PCG-MG	PCG-AMG	PCG-ILU	PCG-SGS
1	11	0.000	5	5	3	5
2	34	0.320	4	4	5	11
3	107	0.225	8	8	8	21
4	342	0.214	12	12	17	40
5	1069	0.199	18	18	30	61
6	3095	0.190	22	23	63	120
7	8331	0.181	36	34	122	216
8	18608	0.173	54	52	420	386

where the right-hand side F contains the required data about $u^{(i-1)}$ from the previous time level. The numerical solution of this PDE serves then as the solution $u^{(i)}$ at the current time level, and the process can proceed to march to the next time level.

The stability and accuracy of the time-marching schemes are analyzed in the exercises at the end of Chapter 12.

The time-dependent problem has thus been reduced to a sequence of time-independent problems with an extra term. In the finite-element scheme, this term contributes extra integrals of the form

$$(\Delta t)^{-1} \int_e \phi_j \phi_i dxdy$$

to the element $A_{i,j}$ in the stiffness matrix, where e is any finite element and both nodal basis functions ϕ_i and ϕ_j are nonzero. Since ϕ_i and ϕ_j are polynomials in x and y, this integral can be calculated as in Chapter 5, Section 14. This is the only change required in the above adaptive-refinement algorithm.

Thus, we repeat the above numerical experiments, with the only change being the above extra term. For simplicity, we assume that $u^{(i-1)} \equiv 0$, so the right-hand side is 0. We also assume that $\Delta t = 1$.

First, we apply this algorithm with zero threshold in Chapter 14, Section 3, which actually produces global refinement. (Automatic boundary refinement as in Chapter 14, Section 8, is also used to make sure that the mesh also refines next to the circular boundary.) With 107000 nodes in the eighth level of refinement, the value at the rightmost point in the domain is

$$u(1,0) = 0.1802.$$

To approximate this value well, the threshold of 0.05 is too large: although it uses at most 2000 nodes, it gives the inaccurate value of $u(1,0) = 0.243$. Similarly, the smaller threshold of 0.025, although it uses at most 7000 nodes, produces the value $u(1,0) = 0.141$, which is still not sufficiently accurate. Therefore, we use a threshold of 0.01, which gives better accuracy in a still moderate number of nodes, as is apparent from Table 19.2.

The numbers of iterations reported in Table 19.2 are similar to those in Table 19.1. This shows that the present approach can also be used in the stable and accurate solution of time-dependent problems (see exercises at the end of Chapter 12).

19.5 Exercises

1. Construct the initial coarse mesh in Figure 19.2. The solution can be found in Section A.16 of the Appendix.

2. Write the code that implements the adaptive-refinement algorithm for the diffusion problem. Remember to modify the code in Chapter 14, Section 8, so that two small triangles are added between the boundary edge and the boundary segment that lies next to it only if this is a circular boundary segment. The solution can be found in Section A.16 of the Appendix.

3. Repeat the above exercises with the extra free term introduced in Section 19.4.

Chapter 20

The Linear Elasticity Equations

In this chapter, we apply the adaptive-refinement algorithm to the linear elasticity system of PDEs in a circular domain. We show the efficiency of this algorithm, particularly when used in conjunction with the multigrid preconditioner to solve the individual linear systems in the individual refinement levels. For this purpose, the multigrid algorithm must be modified in such a way that the transfer of data between fine and coarse grids is done separately for each unknown function.

20.1 System of PDEs

So far, we have considered scalar PDEs, with scalar unknown function $u \equiv u(x, y)$. In this chapter, we consider the more complicated case of a system of PDEs in which two or more unknown functions, say $u \equiv u(x, y)$ and $v \equiv v(x, y)$, need to be solved for. The unknown functions u and v are coupled in the system of PDEs in the sense that they depend on each other and cannot be solved for separately.

The system of coupled PDEs is also called a "vector PDE," because it can be rewritten as a single equation in the unknown (two-dimensional) vector $\vec{u} \equiv (u, v)$ and the corresponding derivatives $\vec{u}_x = (u_x, v_x)$ and $\vec{u}_y = (u_y, v_y)$. In this style, the coefficients in the PDE are placed in a 2×2 matrix.

20.2 The Strong Formulation

Here, we consider the system of linear elasticity equations in the two-dimensional domain Ω. The system is characterized by the Poisson ratio ν, $0 \le \nu < 1$, which is a typical constant that depends on the particular material under consideration. For most materials, $0.25 \le \nu \le 0.35$; the prominent exception is rubber, for which $\nu = 0.5$.

The system of linear elasticity equations in its strong form is given by

$$u_{xx} + \frac{1-\nu}{2}u_{yy} + \nu v_{yx} + \frac{1-\nu}{2}v_{xy} = \mathcal{F},$$

$$\frac{1-\nu}{2}u_{yx} + \nu u_{xy} + \frac{1-\nu}{2}v_{xx} + v_{yy} = \mathcal{G}$$

at points $(x, y) \in \Omega$. Here, $\mathcal{F}$ and $\mathcal{G}$ are given right-hand-side functions in Ω.

Because the system consists of two equations, we also have two boundary conditions at each boundary point $(x, y) \in \partial\Omega$. The boundary conditions can be of different types at different points in $\partial\Omega$. In fact, the boundary can be written as the union

$$\partial\Omega = \Gamma_D \cup \Gamma_N,$$

where Γ_D and Γ_N are disjoint subsets of $\partial\Omega$. Dirichlet boundary conditions are imposed on Γ_D, and Neumann boundary conditions are imposed on Γ_N. More specifically, the boundary conditions in Γ_D are given by

$$u = \gamma_1 \quad \text{and} \quad v = \gamma_2$$

(where γ_1 and γ_2 are given functions in Γ_D), and the boundary conditions in Γ_N are given by

$$\begin{aligned} u_x n_1 + \frac{1-\nu}{2} u_y n_2 + \nu v_y n_1 + \frac{1-\nu}{2} v_x n_2 &= 0, \\ \frac{1-\nu}{2} u_y n_1 + \nu u_x n_2 + \frac{1-\nu}{2} v_x n_1 + v_y n_2 &= 0 \end{aligned}$$

(where (n_1, n_2) is the outer unit normal vector at the relevant point in Γ_N).

In the next section, we rewrite this system in its weak formulation. In this formulation, the system is well posed in the sense that it has a unique solution. The weak formulation is also the basis for the finite-element method used in the discretization.

20.3 The Weak Formulation

The weak formulation can be obtained from the strong one by integrating the equations over Ω and using Green's formula and the boundary conditions. As a consequence, every solution to the strong formulation must also solve the weak formulation. However, a solution to the weak formulation does not necessarily solve the strong formulation. In fact, it may well have nondifferentiable derivatives, with which the strong formulation is not at all well defined. Thus, the weak formulation is better posed than the strong one.

The weak formulation is defined as follows. Find functions $u \equiv u(x, y)$ and $v \equiv v(x, y)$ that agree with γ_1 and γ_2 (respectively) on Γ_D and have square-integrable derivatives in Ω such that, for every pair of functions $s \equiv s(x, y)$ and $t \equiv t(x, y)$ that vanish on Γ_D and have square-integrable derivatives in Ω,

$$\begin{aligned} \int_\Omega \left(u_x s_x + \frac{1-\nu}{2} u_y s_y + \nu v_y s_x + \frac{1-\nu}{2} v_x s_y \right) dxdy &= -\int_\Omega \mathcal{F} s dxdy, \\ \int_\Omega \left(\frac{1-\nu}{2} u_y t_x + \nu u_x t_y + \frac{1-\nu}{2} v_x t_x + v_y t_y \right) dxdy &= -\int_\Omega \mathcal{G} t dxdy. \end{aligned}$$

An equivalent form of the weak formulation can be obtained by taking the sum of the above two equations [20]. Indeed, since s or t may also be just the zero function, the above two equations can be retrieved from their sum. The sum equation can be written as

$$\begin{aligned} \int_\Omega \left((1-\nu)(u_x s_x + v_y t_y) + \nu(u_x + v_y)(s_x + t_y) + \frac{1-\nu}{2}(u_y + v_x)(s_y + t_x) \right) dxdy \\ = -\int_\Omega (\mathcal{F}s + \mathcal{G}t) dxdy. \end{aligned}$$

In fact, if the left-hand side of this equation is denoted by $a((u, v), (s, t))$ and the right-hand side is denoted by $f((s, t))$, then the equation takes the form

$$a((u, v), (s, t)) = f((s, t)).$$

This formulation is used below to define an equivalent, well-posed minimization problem.

20.4 The Minimization Problem

This section and the next one are related to the theory in Chapter 11. Readers who are mainly interested in the practical aspects can skip them and proceed to Section 20.6. More advanced readers may find them interesting and relevant to understand better the finite-element scheme.

Let's define the minimization problem that is equivalent to the weak formulation. Define the functional

$$g((s, t)) \equiv \frac{1}{2}a((s, t), (s, t)) - f((s, t)).$$

The minimization problem is defined as follows: from the family of pairs of functions that agree with γ_1 and γ_2 on Γ_D and have square-integrable derivatives in Ω, pick the pair (u, v) that minimizes g.

By following the proof in Chapter 11, Section 3, one can show that here also the minimization problem is equivalent to the weak formulation. Therefore, in order to show that the weak formulation is well posed, it is sufficient to show that the minimization problem is well posed (has a unique solution). As shown in Chapter 11, Sections 5 and 6, a sufficient condition for this is the coercivity of the quadratic form $a((s, t), (s, t))$. This property is shown below.

20.5 Coercivity of the Quadratic Form

In order to show the coercivity of the quadratic form $a((s, t), (s, t))$, we assume that the domain Ω is connected and Γ_D is (or contains) a curve in $\partial\Omega$. Let s and t be some functions that vanish on Γ_D and have square-integrable derivatives in Ω. Assume also that

$$a((s, t), (s, t)) = 0.$$

Coercivity means that, under the above assumptions, s and t vanish throughout Ω. Indeed, from the latter assumption it follows that

$$s_x = t_y = s_y + t_x = 0$$

throughout Ω. Let (x, y) be some point in Ω, and let us show that both s and t indeed vanish at (x, y). Let us draw the "staircase" in Ω (Figure 20.1) that leads from the curve in Γ_D to (x, y). Let us show that both s and t vanish in the entire staircase. Let us consider the first, horizontal strip in the staircase. Because s vanishes in Γ_D and s_x vanishes in Ω, we have that s (and, hence, also s_y) vanishes in the strip. From the last assumption above, t_x also vanishes in the strip. Because $t = 0$ in Γ_D, we have that t also vanishes in the strip. Now,

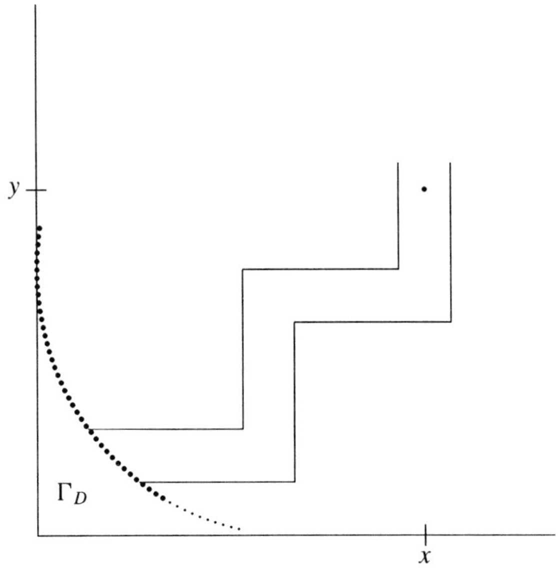

Figure 20.1. *The staircase in Ω that leads from Γ_D to $(x, y) \in \Omega$ in the coercivity proof.*

consider the next, vertical strip in the staircase. Since $t_y = 0$ in Ω, t (and, hence, also t_x) vanishes in this strip. From the last assumption, s_y also vanishes in this strip, which implies that s also vanishes there. By repeating this process, one can "climb" up the stairs and reach (x, y), concluding that both s and t indeed vanish there. This completes the proof of the coercivity property.

As in Chapter 11, Sections 5 and 6, the coercivity property can be used to show that the minimization problem has a unique solution. As a consequence, it follows that the weak formulation is also well posed.

20.6 The Finite-Element Discretization

The finite-element discretization method for the above weak formulation uses the same triangulation as in Chapter 12, Section 1. The discrete problem that can be solved numerically is obtained by restricting the weak formulation to a finite-dimensional subspace of functions, that is, the subspace of functions that are continuous in the entire mesh and linear in each and every triangle in it. This discrete problem or discrete weak formulation is as in Section 20.3, except that the functions u, v, s, and t must also be linear in each triangle in the mesh.

Assume that there are $N/2$ nodes in the mesh that don't lie on Γ_D. The numerical solution (u, v) can then be written as

$$u = \sum_{j=0}^{N/2-1} x_j \phi_j \quad \text{and} \quad v = \sum_{j=N/2}^{N-1} x_j \phi_j,$$

where $\phi_j = \phi_{j+N/2}$ $(0 \le j < N/2)$ is the nodal basis function that has the value 1 at the jth node and vanishes at all the other nodes, and x_j $(0 \le j < N)$ is the unknown coefficient of ϕ_j in the expansion.

In order to obtain the ith equation in the discrete system, $0 \le i < N/2$, we pick $s = \phi_i$ and $t \equiv 0$. Similarly, the ith equation $(N/2 \le i < N)$ is obtained by picking $s \equiv 0$ and $t = \phi_{i-N/2}$. The discrete stiffness system is, thus,

$$Ax = f,$$

where A is the stiffness matrix of order N defined below, x is the N-dimensional vector of unknowns, and f is the given N-dimensional vector defined by

$$f_i = \begin{cases} -\int_\Omega \mathcal{F}\phi_i dxdy & \text{if } 0 \le i < N/2, \\ -\int_\Omega \mathcal{G}\phi_{i-N/2} dxdy & \text{if } N/2 \le i < N, \end{cases}$$

plus a contribution from the boundary conditions.

20.7 The Stiffness Matrix

The stiffness matrix A in the above discrete system can be split into four blocks of order $N/2$:

$$A = \begin{pmatrix} A^{(0,0)} & A^{(0,1)} \\ A^{(1,0)} & A^{(1,1)} \end{pmatrix}.$$

The four blocks can be calculated as in Chapter 12, Section 5. The only change is that the diagonal matrix $diag(P, Q)$ used there should be replaced here by

$$\begin{pmatrix} 1 & 0 \\ 0 & \frac{1-\nu}{2} \end{pmatrix}$$

to calculate $A^{(0,0)}$,

$$\begin{pmatrix} \frac{1-\nu}{2} & 0 \\ 0 & 1 \end{pmatrix}$$

to calculate $A^{(1,1)}$,

$$\begin{pmatrix} 0 & \frac{1-\nu}{2} \\ \nu & 0 \end{pmatrix}$$

to calculate $A^{(0,1)}$, and

$$\begin{pmatrix} 0 & \nu \\ \frac{1-\nu}{2} & 0 \end{pmatrix}$$

to calculate $A^{(1,0)}$. This completes the definition of the stiffness matrix.

As explained in Chapter 13, Section 5, one can use symmetry to avoid recalculating contributions from triangles to the stiffness matrix. This applies to the blocks $A^{(0,0)}$ and $A^{(1,1)}$, which may be viewed as discrete diffusion problems. In the calculation of the $A^{(0,1)}$

block, however, this trick can no longer be used, and every contribution to the stiffness matrix from any triangle must be calculated explicitly. Fortunately, the symmetry property can be used to avoid calculating the $A^{(1,0)}$ block. Because it is the transpose of $A^{(0,1)}$, it is obtained immediately from the substitution

$$A \leftarrow A + A^t - diag(A),$$

carried out after the call to the assembling function.

20.8 The Adaptive-Refinement Criterion

The adaptive-refinement algorithm is as in Chapter 14, Section 3, except for a slight change in the refinement criterion. In Chapter 14, Section 3, the criterion for including the midpoint $(i + j)/2$ in the next finer mesh is

$$|x_i - x_j| > \text{threshold},$$

where x is the numerical solution of the stiffness system at the current level of refinement, and i and j correspond to some nodes in the current mesh. This criterion indicates that the solution may indeed have large variation between i and j, so their midpoint is also needed to provide better accuracy there.

In the present case, both solution functions u and v must be solved for with good accuracy. Therefore, if either of them has large variation between i and j, the midpoint $(i + j)/2$ must be added in the next level of refinement. Thus, the criterion for refinement should read

$$\max(|x_i - x_j|, |x_{i+N/2} - x_{j+N/2}|) > \text{threshold},$$

where $0 \leq i, j < N/2$ are indices corresponding to nodes in the current mesh. This criterion indicates that either u or v may have large variation between the nodes i and j in the current mesh, so the midpoint $(i + j)/2$ should be included in the next, finer, mesh.

20.9 The Modified Multigrid Algorithm

The adaptive-refinement algorithm requires the solution of the stiffness system at each level of refinement. The numerical solution is then used to check the above criterion and decide where to refine in the next level of refinement. Thus, an efficient iterative method for solving the stiffness system is required.

As discussed in Chapter 12, Section 4, the coercivity property implies that the stiffness matrix is SPD. Therefore, one would naturally like to use the PCG iterative method, with a good preconditioner like multigrid. Unfortunately, the multigrid algorithms described in Chapter 17, Sections 8 and 9, don't work well for the elasticity equations. This is because the prolongation operator defined there uses information stored in the matrix elements in A. Because some matrix elements couple the numerical approximation to u with the numerical approximation to v, the prolongation operator may mix u-values with v-values.

The cure is to use, in Algorithms 17.2–17.3, not the original matrix A but rather its block-diagonal part

$$\begin{pmatrix} A^{(0,0)} & 0 \\ 0 & A^{(1,1)} \end{pmatrix}.$$

Since this part doesn't couple unknowns corresponding to u with unknowns corresponding to v, the prolongation operator resulting from it prolongs u-values separately from v-values, preserving continuity in u and v. (A similar approach is introduced in [12] for the rectangular grid.)

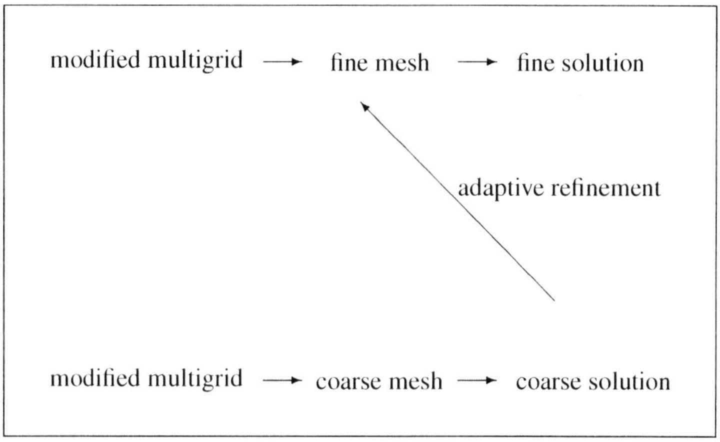

Figure 20.2. *The adaptive-refinement algorithm for the linear elasticity equations: the numerical approximations to u and v obtained by the modified multigrid preconditioner at a particular mesh are used to refine it further and produce the next level of refinement.*

The rest of the multigrid algorithm is as before ($R = P^t$ and $Q = RAP$). This completes the definition of the iterative solver for the individual stiffness systems in the adaptive-refinement algorithm. The entire adaptive-refinement algorithm is displayed schematically in Figure 20.2.

20.10 Numerical Examples

Here, we apply the adaptive-refinement algorithm to the above linear elasticity equations with the Poisson ratio

$$\nu = 1/3.$$

The domain Ω is the unit circle:

$$\Omega = \left\{(x, y) \mid x^2 + y^2 < 1\right\}.$$

The portion of the boundary on which Dirichlet boundary conditions are imposed is

$$\Gamma_D = \left\{(x, y) \mid x^2 + y^2 = 1,\ x \le -0.5\right\}.$$

The portion of the boundary on which Neumann boundary conditions are imposed is the rest of the boundary:

$$\Gamma_N = \left\{(x, y) \mid x^2 + y^2 = 1,\ x > -0.5\right\}$$

(Figure 20.3).

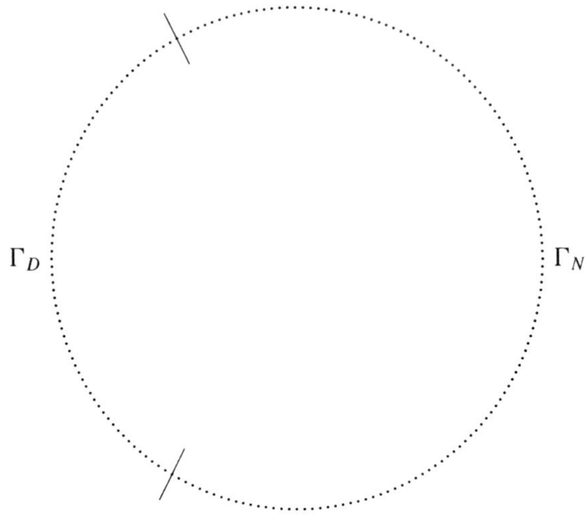

Figure 20.3. *The boundary segments Γ_D (where Dirichlet boundary conditions are given) and Γ_N (where Neumann boundary conditions are given) in the linear elasticity example.*

The Dirichlet boundary conditions are homogeneous in Γ_D:

$$\gamma_1 = \gamma_2 \equiv 0.$$

The right-hand-side functions are

$$\mathcal{F} = \mathcal{G} = -0.1 \cdot \delta_{(1,0)},$$

where the δ-function (centered at $(1, 0)$) is defined by the weak formula

$$\int_{(x,y)\in\Omega} \delta_{(1,0)} w(x, y) dx dy = w(1, 0)$$

for every smooth function $w(x, y)$ defined in the domain Ω. For this example, the right-hand-side vector f in the stiffness system is, thus,

$$f_i = \begin{cases} 0.1 & \text{if } i = 0 \text{ or } i = N/2, \\ 0 & \text{otherwise.} \end{cases}$$

The adaptive-refinement algorithm is used with the criterion in Section 20.8 above (with threshold $= 0.01$). Automatic boundary refinement as in Chapter 14, Section 8, is also used. In each refinement level, the numerical solution is calculated by the PCG iteration in Chapter 17, Section 11. The multigrid algorithm in Chapter 17, Sections 8 and 9, is used as a preconditioner. The multigrid V-cycle uses symmetric Gauss–Seidel as the relaxation method ("useILU" $= 0$ and "gridRatio" $= 0.95$ in Chapter 17, Section 10). The

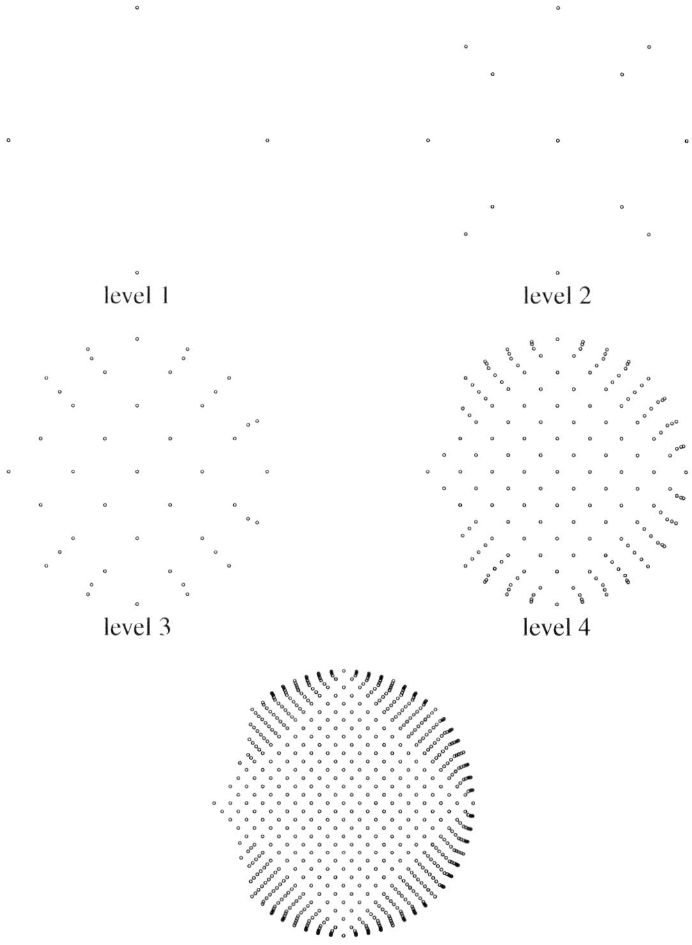

Figure 20.4. *Distribution of nodes at the first five levels of refinement (with automatic boundary refinement). The Poisson ratio is* $\nu = 1/3$.

modification in Section 20.9 is also used. The total cost of a multigrid V-cycle is the same as that of five symmetric Gauss–Seidel iterations.

We don't report results with the ILU preconditioner, because it is much inferior to the other preconditioners and may not converge in reasonable time. In general, ILU is a good preconditioner for nonsymmetric problems but not sufficiently reliable for symmetric problems such as the present one. (ILU with no fill-in could serve as a good relaxation method within multigrid but is not used here.)

The resulting meshes are displayed in Figure 20.4. The initial mesh contains only four nodes, as in Figure 12.9. In the finer meshes, there are slightly more nodes in the right part of the circle, near the Neumann boundary. This is probably because Neumann boundary

Table 20.1. *The adaptive-refinement algorithm (with automatic boundary refinement) applied to the linear elasticity equations in the circle (with Poisson ratio $\nu = 1/3$). The modified multigrid preconditioner costs the same as five symmetric Gauss–Seidel iterations. (The setup time is negligible.)*

level	nodes	PCG-MG	PCG-AMG	PCG-SGS
1	4	13	13	54
2	13	9	9	13
3	45	12	14	32
4	173	17	16	73
5	679	33	32	177
6	2690	57	52	398
7	10329	93	85	895

conditions allow greater variation in the solution, because they don't fix it like Dirichlet boundary conditions.

The numbers of iterations required to solve the stiffness systems are reported in Table 20.1. It turns out that the multigrid preconditioners are more efficient than the symmetric Gauss–Seidel preconditioner. This advantage becomes clearer as finer and finer meshes are used.

20.11 Exercises

1. Write the code that assembles the stiffness matrix for the linear elasticity equations. The solution can be found in Section A.17 of the Appendix.

2. Modify the "refine" function in Chapter 14, Section 6, to implement the criterion for refinement in Section 20.8. The solution can be found in Section A.17 of the Appendix.

3. Modify the "createTransfer" function in Section A.12 of the Appendix to implement the modified multigrid algorithm in Section 20.9. The solution can be found in Section A.17 of the Appendix.

4. Use your answers to the previous exercises to implement the adaptive-refinement algorithm for the linear elasticity equations. Use automatic boundary refinement, as in Chapter 14, Section 8. Print the resulting meshes and verify that they are indeed conformal. How does the number of nodes change from level to level?

Chapter 21

The Stokes Equations

In this chapter, we present the Stokes equations and their relation to the general linear elasticity equations. The present adaptive-refinement algorithm uses modified multigrid to solve linear elasticity problems that approximate the original Stokes equation. On the finest mesh, a Schur-complement preconditioner is used to solve the original Stokes equations. Other algorithms to solve the Stokes and Navier–Stokes equations are also described.

21.1 The Nabla Operator

In order to present the Stokes and Navier–Stokes equations that model fluid dynamics, we need the Nabla operator, denoted by ∇. This operator acts differently on scalar and vector functions. Furthermore, its interpretation depends on the arithmetic symbol that follows it.

The Nabla operator acts upon scalar and vector functions of three spatial variables x, y, and z. For example, let's consider the scalar function

$$s \equiv s(x, y, z)$$

and the vector function

$$v \equiv (v_1(x, y, z), v_2(x, y, z), v_3(x, y, z)).$$

In what follows, we assume that the scalar functions s, v_1, v_2, and v_3 are differentiable to the second order; that is, they have well-defined second derivatives, including mixed ones.

The Nabla operator may be interpreted in three different ways. The interpretation depends on the symbol that follows the ∇ symbol and the type of function that follows it.

When the ∇ symbol is followed by the name of a function with no arithmetic symbol in between, it is interpreted as the gradient operator:

$$\nabla = \begin{pmatrix} \partial/\partial x \\ \partial/\partial y \\ \partial/\partial z \end{pmatrix}.$$

When applied to a scalar function, this operator produces the vector of derivatives

$$\nabla s = \begin{pmatrix} \partial/\partial x \\ \partial/\partial y \\ \partial/\partial z \end{pmatrix} s = \begin{pmatrix} s_x(x, y, z) \\ s_y(x, y, z) \\ s_z(x, y, z) \end{pmatrix}.$$

In the above, the Nabla operator is applied to a scalar function with no mathematical symbol in between. Actually, it can also be applied in this way to a vector function v. The result is a 3×3 matrix whose columns are the gradients of the individual components v_1, v_2, and v_3:

$$\nabla v = (\nabla v_1 \mid \nabla v_2 \mid \nabla v_3).$$

When the ∇ operator is followed by the dot symbol $\cdot$, it is interpreted as the divergence operator. In this case, it can be written as the row vector

$$\nabla \cdot = (\partial/\partial x, \partial/\partial y, \partial/\partial z).$$

This operator acts upon vector functions to produce their divergence (sum of partial derivatives):

$$\begin{aligned} \nabla \cdot v &= (\partial/\partial x, \partial/\partial y, \partial/\partial z)\, v \\ &= v_1(x, y, z)_x + v_2(x, y, z)_y + v_3(x, y, z)_z. \end{aligned}$$

The above symbolic forms allow one also to multiply operators easily. For example, the Laplacian operator $\triangle$ (the divergence of a gradient) can be written as

$$\triangle = \nabla \cdot \nabla = \partial^2/\partial x^2 + \partial^2/\partial y^2 + \partial^2/\partial z^2.$$

This operator acts upon scalar functions and produces the sum of their (nonmixed) second derivatives:

$$\begin{aligned} \triangle s &= \nabla \cdot \begin{pmatrix} s(x, y, z)_x \\ s(x, y, z)_y \\ s(x, y, z)_z \end{pmatrix} \\ &= s(x, y, z)_{xx} + s(x, y, z)_{yy} + s(x, y, z)_{zz}. \end{aligned}$$

This definition can also be extended to produce the vector Laplacian operator for vector functions $v = (v_1, v_2, v_3)$:

$$\triangle v = \begin{pmatrix} \triangle(v_1) \\ \triangle(v_2) \\ \triangle(v_3) \end{pmatrix}.$$

This vector Laplacian operator can also be written symbolically as the diagonal 3×3 matrix

$$\triangle = \begin{pmatrix} \triangle & & \\ & \triangle & \\ & & \triangle \end{pmatrix},$$

where the $\triangle$ in the left-hand side is interpreted as the vector Laplacian, and the $\triangle$'s on the diagonal of the matrix in the right-hand side are interpreted as scalar Laplacians.

When the above Nabla operators are multiplied in the reverse order (gradient of divergence), we get the following 3×3 matrix of second derivatives:

$$\nabla\nabla\cdot = \begin{pmatrix} \partial/\partial x \\ \partial/\partial y \\ \partial/\partial z \end{pmatrix} (\partial/\partial x, \partial/\partial y, \partial/\partial z)$$

$$= \begin{pmatrix} \frac{\partial^2}{\partial x^2} & \frac{\partial^2}{\partial x \partial y} & \frac{\partial^2}{\partial x \partial z} \\ \frac{\partial^2}{\partial y \partial x} & \frac{\partial^2}{\partial y^2} & \frac{\partial^2}{\partial y \partial z} \\ \frac{\partial^2}{\partial z \partial x} & \frac{\partial^2}{\partial z \partial y} & \frac{\partial^2}{\partial z^2} \end{pmatrix}.$$

When the ∇ operator is followed by the $\times$ symbol, then it represents the curl or rotor operator. This operator can be represented by the 3×3 matrix

$$\nabla\times = \begin{pmatrix} 0 & -\partial/\partial z & \partial/\partial y \\ \partial/\partial z & 0 & -\partial/\partial x \\ -\partial/\partial y & \partial/\partial x & 0 \end{pmatrix}.$$

This operator acts upon vector functions to produce their curl (rotor) vector functions:

$$\nabla \times v = \begin{pmatrix} 0 & -\partial/\partial z & \partial/\partial y \\ \partial/\partial z & 0 & -\partial/\partial x \\ -\partial/\partial y & \partial/\partial x & 0 \end{pmatrix} v$$

$$= \begin{pmatrix} v_3(x, y, z)_y - v_2(x, y, z)_z \\ v_1(x, y, z)_z - v_3(x, y, z)_x \\ v_2(x, y, z)_x - v_1(x, y, z)_y \end{pmatrix}.$$

If a vector function v satisfies

$$\nabla \times v = 0$$

at every spatial point (x, y, z), then v is referred to as a conservative vector field. In this case, v has a scalar potential function $\Phi(x, y, z)$, of which v is the negative of the gradient, i.e.,

$$v = -\nabla\Phi,$$

at every spatial point (x, y, z).

Conversely, if a vector function v has a scalar potential function ϕ, then it has zero curl:

$$\nabla \times \nabla\phi = 0,$$

where the 0 on the right stands for the zero three-dimensional vector. As a result, we also have

$$\nabla \times \nabla v = 0,$$

where the curl operator is interpreted to act separately on each column in the 3×3 matrix ∇v, and the 0 on the right stands for the zero 3×3 matrix. In summary, one can write in operator form

$$\nabla \times \nabla = 0.$$

If a vector function v has zero divergence, i.e.,

$$\nabla \cdot v = 0,$$

at every spatial point (x, y, z), then it can be written as the curl of another vector function $w(x, y, z)$, i.e.,

$$v = \nabla \times w$$

at every spatial point (x, y, z).

Conversely, if a vector field v can be written as the curl of another vector field w, then it is divergence-free:

$$\nabla \cdot \nabla \times w = 0.$$

In operator form, one can equivalently write

$$\nabla \cdot \nabla \times = 0.$$

Finally, from the above symbolic forms, one can also easily verify the formula

$$\nabla \times \nabla \times = -\Delta + (\nabla \nabla \cdot)^t.$$

Here, Δ is interpreted as the vector Laplacian, so the right-hand side is the sum of two 3×3 matrices: the negative of the vector Laplacian and the transpose of the gradient of the divergence. Note that although the gradient of the divergence is a symmetric matrix when operating on vector functions that are differentiable to the second order, it may act nonsymmetrically on vector functions with discontinuous second derivatives. This is why we bother to take its transpose in the above formula. This is particularly important in the weak formulation of the linear elasticity equations below, where derivatives are assumed to be square-integrable but not necessarily continuous.

21.2 General Linear Elasticity

Here, we use the above notation to write the linear elasticity equations in a more general and compact form. This form uses a three-dimensional setting, with the four unknown functions v_1, v_2, v_3, and p defined in a three-dimensional domain Ω in the (x, y, z) Cartesian space. (The two-dimensional case can be obtained from this general form as a special case by assuming that $v_3 \equiv 0$ and that the functions are independent of z. The system of PDEs in Chapter 20, Section 2, can then be obtained from the present one by eliminating p, as discussed below.)

The unknown functions in the system of PDEs can be referred to as the scalar function $p(x, y, z)$ and the vector function $v(x, y, z)$, which contains the components $v_1(x, y, z)$, $v_2(x, y, z)$, and $v_3(x, y, z)$. Similarly, the right-hand-side functions are given as the scalar function $g(x, y, z)$ and the vector function $f(x, y, z)$.

Let μ_0 be a given positive constant. Let $0 < \lambda \leq \infty$ and $0 \leq \mu \leq \mu_0$ be given parameters. The general linear elasticity equations are as follows:

$$\begin{pmatrix} -\lambda^{-1} & -\nabla\cdot \\ \nabla & -\mu\left(\Delta + (\nabla\nabla\cdot)^t\right) \end{pmatrix} \begin{pmatrix} p \\ v \end{pmatrix} = \begin{pmatrix} g \\ f \end{pmatrix}.$$

Let us now impose suitable boundary conditions on the above system. Let's assume that the boundary of Ω can be written as the union of two disjoint subsets:

$$\partial\Omega = \Gamma_D \cup \Gamma_N,$$

where Dirichlet boundary conditions of the form

$$v(x, y, z) = \gamma(x, y, z), \qquad (x, y, z) \in \Gamma_D,$$

are imposed on Γ_D (where γ is a vector function given in Γ_D), and mixed boundary conditions of the form

$$-p\vec{n} + \mu\left((\nabla v)^t + \nabla v\right)\vec{n} + \alpha v = \beta$$

are imposed on Γ_N (where the nonnegative function α and the vector function β are given in Γ_N, and $\vec{n} \equiv \vec{n}(x, y, z)$ is the outer unit vector normal to Γ_N at (x, y, z)).

With these boundary conditions, the system can be reduced to a system of only three PDEs, which is equivalent to a well-posed minimization problem. This is done below.

21.3 Reduction to the Linear Elasticity Equations

The 2×2 matrix of differential operators used in the general linear elasticity equations above has the LU decomposition

$$\begin{pmatrix} -\lambda^{-1} & -\nabla\cdot \\ \nabla & -\mu\left(\Delta + (\nabla\nabla\cdot)^t\right) \end{pmatrix}$$
$$= \begin{pmatrix} 1 & 0 \\ -\lambda\nabla & 1 \end{pmatrix} \begin{pmatrix} -\lambda^{-1} & 0 \\ 0 & -\mu\left(\Delta + (\nabla\nabla\cdot)^t\right) - \lambda\nabla\nabla\cdot \end{pmatrix} \begin{pmatrix} 1 & \lambda\nabla\cdot \\ 0 & 1 \end{pmatrix}.$$

In this triple product, the leftmost matrix is lower triangular, so its inverse can be obtained immediately by inserting the minus sign just before its lower-left element. Similarly, the rightmost matrix is upper triangular, so its inverse can also be obtained by inserting the minus sign just before its upper-right element. Because inverting the leftmost and rightmost matrices is trivial, the problem of inverting the original 2×2 matrix in the general linear elasticity equations is actually reduced to inverting the middle matrix in the above triple product. Fortunately, this matrix is diagonal, and its upper-left element is just a scalar. Thus, the original problem is transformed to the problem of inverting the lower-right element in the diagonal matrix in the middle of the above triple product. This element is known as the Schur complement of the original matrix.

The differential operator in the above Schur complement is the one that is actually used in the linear elasticity equations. To see this, let us return to the two-dimensional case, obtained by assuming $v_3 \equiv 0$ and no dependence on the third spatial variable z. In order to obtain the linear elasticity equations in Chapter 20, Section 2, as a special case of the present formulation, we first define

$$\nu \equiv \frac{\lambda}{2\mu + \lambda},$$

so

$$\frac{1-\nu}{2} = \frac{\mu}{2\mu + \lambda}.$$

Now, we multiply the first PDE in the general linear elasticity system by $\lambda\nabla$ and add it to the second one. This way, the lower-left block in the general linear elasticity system disappears, and the lower-right block takes the form of the Schur complement:

$$-\mu\left(\Delta + (\nabla\nabla\cdot)^t\right) - \lambda\nabla\nabla\cdot.$$

By multiplying the resulting PDE throughout by $-(2\mu + \lambda)^{-1}$, the differential operator takes the form

$$\frac{1-\nu}{2}\left(\Delta + (\nabla\nabla\cdot)^t\right) + \nu\nabla\nabla\cdot$$

used in Chapter 20, Section 2. The boundary conditions used there can also be obtained from the present ones: the Dirichlet boundary conditions are just the same, and the mixed boundary conditions are also the same in view of the equation

$$-p = \lambda(\nabla \cdot v + g).$$

In the next section, we present the Stokes equations as a special (limiting) case of the general linear elasticity equations, in which the parameter λ is set to ∞.

21.4 The Stokes Equations

The Stokes equations are obtained from the general linear elasticity equations in Section 21.2 above by setting λ to ∞, so the upper-left block vanishes (Figure 21.1). In the formulation in Chapter 20, Section 2, $\lambda \to \infty$ means that

$$\nu = \frac{\lambda}{2\mu + \lambda} \to 1.$$

It is thus interesting to try to solve the linear elasticity equations with ν close to 1. We have applied the adaptive-refinement algorithm used in Chapter 20, Section 10, with ν as large as 0.9 with good performance. Unfortunately, when ν is as large as 0.99, the convergence rate of the modified multigrid preconditioner deteriorates (presumably due to the large anisotropy). A special trick is required to get around this problem.

21.5 Continuation Process

A possible cure to the above problem is to use a continuation process in the refinement algorithm. This means that the stiffness matrix at the coarsest mesh is assembled using $\nu = 0.9$, and the subsequent stiffness matrices at finer and finer meshes are assembled

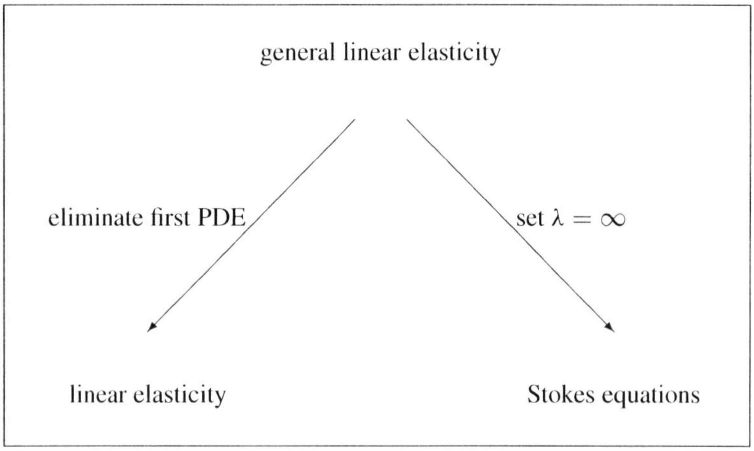

Figure 21.1. *Both the Stokes and the linear elasticity equations can be viewed as special cases of general linear elasticity.*

using larger and larger ν, until $\nu = 0.99$ is used at the finest mesh. Automatic boundary refinement is also used, as in Chapter 14, Section 8. In this process, the convergence factor of the modified multigrid preconditioner is on average 0.95; that is, the energy norm of the preconditioned residual is reduced by 5% on average in each call to the modified multigrid algorithm in the PCG iteration. Although this is not considered rapid convergence, it is still acceptable for such a difficult problem. (See Figure 21.2 for the distribution of nodes in this numerical experiment.)

21.6 Adequacy Consideration

As ν approaches 1 in the above continuation process, the linear elasticity equations in Section 21.3 become more and more anisotropic. Fortunately, unlike in Chapter 12, Section 8, this produces no inadequacy even when the mesh is not aligned with the Cartesian x- and y-axes. Indeed, thanks to the fact that the reduced system in Section 21.3 is obtained by multiplying throughout by $-(2\mu+\lambda)^{-1}$, its right-hand side is bounded independently of ν. This implies that the solution cannot oscillate too rapidly, or the mixed-derivative terms (and, hence, also the right-hand side) would be too large. In other words, the solution is smooth independent of ν, and the discretization is indeed adequate.

21.7 Preconditioner for the Stokes Equations

In the above continuation process, we have solved the linear elasticity equations with ν pretty close to 1. This produces the scalar unknown functions u and v on the finest mesh. The scalar unknown function p can then be obtained from the first equation in the general linear elasticity system in Section 21.2. In other words, p is obtained from the LU decomposition of the original general linear elasticity equations. This completes the definition of the general linear elasticity solver on the finest mesh. Below, we use a discrete version of this

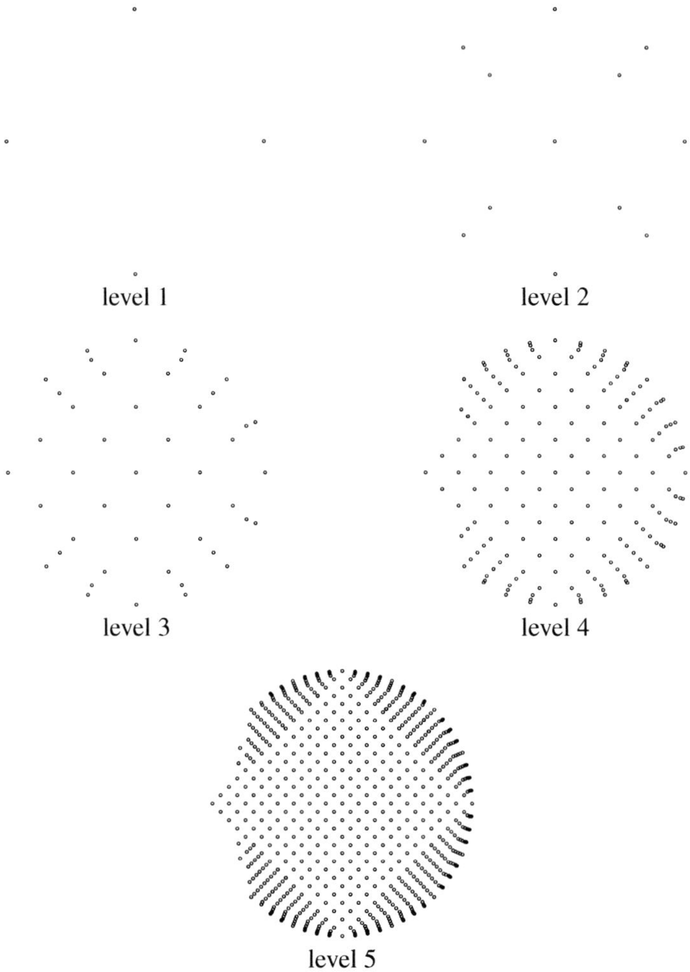

Figure 21.2. *Distribution of nodes at the first five levels of refinement (with automatic boundary refinement). The Poisson ratio increases gradually from* 0.9 *at the first level to* 0.99 *at the seventh level (*11000 *nodes).*

solver as a preconditioner for the original Stokes equations. For this purpose, however, we must first discretize the Stokes equations on the finest mesh or have their stiffness matrix assembled.

In both the general linear elasticity and Stokes systems, the unknown function p is different from the other unknown functions in that it may be discontinuous. Indeed, its derivatives are never used in the weak formulation, thanks to integration by parts in the second (vector) PDE. Therefore, p is approximated in the finest mesh by a piecewise-constant function, which is constant in each individual triangle in the mesh. Actually, the numerical approximation to p can be written uniquely as a linear combination of "elemental" basis functions, each of which has the value 1 in a particular triangle and 0 elsewhere.

The coefficients of these basis functions in the above linear combination are stored in the corresponding components x_i in the unknown vector x that solves the stiffness system. In order to specify the index i that corresponds to this basis function, we must have an indexing scheme that allows access from a particular finite element to the index i associated with it. For this purpose, the "finiteElement" class in Chapter 13, Section 3, must also have an extra (private) integer field named "index" to contain the relevant index i.

Because p is approximated by a piecewise-constant function, there is no point in taking its derivatives, which vanish almost everywhere in Ω. Therefore, the term ∇p must be replaced in the integration by parts that defines the weak formulation by a term in which ∇ is applied not to p but rather to the test function that multiplies it:

$$\int_\Omega \nabla p \cdot w d\Omega = -\int_\Omega p(\nabla \cdot w) d\Omega + \int_{\Gamma_N} p(w \cdot \vec{n}) d\Gamma_N.$$

This way, the weak formulation contains no derivative of p, as required. Furthermore, the bilinear form and stiffness matrix produced from it are symmetric.

The unknowns in the stiffness system can actually be split into two blocks: the first block contains the unknowns $x_0, x_1, \ldots, x_{t-1}$ (where t is the number of triangles in the finest mesh) corresponding to the elemental basis functions, and the second block contains the unknowns $x_t, x_{t+1}, \ldots, x_{t+2n-1}$ (where n is the number of nodes in the finest mesh), which are the coefficients of nodal basis functions in the numerical approximations to u and v. This partitioning induces the following block form of the stiffness matrix for the Stokes equations:

$$A = \begin{pmatrix} A_{00} & A_{01} \\ A_{10} & A_{11} \end{pmatrix},$$

where the index 0 refers to the first block and the index 1 refers to the second block. Similarly, the stiffness matrix for the general linear elasticity equations (with $\lambda < \infty$) has the block form

$$A^{(\lambda)} = \begin{pmatrix} A_{00}^{(\lambda)} & A_{01}^{(\lambda)} \\ A_{10}^{(\lambda)} & A_{11}^{(\lambda)} \end{pmatrix}.$$

Because $A_{00}^{(\lambda)}$ is diagonal, its inverse is immediately available. The Schur complement of $A^{(\lambda)}$ with respect to the above partitioning is defined by

$$S^{(\lambda)} \equiv A_{11}^{(\lambda)} - A_{10}^{(\lambda)} \left(A_{00}^{(\lambda)}\right)^{-1} A_{01}^{(\lambda)}.$$

Using the Schur complement, the block LU decomposition of $A^{(\lambda)}$ is given by

$$A^{(\lambda)} = \begin{pmatrix} I & 0 \\ A_{10}^{(\lambda)}\left(A_{00}^{(\lambda)}\right)^{-1} & I \end{pmatrix} \begin{pmatrix} A_{00}^{(\lambda)} & 0 \\ 0 & S^{(\lambda)} \end{pmatrix} \begin{pmatrix} I & \left(A_{00}^{(\lambda)}\right)^{-1} A_{01}^{(\lambda)} \\ 0 & I \end{pmatrix}$$

(where I stands for the identity matrix of suitable order).

Note that both A and $A^{(\lambda)}$, although symmetric, are no longer positive definite. In fact, they are indefinite: the Schur complement is positive definite, whereas $A_{00}^{(\lambda)}$ is (diagonal and) negative definite.

The Stokes stiffness system $Ax = f$ is now solved by the general minimal residual (GMRES) iteration, with $A^{(\lambda)}$ (for some fixed large λ) as preconditioner. Each GMRES iteration requires the inversion of $A^{(\lambda)}$, namely, solving a system of the form $A^{(\lambda)}e = r$. This is done using the above block LU decomposition; the Schur-complement subsystem is solved approximately by an inner PCG iteration with the modified multigrid preconditioner.

In our numerical experiments, we use a finest mesh of 2713 nodes. The $A^{(\lambda)}$-preconditioner uses $\mu = 1$ and $\lambda = 20$, and the Schur-complement subsystem in it is solved approximately by 20 to 50 inner PCG iterations (with the modified multigrid preconditioner), which reduce the energy norm of the preconditioned residual of this subsystem by two orders of magnitude. The outer acceleration method is GMRES(20, 10) in [40] (which proves here to be better than standard GMRES), restarted 10 times, so a total of 300 outer iterations (with the $A^{(20)}$ preconditioner) are used to reduce the norm of the residual of the original Stokes equations by six orders of magnitude.

Another possible approach to discretizing and solving the general linear elasticity and Stokes equations uses a new formulation of the original PDEs. This approach is described next.

21.8 First-Order System Least Squares

In systems of PDEs such as the Stokes and linear elasticity equations, the number of unknown functions is the same as the number of equations. For example, the system in Section 21.2 contains two equations, a scalar equation and a vector equation, for two unknown functions, scalar and vector. Since a vector consists of three scalar components, the system can also be viewed as a system of four equations in four unknown scalar functions. The property that the number of equations is the same as the number of unknowns is, of course, a necessary condition for well-posedness.

In some cases, it is worthwhile to introduce new unknown functions, along with new equations. In this process, the number of new equations may exceed the number of new unknown functions, so the system becomes overdetermined and may have no solution at all. Still, the extended system does have a unique least-square solution, that is, a vector function for which the norm of the residual is minimized. This solution is obtained from the normal equations.

Consider, for instance, the general linear elasticity equations in Section 21.2 above. This system has the solution (p, v), with p a scalar function and v a three-dimensional vector function. Recall from Section 21.1 that ∇v is a 3×3 matrix and that $\nabla \times \nabla v$ is the zero matrix. Define nine more unknown functions by

$$U \equiv \nabla v,$$

along with nine more equations:

$$\nabla \times U = 0.$$

The first-order-system-least-squares (FOSLS) algorithm [7] consists of two steps. First, solve for the 10 unknown functions (p, U) the system of 13 equations consisting of the 4 original equations together with the 9 new ones in the least-square sense. Then, after p and U have been found and fixed, solve for the original unknown vector function v the system of 9 equations

$$\nabla v = U,$$

again in the least-square sense. (The normal equations associated with this least-square problem are just three independent Poisson equations for the scalar components in v.) It is optional to add to the latter system a 10th equation, which is just the first equation in the original system. The resulting pair (p, v) is referred to as the FOSLS solution to the general linear elasticity equations.

The above algorithm assumes that $\alpha \equiv 0$ in the mixed boundary conditions in Section 21.2. Otherwise, the quadratic form that is minimized in the weak formulation (see Chapter 20, Sections 3 and 4) must also contain an integral of $\alpha\|v\|^2$ over Γ_N, which makes it impossible to solve for U alone, as required in the first step above. In the more general case in which $\alpha \neq 0$, one must solve in the first step also for the restriction of v to Γ_N.

The FOSLS systems in both the above steps seem more stable and less anisotropic than the original system. Thus, they have a better chance of having stable and accurate finite-difference and finite-element discretizations. The adaptive-refinement algorithm can also be used to produce finer and finer meshes to be used in both steps above.

21.9 The Navier–Stokes Equations

Here, we present the Navier–Stokes equations that govern the dynamics of fluids. We assume that the equations are stationary; that is, the process has already converged to a steady state, so no time derivative or time variable is present. The functions depend only on the spatial variables x, y, and z.

The unknown and right-hand-side functions have the same form as in the Stokes equations above. In fact, the only difference between the Stokes and Navier–Stokes systems is in the lower-right block in the coefficient matrix that contains the differential operators. In the following, we describe this difference in detail.

Let Re be a (large) positive parameter that characterizes the problem (the Reynolds number). Define

$$\varepsilon \equiv 1/Re.$$

The stationary Navier–Stokes equations are given by

$$\begin{pmatrix} 0 & -\nabla\cdot \\ \nabla & -\varepsilon\Delta + (\nabla v)^t \end{pmatrix} \begin{pmatrix} p \\ v \end{pmatrix} = \begin{pmatrix} g \\ f \end{pmatrix},$$

where Δ stands for the vector Laplacian (the 3×3 diagonal matrix with scalar Laplacians on its main diagonal).

Note that the lower-right block in the coefficient matrix that contains the differential operators depends on the yet unknown vector v. Thus, the equations are no longer linear, like the Stokes equations, but rather quasi-linear, as in Chapter 9. In fact, the lower-right block is the three-dimensional analogue of the singularly perturbed conservation law in Chapter 9, Section 11. Thus, the solution may exhibit all sorts of irregular behavior, such as shock and rarefaction waves.

Because the problem is nonlinear, it is natural to use Newton's iteration, in which the original system is linearized successively around better and better approximate solutions. The iteration must start from an initial guess that already incorporates the above nonlinear waves. A possible initial guess is the numerical solution produced by the alternating-direction Godunov scheme in Chapter 9, Section 11.

Another possible initial guess can be obtained from the adaptive-refinement algorithm. In this approach, the numerical solution v at a particular mesh is used not only to refine the mesh but also as the initial guess in Newton's iteration on the next, finer, mesh. In other words, v is used to linearize the system of PDEs around it, refine the current mesh, and initialize the Newton iteration for the corresponding stiffness system on the next, finer, mesh [22].

The stiffness system is recalculated at each level, with piecewise-constant approximation to p as in Section 21.7 above. In the integration by parts used to define the weak formulation, one should bear in mind that the nonlinear term contributes two terms to the bilinear form:

$$\begin{aligned}\int_\Omega v \cdot (\nabla v) \cdot w d\Omega &= -\int_\Omega v \cdot (\nabla w) \cdot v d\Omega - \int_\Omega (\nabla \cdot v) v \cdot w d\Omega + \int_{\Gamma_N} (v \cdot w)(v \cdot \vec{n}) d\Gamma_N \\ &= -\int_\Omega v \cdot (\nabla w) \cdot v d\Omega - \int_\Omega g v \cdot w d\Omega + \int_{\Gamma_N} (v \cdot w)(v \cdot \vec{n}) d\Gamma_N.\end{aligned}$$

(usually, only half of this term is integrated by parts, to account for the extra term $(v \cdot \vec{n})v/2$ in the mixed boundary conditions). The detailed formulation is given in [22].

The original Navier–Stokes system can be perturbed by replacing the zero element in the upper-left corner of the matrix of differential operators by $-\lambda^{-1}$ for some fixed (large) parameter λ. The stiffness matrix $A^{(\lambda)}$ produced from the linearized perturbed system can then serve as preconditioner for the corresponding linear problem in each Newton iteration (in much the same way as in Section 21.7 above).

21.10 Exercises

1. Modify your code from the exercises at the end of Chapter 20 to implement the continuation process in Section 21.5 for the solution of the two-dimensional linear elasticity equations with $\nu \doteq 1$. Actually, all you have to do is increase ν gradually in the loop in Section A.15 of the Appendix until it reaches a value sufficiently close to 1 at the finest mesh.

2. Print the meshes resulting from your code, and verify that they are indeed conformal.

3. Assemble the stiffness matrix for general linear elasticity in two dimensions by adding the unknown function p, which takes constant values in the individual triangles. Verify that the stiffness matrix is indeed symmetric. Use this matrix as a preconditioner for the Stokes equations. The solution can be found on the Web page http://www.siam.org/books/cs01.

Chapter 22

Electromagnetic Waves

In this chapter, we consider the Maxwell equations that model electromagnetic waves. In certain cases, these equations can be reduced to the scalar Helmholtz equation. We discuss the adequacy of finite-difference and finite-element discretization methods for this problem. Finally, we discuss models for the measuring problem, which can also be formulated as Maxwell or Helmholtz equations.

22.1 The Wave Equation

We start the discussion with the second-order linear wave equation, which governs the propagation of waves in a homogeneous medium. The domain is the unit square $0 < x, y < 1$, and the time variable is $-\infty < t < \infty$. The constant c is the wave speed. The functions $f(t, x, y)$ and $g(t, y)$ are given in advance, and the function $u(t, x, y)$ is the unknown solution. The wave equation is

$$u(t, x, y)_{tt} - c^2 \left(u(t, x, y)_{xx} + u(t, x, y)_{yy}\right) = f(t, x, y).$$

We consider the following boundary conditions: on the left edge of the square, the boundary conditions are of Dirichlet type:

$$u(t, 0, y) = g(t, y).$$

On the top and bottom edges, the boundary conditions are of homogeneous Neumann type:

$$u(t, x, 0)_y = u(t, x, 1)_y = 0.$$

Finally, on the right edge, the boundary conditions are of mixed type:

$$u(t, 1, y)_x + \frac{1}{c} u(t, 1, y)_t = 0.$$

These boundary conditions imply that the wave issues from the left edge and leaves the square only through the right edge (with speed c) but not through the top or bottom edge.

Assume that the given functions $f(t, x, y)$ and $g(t, y)$ are square-integrable and also integrable in absolute value with respect to the time variable t:

$$\int_{-\infty}^{\infty} |f(t, x, y)|^2 dt < \infty,$$

$$\int_{-\infty}^{\infty} |f(t, x, y)| dt < \infty,$$

$$\int_{-\infty}^{\infty} |g(t, y)|^2 dt < \infty,$$

$$\int_{-\infty}^{\infty} |g(t, y)| dt < \infty.$$

Then, these functions, as well as the solution $u(t, x, y)$, can be written in the Fourier form

$$f(t, x, y) = \int_{-\infty}^{\infty} \exp(i\omega t) f_\omega(x, y) d\omega,$$

$$g(t, y) = \int_{-\infty}^{\infty} \exp(i\omega t) g_\omega(y) d\omega,$$

$$u(t, x, y) = \int_{-\infty}^{\infty} \exp(i\omega t) u_\omega(x, y) d\omega,$$

where $f_\omega(x, y)$ and $g_\omega(y)$ are the given Fourier coefficients of the ω-frequency in the Fourier expansion (with respect to the time variable t), and $u_\omega(x, y)$ is the unknown Fourier coefficient for the unknown solution $u(t, x, y)$. In order to find $u_\omega(x, y)$, we restrict the wave equation and the above boundary conditions to the ω-frequency:

$$-\omega^2 u_\omega(x, y) - c^2 \left(u_\omega(x, y)_{xx} + u_\omega(x, y)_{yy} \right) = f_\omega(x, y),$$

with the boundary conditions

$$u_\omega(0, y) = g_\omega(y)$$

on the left edge of the square;

$$u_\omega(x, 0)_y = u_\omega(x, 1)_y = 0$$

on the top and bottom edges; and

$$u_\omega(1, y)_x + i\frac{\omega}{c} u_\omega(1, y) = 0$$

on the right edge, where the wave exits the square.

Thus, we have reduced the original wave equation to a time-independent equation for $u_\omega(x, y)$, known as the Helmholtz equation. In what follows, we consider numerical algorithms to solve this equation.

22.2 The Helmholtz Equation

The Helmholtz equation in the unit square is defined as follows:

$$-u_{xx}(x, y) - u_{yy}(x, y) - 4\pi^2 K^2 u(x, y) = F(x, y), \quad 0 < x, y < 1,$$

where $F(x, y)$ is a given function and K is a given positive constant. The PDE is also accompanied by boundary conditions on the edges of the unit square. Here, we consider the particular case in which homogeneous Neumann boundary conditions of the form

$$u_y(x, 0) = u_y(x, 1) = 0$$

are given on the upper and lower edges, Dirichlet boundary conditions are given on the left edge (where $x = 0$), and mixed complex boundary conditions of the form

$$u_x(1, y) + 2\pi i K u(1, y) = 0$$

(with $i = \sqrt{-1}$ being the imaginary number) are given on the right edge of the unit square.

In what follows, we consider some finite-difference and finite-element discretization methods for the numerical solution of the Helmholtz equation and discuss their adequacy.

22.3 Finite-Difference Discretization

The finite-difference scheme uses a uniform two-dimensional grid to approximate the unit square (see Figure 7.4). Let i be the row index and j be the column index in this grid. Let

$$u_{i,j} \doteq u(jh, ih)$$

be the numerical approximation to the solution of the PDE at the corresponding grid point (where h is the meshsize in both the x and y spatial directions). Using finite-difference approximations for the spatial derivatives as in Chapter 12, Section 8, we have the following finite-difference approximation to the Helmholtz equation:

$$h^{-2}\left(4u_{i,j} - u_{i,j-1} - u_{i,j+1} - u_{i-1,j} - u_{i+1,j}\right) - 4\pi^2 K^2 u_{i,j} = F(jh, ih).$$

The boundary conditions are also incorporated in the finite-difference operator, as in Chapter 7, Section 4.

In the next section, we discuss the adequacy of this scheme for the Helmholtz equation.

22.4 Adequacy in Finite Differences

The accuracy of the finite-difference scheme means that the discretization error approaches 0 as $h \to 0$. Here, however, accuracy doesn't tell the whole story. Indeed, because we are particularly interested in a very large Helmholtz parameter K, a much more relevant limit case is the one in which both $h \to 0$ and $K \to \infty$ at the same time. When the discretization error approaches 0 in this limit process, we say that the scheme is adequate (see Chapter 8, Section 4).

In order to estimate the adequacy of the finite-difference scheme, we consider a typical model solution of the Helmholtz equation of the form

$$u(x, y) = \exp(2\pi i(k_1 x + k_2 y)),$$

where k_1 and k_2 are integers satisfying

$$k_1^2 + k_2^2 = K^2.$$

Because this is just a model-case analysis, we take the liberty of assuming that K^2 can indeed be written as a sum of squares of integers and also disregard the boundary conditions for the moment. (Actually, this model solution satisfies periodic boundary conditions rather than the original ones.) The truncation error for this solution is

$$\begin{aligned}\frac{1}{12}|h^2(u_{xxxx}(x,y)+u_{yyyy}(x,y))| &= \frac{1}{12}h^2(4\pi^2)^2(k_1^4+k_2^4)\\ &\geq \frac{1}{24}h^2(4\pi^2)^2(k_1^2+k_2^2)^2\\ &= \frac{1}{24}h^2(4\pi^2)^2K^4.\end{aligned}$$

A necessary condition for the truncation error approaching 0 when $h \to 0$ and $K \to \infty$ at the same time is, thus,

$$h \ll K^{-2}$$

(that is, h approaches 0 faster than K^{-2}). This means that the grid must contain many points, and the computational cost is large. In what follows, we consider finite-element schemes and discuss their adequacy.

22.5 Bilinear Finite Elements

As we have seen above, the finite-difference scheme requires a very small meshsize h and, hence, large computational resources to have sufficient adequacy. Therefore, we switch here to finite elements in the hope that they will provide adequacy for a lower cost.

For simplicity, we consider first a mesh of square finite elements as in Figure 22.1. The discretization on this mesh is in principle as in Chapter 12, Section 3, except that square finite elements are used rather than triangles. The stiffness matrix is calculated basically as in Chapter 12, Section 5, except that the reference element r is no longer a triangle, as in Figure 12.2, but rather a square, as in Figure 22.2. The four typical (standard) nodal functions in this square are defined as follows:

$$\begin{aligned}\phi_{0,0} &= (1-x)(1-y),\\ \phi_{1,0} &= x(1-y),\\ \phi_{0,1} &= (1-x)y,\\ \phi_{1,1} &= xy.\end{aligned}$$

Each of these bilinear functions has the value 1 at one of the corners of the square and 0 at the other three corners. This is why these typical nodal functions are so helpful in the calculation of the stiffness matrix using the guidelines in Chapter 12, Section 5.

Because the finite-element mesh in Figure 22.1 is rectangular and uniform, there is no need to use the sophisticated data structures in Chapter 13 that implement unstructured meshes. It is easier and more efficient to use the objects in Chapter 7 that implement rectangular grids.

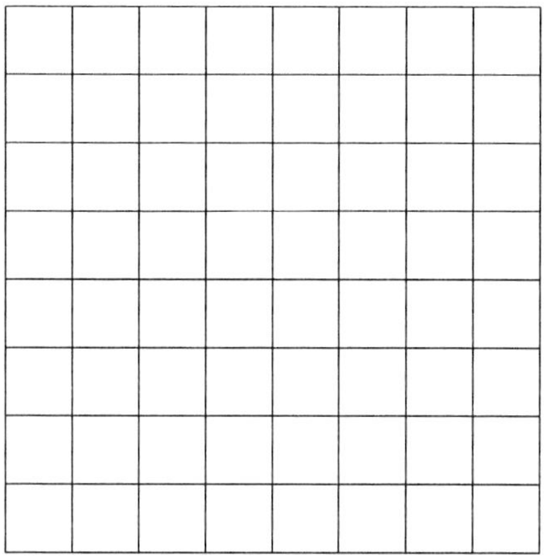

Figure 22.1. *The bilinear finite-element mesh.*

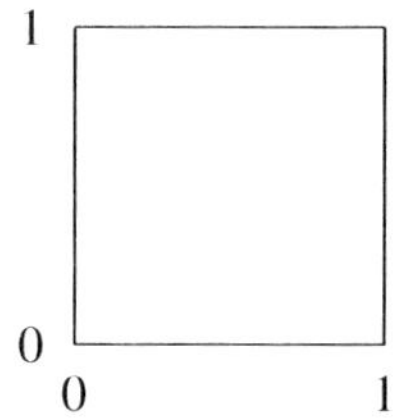

Figure 22.2. *The reference element for the bilinear finite-element scheme.*

22.6 Adequacy in Bilinear Finite Elements

In order to test the adequacy of the bilinear finite-element scheme for the Helmholtz equation, we consider a model case for which the solution $u(x, y)$ is known, and, hence, the discretization error is also available. In particular, we consider the Helmholtz equation and boundary conditions in Section 22.2. The Dirichlet boundary conditions on the left edge are given by

$$u(0, y) = 1.$$

Thus, the solution to the boundary-value problem is

$$u(x, y) = \exp(-2\pi i K x).$$

With the solution available, we also have the discretization error for the present numerical scheme. For this, we apply the bilinear finite-element scheme to the above problem. The difference between the solution of the original PDE and the numerical solution of the numerical scheme is the discretization error at the grid points.

It turns out that, with Helmholtz parameter $K = 5$, a 100×100 grid is required to have a sufficiently small discretization error. In fact, the maximal discretization error on this grid is 0.02. However, when K increases to $K = 10$, the solution is much more oscillatory, and the above grid is too coarse. Indeed, the maximal discretization error on it is as large as 0.2. The minimal grid on which the discretization error is sufficiently small is a 400×400 grid. Indeed, on this grid, the discretization error is at most 0.06.

The conclusion is, thus, that the bilinear finite-element scheme is in principle the same as the finite-difference scheme in terms of adequacy. Indeed, when K is doubled, the meshsize h must decrease fourfold to preserve adequacy. In other words,

$$h \sim K^{-2}$$

is necessary to have a sufficiently small discretization error.

A more adequate FOSLS discretization is proposed in [23]. However, it is tested only for K as small as $K \leq 8/(2\pi)$, and with boundary conditions of the third kind only.

22.7 The Measuring Problem

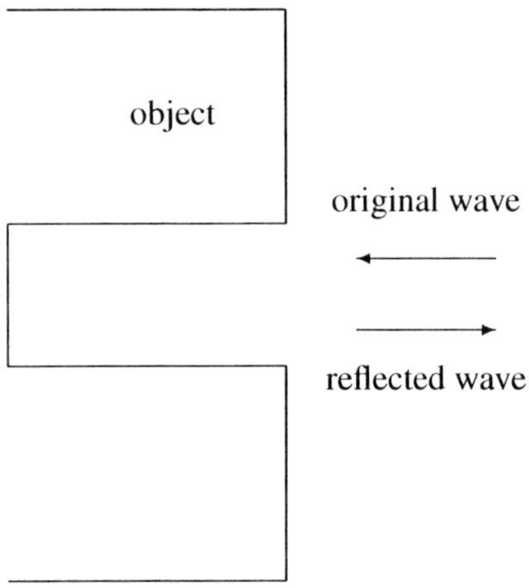

Figure 22.3. *The measuring problem: find the depth of the slit in the object on the left from the wave reflected from it.*

Here, we consider the problem of measuring the depth of a slit in an object. This is done by sending a wave toward the object and measuring the wave reflected from it by solving the Helmholtz equation.

The procedure is displayed in Figure 22.3. The object on the left has a slit in the middle of its right side. The original wave is sent toward the object from a source on the far right. This wave hits the object on its right side (where the slit is) and then is reflected back to the right. Thus, the Helmholtz equation governs the behavior of the reflected wave and should be solved for it. If the solution matches the observed data about the reflected wave, then one may conclude that the depth of the slit is indeed as in Figure 22.3. Otherwise, the depth should be changed. When a depth is chosen for which the observed data about the reflected wave matches the data expected from the Helmholtz equation, we conclude that this is the correct depth, and the measuring problem is indeed solved.

The domain in which the Helmholtz equation is solved for the reflected wave is the area to the right of the object. This domain, however, has only a left boundary (which is also the right side of the object); the top, bottom, and right boundary segments are missing and have to be completed artificially. Two possible ways to do this are described below.

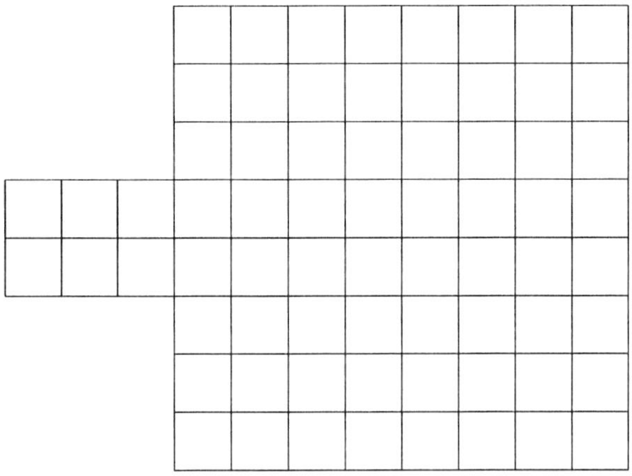

Figure 22.4. *The bilinear finite-element mesh for the measuring problem.*

22.8 The Nearly Rectangular Model

It seems natural to solve the Helmholtz equation for the reflected wave in a nearly rectangular domain as in Figure 22.4. The left boundary of this domain is the right side of the object (with the slit in it). The top and bottom boundary segments are just straight horizontal lines. The right boundary segment is just a straight vertical line.

The boundary conditions are as in Section 22.2. Dirichlet boundary conditions are imposed on the left, from which the reflected wave issues. These boundary conditions are available from the original wave sent into the object from a source at the far right. Because all the information on this original wave is known, the reflected wave has the same data at the right side of the object (including the slit). Because the reflected wave is parallel to the top and bottom edges, homogeneous Neumann boundary conditions are imposed on them. Finally, mixed complex boundary conditions are imposed on the right edge, through which the reflected wave leaves the domain. All that is left to do is to solve the Helmholtz equation numerically.

The mesh illustrated in Figure 22.4 can be used in both finite-difference and bilinear finite-element discretization methods. We have used the bilinear finite-element scheme on a 400×400 grid, to which a 10×40 subgrid is added on the left, as in Figure 22.4, to discretize the area inside the slit.

In order to use efficient data structures, as in Chapter 7, that are based on arrays, the grid is completed to a 400×440 computational grid by adding fictitious points above and below the subgrid in the slit. For every fictitious point i, j, we have used a trivial equation of the form $u_{i,j} = 0$.

In our numerical experiments, we have considered the above problem with $K = 10$. To the discrete linear system, we have applied the multigrid method in Chapter 8 of [39] (V(1,1) with up to 50 Kacmarz relaxations on the fourth level), accelerated by outer conjugate gradient squared (CGS) iteration. The iteration converges at a convergence rate of 0.95; that is, the norm of the preconditioned residual is reduced on average by 5% at each call to multigrid within CGS. Although this rate is not considered very good, it is still acceptable for this highly indefinite and large system. (For comparison, the convergence rate for the corresponding model problem in Section 22.6 in a 400×400 grid is 0.9, which is only twice as good.)

We decline to use here more expensive multigrid algorithms, such as the oblique-projection method in [3], which requires an exact coarse-grid solve in each iteration, and the V(2,40)-cycle in [14], which requires up to 40 inner GMRES postrelaxations.

22.9 The Nearly Circular Model

The nearly rectangular model used above, although easy to implement, suffers from two major drawbacks. The first drawback has to do with the suitability of the mathematical model to the physical phenomenon. In fact, the nearly rectangular domain in which the PDE is defined may be unsuitable for the wave reflected from the object. Indeed, the corners at the entrance to the slit reflect the wave not only to the right but also in oblique directions. Therefore, the wave reflected from these corners may exit the domain not only through the right edge but also through the top and bottom edges. This possibility is ignored in the above model, where the homogeneous Neumann boundary conditions on the top and bottom edges assume no exit. Furthermore, the wave reflected from these corners may cross the right edge of the domain at all sorts of angles. Again, this possibility is ignored in the above model, where the mixed boundary conditions on the right allow exit in the normal direction only.

The second drawback in the nearly rectangular model has to do with the numerical scheme. In order to be able to use efficient computer arrays as in Chapter 7, one would like to use a rectangular mesh, as in Figure 22.4. However, this uniform mesh cannot guarantee sufficient accuracy at places where the solution is expected to have large variation. For example, in and around the slit, one would like to have extra resolution. This is impossible with the standard uniform mesh in Figure 22.4; an unstructured mesh as in Chapter 13 is required.

Thus, we turn here to the nearly circular model for the measuring problem. In this model, the object is surrounded by a big (almost complete) circle to complete the missing boundary on the upper, lower, and right sides of the domain (Figure 22.5). On this circular boundary, mixed complex boundary conditions of the form

$$u_n(x, y) + 2\pi i K u(x, y) = 0$$

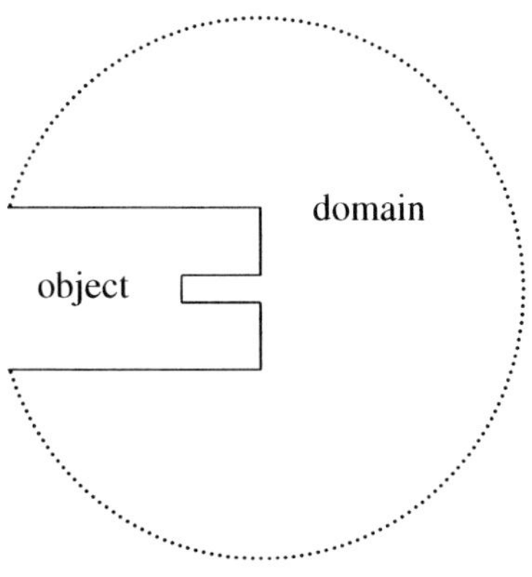

Figure 22.5. *The nearly circular model for the measuring problem.*

are imposed (where $\vec{n} \equiv \vec{n}(x, y)$ is the outer normal unit vector at the corresponding point (x, y)). These boundary conditions allow the reflected wave to exit only in the direction perpendicular to the circular boundary, as required.

The left part of the boundary is shaped by the object, which lies to the left of the domain (Figure 22.5). Homogeneous Neumann boundary conditions are imposed on the top and bottom of the object to prevent the wave from exiting. Dirichlet boundary conditions are imposed on the right side of the object, with the slit in it. This completes the definition of the boundary-value problem.

Note that the wave reflected from the corners at the entrance to the slit meets the circular boundary at an angle of about $\pi/2$. The bigger the circle, the closer are the angles to $\pi/2$. Therefore, in order to make the mixed complex boundary conditions at the circular boundary make sense, one would like to use a rather big circle; the mixed complex boundary conditions, which allow the wave to exit in the normal direction only, will then be suitable for the reflected wave.

The circular boundary should thus be approximated well by a high-resolution finite-element mesh (as in Chapter 12, Section 9). Farther away from the circular boundary or slit, lower resolution may be sufficient. The adaptive-refinement algorithm can start from an initial mesh that is already rather fine at the circular boundary, as in Chapter 12, Section 9. The mesh is then refined further where the numerical solution has large variation (e.g., at the corners of the slit). This procedure is repeated until the required accuracy is achieved. Because only the absolutely necessary nodes are added to the mesh in each refinement step, the total number of nodes in the final (finest) mesh is kept moderate.

22.10 Nonlinear Finite Elements

Here we modify the circular domain in Figure 22.5 into a half-circular domain. This domain can be discretized more accurately by nonlinear finite elements. The mesh is illustrated in Figure 22.6. Mixed complex boundary conditions are imposed not only on the circular boundary on the right to allow the reflected wave to exit but also on the vertical edges above and below the object on the left to allow the original wave to exit.

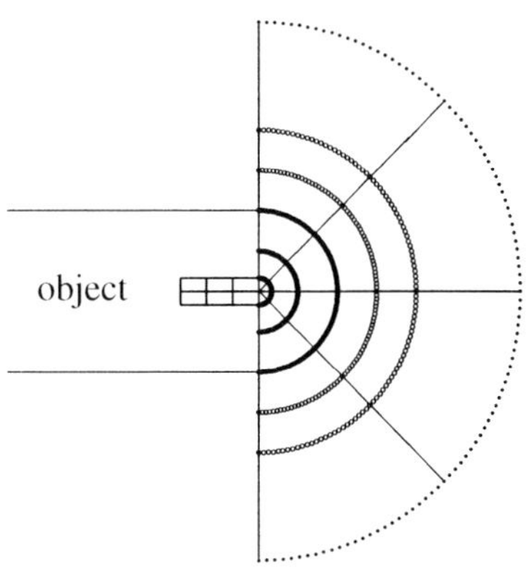

Figure 22.6. *The nonlinear finite-element mesh in the half-circular domain.*

In the mesh, only the finite elements in the slit are linear; in fact, they are small squares. The rest of the finite elements are nonlinear: they have two circular sides and two linear edges. Still, they can be transformed into rectangles using the standard polar coordinates. Thus, the mapping from the square reference element in Figure 22.2 to these nonlinear finite elements can be used to assemble the stiffness matrix. Here, however, the situation is not as straightforward as in Chapter 12, Section 5: because the mapping is nonlinear, its Jacobian S is no longer constant. Therefore, numerical integration may be necessary in some cases.

In this approach, the circular boundary on the right is approximated better than before. As discussed above, the circular shape is essential for the correct modeling of the physical phenomenon. Therefore, the nonlinear finite-element mesh may be more accurate and adequate. (Of course, a suitable nonlinear finite-element mesh must use a meshsize much smaller than in Figure 22.6 to guarantee adequacy. This can be achieved by adding more internal half-circles and radial lines in the mesh.)

In the rest of this chapter, we show that the Helmholtz equation can also be derived from the Maxwell equations that govern the connection between electric and magnetic fields.

22.11 The Maxwell Equations

The phenomenon of electromagnetic waves is described using the Maxwell equations, which govern the connection between electric and magnetic fields. In this system, the unknowns are the electric vector field E and the magnetic vector field B. These vector functions depend on the three spatial variables x, y, and z, as well as the time variable t. The given data in the right-hand side of the system are the scalar function $\rho(t, x, y, z)$ representing the density of electric charges and the vector function $J(t, x, y, z)$ representing the vector current density (the amount of current that flows through a unit area in unit time in the direction perpendicular to that area.)

We use here the c.g.s. system of units (centimeters, grams, seconds). In this system, the constant c that represents the speed of light is 3×10^{10} cm/s.

Note that the Nabla operator used below contains only the spatial derivatives as in Chapter 21, Section 1, whereas the time derivative is represented by a subscript t.

The Maxwell equations give the unknown electric and magnetic fields in terms of the given data about the charge and current densities:

$$
\begin{aligned}
\nabla \cdot E(t, x, y, z) &= 4\pi\rho(t, x, y, z), \\
\nabla \times E(t, x, y, z) &= -\frac{1}{c}B(t, x, y, z)_t, \\
\nabla \cdot B(t, x, y, z) &= 0, \\
\nabla \times B(t, x, y, z) &= \frac{4\pi}{c}J(t, x, y, z) + \frac{1}{c}E(t, x, y, z)_t.
\end{aligned}
$$

The drawback in this setting is that it contains more equations than unknowns. Indeed, it has two vector unknowns, E and B, which can be viewed as six scalar unknowns. The number of equations, however, is greater than six: there are two vector equations (the second and fourth) and two scalar equations (the first and third), which amount to eight scalar equations. In what follows, we reduce the system to an equivalent system in which the number of equations is the same as the number of unknowns.

Let us use the facts mentioned at the end of Chapter 21, Section 1, to introduce scalar and vector potential functions for the electric and magnetic vector fields. From the third equation in the above system, we have that the magnetic field B has zero divergence at every given time and every spatial point. Therefore, there exists a vector potential function $V(t, x, y, z)$ such that

$$B = \nabla \times V$$

at every fixed time t and every spatial point. The second equation in the Maxwell system can thus be rewritten as

$$\nabla \times \left(E + \frac{1}{c} V_t \right) = 0.$$

Thus, at every given time, the vector field $E + V_t/c$ is conservative, and, hence, has a scalar potential function $\Phi(t, x, y, z)$ for which

$$E + \frac{1}{c} V_t = -\nabla \Phi$$

at every fixed time t and every spatial point. By substituting this equation in the first equation in the Maxwell system, we have

$$-\Delta \Phi - \frac{1}{c} \nabla \cdot V_t = 4\pi \rho(t, x, y, z).$$

Let us now take the time derivative of the previous equation:

$$E_t + \frac{1}{c} V_{tt} = -\nabla \Phi_t.$$

By substituting this result in the fourth equation in the Maxwell system, we have

$$\nabla \times \nabla \times V = \frac{4\pi}{c} J(t, x, y, z) - \frac{1}{c} \left(\frac{1}{c} V_{tt} + \nabla \Phi_t \right).$$

Using the formula at the end of Chapter 21, Section 1, this equation can be rewritten as

$$\left(-\Delta + (\nabla \nabla \cdot)^t + \frac{1}{c^2} \frac{\partial^2}{\partial t^2} \right) V + \frac{1}{c} \nabla \Phi_t = \frac{4\pi}{c} J(t, x, y, z).$$

Thus, the original Maxwell system has been reduced to an equivalent system of two equations:

$$-\Delta \Phi - \frac{1}{c} \nabla \cdot V_t = 4\pi \rho(t, x, y, z),$$
$$\left(-\Delta + (\nabla \nabla \cdot)^t + \frac{1}{c^2} \frac{\partial^2}{\partial t^2} \right) V + \frac{1}{c} \nabla \Phi_t = \frac{4\pi}{c} J(t, x, y, z).$$

This system can also be written more compactly as

$$\begin{pmatrix} -\Delta & -\frac{1}{c} \nabla \cdot \dfrac{\partial}{\partial t} \\ \dfrac{1}{c} \nabla \dfrac{\partial}{\partial t} & -\Delta + (\nabla \nabla \cdot)^t + \dfrac{1}{c^2} \dfrac{\partial^2}{\partial t^2} \end{pmatrix} \begin{pmatrix} \Phi \\ V \end{pmatrix} = \begin{pmatrix} 4\pi \rho \\ \dfrac{4\pi}{c} J \end{pmatrix}.$$

We refer to this system as the reduced Maxwell system. In this system, the number of equations is the same as the number of unknowns. Indeed, it contains two equations (a scalar equation and a vector equation) and requires two unknowns (a scalar function and a vector function).

Let us now apply the Fourier transform in the time variable t, as in Section 22.1. Let the subscript ω denote the Fourier coefficient of frequency ω in the Fourier expansion of the corresponding function. Because the time derivative is replaced by multiplication by $i\omega$ in the transformed system, we have

$$\begin{pmatrix} -\Delta & -i\frac{\omega}{c}\nabla\cdot \\ i\frac{\omega}{c}\nabla & -\Delta + (\nabla\nabla\cdot)^t - \frac{\omega^2}{c^2} \end{pmatrix} \begin{pmatrix} \Phi_\omega \\ V_\omega \end{pmatrix} = \begin{pmatrix} 4\pi\rho_\omega \\ \frac{4\pi}{c}J_\omega \end{pmatrix}.$$

In the next section, we use one extra assumption to reduce further the reduced Maxwell system into a system of four independent scalar Helmholtz equations.

22.12 Reduction to Three-Dimensional Helmholtz Equations

Here we further reduce the above reduced Maxwell system to a system of four decoupled scalar Helmholtz equations that can in fact be solved independently of each other. For this purpose, however, we need to assume that

$$\nabla \cdot V = -\frac{1}{c}\Phi_t$$

and that this function has continuous derivatives in space. The reduction is displayed schematically in Figure 22.7.

Using this assumption in the reduced Maxwell system, one can see immediately that certain terms are canceled, and we get

$$\begin{pmatrix} -\Delta + \frac{1}{c^2}\frac{\partial^2}{\partial t^2} & 0 \\ 0 & -\Delta + \frac{1}{c^2}\frac{\partial^2}{\partial t^2} \end{pmatrix} \begin{pmatrix} \Phi \\ V \end{pmatrix} = \begin{pmatrix} 4\pi\rho \\ \frac{4\pi}{c}J \end{pmatrix}.$$

Using the Fourier transform as at the end of Section 22.11 above, we have

$$\begin{pmatrix} -\Delta - \frac{\omega^2}{c^2} & 0 \\ 0 & -\Delta - \frac{\omega^2}{c^2} \end{pmatrix} \begin{pmatrix} \Phi_\omega \\ V_\omega \end{pmatrix} = \begin{pmatrix} 4\pi\rho_\omega \\ \frac{4\pi}{c}J_\omega \end{pmatrix}.$$

Note that the second equation is a vector equation that contains three independent scalar Helmholtz equations. Thus, the original Maxwell system has been reduced to the Helmholtz equation in the three spatial dimensions x, y, and z.

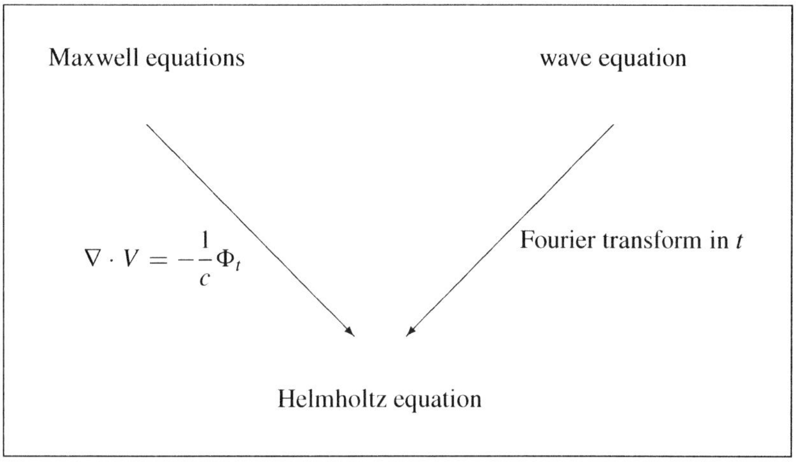

Figure 22.7. *The wave and Maxwell equations can be reduced to the Helmholtz equation by a Fourier transform in the time variable t and the assumption* $\nabla \cdot V = -\frac{1}{c}\Phi_t$, *respectively.*

22.13 Exercises

1. Use the "dynamicVector2" and "difference2" objects in Section A.4 of the Appendix to implement the finite-difference discretization of the Helmholtz equation (Section 22.3). The discrete boundary conditions are incorporated in the coefficient matrix as in Chapter 7, Section 4. Because the boundary conditions on the right edge are complex, you actually need "dynamicVector2<complex>" and "difference2<complex>" objects.

2. Use the above objects to implement the bilinear finite-element discretization of the Helmholtz equation (Section 22.5). Calculate the stiffness matrix according to the guidelines in Chapter 12, Section 5, with the required changes. In particular, keep in mind that square elements are used here rather than triangles.

3. Apply your finite-difference and bilinear finite-element codes to the model problem in Section 22.6. Compare the discretization errors and levels of adequacy in the two methods.

4. Apply your finite-difference and bilinear finite-element codes to the measuring problem in Section 22.8. Use fictitious grid points above and below the slit to complete the grid into a rectangular computational grid.

5. Construct the initial, coarse triangulation that approximates the domain in Figure 22.5 poorly. The solution can be found in Section A.16 of the Appendix.

6. Assemble the stiffness matrix for the Helmholtz equation on the above mesh. Use the code in Chapter 16, Section 5, with the required changes. In particular, use the "polynomial" object in Chapter 5, Sections 13 and 14, to calculate the contribution from the Helmholtz term.

7. Repeat the above exercise, only this time modify the code in Chapter 16, Section 5, to assemble the contribution from the mixed boundary conditions. Note that, because these boundary conditions are complex, you must use a "sparseMatrix<complex>" object to store the stiffness matrix.

8. Use your answer to the previous exercise to implement the adaptive-refinement algorithm to refine the above coarse mesh. Remember to use automatic boundary refinement only at boundary edges that lie next to the circular boundary segment. To loop over refinement levels, use the code in Section A.15 of the Appendix with the required changes. In particular, because the boundary conditions are complex, you need "dynamicVector<complex>", "sparseMatrix<complex>", and "multigrid<complex>" objects. (The implementation of triangles and meshes can remain the same as before.)

Appendix

A.1 Operations with Vectors

Here is the detailed implementation of some arithmetic operators of the "vector" class that have been left as an exercise in Chapter 2, Section 18:

```
template<class T, int N>
const vector<T,N>&
vector<T,N>::operator-=(const vector<T,N>&v){
    for(int i = 0; i < N; i++)
      component[i] -= v[i];
    return *this;
}  //  subtracting a vector from the current vector

template<class T, int N>
const vector<T,N>&
vector<T,N>::operator*=(const T& a){
    for(int i = 0; i < N; i++)
      component[i] *= a;
    return *this;
}  //  multiplying the current vector by a scalar

template<class T, int N>
const vector<T,N>&
vector<T,N>::operator/=(const T& a){
    for(int i = 0; i < N; i++)
      component[i] /= a;
    return *this;
}  //  multiplying the current vector by a scalar

template<class T, int N>
const vector<T,N>
```

```
operator-(const vector<T,N>&u, const vector<T,N>&v){
  return vector<T,N>(u) -= v;
} // vector minus vector

template<class T, int N>
const vector<T,N>
operator*(const vector<T,N>&u, const T& a){
  return vector<T,N>(u) *= a;
} // vector times scalar

template<class T, int N>
const vector<T,N>
operator*(const T& a, const vector<T,N>&u){
  return vector<T,N>(u) *= a;
} // T times vector

template<class T, int N>
const vector<T,N>
operator/(const vector<T,N>&u, const T& a){
  return vector<T,N>(u) /= a;
} // vector times scalar
```

A.2 Operations with Matrices

Here is the actual implementation of the products of vector and matrix, matrix and vector, and matrix and matrix declared in Chapter 2, Section 20.

```
template<class T, int N, int M>
const matrix<T,N,M>&
matrix<T,N,M>::operator*=(const T&a){
  for(int i=0; i<M; i++)
    set(i,(*this)[i] * a);
  return *this;
} // multiplication by scalar

template<class T, int N, int M>
const matrix<T,N,M>&
matrix<T,N,M>::operator/=(const T&a){
  for(int i=0; i<M; i++)
    set(i,(*this)[i] / a);
  return *this;
} // division by scalar

template<class T, int N, int M>
const matrix<T,N,M>
operator*(const T&a,const matrix<T,N,M>&m){
```

```
    return matrix<T,N,M>(m) *= a;
  } // scalar times matrix

  template<class T, int N, int M>
  const matrix<T,N,M>
  operator*(const matrix<T,N,M>&m, const T&a){
    return matrix<T,N,M>(m) *= a;
  } // matrix times scalar

  template<class T, int N, int M>
  const matrix<T,N,M>
  operator/(const matrix<T,N,M>&m, const T&a){
    return matrix<T,N,M>(m) /= a;
  } // matrix divided by scalar

  template<class T, int N, int M>
  const vector<T,M>
  operator*(const vector<T,N>&v,const matrix<T,N,M>&m){
    vector<T,M> result;
    for(int i=0; i<M; i++)
      result.set(i, v * m[i]);
    return result;
  } // vector times matrix

  template<class T, int N, int M>
  const vector<T,N>
  operator*(const matrix<T,N,M>&m,const vector<T,M>&v){
    vector<T,N> result;
    for(int i=0; i<M; i++)
      result += v[i] * m[i];
    return result;
  } // matrix times vector

  template<class T, int N, int M, int K>
  const matrix<T,N,K>
  operator*(const matrix<T,N,M>&m1,const matrix<T,M,K>&m2){
    matrix<T,N,K> result;
    for(int i=0; i<K; i++)
      result.set(i,m1 * m2[i]);
    return result;
  } // matrix times matrix
```

Here are some more functions that compute the determinant, inverse, and transpose of 2×2 matrices of class "matrix2" in Chapter 2, Section 20:

```
  typedef matrix<double,2,2> matrix2;
```

```
double det(const matrix2&A){
  return A(0,0)*A(1,1) - A(0,1)*A(1,0);
} // determinant of 2 by 2 matrix
```

The above "det()" function is now used to compute A^{-1} by Kremer's formula:

$$\begin{pmatrix} A_{0,0} & A_{0,1} \\ A_{1,0} & A_{1,1} \end{pmatrix}^{-1} = \det(A)^{-1} \begin{pmatrix} A_{1,1} & -A_{0,1} \\ -A_{1,0} & A_{0,0} \end{pmatrix}.$$

This is done as follows:

```
const matrix2 inverse(const matrix2&A){
  point column0(A(1,1),-A(1,0));
  point column1(-A(0,1),A(0,0));
  return matrix2(column0,column1)/det(A);
} // inverse of 2 by 2 matrix
```

Finally, the transpose of a 2×2 matrix is computed as follows:

```
const matrix2 transpose(const matrix2&A){
  return
   matrix2(point(A(0,0),A(0,1)),point(A(1,0),A(1,1)));
} // transpose of 2 by 2 matrix
```

A.3 Operations with Dynamic Vectors

Here is the detailed implementation of some arithmetic operators of the "dynamicVector" class that have been left as an exercise in Chapter 3, Section 3 (subtraction, multiplication, and division by scalar, inner product, etc.):

```
template<class T>
const dynamicVector<T>&
dynamicVector<T>::operator-=( const dynamicVector<T>&v){
    for(int i = 0; i < dimension; i++)
      component[i] -= v[i];
    return *this;
} // subtract a dynamicVector from the current one

template<class T>
const dynamicVector<T>&
dynamicVector<T>::operator*=(const T& a){
    for(int i = 0; i < dimension; i++)
      component[i] *= a;
    return *this;
} // multiply the current dynamicVector by a scalar
```

```
template<class T>
const dynamicVector<T>&
dynamicVector<T>::operator/=(const T& a){
    for(int i = 0; i < dimension; i++)
      component[i] /= a;
    return *this;
} // divide the current dynamicVector by a scalar

template<class T>
const dynamicVector<T>
operator-(const dynamicVector<T>&u,
        const dynamicVector<T>&v){
  return dynamicVector<T>(u) -= v;
} // dynamicVector minus dynamicVector

template<class T>
const dynamicVector<T>
operator*(const dynamicVector<T>&u, const T& a){
  return dynamicVector<T>(u) *= a;
} // dynamicVector times scalar

template<class T>
const dynamicVector<T>
operator*(const T& a, const dynamicVector<T>&u){
  return dynamicVector<T>(u) *= a;
} // T times dynamicVector

template<class T>
const dynamicVector<T>
operator/(const dynamicVector<T>&u, const T& a){
  return dynamicVector<T>(u) /= a;
} // dynamicVector divided by scalar

template<class T>
T operator*(const dynamicVector<T>&u,
        const dynamicVector<T>&v){
    T sum = 0;
    for(int i = 0; i < u.dim(); i++)
      sum += u[i] * +v[i];
    return sum;
} // inner product
```

A.4 Two-Dimensional Convection-Diffusion Equation

Here, we present the detailed implementation of the semi-implicit finite-difference scheme for the convection-diffusion equation in two spatial dimensions (Chapter 7, Section 13). The required objects are similar to those used in the analogous one-dimensional case.

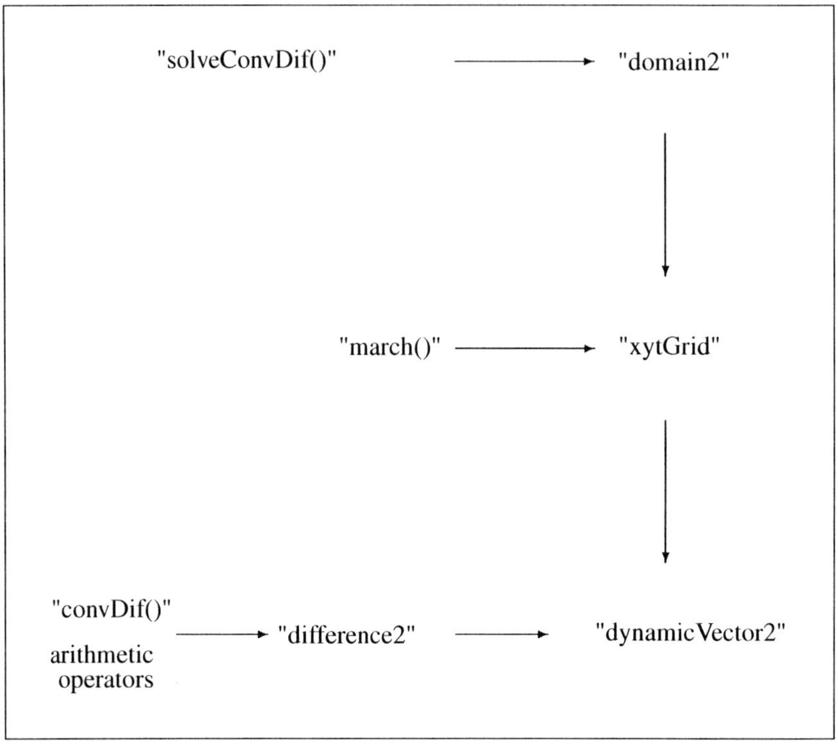

Figure A.1. *Hierarchy of objects for the convection-diffusion equation in two spatial dimensions: the "domain2" object uses an "xytGrid" object, which uses "dynamicVector2" and "difference2" objects.*

The hierarchy of objects used in the workplan is described in Figure A.1. The "domain2" object is in the highest level, with the function "solveConvDif()" that acts upon it. The "xytGrid" object is in the lower level with the function "march()" that acts upon it. In fact, "solveConvDif()" invokes "march()" to act upon the "xytGrid" object contained in the "domain2" object. Finally, the "dynamicVector2" and "difference2" objects are in the lowest level. These objects correspond to a particular time level in the time-space grid and are also connected by arithmetic operations between them and the function "convDif()" that initializes them with the discrete convection-diffusion terms at the time level under consideration.

Here, we introduce the objects in the upward order, from the most elementary ones at the lowest level to the most complex ones at the highest level. First, we introduce the "dynamicVector2" class that implements vectors corresponding to rectangular grids as in

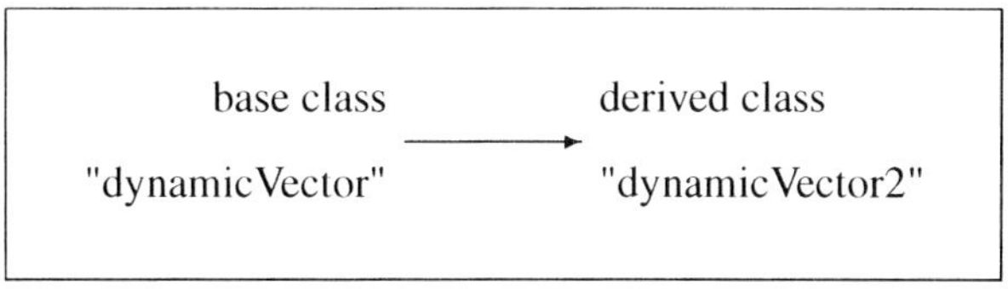

Figure A.2. *Schematic representation of inheritance from the base class "dynamicVector" to the derived class "dynamicVector2".*

Figure 7.4. As described schematically in Figure A.2, the "dynamicVector2" class is derived from the base "dynamicVector" class in Chapter 3, Section 3, by introducing an extra integer data field to indicate the width of the grid:

```
template<class T>
class dynamicVector2 : public dynamicVector<T>{
    int N;
  public:
    dynamicVector2(int, int, const T&);
```

The constructor declared above will be defined later on.

The components in the "dynamicVector2" object can be accessed by "operator()" with two integer arguments to indicate the spatial location in the grid. For example, $u_{i,j}$ can be accessed (for reading or writing) as "u(i,j)":

```
    T& operator()(int i, int j){
      return component[i*N+j];
    } // access (i,j)th component
```

Actually, $u_{i,j}$ can also be read by another version of "operator()", which takes three arguments: two integers as before and a string to indicate that the read-only version is used. With this version, $u_{i,j}$ can be read as, e.g., "u(i,j,"read")":

```
    const T& operator()(int i, int j, char*) const{
      return component[i*N+j];
    } // (i,j)th component (read only)
```

The following member functions return the length and width of the grid:

```
    int length() const{
      return dim()/N;
    } // length of grid

    int width() const{
      return N;
    } // width of grid
```

Since the derived class contains an extra data field that is absent in the base class, it makes no sense to convert the base-class object into a derived-class object, because the extra data field would remain undefined. This is why arithmetic operators like "*=" and "+=" must be redefined in the derived class: the base-class versions return a base-class object that cannot be converted to the required derived-class object. The present versions rewritten here return derived-class objects, as required:

```
    const dynamicVector2& operator+=(const dynamicVector2&);
    const dynamicVector2& operator-=(const dynamicVector2&);
    const dynamicVector2& operator*=(const T&);
    const dynamicVector2& operator/=(const T&);
};
```

This completes the block of the "dynamicVector2" class. The functions that are only declared above are defined explicitly below.

Because the data fields in the base "dynamicVector" class are declared as "protected" rather than "private", they can be accessed and changed in the derived "dynamicVector2" class. This property is useful in the constructor that takes integer arguments. Indeed, this constructor first implicitly calls the default constructor of the base "dynamicVector" class, which creates a trivial "dynamicVector" object with no components at all. The data fields in this object are then reconstructed by the present constructor of the "dynamicVector2" class and assume meaningful values. This can be done thanks to the access privilege that derived classes have to "protected" members of base classes:

```
template<class T>
dynamicVector2<T>::dynamicVector2(
    int m=0, int n=0, const T&t=0){
  dimension = n*m;
  N = n;
  component = dimension ? new T[dimension] : 0;
  for(int i=0; i<dimension; i++)
    component[i] = t;
} // constructor
```

The copy constructor need not be defined, because the default copy constructor works just fine: it first implicitly invokes the copy constructor of the base "dynamicVector" class to copy the inherited data fields and then copies the remaining data field 'N'. The same is true for the assignment operator.

Here is the actual definition of the member arithmetic operators declared above:

```
template<class T>
const dynamicVector2<T>&
dynamicVector2<T>::operator+=(const dynamicVector2<T>&v){
    for(int i = 0; i < dimension; i++)
      component[i] += v[i];
    return *this;
} // add a dynamicVector to the current dynamicVector
```

```
template<class T>
const dynamicVector2<T>&
dynamicVector2<T>::operator-=(const dynamicVector2<T>&v){
    for(int i = 0; i < dimension; i++)
      component[i] -= v[i];
    return *this;
}  //  subtract a dynamicVector from the current one

template<class T>
const dynamicVector2<T>&
dynamicVector2<T>::operator*=(const T& a){
  for(int i = 0; i < dimension; i++)
    component[i] *= a;
  return *this;
}  //  multiply the current dynamicVector by a scalar

template<class T>
const dynamicVector2<T>&
dynamicVector2<T>::operator/=(const T& a){
  for(int i = 0; i < dimension; i++)
    component[i] /= a;
  return *this;
}  //  divide the current dynamicVector by a scalar
```

Here are some nonmember binary arithmetic operators:

```
template<class T>
const dynamicVector2<T>
operator+(const dynamicVector2<T>&u,
        const dynamicVector2<T>&v){
  return dynamicVector2<T>(u) += v;
}  //  dynamicVector2 plus dynamicVector2

template<class T>
const dynamicVector2<T>
operator-(const dynamicVector2<T>&u,
        const dynamicVector2<T>&v){
  return dynamicVector2<T>(u) -= v;
}  //  dynamicVector2 minus dynamicVector2

template<class T>
const dynamicVector2<T>
operator*(const dynamicVector2<T>&u, const T& a){
  return dynamicVector2<T>(u) *= a;
}  //  dynamicVector2 times scalar
```

```
template<class T>
const dynamicVector2<T>
operator*(const T& a, const dynamicVector2<T>&u){
  return dynamicVector2<T>(u) *= a;
}  //  T times dynamicVector2

template<class T>
const dynamicVector2<T>
operator/(const dynamicVector2<T>&u, const T& a){
  return dynamicVector2<T>(u) /= a;
}  //  dynamicVector2 divided by scalar
```

Here is the unary negative operator:

```
template<class T>
const dynamicVector2<T>
operator-(const dynamicVector2<T>&u){
  return dynamicVector2<T>(u) *= -1.;
}  //  negative of a dynamicVector2
```

Here is a function that prints a "dynamicVector2" object onto the screen:

```
template<class T>
void print(const dynamicVector2<T>&v){
  for(int i = 0;i < v.length(); i++){
    for(int j = 0;j < v.width(); j++)
      printf("v[%d,%d]=%f;   ",i,j,v(i,j,"read"));
    printf("\n");
  }
}  //  printing a dynamicVector2
```

We now implement the difference operator that acts on "dynamicVector2" objects. This object is implemented in the "difference2" class below.

The "difference2" class is derived from "list<dynamicVector2<T> >". More precisely, a "difference2" object is actually a list of nine "dynamicVector2" objects. In this list, nine numbers can be stored per grid point. These numbers uniquely define the discrete convection-diffusion terms in two spatial dimensions:

```
template<class T>
class difference2 : public list<dynamicVector2<T> >{
  public:
    difference2(int,int,const T&,const T&,const T&,const T&,
        const T&,const T&,const T&,const T&,const T&);
```

The above constructor takes two integer arguments to specify the size of the grid and nine 'T' arguments to specify the constant coefficients in the constructed "difference2" object. This constructor is only declared above. Here are some more declarations, to be implemented explicitly later on:

```
      const difference2<T>&
          operator+=(const difference2<T>&);
      const difference2<T>&
          operator-=(const difference2<T>&);
      const difference2& operator*=(const T&);
```

The individual elements of the form $D_{i,j,k,l}$ in the difference operator D are accessed (for reading or writing) by the "operator()" with four integer arguments (called as "D(i,j,k,l)"):

```
      T& operator()(int i,int j,int k,int l){
        return (*item[(k-i+1)*3+l-j+1])(i,j);
      } // access (i,j,k,l)th element
```

The above individual element can also be read by the read-only version, invoked by "D(i,j,k,l,"read")":

```
      const T&
      operator()(int i,int j,int k,int l, char*) const{
        return (*item[(k-i+1)*3+l-j+1])(i,j,"read");
      } //  (i,j,k,l)th element (read only)
```

The following member functions return the width and length of the spatial grid:

```
      int width() const{ return item[0]->width(); }
      int length() const{ return item[0]->length(); }
  };
```

This concludes the block of the "difference2" class.

We now define the constructor declared above. When this constructor is called, it first implicitly invokes the default constructor of the base "list" class, which constructs a trivial list with no items. Thanks to the fact that the data fields in the "list" class are declared "protected" rather than "private", they can be reconstructed in the present constructor to implement the actual spatial difference operator:

```
  template<class T>
  difference2<T>::difference2(
      int m=0,int n=0,const T&a=0,const T&b=0,
      const T&c=0,const T&d=0,const T&e=0,
      const T&f=0,const T&g=0,const T&h=0,const T&i=0){
    number = 9;
    item = new dynamicVector2<T>*[9];
    item[0] = new dynamicVector2<T>(m,n,a);
    item[1] = new dynamicVector2<T>(m,n,b);
    item[2] = new dynamicVector2<T>(m,n,c);
    item[3] = new dynamicVector2<T>(m,n,d);
    item[4] = new dynamicVector2<T>(m,n,e);
    item[5] = new dynamicVector2<T>(m,n,f);
```

```
    item[6] = new dynamicVector2<T>(m,n,g);
    item[7] = new dynamicVector2<T>(m,n,h);
    item[8] = new dynamicVector2<T>(m,n,i);
  } // constructor
```

No copy constructor needs to be defined, because the default copy constructor of the base "list" class (invoked implicitly upon copying a "difference2" object) works just fine. The same is true for the assignment operator.

We now proceed to the definition of the member arithmetic operators declared above:

```
  template<class T>
  const difference2<T>&
  difference2<T>::operator+=(const difference2<T>&d){
    for(int i=0; i<number; i++)
      *item[i] += d[i];
    return *this;
  } // add a difference2 to the current one

  template<class T>
  const difference2<T>&
  difference2<T>::operator-=(const difference2<T>&d){
    for(int i=0; i<number; i++)
      *item[i] -= d[i];
    return *this;
  } // subtract a difference2 from the current one

  template<class T>
  const difference2<T>&
  difference2<T>::operator*=(const T&t){
    for(int i=0; i<number; i++)
      *item[i] *= t;
    return *this;
  } // multiply the difference2 by a scalar T
```

Here are some nonmember binary arithmetic operators:

```
  template<class T>
  const difference2<T>
  operator+(const difference2<T>&d1,
      const difference2<T>&d2){
    return difference2<T>(d1) += d2;
  } // addition of two difference2s

  template<class T>
  const difference2<T>
  operator-(const difference2<T>&d1,
      const difference2<T>&d2){
```

```
    return difference2<T>(d1) -= d2;
  } // subtraction of two difference2s

  template<class T>
  const difference2<T>
  operator*(const T&t, const difference2<T>&d){
    return difference2<T>(d) *= t;
  } // scalar times difference2

  template<class T>
  const difference2<T>
  operator*(const difference2<T>&d, const T&t){
    return difference2<T>(d) *= t;
  } // difference2 times scalar

  template<class T>
  const dynamicVector2<T>
  operator*(const difference2<T>&d,
      const dynamicVector2<T>&v){
    dynamicVector2<T> dv(v.length(),v.width(),0);
    for(int i=0; i<v.length(); i++)
      for(int j=0; j<v.width(); j++)
        for(int k=max(0,i-1);
            k<=min(v.length()-1,i+1); k++)
          for(int l=max(0,j-1);
          l<=min(v.width()-1,j+1); l++)
            dv(i,j) += d(i,j,k,l,"read")*v(k,l,"read");
    return dv;
  } // difference2 times dynamicVector2
```

The following "operator/" solves approximately the linear system "D*x=f" using 100 consecutive Gauss–Seidel iterations. Of course, the multigrid linear-system solver (Chapter 17, Section 8) is much more efficient. Still, the Gauss–Seidel iteration is good enough for our main purpose: to write and debug the entire convection-diffusion solver:

```
  template<class T>
  const dynamicVector2<T>
  operator/(const dynamicVector2<T>&f,
      const difference2<T>&d){
    dynamicVector2<T> x(f);
    for(int iteration=0; iteration < 100; iteration++)
      for(int i=0; i<f.length(); i++)
        for(int j=0; j<f.width(); j++){
          double residual = f(i,j,"read");
          for(int k=max(0,i-1);
              k<=min(f.length()-1,i+1); k++)
```

```
            for(int l=max(0,j-1);
                l<=min(f.width()-1,j+1); l++)
              residual -= d(i,j,k,l,"read")*x(k,l,"read");
          x(i,j) += residual/d(i,j,i,j,"read");
        }
    return x;
  } //  solving d*x=f approximately by 100 GS iterations
```

Next, we define the "xytGrid" class, which implements the time-space grid. This class is analogous to the "xtGrid" class in Chapter 7, Section 11. It is implemented as a list of "dynamicVector2" objects that contain the individual time levels (Figure A.3).

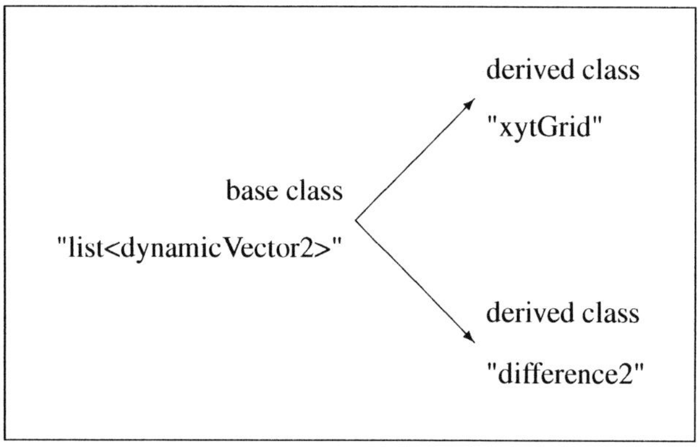

Figure A.3. *Schematic representation of inheritance from the base class "list" (list of "dynamicVector2" objects) to the derived classes "xytGrid" and "difference2".*

Again, no copy constructor or assignment operator needs to be defined, because the corresponding default operators of the base "list" class (invoked implicitly upon copying or assigning an "xytGrid" object) work just fine:

```
  template<class T>
  class xytGrid : public list<dynamicVector2<T> >{
    public:
      xytGrid(int,int,int,const T&);
      int timeSteps() const{
        return size();
      } // number of time levels

      int width() const{
        return item[0]->width();
      } // width of grid

      int length() const{
        return item[0]->length();
      } // length of grid
```

A grid point in the time-space grid can be accessed (for reading or writing) by the "operator()" with three integer arguments to indicate its time level and spatial location. For example, "g(i,j,k)" refers to the "(j,k)"th grid point in the 'i'th time level in the "xytGrid" object 'g':

```
    T& operator()(int i,int j,int k){
      return (*item[i])(j,k);
    } //  (i,j,k)th grid point
```

The entire time level is also returned by another version of "operator()" with only one argument to indicate the index of the time level. For example, "g(i)" returns the entire 'i'th time level in the "xytGrid" object 'g'. Although this operator is already implemented in the base "list" class, it must be rewritten here explicitly to prevent confusion with the other version of "operator()" defined above:

```
    dynamicVector2<T>& operator()(int i){
      if(item[i])return *item[i];
    } //  ith time level
};
```

This concludes the block of the "xytGrid" class.

The constructor that takes integer arguments to specify the grid size is only declared above. In the following, we define it explicitly. When this constructor is actually called, the default constructor of the base "list" class is invoked implicitly to construct a trivial list with no items in it. Thanks to the fact that the items in the base "list" class are declared "protected" rather than "private", they can be accessed from the present constructor and set to contain the individual time levels in the time-space grid:

```
template<class T>
xytGrid<T>::xytGrid(int m=0,
  int n=0,int l=0,const T&a=0){
  number = m;
  item = m ? new dynamicVector2<T>*[m] : 0;
  for(int i=0; i<m; i++)
    item[i] = new dynamicVector2<T>(n,l,a);
} //  constructor
```

Next, we implement the semi-implicit time-marching scheme in an analogous way to that in Chapter 7, Section 8. For simplicity, a homogeneous problem is considered; that is, the external functions that define the initial-boundary-value problem are equal to 0, except the initial condition, which is a nonzero function:

```
double F(double, double, double){return 0.;}
double C1(double, double, double){return 0.;}
double C2(double, double, double){return 0.;}
double Alpha(double, double, double){return 0.;}
double G(double, double, double){return 0.;}
double Initial(double x, double y){
```

```
    return (1.-x*x)*(1.-y*y);
}
const double Epsilon=1.;
```

Here is the "convDif()" function that places the discrete convection-diffusion spatial derivatives in a "difference2" object and the corresponding right-hand-side vector in a "dynamicVector2" object:

```
template<class T>
void convDif(difference2<T>&d,dynamicVector2<T>&f,
  double hx,double hy,double deltaT,double t){
  for(int k=0; k<d.length(); k++)
   for(int j=0; j<d.width(); j++){
     if(t>deltaT/2)f(k,j)=F(j*hx,k*hy,t-deltaT/2);
    double c1=C1(j*hx,k*hy,t)/hx;
    if(c1>0.){
      d(k,j,k,j)=c1;
      d(k,j,k,j-1)=-c1;
    }
    else{
      d(k,j,k,j)=-c1;
      d(k,j,k,j+1)=c1;
    }
    double c2=C2(j*hx,k*hy,t)/hy;
    if(c2>0.){
      d(k,j,k,j)+=c2;
      d(k,j,k-1,j)=-c2;
    }
    else{
      d(k,j,k,j)-=c2;
      d(k,j,k+1,j)=c2;
    }
   }
  d += Epsilon * difference2<T>(d.length(),
            d.width(),0,-1/hy/hy,0,-1/hx/hx,
          2/hx/hx+2/hy/hy,-1/hx/hx,0,-1/hy/hy,0);
  for(int k=0; k<d.length(); k++){
    d(k,0,k,0) += d(k,0,k,-1);
    d(k,0,k,0) -= d(k,0,k,-1) * hx * Alpha(0.,k*hy,t);
    if(t>deltaT/2){
      f(k,0) -= d(k,0,k,-1) * hx * G(0,k*hy,t-deltaT/2);
      f(k,d.width()-1) -= d(k,d.width()-1,k,d.width())
          * G(d.width()*hx,k*hy,t-deltaT/2);
    }
  }
  for(int j=0; j<d.width(); j++){
```

```
      d(0,j,0,j) += d(0,j,-1,j);
      d(0,j,0,j) -= d(0,j,-1,j) * hy * Alpha(j*hx,0.,t);
      if(t>deltaT/2){
        f(0,j) -=
          d(0,j,-1,j) * hy * G(j*hx,0.,t-deltaT/2);
        f(d.length()-1,j) -=
            d(d.length()-1,j,d.length(),j)
            * G(j*hx,d.length()*hy,t-deltaT/2);
      }
    }
  } // discrete convection-diffusion term+right-hand side
```

The above function places the discrete convection-diffusion spatial derivatives at the relevant time level in its first argument (the "difference2" object 'd') and the corresponding right-hand side in its second argument (the "dynamicVector2" object 'f'). This function is now used in the actual semi-implicit time-marching scheme:

```
  template<class T>
  void march(xytGrid<T>&g, double hx,
          double hy, double deltaT){
    difference2<T> I(g.length(),g.width(),0,0,0,0,1,0,0,0,0);
    for(int k=0; k<g.length(); k++)
      for(int j=0; j<g.width(); j++)
        g(0,k,j) = Initial(j*hx,k*hy);
    dynamicVector2<T> f(g.length(),g.width());
    difference2<T> previous(g.length(),g.width());
    convDif(previous,f,hx,hy,deltaT,0);
    for(int time=1; time < g.timeSteps(); time++){
      difference2<T> current(g.length(),g.width());
      convDif(current,f,hx,hy,deltaT,time*deltaT);
      g(time) =
       ((I-.5 * deltaT * previous) * g[time-1]+deltaT * f)
            / (I + 0.5 * deltaT * current);
      previous = current;
    }
    print(g[g.timeSteps()-1]);
  }  //  semi-implicit time marching
```

The "domain2" class below is analogous to the "domain" class in Chapter 7, Section 8:

```
  class domain2{
      xytGrid<double> g;
      double Time;
      double Width;
      double Length;
    public:
      domain2(double T, double Lx,
```

```
        double Ly, double accuracy)
    : g((int)(T/accuracy)+1,(int)(Ly/accuracy)+1,
        (int)(Lx/accuracy)+1),
      Time(T),Width(Lx),Length(Ly){
    } // constructor

    void solveConvDif(){
      march(g,Width/g.width(),Length/g.length(),
          Time/g.timeSteps());
    } // solve convection-diffusion equation
};
```

This concludes the definition of the "domain2" class. The "main()" program that uses it to numerically solve the convection-diffusion equation in two spatial dimensions looks like this:

```
int main(){
  domain2 D2(10.,1.,1.,.1);
  D2.solveConvDif();
  return 0;
}
```

This completes the numerical solution of the time-dependent convection-diffusion equation in two spatial dimensions.

A.5 Stability in the Explicit Scheme

The condition that guarantees that the explicit time-marching scheme in Chapter 7, Section 5, is also stable is that Δt is so small that all the main-diagonal elements in $\Delta t A_i$ are smaller than 1. Indeed, because A_i is assumed to be diagonally dominant, it follows from Gersgorin's theorem that all the eigenvalues of $\Delta t A_i$ lie strictly in a circle of radius 1 around 1 in the complex plane. Now, for the explicit scheme, we have, in Chapter 8, Section 3,

$$Q_i B_i^{-1} = I - \Delta t A_i.$$

From the above discussion, all the eigenvalues of these matrices are smaller than 1 in magnitude. The rest of the stability analysis is as in Chapter 8, Section 3.

A.6 Jordan Form of a Tridiagonal Matrix

Let

$$T = \begin{pmatrix} p_1 & o_2 & & \\ q_2 & p_2 & o_3 & \\ & q_3 & p_3 & \ddots \\ & & \ddots & \ddots \end{pmatrix}$$

be a tridiagonal matrix with $p_1, p_2, \ldots$ on the main diagonal, $o_2, o_3, \ldots$ on the diagonal just above it, $q_2, q_3, \ldots$ on the diagonal just below it, and zero elsewhere. Let us define a diagonal matrix E in such a way that ETE^{-1} is symmetric:

$$E = \begin{pmatrix} e_1 & & \\ & e_2 & \\ & & \ddots \end{pmatrix},$$

where

$$e_1 = 1,$$
$$e_j = e_{j-1}\sqrt{o_j/p_j}.$$

(It is assumed that the above square root is real and positive.) It is easy to see that ETE^{-1} is indeed symmetric, so it has an orthogonal Jordan matrix R for which $\|R\| = \|R^{-1}\| = 1$ and $RETE^{-1}R^{-1}$ is diagonal. As a result, T has the Jordan matrix

$$J \equiv RE.$$

Now, we add the subscript i to denote the ith time level. By repeating the above for each time level, we obtain the Jordan matrix for the tridiagonal matrix A_i in Chapter 8, Section 2 (also denoted by $D^{(i)}$ in Chapter 7, Section 3) in the form

$$J_i = R_i E_i,$$

where E_i is a diagonal matrix and R_i is an orthogonal matrix.

Assume now that the convection coefficient $C(t, x)$ in the convection-diffusion equation is a rarefaction or shock wave as in Figures 9.1 to 9.3. Consider a particular main-diagonal element in E_i (say, the jth one). It is easy to see that this element can only decrease as the wave progresses and i increases. Indeed, if the convection coefficient $C(t, x)$ is a rarefaction or a shock wave as in Figure 9.1 or 9.3, then

$$C((i+1)\Delta t, x) \geq C(i\Delta t, x)$$

for every x. Therefore, for fixed j, o_j/q_j can only decrease as i increases. As a result, e_j (for fixed j) also can only decrease as i increases.

Let us show that e_j (for fixed j) is a monotonically decreasing function of i also when $C(t, x)$ is the rarefaction wave in Figure 9.2. In this case, the above inequality holds only for $x \leq 0$ but not for $x > 0$. Still, for a fixed grid point j on the x-interval, e_j is calculated from data at points to its left only. Therefore, when j lies to the left of the origin, the situation is as before, and e_j indeed decreases as i increases. Even when j lies to the right of the origin, there are to the left of it more points where the above inequality holds than points where it is violated. Furthermore, thanks to symmetry considerations around the origin in Figure 9.2, the effect on e_j from grid points to the left of j where the above inequality is violated is canceled by the effect of their symmetric counterparts to the left of the origin, where it holds. Thus, here also e_j (for fixed j) can only decrease as i increases.

As a result, we have

$$\begin{aligned}\|J_{i+1}J_i^{-1}\| &= \|R_{i+1}E_{i+1}E_i^{-1}R_i^{-1}\| \\ &\le \|R_{i+1}\| \cdot \|E_{i+1}E_i^{-1}\| \cdot \|R_i^{-1}\| \\ &= \|E_{i+1}E_i^{-1}\| \\ &\le 1.\end{aligned}$$

This completes the proof of the stability condition in Chapter 8, Section 3, for the convection-diffusion equation with the convection coefficient $C(t, x)$ being a rarefaction or shock wave as in Figures 9.1 to 9.3.

Note also that, when the convection coefficient $C(t, x)$ is a shock wave, as in Figure 9.3, with $a \ne 0$ and $b \ne 0$, and

$$\Delta t \sim h \sim \varepsilon$$

are all of the same order as they approach zero, $\Delta t E_i A_i E_i^{-1}$ is strictly diagonally dominant, with eigenvalues well away from the origin. Therefore, in this case, the eigenvalues of $Q_i B_i^{-1}$ are much smaller than 1 in magnitude, and we have the improved upper bounds in Chapter 8, Section 3.

A.7 Denoising Digital Images

The function "convDif()" in Section A.4 above sets the difference operator to contain the discrete spatial derivatives in the convection-diffusion equation. Here, we introduce the "nonlinearDif()" function, which produces the difference operator that contains the discrete nonlinear diffusion terms used in the denoising algorithm in Chapter 10.

In order to treat both grayscale and RGB color images in the same code, we use the global integer "Colors". If "Colors" is set to 3, then an RGB color image is considered. If, on the other hand, "Colors" is set to 1, then a grayscale image is considered.

The color vectors in the digital image are stored in an array of dimension "Colors". More precisely, this array contains the addresses of the "dynamicVector2" objects that actually store the color vectors. For instance,

```
const int Colors = 3;
dynamicVector2<double>* u[Colors];
```

declares an array 'u' that may contain the three addresses of vector colors in an RGB digital image.

The function "nonlinearDif()" defined below takes several arguments, the first of which is the difference operator 'd' in which the discrete diffusion terms should be stored. It is assumed that 'd' already has dimensions corresponding to the number of pixels in the digital image and initially contains zero values only. Then, the function "addSpatialDerivative()" (to be defined later) is used to add to 'd' the discrete nonlinear diffusion terms in the x and y spatial directions. This function takes two integer arguments: "dir1" and "dir2". These arguments indicate the spatial direction used in the particular discrete derivative that is added to 'd': when "dir1" is 1 and "dir2" is 0, the y-derivative is added; when "dir1" is 0 and "dir2" is 1, the x-derivative is added. Since the "addSpatialDerivative()" function is called twice, both the x- and y-derivatives are added to 'd', as required:

```
template<class T>
void nonlinearDif(difference2<T>&d, dynamicVector2<T>&f,
    dynamicVector2<T>** u, double deltaT,double t){
  for(int k=0; k<d.length(); k++)
   for(int j=0; j<d.width(); j++)
     if(t>deltaT/2)f(k,j)=0.;
  T Eta = 1000.;
  T hy = 1./d.length();
  T hx = 1./d.width();
  addSpatialDerivative(d,hy,hx,1,0,u,Eta);
  addSpatialDerivative(d,hx,hy,0,1,u,Eta);
} // set the difference operator for denoising
```

In the above code, we have used for simplicity a constant parameter "Eta". This parameter may also take a value that depends on the digital image in 'u', as in the definition of η at the end of Chapter 10, Section 7.

The above "nonlinearDif()" function can replace the "convDif()" function used in Section A.4 above to produce a denoising code. A slight modification should be made to account for the fact that 'u' is actually an array of pointers-to-dynamicVector2 rather than just a reference-to-dynamicVector2. Furthermore, the code in Section A.4 is currently set to use a single Newton iteration per time step; an inner loop should be introduced to allow more Newton iterations in each time step.

Let us now define the function "addSpatialDerivative()" that adds to the difference operator 'd' the discrete second derivative in the spatial direction ("dir2","dir1") (see Chapter 10, Section 7):

```
template<class T>
void addSpatialDerivative(difference2<T>&d,
    const T&h, const T&h2, int dir1, int dir2,
    dynamicVector2<T>** u, const T&Eta){
  for(int i=0;i<d.length()-dir1;i++)
    for(int j=0;j<d.width()-dir2;j++){
      T ux2 = 0.;
      T uy2 = 0.;
      T uxuy = 0.;
      for(int c=0;c<Colors;c++){
          T ux = ((*u[c])(i+dir1,j+dir2,"read")
              - (*u[c])(i,j,"read"))/h;
          T uy =
            ((*u[c])(min(d.length()-1,i+1),
                min(d.width()-1,j+1),"read")
            + (*u[c])(min(d.length()-1,i+1-dir1),
                min(d.width()-1,j+1-dir2),"read")
            - (*u[c])(max(0,i+2*dir1-1),
                max(0,j+2*dir2-1),"read")
            - (*u[c])(max(0,i+dir1-1),
```

```
                    max(0,j+dir2-1),"read")
                )/(4*h2);
              ux2 += ux * ux;
              uy2 += uy * uy;
              uxuy += ux * uy;
          }
          ux2 /= Eta;
          uy2 /= Eta;
          uxuy /= Eta;
          ux2 += 1.;
          uy2 += 1.;
          T coef = 1./(h*h)/(ux2*uy2-uxuy*uxuy);
          d(i,j,i+dir1,j+dir2) -= coef;
          d(i,j,i,j) += coef;
          d(i+dir1,j+dir2,i,j) -= coef;
          d(i+dir1,j+dir2,i+dir1,j+dir2) += coef;
        }
  }  //  add discrete nonlinear diffusion
```

The above code has not yet been tested on an actual image, because the examples in Chapter 10, Section 8, were actually produced from a slightly different code that also uses a multigrid linear-system solver in each Newton iteration. You are welcome to test it in the exercises at the end of Chapter 10.

A.8 Members of the Mesh Class

The "maxNorm" function defined below calculates the maximum modulus of a given function at nodes that lie away from the origin (0, 0). The "maxNorm" function is a member of the "mesh" class in Chapter 13, Section 4, and as such can use the recursive structure of the "mesh" object, which is actually a connected list of triangles. In fact, the "maxNorm" function first calculates the maximum modulus at the vertices of the first triangle in the mesh before being called recursively for the rest of the triangles in the mesh.

The maximum in the "maxNorm" function is taken over all nodes located away from the singularity at the origin. Nodes that are too close to the origin are excluded by an "if" question that checks the actual location of the 'i'th vertex in the first triangle, "item[i]()". (This actual location is returned by the "operator()" of the "node" class in Chapter 13, Section 2.) If this location is indeed too close to the origin in terms of sum-of-squares, then the 'i'th vertex in the first triangle is indeed excluded from the evaluation of the maximum:

```
double
mesh<triangle>::maxNorm(const dynamicVector<double>&v){
  double result=0.;
  for(int i=0; i<3; i++)
    if(squaredNorm(item[i]())>0.01)
      result=max(result,abs(v[item[i].getIndex()]));
  if(next)
```

```
    result =
      max(result,((mesh<triangle>*)next)->maxNorm(v));
  return result;
}  //  maximum modulus at nodes away from singularity
```

The "refineBoundary" function is a member of the "mesh" class in Chapter 13, Section 4, so it has access to the protected fields in the base "connectedList" class. This privilege is used to define a local pointer-to-mesh variable, named "runner", which scans the triangles in the mesh. For each triangle encountered, two smaller triangles are added to the mesh between this triangle and the circular boundary next to it (as in Figure 12.10). Actually, this procedure adds another "layer" of triangles next to the circular boundary. The triangles in this layer are ordered counterclockwise. Once the layer is complete, the "runner" pointer points to the start of it and is ready to construct an even finer layer of triangles to approximate the boundary even better. In the present example, the "runner" pointer scans 10 such layers, producing 10 finer and finer layers of triangles closer and closer to the circular boundary. Once the final (finest) layer has been created, 2048 nodes are placed along the circular boundary to approximate it quite well.

Here is the detailed implementation of the "refineBoundary" function. In the first code line, the "math.h" library is included, which provides the mathematical functions "sqrt()" (square root), "sin()" (sine), "cos()" (cosine), and "acos()" ($\cos^{-1}$, the inverse of the cosine function). These functions are useful in calculating the locations of the new nodes created in each layer of triangles:

```
#include<math.h>
void mesh<triangle>::refineBoundary(int levels){
  mesh<triangle>* runner = this;
  for(int i=1; i<levels; i++)
    for(int j=0; j<power(2,i); j++){
      point vertex0 = (*runner)()[0]();
      point vertex1 = (*runner)()[1]();
      point midpoint = (vertex0 + vertex1)/2.;
      double angle1 = acos(vertex1[0]);
      if(j >= power(2,i-1))angle1 = -angle1;
      double angleIncrement =
          acos(sqrt(squaredNorm(midpoint)));
      double angleMidpoint = angle1 - angleIncrement;
      point newPoint(cos(angleMidpoint),sin(angleMidpoint));
      node<point> newVertex(newPoint);
      triangle t1(runner->item(0),newVertex,runner->item(1));
      append(t1);
      angleMidpoint = angle1 + angleIncrement;
      newVertex =
       node<point>(point(cos(angleMidpoint),
               sin(angleMidpoint)));
      triangle t2(runner->item(1),newVertex,runner->item(2));
      append(t2);
```

```
      runner = (mesh<triangle>*)runner->next;
    }
} // refine at the boundary of a circular domain
```

A.9 Operations with Sparse Matrices

Here is the detailed implementation of the member arithmetic operators and functions of the "sparseMatrix" class that have been left as an exercise in Chapter 16, Section 4.

```
template<class T>
int sparseMatrix<T>::columnNumber() const{
  int maxColumn = -1;
  for(int i=0; i<rowNumber(); i++)
    if(item[i])maxColumn =
      max(maxColumn, item[i]->last()().getColumn());
    return maxColumn + 1;
} // number of columns

template<class T>
const sparseMatrix<T>&
sparseMatrix<T>::operator+=(const sparseMatrix<T>&M){
  for(int i=0; i<rowNumber(); i++)
    *item[i] += *M.item[i];
  return *this;
} // add a sparse matrix

template<class T>
const sparseMatrix<T>&
sparseMatrix<T>::operator-=(const sparseMatrix<T>&M){
  for(int i=0; i<rowNumber(); i++){
    row<T> minus = -1. * *M.item[i];
    *item[i] += minus;
  }
  return *this;
} // subtract a sparse matrix

template<class T>
const sparseMatrix<T>&
sparseMatrix<T>::operator*=(const T&t){
  for(int i=0; i<rowNumber(); i++)
    *item[i] *= t;
  return *this;
} // multiply by T
```

Here are some nonmember arithmetic operators:

```
template<class T>
const sparseMatrix<T>
operator+(const sparseMatrix<T>&M1,
        const sparseMatrix<T>&M2){
  return sparseMatrix<T>(M1) += M2;
}  //  matrix plus matrix

template<class T>
const sparseMatrix<T>
operator-(const sparseMatrix<T>&M1,
        const sparseMatrix<T>&M2){
  return sparseMatrix<T>(M1) -= M2;
}  //  matrix minus matrix

template<class T>
const sparseMatrix<T>
operator*(const T&t, const sparseMatrix<T>&M){
  return sparseMatrix<T>(M) *= t;
}  //  scalar times sparse matrix

template<class T>
const sparseMatrix<T>
operator*(const sparseMatrix<T>&M, const T&t){
  return sparseMatrix<T>(M) *= t;
}  //  sparse matrix times scalar

template<class T>
const dynamicVector<T>
operator*(const sparseMatrix<T>&M,
        const dynamicVector<T>&v){
  dynamicVector<T> result(M.rowNumber(),0.);
  for(int i=0; i<M.rowNumber(); i++)
    result(i) = M[i] * v;
  return result;
}  //  matrix times vector
```

Here is the implementation of some friend functions of the "sparseMatrix" class. The "operator*" function returns the product of two sparse matrices. The calculation uses the algorithm described in Chapter 16, Section 4. The "diagonal" function returns the main diagonal of a sparse matrix. The "transpose" function returns the transpose of a sparse matrix:

```
template<class T>
const sparseMatrix<T>
operator*(const sparseMatrix<T>&M1,
        const sparseMatrix<T>&M2){
```

```
    sparseMatrix<T> result(M1.rowNumber());
    for(int i = 0; i < M1.rowNumber(); i++){
      result.item[i] =
          new row<T>(M1.item[i]->getValue() *
          *M2.item[M1.item[i]->getColumn()]);
      for(const row<T>* runner =
          (const row<T>*)M1.item[i]->readNext();
          runner; runner =
          (const row<T>*)runner->readNext()){
        row<T> r =
            runner->getValue() *
            *M2.item[runner->getColumn()];
        *result.item[i] += r;
      }
    }
    return result;
}  //  matrix times matrix

template<class T>
const sparseMatrix<T>
diagonal(const sparseMatrix<T>&M){
  sparseMatrix<T> diag(M.rowNumber());
  for(int i=0; i<M.rowNumber(); i++)
    diag.item[i] = new row<T>(M(i,i),i);
  return diag;
}  //  return the main diagonal part

template<class T>
const sparseMatrix<T>
transpose(const sparseMatrix<T>&M){
  sparseMatrix<T> Mt(M.columnNumber());
  for(int i=0; i<M.rowNumber(); i++)
    for(const row<T>* runner = M.item[i]; runner;
        runner = (const row<T>*)runner->readNext()){
      if(Mt.item[runner->getColumn()])
        Mt.item[runner->getColumn()]->
            append(runner->getValue(),i);
      else
        Mt.item[runner->getColumn()] =
            new row<T>(runner->getValue(),i);
    }
  return Mt;
}  //  transpose of matrix
```

A.10 Kacmarz Iteration

Here, we implement the Kacmarz iteration defined in Chapter 17, Section 6. The naive implementation is as follows:

```
for(int i=0; i<100; i++)
  GaussSeidel(transpose(A) * A, transpose(A) * f, x);
print(x);
```

In each iteration, the transpose of 'A' is recalculated and used to multiply 'A' and 'f'. Of course, this approach is extremely inefficient. It makes much more sense to form the normal equations once and for all before the iteration starts:

```
sparseMatrix<double> At = transpose(A);
sparseMatrix<double> AtA = At * A;
dynamicVector<double> Atf = At * f;
for(int i=0; i<100; i++)
  GaussSeidel(AtA, Atf, x);
print(x);
```

A.11 ILU Iteration

Here, we consider the ILU iteration in Chapter 17, Section 7. First, we show how the "factorize()" function can be modified in such a way that no fill-in is ever used. (This version is particularly useful as a relaxation method within the multigrid V-cycle.) To this end, only three changes should be made in the original "factorize()" function in Chapter 17, Section 7. First, the "if" question that checks whether or not "factor" is sufficiently large in magnitude should be dropped, and the block that follows it should always be carried out. Second, the "+=" operator used in this block should be replaced by an "&=" operator. This operator should be defined in the body of the "row" class. It adds to the current "row" object certain elements from the "row" argument passed to it by reference. More precisely, only those elements in the "row" argument for which corresponding elements already exist in the current "row" object are added. This way, extra fill-in is never introduced in the current "row" object, and the original sparsity pattern is preserved. This is also why the call to the "truncateItems()" function at the end of the "factorize()" function is unnecessary and should be dropped as well: because there is no fill-in, there is no need to truncate matrix elements. This completes the definition of the "factorize()" version that uses no fill-in at all.

Next, we implement forward elimination in a lower triangular matrix 'L' with main-diagonal elements that are all equal to 1 (solving $Lx = f$) and back substitution in an upper triangular matrix 'U' (solving $Ux = f$). The functions that do these tasks are members of the "sparseMatrix" class, so they have access to the "item" data field of "sparseMatrix" objects, which contains the addresses of their rows. Using the row-times-vector product available in the "row" class in Chapter 16, Section 3, forward elimination and back substitution are implemented easily and elegantly.

Note that the local vector "result" that contains the solution cannot be returned by reference, because it disappears at the end of the function. It must be returned by value, as indicated by the word "dynamicVector<T>" (rather than "dynamicVector<T>&") that precedes the function name. This way, the local vector "result" is copied to a temporary, unnamed vector that stores the output even after the function terminates. An even better idea is to declare the output as a constant by adding the reserved word "const" before the function name. This guarantees that the returned vector can serve as an argument in subsequent calls to other functions:

```
template<class T>
const dynamicVector<T>
sparseMatrix<T>::forwardElimination(
        const dynamicVector<T>&f) const{
  dynamicVector<T> result(f.dim(),0.);
  result(0) = f[0];
  for(int i=1; i<f.dim(); i++)
    result(i) =
      item[i] ? f[i] - *item[i] * result : f[i];
  return result;
} // forward elimination in L

template<class T>
const dynamicVector<T>
sparseMatrix<T>::backSubstitution(
    const dynamicVector<T>&f) const{
  dynamicVector<T> result(f.dim(),0.);
  for(int i = f.dim() - 1; i>=0; i--){
    result(i) = item[i]->readNext() ?
      f[i]-*(row<T>*)item[i]->readNext() * result : f[i];
    result(i) /= (*item[i])().getValue();
  }
  return result;
} // back substitution in U
```

A.12 Multigrid Iteration

Here, we implement the "coarsen" member function of the "sparseMatrix" class, which constructs the coarse grid c in Chapter 17, Section 8. This coarse grid is stored in a vector of integers named "coarse" according to the rule that the ith component in "coarse" is nonzero if and only if $i \in c$. In fact, the indices i for which $i \in c$ are ordered in a sequence, and the ith component in "coarse" takes their index in this sequence: $1, 2, 3, \ldots, |c|$ (where $|c|$ is the number of coarse-grid points). This is done in the final loop at the end of the "coarsen" function.

The vector of integers "coarse" is local and disappears when the function terminates. It must be returned by value rather than reference, as is indeed indicated by the word "dynamicVector<int>" (rather than "dynamicVector<int>&") before the function name. Actually, the returned vector is also declared constant by writing the reserved word "const" before the function name. This way, it can be passed as a concrete argument to subsequent functions:

```
template<class T>
const dynamicVector<int>
sparseMatrix<T>::coarsen(double threshold) const{
  dynamicVector<int> coarse(rowNumber(), 1);
  for(int i=0; i<rowNumber(); i++)
    if(coarse[i])
      for(const row<T>* runner = item[i]; runner;
        runner=(const row<T>*)runner->readNext())
        if((runner->getColumn() != i)&&(
             abs(runner->getValue()) >
             threshold * abs((*this)(i,i))))
          coarse(runner->getColumn()) = 0;
  for(int i=0; i<rowNumber(); i++)
    if(!coarse[i]){
      int drop=1;
      for(const row<T>* runner = item[i]; runner;
        runner=(const row<T>*)runner->readNext())
        if((coarse[runner->getColumn()])&&
            (runner->getValue() / (*this)(i,i)
                <= -threshold))
          drop=0;
      if(drop) coarse(i) = 1;
    }
  int count = 1;
  for(int i=0; i<rowNumber(); i++)
    if(coarse[i])coarse(i) = count++;
  return coarse;
} // define the coarse grid
```

Before we implement the "createTransfer" function using the guidelines in Algorithm 17.3 in Chapter 17, Section 8, we must define the "dropPositiveItems" member function of the "row" class, which drops positive off-diagonal elements and elements that are too small in magnitude. Like the "dropItems" function in Chapter 16, Section 3, "dropPositiveItems" also uses the recursive nature of the "row" class. First, the second row element is checked and dropped if it is indeed positive or too small in magnitude. Then, the function is called recursively for the remainder of the row. Finally, the first element in the row is also checked and dropped if it is positive or too small in magnitude:

```
template<class T>
void row<T>::dropPositiveItems(int i,
    const T&center, double threshold){
  if(next){
    if(((*next)().getColumn() != i)&&
      ((*next)().getValue()/center >= -threshold)){
      dropNextItem();
      dropPositiveItems(i, center, threshold);
```

```
    }
    else
      (*(row<T>*)next).dropPositiveItems(i,center, threshold);
    if((getColumn() != i)&&(getValue()/center >= -threshold))
      dropFirstItem();
  }
} // drop positive off-diagonal elements
```

The latter "if" question and the command that follows it at the end of the above code can actually be removed from this function and placed in another function that drops only the first element, if appropriate. This might increase efficiency, because as it stands there are some unnecessary repetitive checks.

We are now ready to define the "createTransfer" member function of the "sparseMatrix" class. First, this function calls the above "coarsen" function to define the coarse grid c, which is stored in a vector of integers named "coarse" (the ith component in "coarse" is nonzero if and only if $i \in c$). This vector is then used to define the prolongation matrix P, as in Chapter 17, Section 8. In fact, P is obtained from the original coefficient matrix A (the current "sparseMatrix" object) by dropping the columns in A with index j satisfying $j \notin c$. This is done by calling the "dropItems" function in Chapter 16, Section 3. In other words, the rows in the current "sparseMatrix" object are "masked" by the integer vector "coarse", and all matrix elements with column index j for which the jth component in "coarse" vanishes are dropped. Further, "coarse" is also used to renumber the columns left in the current "sparseMatrix" object by the numbers $1, 2, 3, \ldots, |c|$ using the "renumberColumns" function in Chapter 16, Section 3. Once this has been done, the current "sparseMatrix" object is the required prolongation matrix P. The function also returns as output the restriction matrix $R = P^t$ for further use in the multigrid algorithm:

```
template<class T>
const sparseMatrix<T>
sparseMatrix<T>::createTransfer(){
  const dynamicVector<int> coarse = coarsen();
  for(int i=0; i<rowNumber(); i++){
    if(coarse[i]){
      delete item[i];
      item[i] = new row<T>(1., coarse[i] - 1);
    }
    else{
      item[i]->dropPositiveItems(i, (*item[i])[i], 0.05);
      item[i]->dropItems(coarse);
      item[i]->renumberColumns(coarse);
      *item[i] /= item[i]->rowSum();
    }
  }
  return transpose(*this);
} // create transfer operators
```

One can easily modify this code to implement the algebraic multigrid (AMG) method in Chapter 17, Section 9, by replacing the call to the "dropPositiveItems" function at the beginning of the above "else" block by the following loop:

```
for(const row<T>* runner = item[i]; runner;
      runner=(const row<T>*)runner->readNext()){
  int j = runner->getColumn();
  if((j != i)&&(!coarse[j])){
    T Pij = runner->getValue();
    T AjSum =
      item[i]->rowSumCoarse(*A.item[j],coarse);
    if(fabs(AjSum) > 1.e-10)
      item[i]->addCoarse(
          (Pij / AjSum) * *A.item[j],coarse);
  }
}
```

where 'A' is the original current "sparseMatrix" object without its positive off-diagonal elements, defined at the beginning of the function by

```
for(int i=0; i<rowNumber(); i++)
  if(!coarse[i])
    item[i]->dropPositiveItems(i,
        (*item[i])[i], thresholdMG);
sparseMatrix<T> A = *this;
```

"rowSumCoarse" and "addCoarse" are member functions of the "row" class. "rowSumCoarse" sums only those elements in the "row" argument that lie in columns $j \in c$ for which the corresponding element in the current "row" object is also nonzero:

```
const T rowSumCoarse(const row<T>&r,
    const dynamicVector<int>&coarse)const{
  T contribution =
    coarse[getColumn()] ? r[getColumn()] : 0.;
  return next ? contribution +
        ((row<T>*)next)->rowSumCoarse(r,coarse)
        : contribution;
}  //  row sum at coarse points
```

"addCoarse" adds values to nonzero elements in the current "row" object that lie in columns j for which $j \in c$:

```
void addCoarse(const row<T>&r,
    const dynamicVector<int>&coarse){
  if(coarse[getColumn()])
    item += r[getColumn()];
  if(next) ((row<T>*)next)->addCoarse(r,coarse);
}  //  add values at coarse points
```

Note that the vector of integers "coarse" that indicates the coarse grid is passed to the above functions by reference rather than by name to avoid highly expensive and unnecessary reconstructions, particularly in the inner recursive calls.

Next, we implement the assignment operator of the "multigrid" class in Chapter 17, Section 10. This operator uses the recursive nature of the "multigrid" class: it first assigns the matrices corresponding to the finest grid and is then applied recursively to the coarser grids in the "next" field:

```
template<class T>
const multigrid<T>&
multigrid<T>::operator=(const multigrid<T>&mg){
  if(this != &mg){
    A = mg.A;
    U = mg.U;
    L = mg.L;
    P = mg.P;
    R = mg.R;
    if(next){
      if(mg.next)
        *next = *mg.next;
      else{
        delete next;
        next = 0;
      }
    }
    else
      if(mg.next)next = new multigrid(*mg.next);
  }
  return *this;
} // assignment operator
```

Similarly, the "print" friend function first prints the matrices in the "multigrid" object and then is applied recursively to the coarser grids in the "next" field:

```
template<class T>
void print(const multigrid<T>&mg){
  print(mg.A);
  print(mg.P);
  print(mg.R);
  if(mg.next)print(*mg.next);
} // print the multigrid object
```

A.13 Acceleration Techniques

Here, we implement several Krylov-subspace methods to accelerate the convergence of the basic multigrid iteration or any other iteration. The preconditioned conjugate gradient

(PCG) acceleration method is limited to symmetric and positive definite (SPD) problems and SPD preconditioners. Therefore, we also implement the conjugate gradient squared (CGS), transpose-free quasi-minimal residual (TFQMR), and general minimal residual (GMRES) acceleration methods, which can be used in more general cases as well.

The PCG algorithm in Chapter 17, Section 11, works well for linear systems with SPD coefficient matrix A and iterative methods with SPD preconditioner $\mathcal{P}$. Here, we assume that the preconditioner is that of the multigrid iteration in Chapter 17, Section 8. (The ILU and symmetric Gauss–Seidel iterations can also be obtained from it by setting "gridRatio" $= -1$ in the implementation in Chapter 17, Section 10, which means that no coarse grids are actually used.) For this iterative method, the preconditioner is SPD whenever A is (see Chapter 10, Section 7, of [39]), so PCG is indeed applicable.

Here is the detailed implementation of the PCG iteration. It terminates when the initial error has been reduced (in terms of the energy norm) by six orders of magnitude:

```
template<class T>
void
PCG(const multigrid<T>& MG, const sparseMatrix<T>&A,
  const dynamicVector<T>&f, dynamicVector<T>&x){
  const double eps=1.e-15, threshold=1.e-12;
  const int iterationnumber = 1000;
  dynamicVector<T> zero(x.dim(),0.);
  dynamicVector<T> keep(x);
  dynamicVector<T> rr(MG.Vcycle(f,keep) - x);
  dynamicVector<T> r = f - A * x;
  dynamicVector<T> pp(rr);
  dynamicVector<T> p = r;
  double gamma = r * rr;
  double gammainit = gamma;
  int count = 0;
  while((abs(gamma/gammainit) >=
        threshold * threshold)&&
         (count <= iterationnumber)){
    keep = pp;
    dynamicVector<T> ww = pp - MG.Vcycle(zero,keep);
    dynamicVector<T> w = A * pp;
    double alpha = gamma / (pp * w);
    x += alpha * pp;
    rr -= alpha * ww;
    r -= alpha * w;
    double gammaold = gamma;
    gamma = r * rr;
    double beta = gamma/gammaold;
    pp = rr + beta * pp;
    count++;
    printf("at MG it. %d in PCG, (r,Pr)=%f\n",count,gamma);
  }
```

```
    printf("total MG it. in PCG=%d\n", count + 1);
} // PCG acceleration
```

The "PCG()" function is called as follows:

```
PCG(MG,A,f,x);
print(x);
```

Next, we implement the CGS algorithm in [43] for the preconditioned system

$$\mathcal{P}^{-1}Ax = \mathcal{P}^{-1}f,$$

where $\mathcal{P}$ is the preconditioner of the multigrid iteration in Chapter 17, Section 8. In this system, the application of the coefficient matrix $\mathcal{P}^{-1}A$ to a vector is done as in Chapter 17, Section 11.

If one wants to apply CGS to the ILU or symmetric Gauss–Seidel iteration, then one should just use a multigrid object with zero "next" field.

The initial direction vector is the initial preconditioned residual. The code also uses the "fabs" and "sqrt()" functions available in the included "math.h" library. The "fabs()" function (absolute value) is actually equivalent to the "abs()" function in Chapter 1, Section 9. The "sqrt()" function returns the square root of a real number.

One of the arguments in the "CGS()" function is the integer "TFQMR". If this argument is zero, then CGS is indeed used. Otherwise, TFQMR [16] is used instead:

```
#include<stdio.h>
#include<math.h>
template<class T>
void CGS(const multigrid<T>& MG,
      const sparseMatrix<T>&A, int TFQMR,
     const dynamicVector<T>&f, dynamicVector<T>&x){
  const double eps=1.e-15;
  const int iterationnumber = 1000;
  T omegainit,omega0,omega1,tau,vv;
  T rho0,sigma,eta,alpha,rho1,beta;
  dynamicVector<T> keep(x);
  dynamicVector<T> rr(MG.Vcycle(f,keep) - x);
  dynamicVector<T> rbar(rr);
  dynamicVector<T> rcgs(rr),u(rr),pp(rr);
  tau = omegainit = omega1 = omega0 = sqrt(rcgs * rcgs);
  printf("res0=");
  print(omegainit);
  printf("\n");
  eta=0.;
  vv=0.;
  if(fabs(rho0 = rbar * rr) < eps)
    printf("rho0=%f,mg it.=1\n",fabs(rho0));
  dynamicVector<T> zero(x.dim(),0.);
```

```
dynamicVector<T> d(zero),v(zero),q(zero);
int count = 1;
do{
  keep = pp;
  v = pp - MG.Vcycle(zero,keep);
  if(fabs(sigma = rbar * v) < eps)
    printf("sigma=%f,mg it.=%d\n", fabs(sigma),2 * count);
  if(fabs(alpha=rho0/sigma) < eps)
    printf("alpha=%f,mg it.=%d\n", fabs(alpha),2 * count);
  q = u - alpha * v;
  dynamicVector<T> uq = u + q;
  keep = uq;
  rcgs -= alpha * (uq - MG.Vcycle(zero,keep));
  omega1=sqrt(rcgs * rcgs);
  if(!TFQMR){
    x += alpha * uq;
    printf("res=%f,it.=%d\n", fabs(omega1),2 * count + 1);
  }
  else{
    for(int m=2*count+1; m<=2*count+2;m++){
      T omega;
      if(m==2*count+1){
        omega=sqrt(omega0*omega1);
        keep=u;
      }
      else{
        omega=omega1;
        keep=q;
      }
      T scala=vv*vv*eta/alpha;
      d = keep + scala * d;
      vv=omega/tau;
      T c=1./sqrt(1.+vv*vv);
      tau *= vv*c;
      eta=c*c*alpha;
      x += eta * d;
      printf("res=%f,it.=%d\n",
        sqrt((A*x-f)*(A*x-f)),2 * count+1);
    }
  }
  omega0=omega1;
  if(fabs(rho1=rbar*rcgs)<eps)
    printf("rho1=%f,mg it.=%d\n", fabs(rho1),2*count+1);
  beta=rho1/rho0;
  rho0=rho1;
  u = rcgs + beta * q;
```

```
      pp = u + beta * (q + beta * pp);
    } while((fabs(omegal/omegainit) >= thresholdCG)&&
            (++count <= iterationnumber));
    printf("total MG it. in CGS=%d\n",2 * count + 1);
  } // CGS or TFQMR acceleration
```

Finally, we also implement the GMRES acceleration method [34]. Here, we use two extra integer arguments. The argument "preIterations" denotes the number of multigrid iterations used before GMRES starts (as in [40]). The argument 'K' denotes the dimension of the Krylov subspace used in GMRES:

```
template<class T>
const dynamicVector<T>&
GMRES(const multigrid<T>&MG, const sparseMatrix<T>&A,
        int preIterations, int K,
        const dynamicVector<T>&f, dynamicVector<T>&x){
  for(int i=0;i<preIterations;i++)
    MG.Vcycle(f,x);
  dynamicVector<T> s = x;
  dynamicVector<T> r = x;
  T R[K+1][K+1];
  for(int i=0;i<=K;i++)
    for(int j=0;j<=K;j++)
      R[i][j]=0.0;
  T Givens[2][K];
  T xi[K];
  for(int i=0;i<K;i++){
    Givens[0][i] = 0.0;
    Givens[1][i] = 0.0;
    xi[i] = 0.0;
  }
  dynamicVector<T>* Q[K+1];
  dynamicVector<T> zero(x.dim(),0.0);
  dynamicVector<T> keep(x);
  double res=sqrt((A*x-f)*(A*x-f));
  for(int k=0;k<=K;k++){
    if(k)keep = *Q[k-1];
    Q[k] = k ?
      new dynamicVector<T>(*Q[k-1]-MG.Vcycle(zero,keep))
      : new dynamicVector<T>(MG.Vcycle(f,keep)-x);
    for(int j=0;j<k;j++)
      *Q[k] -= (R[j][k] = *Q[j] * *Q[k]) * *Q[j];
    *Q[k] *= 1.0/(R[k][k] = sqrt(*Q[k] * *Q[k]));
    T givensa;
    T givensb;
    if(k){
```

```
      for(int j=1;j<k;j++){
        givensa=R[j-1][k];
        givensb=R[j][k];
        R[j-1][k] =
          givensa*Givens[0][j-1]+givensb*Givens[1][j-1];
        R[j][k] =
          -givensa*Givens[1][j-1]+givensb*Givens[0][j-1];
      }
      T ab=sqrt(R[k-1][k]*R[k-1][k]+R[k][k]*R[k][k]);
      Givens[0][k-1]=R[k-1][k]/ab;
      Givens[1][k-1]=R[k][k]/ab;
      givensa=R[k-1][k];
      givensb=R[k][k];
      R[k-1][k] =
        givensa*Givens[0][k-1]+givensb*Givens[1][k-1];
      R[k][k]=0.0;
      R[k][0]=-R[k-1][0]*Givens[1][k-1];
      R[k-1][0]=R[k-1][0]*Givens[0][k-1];
    }
    for(int i=k-1;i>=0;i--){
      xi[i]=R[i][0];
      for(int j=i+2;j<=k;j++)
        xi[i] -= R[i][j] * xi[j-1];
      xi[i] /= R[i][i+1];
    }
    s = x;
    for(int j=0;j<k;j++)
      s += xi[j] * *Q[j];
    printf("res. at step k=%d is %f\n",
        k,res=sqrt((r=A*s-f)*r));
  }
  return x = s;
} // GMRES with initial iterations
```

A.14 Parallel Implementation

Here, we show how the functions in the "dynamicVector" class can be rewritten in a form suitable for parallel implementation. In the present version, each loop on the components in the dynamic vector is broken into 'K' subloops, where 'K' is a global integer denoting the number of processors available. Each of these 'K' subloops can then be assigned to a different processor and be carried out independently. Furthermore, each subloop uses data stored continuously in the computer memory, which is particularly efficient.

Doing this in C++ has the important advantage that all these implementation details are completely hidden from users of the "dynamicVector" class. In fact, these users don't need to change their own codes or even know about the new implementation.

For example, the "dynamicVector2" class derived from the "dynamicVector" class also benefits from this parallel implementation, with absolutely no change required in it. In fact, although it actually implements two-dimensional grids, the "dynamicVector2" class also benefits from the efficient data distribution and continuous subloops discussed above.

Here is how the original loop over the components in the "dynamicVector" object is broken into 'K' subloops:

```
for(int k = 0; k < K; k++){
  int size = dim() / K;
  int res = dim() % K;
  int init = min(k,res) * (size + 1) + max(k-res,0) * size;
  int end = min(k+1,res) * (size+1) + max(k+1-res,0) * size;
  for(int i=init; i<end; i++){

    ...

  }
}
```

In the above code, 'k' is the index in the outer loop over the processors. For each fixed 'k', a continuous subset of vector components is specified. This subset starts from the component indexed by the integer "init" and ends just before the component indexed by the integer "end". The subloop over this subset is ready to be carried out on the 'k'th processor in the actual parallel implementation.

The same approach can be used in the parallel implementation of sparse matrices. The "broken" loop in the "sparseMatrix" class is basically the same as in the above code, except that the function "dim()" is replaced by "rowNumber()", and the subloop carries out some operations on the matrix row rather than on the vector component.

The parallel implementation of the "sparseMatrix" class, however, is not always well balanced. For example, some rows in the matrix may be denser (contain more nonzero elements) than others, thus requiring more computational work than others. A more pedantic implementation would detect such a row in advance and divide the work between two processors. For simplicity, we assume here that the number of processors is far smaller than the number of rows in the matrix, so the loop over the rows can be divided rather evenly among the processors without breaking individual rows.

A.15 The Adaptive-Refinement Code

Here, we implement the adaptive-refinement algorithm in Chapter 14, Section 3, for the Poisson equation in the unit circle. For simplicity, we assume that homogeneous Neumann boundary conditions are imposed, so the code in Chapter 16, Section 5, assembles the stiffness matrix properly. It is also assumed that the right-hand side is the δ-function (centered at $(1,0)$), as in Chapter 20, Section 10, so the right-hand-side vector f in the stiffness system has the value 1 at its first component and 0 at all the others:

$$f_i = \begin{cases} 1 & \text{if } i = 0, \\ 0 & \text{otherwise.} \end{cases}$$

First, the initial mesh in Figure 12.9 is constructed. This mesh contains only two triangles, "t1" and "t2":

$$"t1" = \triangle((1,0),(0,1),(-1,0)),$$
$$"t2" = \triangle((-1,0),(0,-1),(1,0)).$$

First, the nodes that are used as vertices in these triangles are defined:

```
node<point> a(point(1,0));
node<point> b(point(0,1));
node<point> c(point(-1,0));
node<point> d(point(0,-1));
```

Then, the upper triangle "t1" is defined from these nodes:

```
triangle t1(a,b,c);
```

Note that, once the nodes $(1,0)$ and $(-1,0)$ are defined and used as vertices in "t1", they must be referred to as "t1(0)" and "t1(2)" rather than 'a' and 'c' (respectively), to make sure that their "sharingElements" fields indeed increase properly whenever new triangles like "t2" are defined:

```
triangle t2(t1(2),d,t1(0));
```

These triangles are now placed in a "mesh" object:

```
mesh<triangle> m(t1);
m.append(t2);
```

Once the mesh is constructed, the original triangles can be deleted:

```
t1.~triangle();
t2.~triangle();
```

The initial, coarse mesh in Figure 12.9 is now complete and ready for the adaptive-refinement algorithm. It is assumed that automatic boundary refinement is also used, as in Chapter 14, Section 8, so the mesh refines not only in the interior of the circle but also at the boundary.

The code uses a loop over 10 refinement levels:

```
for(int i=0; i<10; i++){
```

In the beginning of the loop, the right-hand-side vector f is defined. The dimension of this vector (the number of nodes in the mesh) is returned by the "indexing()" function, which also assigns indices to the nodes in a continuous order:

```
    dynamicVector<double> f(m.indexing(),0.);
    f(0) = 1.;
```

These indices are then used to refer to nodes when the stiffness matrix A is assembled:

```
sparseMatrix<double> A(m);
A += transpose(A) - diagonal(A);
```

The PCG iteration is now used to solve the stiffness system:

```
    dynamicVector<double> x(A.order(),0.);
    multigrid<double> MG(A);
    PCG(MG,A,f,x);
    printf("value at (1,0)=%f\n",x[0]);
    printf("at level %d number of nodes=%d\n",i,x.dim());
    if(i<9)m.refine(x, 0.01);
}
```

The final command in the loop refines the mesh using the numerical solution x. This completes the adaptive-refinement code.

A.16 The Diffusion Solver

Here, we show how to implement the adaptive-refinement algorithm to solve the diffusion problem in Chapter 19. First, we construct the initial coarse mesh in Figure 19.2. For this purpose, we need some constant parameters.

Note that the part of the boundary where Dirichlet boundary conditions are imposed in Figure 19.1 has a small slit in it. For simplicity, we assume here that the slit is four times as narrow as the entire rectangle that contains it (on which homogeneous Neumann boundary conditions are imposed). The code below uses two positive parameters, "SlitWidth" and "SlitLength", so that the slit is of size "SlitWidth"/4 by "SlitLength".

The coarse mesh contains 9 triangles and 11 nodes (Figure A.4). In the code below, the nodes are defined and then used to define the triangles as well. Note that once a triangle is defined, the nodes used as vertices in it are no longer referred to by their original names but rather as vertices in this triangle. This way, their "sharingElements" fields increase properly each time they are used as vertices in a new triangle. In fact, the original "node" objects are deleted once they have been placed in the required triangles:

```
const double SlitWidth=.2;
const double SlitLength=.5;
double angle = asin(SlitWidth);
double sinAngle=sin(angle);
double sinAngle2=sin(angle/2);
double cosAngle=cos(angle);
double cosAngle2=cos(angle/2);
node<point> a1(point(1,0));
node<point> a2(point(sinAngle2,cosAngle2));
node<point> a3(point(-cosAngle,sinAngle));
node<point> a4(point(-cosAngle,-sinAngle));
node<point> a5(point(sinAngle2,-cosAngle2));
```

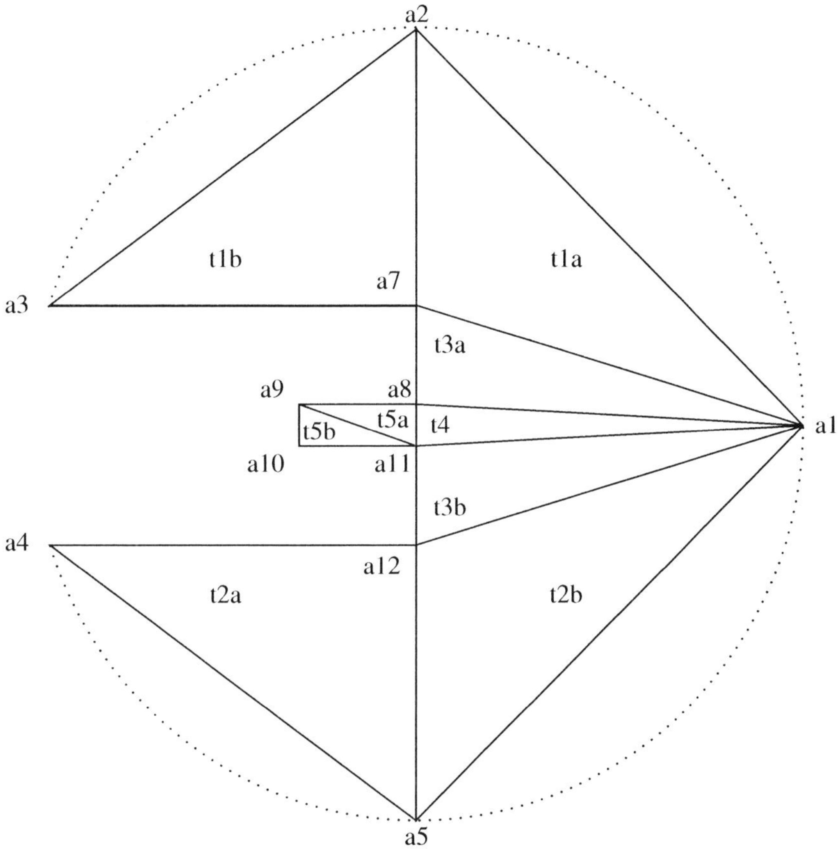

Figure A.4. *The coarse mesh that contains only* 11 *nodes and* 9 *triangles and serves as input for the adaptive-refinement algorithm to solve the diffusion problem.*

```
node<point> a7(point(0,SlitWidth));
node<point> a8(point(0,SlitWidth/4));
node<point> a9(point(-SlitLength,SlitWidth/4));
node<point> a10(point(-SlitLength,-SlitWidth/4));
node<point> a11(point(0,-SlitWidth/4));
node<point> a12(point(0,-SlitWidth));
```

This completes the definition of the 11 nodes in Figure A.4. Now, we use these nodes to define the required triangles:

```
triangle t1a(a1,a7,a2);
triangle t1b(t1a(2),t1a(1),a3);
triangle t2a(a4,a12,a5);
triangle t2b(t2a(2),t2a(1),t1a(0));
triangle t3a(t1a(0),a8,t1a(1));
triangle t3b(t2b(1),a11,t2b(2));
```

```
triangle t4(t3a(1),t3b(1),t2b(2));
triangle t5a(a9,t3a(1),t3b(1));
triangle t5b(t5a(0),a10,t3b(1));
```

These nine triangles are now used to form the coarse mesh:

```
mesh<triangle> m(t1a);
m.append(t1b);
m.append(t2a);
m.append(t2b);
m.append(t3a)
m.append(t3b)
m.append(t4)
m.append(t5a)
m.append(t5b)
```

Once they have been placed in a "mesh" object, the original triangles can be removed:

```
t1a.~triangle();
t1b.~triangle();
t2a.~triangle();
t2b.~triangle();
t3a.~triangle();
t3b.~triangle();
t4.~triangle();
t5a.~triangle();
t5b.~triangle();
```

This completes the construction of the coarse mesh in Figure A.4.

The adaptive-refinement algorithm uses a loop over the refinement levels as in Section A.15. In each refinement level, the stiffness matrix must be assembled. To do this, we need some extra member functions in the "finiteElement" class. These functions return the diffusion coefficients P and Q in the current finite element in the mesh:

```
double p() const{
  for(int i=0; i<N; i++)
    if(((*vertex[i])()[0]>0.)||((*vertex[i])()[1]>0.))
      return 1.;
  return 100.;
} // coefficient of x-derivative

double q() const{
  for(int i=0; i<N; i++)
    if(((*vertex[i])()[0]>0.)||((*vertex[i])()[1]>0.))
      return 1.;
  return 1000.;
} // coefficient of y-derivative
```

These functions should be used in the calculation of the stiffness matrix according to the guidelines in Chapter 12, Section 5. Actually, the only change is that the 2×2 matrix "weight" in the constructor that assembles the stiffness matrix in Chapter 16, Section 5, should also contain the factor $diag(P, Q)$ in the middle of it:

```
point diffX((*runner)().p(),0.);
point diffY(0.,(*runner)().q());
matrix2 diffCoef(diffX,diffY);
matrix2 S((*runner)()[1]() - (*runner)()[0](),
          (*runner)()[2]() - (*runner)()[0]());
matrix2 Sinverse = inverse(S);
matrix2 weight = abs(det(S)/2) *
   Sinverse * diffCoef * transpose(Sinverse);
```

We leave to the reader the details of incorporating the mixed boundary conditions and eliminating the equations corresponding to nodes on the Dirichlet boundary (see Section A.17). This completes the definition of the stiffness matrix.

In order to refine not only in the interior of the domain but also at the circular part of its boundary, we also use automatic boundary refinement (Chapter 14, Section 7). For this purpose, we introduce an extra "if" question in the code in Chapter 14, Section 8, to make sure that automatic boundary refinement is used only at boundary edges that lie next to circular boundary segments.

The adaptive-refinement loop is as in Section A.15 above. This completes the definition of the diffusion solver.

A.17 The Linear Elasticity Solver

Here, we show how to implement the adaptive-refinement algorithm in Chapter 20, Section 8, to solve the linear elasticity equations. First, we present the constructor that assembles the stiffness matrix described in Chapter 20, Section 7. The code is similar to the code in Chapter 16, Section 5, except that the three blocks $A^{(0,0)}$, $A^{(1,1)}$, and $A^{(0,1)}$ in Chapter 20, Section 7, need to be constructed. Thus, the body of the inner loop in the code in Chapter 16, Section 5, is actually repeated here three times, using three different 2×2 matrices: "diffCoef" to construct $A^{(0,0)}$, "diffCoef2" to construct $A^{(1,1)}$, and "mixed" to construct $A^{(0,1)}$.

In order to impose Dirichlet boundary conditions at points in Γ_D, we define a vector of integers named "DirichletBoundary". For each node i in Γ_D, the ith component in "DirichletBoundary" vanishes. The rest of the components in "DirichletBoundary" have the value 1. Now, the matrix A is “masked” by the "DirichletBoundary" vector in the sense that the ith row and ith column are dropped for every $i \in \Gamma_D$ and replaced with the standard unit vector $e^{(i)}$, which has the value 1 at its ith component and 0 at all the others. This results in a trivial equation for the unknown x_i in the stiffness system for every $i \in \Gamma_D$.

The constructor that assembles the stiffness matrix for the elasticity equations is thus as follows:

```
template<class T>
sparseMatrix<T>::sparseMatrix(mesh<triangle>&m){
```

```
item = new row<T>*[number = 2 * m.indexing()];
for(int i=0; i<number; i++)
  item[i] = 0;
point gradient[3];
gradient[0] = point(-1,-1);
gradient[1] = point(1,0);
gradient[2] = point(0,1);
dynamicVector<int> DirichletBoundary(number,1);
for(const mesh<triangle>* runner = &m; runner;
   runner=(const mesh<triangle>*)runner->readNext()){
  point diffX(1.,0.);
  point diffY(0.,(1.-NU)/2.);
  point diffX2((1.-NU)/2.,0.);
  point diffY2(0.,1.);
  matrix2 diffCoef(diffX,diffY);
  matrix2 diffCoef2(diffX2,diffY2);
  point mixedX(0.,NU);
  point mixedY((1.-NU)/2.,0.);
  matrix2 mixed(mixedX, mixedY);
  matrix2 S((*runner)()[1]() - (*runner)()[0](),
          (*runner)()[2]() - (*runner)()[0]());
  matrix2 Sinverse = inverse(S);
  matrix2 weight = abs(det(S)/2) *
      Sinverse * diffCoef * transpose(Sinverse);
  matrix2 weight2 = abs(det(S)/2) *
      Sinverse * diffCoef2 * transpose(Sinverse);
  matrix2 weightMixed = abs(det(S)/2) *
      Sinverse * mixed * transpose(Sinverse);
  for(int i=0; i<3; i++){
    int I = (*runner)()[i].getIndex();
    if((abs(squaredNorm((*runner)()[i]())-1.)<1.e-7)&&
       ((*runner)()[i]()[0] <= -0.5)){
      DirichletBoundary(I) = 0;
      DirichletBoundary(I+number/2) = 0;
    }
    for(int j=0; j<3; j++){
      int J = (*runner)()[j].getIndex();
      if(j>=i){
        if(item[I]){
          row<T> r(gradient[j]*weight*gradient[i],J);
          *item[I] += r;
        }
        else
          item[I] =
           new row<T>(gradient[j]*weight*gradient[i],J);
        if(item[I+number/2]){
```

```
              row<T> r(gradient[j]*weight2*gradient[i],
                    J+number/2);
              *item[I+number/2] += r;
            }
            else
              item[I+number/2] =
                new row<T>(gradient[j]*weight2*gradient[i],
                    J+number/2);
          }
          if(item[I]){
            row<T> r(gradient[j]*weightMixed*gradient[i],
                    J+number/2);
            *item[I] += r;
          }
          else
            item[I] = new
             row<T>(gradient[j]*weightMixed*gradient[i],
                     J+number/2);
        }
      }
    }
    for(int i=0; i<number; i++){
      if(DirichletBoundary[i])
        item[i]->dropItems(DirichletBoundary);
      else
        *item[i] = row<T>(1.,i);
    }
  } //  assemble stiffness matrix for linear elasticity
```

As explained in Chapter 20, Section 7, after the matrix A has been constructed using the above constructor, it should be further modified by

```
    A += transpose(A) - diagonal(A);
```

This completes the definition of the stiffness matrix.

Next, we implement the adaptive-refinement algorithm in Chapter 20, Section 8. For this purpose, the "if" question in the "refine" function in Chapter 14, Section 6, should be modified to read

```
    if((item[i].getIndex() >= 0)&&
      (item[j].getIndex() >= 0)&&
      ((abs(v[item[i].getIndex()] -
      v[item[j].getIndex()])>threshold)||
      (abs(v[item[i].getIndex()+v.dim()/2] -
      v[item[j].getIndex()+v.dim()/2])>threshold))){

        ...
```

This way, the criterion in Chapter 20, Section 8, is implemented.

Next, we implement the modified multigrid algorithm in Chapter 20, Section 9. The algorithm is in principle the same as in Chapter 17, Section 8, except that the block-diagonal part of A,

$$blockdiag(A^{(0,0)}, A^{(1,1)}),$$

is used to construct the coarse grid and the prolongation matrix P and restriction matrix R. In order to obtain this block-diagonal matrix, we insert a few code lines at the beginning of the "createTransfer" function in Section A.12. These code lines delete the $A^{(0,1)}$ and $A^{(1,0)}$ blocks from the current copy of the stiffness matrix A. (Of course, it is assumed that A is also stored elsewhere for safekeeping, so these changes don't affect it.) The result is the required block-diagonal part of A, which is then used in the remainder of the "createTransfer" function to create the coarse grid and prolongation and restriction matrices.

The code lines that should be inserted at the beginning of the "createTransfer" function are as follows:

```
dynamicVector<int> one(rowNumber(),1);
dynamicVector<int> two(rowNumber(),1);
for(int i=0; i<rowNumber()/2; i++)
  two(i) = 0;
for(int i=rowNumber()/2; i<rowNumber(); i++)
  one(i) = 0;
for(int i=0; i<rowNumber()/2; i++)
  item[i]->dropItems(one);
for(int i=rowNumber()/2; i<rowNumber(); i++)
  item[i]->dropItems(two);
```

This completes the changes required in the linear elasticity solver. The loop that implements the adaptive-refinement algorithm is basically as in Section A.15.

Bibliography

[1] Bank, R.E., Dupont, T.F., and Yserentant, Y.: The Hierarchical Basis Multigrid Method. *Numer. Math.* 52 (1988), pp. 427–458.

[2] Ben-Artzi, M., and Falcovitz, J.: A Second-Order Godunov-Type Scheme for Compressible Fluid Dynamics. *J. Comput. Phys.* 55 (1984), pp. 1–32.

[3] Bramble, J.H., Leyk, Z., and Pasciak, J.E.: Iterative Schemes for Non-Symmetric and Indefinite Elliptic Boundary Value Problems. *Math. Comp.* 60 (1993), pp. 1–22.

[4] Bramble J.H., Pasciak J.E., and Schatz A.H.: The Constructing of Preconditioners for Elliptic Problems on Regions Partitioned into Substructures I. *Math. Comp.* 46 (1986), pp. 361–369.

[5] Brandt, A., and Yavneh, I.: Inadequacy of First-Order Upwind Difference Schemes for some Recirculating Flows. *J. Comput. Phys.* 93 (1991), pp. 128–143.

[6] Brenner, S.C.; and Scott, L.R.: The Mathematical Theory of Finite Element Methods. Texts in Applied Mathematics, 15, Springer-Verlag, New York, 2002.

[7] Cai, Z., Lee, C.-O., Manteuffel, T.A., and McCormick, S.F.: First-Order System Least Squares for the Stokes and Linear Elasticity Equations: Further Results. *SIAM J. Sci. Comput.* 21 (2000), pp. 1728–1739.

[8] Chan, T.F., and Vanek, P.: Detection of Strong Coupling in Algebraic Multigrid Solvers. In *Multigrid Methods* VI, Vol. 14, Springer-Verlag, Berlin, 2000, pp. 11–23.

[9] Chang, Q., and Huang, Z.: Efficient Algebraic Multigrid Algorithms and Their Convergence. *SIAM J. Sci. Comput.* 24 (2002), pp. 597–618.

[10] D'Azevedo, E.F.; Romine, C.H.; and Donato, J.H.: Coefficient Adaptive Triangulation for Strongly Anisotropic Problems. In *Preproceedings of the 5th Copper Mountain Conference on Iterative Methods*, Manteuffel, T.A. and McCormick, S.F. (eds.), 1998. Available online at www.mgnet.org

[11] Dendy, J.E.: Black Box Multigrid for Nonsymmetric Problems. *Appl. Math. Comput.*, 13 (1983), pp. 261–283.

[12] Dendy, J.E.: Semicoarsening Multigrid for Systems. *Electron. Trans. Numer. Anal.* 6 (1997), pp. 97–105.

[13] Douglas, C.C.: Cache Based Multigrid Algorithms. In the *MGnet Virtual Proceedings of the 7th Copper Mountain Conference on Multigrid Methods*, 1997. Available online at www.mgnet.org

[14] Elman, H.C., Ernst, O.G., and O'Leary, D.P.: A Multigrid Method Enhanced by Krylov Subspace Iteration for Discrete Helmholtz Equations. *SIAM J. Sci. Comput.* 23 (2001), pp. 1291–1315.

[15] Evans, D.J.: Preconditioning Methods. Gordon and Breach, New York, 1983.

[16] Freund R.W.: A Transpose Free Quasi-Minimal Residual Algorithm for Non-Hermitian Linear Systems. *SIAM J. Sci. Comput.* 14 (1993), pp. 470–482.

[17] Gottlieb, A., and Almasi, G.S.: Highly Parallel Computing. Benjamin Cummings, Redwood City, CA, 1989.

[18] Gustafsson, I.: On Modified Incomplete Factorization Methods. In *Numerical Integration of Differential Equations and Large Linear Systems*, Lecture Notes in Mathematics, 968, Springer, Berlin, 1982, pp. 334–351.

[19] Hartman, P.: Ordinary Differential Equations, 2nd ed. Birkhäuser, New York, 1982.

[20] Klawonn, A., and Widlund, O.B.: A Domain Decomposition Method with Lagrange Multipliers and Inexact Solvers for Linear Elasticity. *SIAM J. Sci. Comput.* 22 (2000), pp. 1199–1219.

[21] Kraus, J.K., and Schicho, J.: Algebraic Multigrid Based on Computational Molecules, 1: Scalar Elliptic Problems. RICAM report, Austrian Academy of Science, Linz, Austria, March 2005.

[22] Layton, W., Lee, H.K., and Peterson, J.: Numerical Solution of the Stationary Navier–Stokes Equations Using a Multilevel Finite Element Method. *SIAM J. Sci. Comput.* 20 (1998), pp. 1–12.

[23] Lee, B., Manteuffel, T.A., McCormick, S.F., and Ruge, J.: First-Order System Least-Squares for the Helmholtz Equation. *SIAM J. Sci. Comput.* 21 (2000), pp. 1927–1949.

[24] Mavriplis, D.J.: Directional Coarsening and Smoothing for Anisotropic Navier-Stokes Problems. In the *MGnet Virtual Proceedings of the 7th Copper Mountain Conference on Multigrid Methods*, 1997. Available online at www.mgnet.org.

[25] McCormick, S. F.; and Quinlan, D.: Asynchronous Multilevel Adaptive Methods for Solving Partial Differential Equations: Performance Results. *Parallel Comput.* 12 (1990), pp. 145–156.

[26] Meijerink, J.A., and Van der Vorst, H.A.: An Iterative Solution Method for Linear Systems of which the Coefficients Matrix is a Symmetric M-matrix. *Math. Comp.* 31 (1977), pp. 148–162.

[27] Michelson, D.: Bunsen Flames as Steady Solutions of the Kuramoto–Sivashinsky Equation. *SIAM J. Math. Anal.* 23 (1992), pp. 364–386.

[28] Mitchell, W.F.: Optimal Multilevel Iterative Methods for Adaptive Grids. *SIAM J. Sci. Stat. Comput.* 13 (1992), pp. 146–167.

[29] Morton, W.K.: *Numerical Solution of Convection-Diffusion Problems*. Applied Mathematics and Mathematical Computation, 12. Chapman and Hall, London, 1996.

[30] Ortega J.M.: Introduction to Parallel and Vector Solution of Linear Systems. Plenum Press, New York, 1988.

[31] Perona, P., and Malik, J.: Scale Space and Edge Detection Using Anisotropic Diffusion. *IEEE Trans. Pattern Anal.* 12 (1990), pp. 629–639.

[32] Roos, H.G., Stynes, M., and Tobiska, L.: Numerical Methods for Singularly Perturbed Differential Equations: Convection-Diffusion and Flow Problems. Springer, Berlin, New York, 1996.

[33] Ruge, J.W., and Stüben, K.: Algebraic Multigrid. In *Multigrid Methods*, McCormick, S.F. (ed.), SIAM, Philadelphia, PA, 1987, pp. 73–130.

[34] Saad, Y., and Schultz, M.H.: GMRES: A Generalized Minimal Residual Algorithm for Solving Nonsymmetric Linear Systems. *SIAM J. Sci. Stat. Comput.* 7 (1986), pp. 856–869.

[35] Sapiro, G., and Ringach, D.L.: Anisotropic Diffusion of Multivalued Images with Applications to Color Filtering. *IEEE Trans. Image Process.* 5 (1996), pp. 1582–1586.

[36] Shahinyan, M.: Algorithms in Graph Theory with Application in Computer Design. M.Sc. thesis, Faculty of Applied Mathematics, Yerevan State University, Yerevan, Armenia (1986). (Advisor: Dr. S. Markossian.)

[37] Shapira, Y.: Asymptotic Solutions for the Kuramoto-Sivashinsky Equation. M.Sc. Thesis, Department of Mathematics, Hebrew University, Jerusalem, Israel (1988). (Advisor: Dr. D. Michelson.)

[38] Shapira, Y.: Adequacy of Finite Difference Schemes for Convection-Diffusion Equations. *Numer. Methods Partial Differential Equations* 18 (2002), pp. 280–295.

[39] Shapira, Y.: Matrix-Based Multigrid. Kluwer Academic Publishers, Boston, 2003.

[40] Sidi, A., and Shapira, Y.: Upper Bounds for Convergence Rates of Acceleration Methods with Initial Iterations. *Numer. Algorithms* 18 (1998), pp. 113–132.

[41] Smoller, J.: Shock Waves and Reaction-Diffusion Equations 2nd ed. Springer-Verlag, New York, 1994.

[42] Sochen, N., Kimmel, R., and Malladi, R.: A General Framework for Low Level Vision. *IEEE Trans. Image Process.* 7 (1998), pp. 310–318.

[43] Sonneveld, P.: CGS, A Fast Lanczos-Type Solver for Nonsymmetric Linear Systems. *SIAM J. Sci. Statist. Comput.* 10 (1989), pp. 36–52.

[44] Strang, G. and Fix, G.: An Analysis of the Finite Element Method. Prentice–Hall, Englewood Cliffs, NJ, 1973.

[45] Tanabe, K.: Projection Methods for Solving a Singular System of Linear Equations and Its Applications. *Numer. Math.* 17 (1971), pp. 203–214.

[46] Varga, R.: Matrix Iterative Analysis. Prentice–Hall, Englewood Cliffs, NJ, 1962.

[47] Ward, R.C.: Numerical Computation of the Matrix Exponential with Accuracy Estimate, *SIAM J. Numer. Anal.* 14 (1977), pp. 600–610.

[48] Young, D.: Iterative Solution of Large Linear Systems. Academic Press, New York, 1971.

Index

"++", 10
"+ =", 10
  in complex, 69
  in dynamic vector, 443
  in matrix, 441
  in point, 64
  in sparse matrix, 462
"−−", 10
"− =", 10
  in dynamic vector, 442
  in sparse matrix, 462
  in vector, 439
"− >", 56
"//", 8
"/ =", 10
  in complex, 82
"/*", 8
'<', 14
"<=", 14
'=', 8, 14
"==", 14
'>', 14
">=", 14
"&&", 14
"*/", 8
"*=", 10
  in complex, 82
  in sparse matrix, 462
'*', 22, 121
"| |", 15
'∼', 55
'!', 14
"!=", 14
"( )", 11, 13
'.', 49
';', 6
"[ ]", 23
'%', 9, 38, 146
"%c", 20, 37
"%d", 13
"%f", 13
'&', 58, 121
": :", 50
':', 52, 80
"? :", 14

accuracy, 214
adaptive
  discretization, 194
  mesh (see refinement), 194
  refinement (see refinement), 299
address, 22
  of nodes, 283
  to pass an, 25, 26, 54, 57
    to temporary object, 55
  virtual, 121
    in Harwell–Boeing, 385
    of nodes, 125, 129
    of nonzeroes, 385
adequacy, 214, 272
  in Helmholtz, 427, 429
  in Riemann, 230
  in Stokes, 419
ADI, 234
algorithm, 3, 136, 143
  inherently sequential, 370
  parallelizable, 370
allocate (see memory), 5, 51
alternating
  direction, 234
  Schwarz relaxation, 373
AMG (see multigrid), 354
AMGe, 365

AMGM, 365
anisotropic, 249, 271
  in Stokes, 419
architecture, 370, 380
  parallel, 373, 385
  sequential, 385
argument
  by address, 25, 54
  by name, 24, 57, 470
  by reference, 59, 73
  by value, 24, 54, 57, 59, 73
  concrete, 12, 16
    constant, 61
    local copy, 24, 54
    local reference, 58
    in template, 75
  dummy, 11, 17, 121
  local (see variable), 12
  that is a pointer, 25
arithmetic
  expression, 36
  operation, 9, 196
    on complex, 68
    on dynamic vector, 98
    on matrix, 90
    on polynomial, 138
    priority order of, 9
    on sparse matrix, 462
    on vector, 75
array, 23, 121
  of addresses, 282
  of edges, 125
  of nodes, 124
  of same-size objects, 111
  two-dimensional, 24, 29
assembling, 265, 293, 337
  in elasticity, 481
  quadratic, 316
assignment, 8
assignment operator (see operator), 60
automaton, 12, 51, 121

back substitution, 89, 347, 465
base class, 80, 199, 445
bilinear form, 251, 261, 295, 399
  in Navier-Stokes, 424
binary representation, 22, 27, 376
bisection, 30
block
  of class, 71
  of a function, 11
  of instructions, 4, 15
  in a matrix, 212
  relaxation, 373, 385, 388
boundary
  condition, 190, 250, 478
    Dirichlet (first kind), 181, 190, 481
    in elasticity, 481
    mixed (third kind), 182, 187, 425, 427
    Neumann (second kind), 182
  curved, 133, 309
  irregular, 133, 298, 397
  layer, 184
  refinement, 307, 398
    in elasticity, 411
    in Stokes, 419
boundary-value problem, 251

cache, 371
  oriented, 371, 373
CG (see conjugate gradient), 359
chain rule, 265
  in Riemann, 220, 232
"char", 7, 39
character (see "char"), 7
characteristic, 220, 228, 232, 234
class, 48
code, 3
coercivity, 253
  in elasticity, 405
column pointer, 385
comment, 8
communication, 375
compilation
  error, 23
    due to ambiguity, 85
    constant, 50
    temporary, 54, 61

compilation (continued)
  time, 4, 5
    in array, 23
    in recursion, 26
    in template, 72
compiler, 9
  in assignment, 60, 61
  in constructor, 51, 52, 54
  in conversion, 63
  in function, 12, 17
  in inheritance, 81
  in operator, 64, 66, 67
  in pointer, 23
  in reference, 58
  in template, 72
complex number, 68
  implementation of, 68
    (has a), 79
    (is a), 80
complexity, 369, 380
  in graph, 123
  in triangulation, 129, 131
computable problem, 3
  in polynomial time, 4
computational
  problem, 3, 135, 369, 380
  task, 3, 369
computer, 370
  digital, 3
  parallel, 369
  sequential, 369
concrete object (see argument), 281
concrete variable (see argument), 17
conformal, 129, 259
conformity, 259, 273, 301, 302
conjugate gradient, 359
  preconditioned, 359, 470
  squared, 361, 472
connected list, 102, 111
  of edges, 126
  flexibility of, 112, 132
  of integers, 127, 282
  of nodes, 282
  in row, 329
  in sparse matrix, 127, 325
  of triangles, 132, 284, 291
  in triangulation, 131, 133, 291
  two-sided, 112
connection machine, 375
conservation law, 219, 232
  system of, 234
"const", 9
  argument, 61
  current object, 50, 56, 63
    in template, 74
  returned variable, 54, 56, 466
constant (see "const"), 9
constructor, 51, 287
  in assembling, 337
    in elasticity, 481
  default, 51
    in dynamic vector, 446
    in inheritance, 81, 202
    in list, 199, 200, 449
    in template, 74
  in dynamic vector, 95, 446
convection-diffusion, 187, 444
conversion, 9
  explicit, 53
    of pointers, 294, 303, 304, 306, 330–332
  implicit, 53
    in inheritance, 85
  inverse, 63
copy constructor, 59, 71, 287
  in connected list, 103
  default, 53, 450
    in dynamic vector, 446
    in template, 74
  in dynamic vector, 97, 446
  in inheritance, 450
  in list, 101
  in template, 74
  in vector, 76
Crank–Nicolson (see semi-implicit scheme), 193
cubic (see function), 316
curl, 415
current object (see object), 54

data access
  cache, 371

data access (continued)
- edge-to-node, 125
- node-to-edge, 126
- node-to-triangle, 131
- primary memory, 371
- secondary memory, 371
- triangle-to-node, 131, 133

data structure, 93, 121
- for mesh, 247
- for sparse matrix, 325
- to store triangles, 131
- uses pointers, 127

debug, 5
declaration, 50, 57, 66, 448
degenerate triangle, 262, 276
delete (see "delete"), 55
"delete", 55
- in template, 73, 287

"delete [ ]", 96
denoising, 237, 458
- color, 241, 458

dereferencing, 121
derived class, 80, 199, 445
destructor, 55, 287
- default, 68, 287
  - in dynamic vector, 96
  - in inheritance, 82
  - in template, 73
  - in vector, 76
- in vector, 76

diagonally dominant, 210, 262, 270
difference operator, 201
- in two dimensions, 448

difference scheme (see finite difference), 166
diffusion
- coefficient, 271, 395
- equation, 395
- nonlinear, 238
- problem, 249
- solver, 478
- strong, 272, 397
- time-dependent, 238, 250, 399
- weak, 272

digital image, 237, 458
- color, 240, 458

discretization (see finite difference), 188, 239
discretization error, 214
- in Helmholtz, 427, 429
- in Riemann, 230

distributed memory, 374
divergence, 414
- free, 416
- zero, 416, 435

do, 18
"do", 18
domain
- circular, 274
- complicated, 297, 395
- irregular, 259, 297
- nonconvex, 309
- rectangular, 205, 238, 266
- time-space, 181, 191
  - the implementation of, 195
  - in Riemann, 220
  - in two dimensions, 234, 455

domain decomposition, 362
double precision (see "double"), 7
"double", 7
"(double)", 9

edge
- array of, 125
- in graph, 114, 122
- in triangulation, 128

elasticity
- linear, 403, 417
  - general, 416, 419
  - solver, 481

electric field, 434
electromagnetic, 425, 434
else, 15
"else", 15
energy inner product, 278
energy norm, 360, 387, 419, 422
- of matrix, 278

error estimate, 272, 343
- computational, 168
- discretization, 216
- Taylor scheme, 172

execute, 3

execution, 12
existence
  in ODE, 161, 170, 175
  in PDE, 254
explicit scheme, 191, 212, 456
exponent function, 40
  of matrix, 90
exponential time, 3

file
  of instructions, 374
  open a, 25
  output, 5
  pointer, 25, 386
  print on, 25
  read from, 25
  static, 40
  variable, 25, 386
"FILE", 25, 386
finite difference, 188, 444
  in Helmholtz, 427
  in ODE, 166
finite element, 247, 281
  bilinear, 428
  cubic, 316
  in elasticity, 406
  high-order, 313
  implementation of, 286
  linear, 259
  mesh (see refinement), 396
  nonlinear, 433
  quadratic, 314
float number (see "float"), 7
"float", 7
"(float)", 43
"fopen", 25, 386
for, 18
"for", 18
forward elimination, 89, 347, 465
FOSLS, 422
"fprintf", 25
friend, 57
"fscanf", 25, 386
function, 4, 8, 11
  argument in, 24
  elemental basis, 420
  friend, 57
  interface, 48, 49
    constructor is an, 51
    current object in, 54
    parallelized, 380
    returned object in, 54
  language, 121
  member, 55
  nodal basis, 264
    three-dimensional, 280
    conformity in, 302
    in elasticity, 407
    linear, 271, 313
    quadratic, 314
    in time marching, 400
  typical nodal
    bilinear, 428
    linear, 263

Gauss elimination, 89, 323
Gauss–Seidel relaxation, 344, 371, 451
  block, 373, 385
  symmetric, 346, 384, 388
Gersgorin theorem, 209, 456
"getchar", 39
global
  array, 152
  function, 196
  integer, 475
  refinement, 297, 312, 396, 400
GMRES, 362, 422, 474
Godunov scheme, 222, 226, 232
  explicit, 228, 234
  implicit, 235
golden ratio, 33
gradient, 238, 413
  of cubic function, 317
  of divergence, 416
  of nodal basis function, 265
  of nodal function, 265, 294, 337
  of quadratic function, 316
  vanishes, 271
graph, 114
  coloring, 122
Green
  formula, 251

Green formula (continued)
in elasticity, 404
in Riemann, 220, 224
function, 262
grid, 133
in ODE, 165
in PDE, 188, 199
in two dimensions, 205, 444
rectangular, 190
time-space, 212
implementation of, 199
in two dimensions, 444

Harwell–Boeing collection, 126, 385
numerical results, 388
read matrix, 387
has a, 47, 79, 291
heat equation, 182, 216
Helmholtz equation, 364, 426, 436
hidden
in inheritance, 79
in parallelism, 383, 475
in polynomial, 141
in recursion, 36
storage, 94, 143
high-level
code
in ODE, 168, 172
in parallelism, 380
language, 4, 370
object, 132
programming, 136
in inheritance, 82
Horner algorithm, 144
in exponent, 41
in ODE, 166, 168, 173
in power, 146
in Taylor series, 149
hypercube, 376, 382

I/O (see input/output), 4
if (see "if"), 14
"if", 14, 15
ILU
factorization, 347
iteration, 465
no fill-in
implementation of, 465
in multigrid, 353, 362, 399, 411
image processing (see digital), 237
implementation, 3, 126, 135, 143, 194
downward, 124, 130
parallel, 369, 370, 475
low-level, 379, 383
upward, 124
implicit scheme, 193, 212, 399
in denoising, 239
include, 13, 20
"include", 13, 20
incomplete factorization (see ILU), 347
indefinite matrix, 421
indexing
direct, 102, 111
of finite elements
in Stokes, 421
indirect, 102, 111
of nodes, 284
in finite element, 289
in mesh, 292
information, 131, 137
inheritance, 79
initial-boundary-value problem, 187
initialization list, 52, 68, 195, 357
in dynamic vector, 96
in inheritance, 81
initialize, 8, 48, 51
input/output, 3, 13, 25, 370
integer (see "int"), 7
"(int)", 9
"int", 7, 11
is a, 47, 79, 291
isotropic, 269
iterative method, 341, 342

Jacobi relaxation, 344, 382, 384, 388
block, 385, 388
Jacobian, 264
Jordan, 163
form, 165, 210, 456
matrix, 216, 236

Kacmarz iteration, 346, 378, 465

Kremer formula, 442
Krylov, 359, 360, 470, 474
Kuramoto–Sivashinsky, 171

L-matrix, 262
Laplace equation, 262
Laplacian, 414
  vector, 414, 416, 423
linear system (see matrix), 323
  in implicit scheme, 193
  solvers, 341, 398
linearization
  in denoising, 240
  in Navier–Stokes, 423
  in Riemann, 228, 230
linked list (see connected list), 102
list, 99, 111, 198
  of connected lists, 131
  of instructions, 3
  of nodes, 288
  of operations, 135
  of rows, 325, 335
  of variable-size objects, 111
local
  argument (see variable), 12
  maximum, 30
  object (see variable), 137
  refinement (see refinement), 297
  variable (see variable), 12
logical
  and, 14, 62
  if, 7
  not, 70
  operator, 14, 62
    priority order of, 15, 62
  or, 15
loop, 17, 19
  nested, 20, 349
    in array, 24, 29
    over edges, 126
    over triangles, 131, 133
    over vertices, 295
  subloop, 390, 475
low-level
  language, 4
  object, 169, 294
  in parallelism, 369, 379, 383
  programming, 136
    in polynomial, 141
LU decomposition, 88, 347
  block, 421
  in PDEs, 417

M-matrix, 262, 269
magnetic field, 434
"main", 13
manifold
  invariant, 170, 174
  stable, 170, 174, 175
"math.h", 17
matrix, 86, 440
  dense, 325
  determinant of, 88, 441
  element, 326
  exponent of, 90
  indefinite, 421
  inverse of, 88, 441
  L-, 262
  M-, 262, 269
  negative definite, 421
  positive definite, 262
  rectangular, 352
  sparse, 325
    coloring of, 124, 127
    implementation of, 334
    multiplication of, 378
    operators on, 462
    transpose of, 463
  SPD, 262
  stiffness, 259, 261
    to assemble, 265, 293, 337
    in elasticity, 407, 481
    symmetric, 261
  symmetric, 122, 261, 262
  transpose of, 441
maximum, 30
Maxwell equations, 434
member, 55
  private, 56, 62, 68, 73, 81
    in tree, 113
  protected, 81, 330
  public, 67, 73

member, public (continued)
 in connected list, 103
 in inheritance, 81, 82
memory, 3, 370
 allocation, 5, 33, 68
 in dynamic vector, 95
 in array, 23
 in constructor, 51, 54
 in destructor, 55
 distributed, 374
 long, 150, 173, 207
 primary, 3, 371
 in recursion, 26
 secondary, 3, 371
 shared, 373
 short, 150, 158, 168, 207
 in template, 74
merging connected lists, 108
mesh, 133
 implementation of, 291
 nonuniform, 259, 281, 297
 in ODE, 163, 166
 refinement, 304
 uniform, 266
 unstructured, 259, 281
meshsize
 in ODE, 163, 166
 in PDE, 188
 in two dimensions, 205
midpoint scheme (see semi-implicit scheme), 193
MIMD, 374
minimization problem, 252
 in elasticity, 405
model
 circular, 432
 physical, 194
 rectangular, 431
modular, 5
molecule, 365
multigrid, 351, 355, 362, 398, 466
 algebraic, 354, 362, 398, 469
 black-Box, 364
 in elasticity, 408
 nonsymmetric, 362
multilevel
 programming, 136

Nabla, 413
Navier–Stokes equations, 423
negative definite, 421
new (see "new"), 95
"new", 134, 287, 342
 connected list, 104
 dynamic vector, 95
 graph, 115
 row, 349, 379
Newton
 binomial, 30, 152
 iteration
 in denoising, 240, 241
 in Navier–Stokes, 423
 in Riemann, 229, 230
 in system, 234
Newton–Rabson, 229
node, 133
 array of, 124
 in graph, 114, 122
 implementation of, 284
 in triangulation, 128
nonlinear PDE (see PDE), 219
nonsymmetric system, 353, 362
nonuniform mesh (see mesh), 297
normal equation, 346, 378, 465
numerical
 approximation, 209
 in Helmholtz, 427
 in ODE, 165, 171
 in PDE, 185, 262, 302, 341
 in Stokes, 421
 example
 in denoising, 242
 in elasticity, 409
 scheme (see finite difference; implicit scheme; semi-implicit scheme), 133
 solution
 of linear system, 323
 in ODE, 165, 176
 in PDE, 396

object, 5, 47, 198
abstract, 281
concrete (see argument), 281
current, 49, 54
"const", 63
in inheritance, 81, 139
nonconstant, 65, 70
reference to, 62
in template, 75
temporary, 67
"this", 56, 59, 64
local (see variable), 137
returned, 54, 65, 144
constructed, 137
nonreference, 70, 71
temporary, 54
type of, 60
temporary, 67, 140
to change, 61, 64, 140
nonconstant, 65
pointer to, 54
returned, 68
object-oriented, 282
analysis, 170
implementation, 135
language, 5, 47, 121
programming, 65, 68, 121
ODE, 161
nonlinear, 170
stable, 162
system of, 162
stable, 163
open a file, 25
operation (see operator), 9
operator, 62
arithmetic (see arithmetic), 68
assignment, 60
of complex, 69
of connected list, 105
of dynamic vector, 97
of list, 101
binary, 65
curl, 415
differential, 417, 423
divergence (see divergence), 414
gradient (see gradient), 413
Laplacian, 414
logical (see logical), 14
Nabla, 413
rotor, 415
unary, 63
ordering connected list, 110
output, 13

Pade, 42, 90
parallel
architecture, 373
computer, 369
implementation, 369, 383, 475
parallelism, 369, 381
parallelizable, 370, 381
Pascal
calculation machine, 3
triangle, 29, 152
PCG (see conjugate gradient), 359
PDE, xxiii, 133, 181
two-dimensional, 205, 417, 444
three-dimensional, 436
elliptic, 182
hyperbolic, 183
nonlinear, 219
in denoising, 238
system, 423
system of, 233, 241
parabolic, 183
system of, 233, 403, 416, 434
time-dependent, 400
Picard, 229
pointer, 22
as argument, 25
in connected list, 102
to connected list
converted from, 303, 304, 306, 330, 332
to character, 23, 24
to constant, 23, 56, 290
to constant point, 56
to double, 22, 24, 95
to file, 25, 386
to integer, 22, 25

pointer (continued)
- to mesh, 303
  - converted to, 294, 303, 304, 306
  - local, 461
- to multigrid, 355
- to node
  - in triangle, 283, 286
- to nonconstant, 23, 290
- to point, 56
- to "point" object
  - is returned, 56
- to pointers, 24
- to row
  - converted to, 330–332
- to static file, 40
- to T, 99
- to temporary object, 55

Poisson
- equation, 182
  - in a circle, 274, 476
  - in FOSLS, 423
  - Jacobi iteration for, 345
  - one-dimensional, 276
  - stiffness matrix in, 294
- ratio, 403, 409, 411
  - in Stokes, 420

polynomial, 138
- calculation of, 143, 144
- integration of, 154
- multiplication of, 141
- of three variables, 159
- time, 3, 4
- of two variables, 152, 400
  - cubic, 316
  - quadratic, 314

positive definite, 262
preconditioner, 343, 361, 383, 395
- in elasticity, 403
- in Stokes, 419

primary memory (see memory), 3, 371
print (see "printf", "fprintf"), 13
"printf", 13
private (see "private"), 49
"private", 49
"private:", 50
private member (see member), 56

problem
- boundary value, 251
- computational, 3, 135, 369, 380
- initial-boundary value, 187
- measuring, 430
- minimization, 252
  - in elasticity, 405

process, 121
processor, 3, 370
- main, 375

processors, 369, 373
program, 3
- executable, 375

program (see "main"), 13
programming, 3
prolongation, 352, 354, 362, 468
protected (see "protected"), 81
"protected"
- in connected list, 103
- in dynamic vector, 95, 446
- in list, 199, 201, 386, 453
- in polynomial, 139

"protected:", 81
protected member (see member), 81
public (see "public"), 49
public member (see member), 67
":public", 80
"public:", 49, 68, 73

quadratic (see function), 314
quadratic form, 253, 255, 261
- in elasticity, 405
- in FOSLS, 423

random-choice scheme, 225
rarefaction wave, 220, 231
read (see "fscanf"), 25
rectangular
- domain, 205, 238, 266
- grid, 190
- matrix, 352

recursion, 26
- in arithmetic expression, 36
- in binary representation, 27
- in connected list, 107
- in local maximum, 35

recursion (continued)
  in mesh refinement, 304
  in multigrid, 355
  in row, 331
reference, 57
  to pass a reference
    to finite element, 291
    to node, 302
    to row, 465
    to vector, 332, 345, 470
  to return a, 59, 66, 70, 350
reference element, 428
reference triangle
  cubic, 316
  linear, 263, 314
  quadratic, 314
referencing, 121
refinement, 304
  adaptive, 299, 476
    in elasticity, 408
  boundary, 297, 307
  global, 297, 312, 396, 400
  local, 297
regularity, 305
relaxation (see Jacobi; Gauss–Seidel relaxation), 371
reserved word, 4
return, 12, 33
  by reference, 59, 66, 70, 350, 466
  by value, 59, 66, 70, 350, 465, 466
"return", 12
returned (see object; "const"), 54
returned object (see object), 54
Reynolds number, 423
Riemann problem, 219
  linearized, 230
  N-wave, 226
  singularly perturbed, 226, 228, 230, 233
rotor, 415
row object, 329
run, 4
  time, 4, 24, 26
    in connected list, 105
    in dynamic vector, 95
    in list, 99
scan (see "fscanf"), 25
"scanf", 25
scheme (see finite difference; implicit scheme; semi-implicit scheme), 133
Schur complement, 421, 422
  in PDEs, 417
Schwarz iteration, 373
scope, 15
secondary memory, 3, 371
semi-implicit scheme, 193, 212, 399
sequential
  algorithm, 370
  computer, 369
shape, 128
shared memory, 373
shock wave, 221, 231
SIMD, 17
singular perturbation, 226
"sizeof", 43
solution
  asymptotic, 174
  irregular, 423
  numerical (see numerical), 165
  unique, 255
sparse matrix (see matrix), 124
sparsity, 318
SPD, 262, 360, 383
stability, 209, 212, 456
stable ODE (see ODE), 162
standard I/O (see "stdio.h"), 13
state machine, 12, 51, 121
"static", 40
static file, 40
"stdio.h", 13
stiffness matrix (see matrix), 259
stiffness system, 260
Stokes equations, 418, 419
storage, 51, 137
  in inheritance, 82
string, 24, 36
  to copy a, 36
  as input, 36
  to print a, 13, 25
strong formulation, 251
  in elasticity, 403

subshape, 128
subspace
invariant, 164
stable, 164
Krylov, 359, 360, 470, 474
symmetric
bilinear form, 251, 256
differencing, 188, 192
matrix, 122
PDE, 247
stiffness matrix, 261, 262, 421
system (see PDE; linear system), 193

task
assigning, 375
computational, 3, 369
modularity in, 4
recursion in, 36
Taylor
approximation, 90
error, 148, 169
expansion, 40, 148
polynomial, 40
scheme, 167, 170, 172, 176
template, 72, 94
in dynamic vector, 95
tetrahedron, 279, 286, 290
TFQMR, 362, 472
this (see "this"), 56
"this", 56, 59
time
exponential, 3
polynomial, 3, 4
time level, 190, 211, 399, 444
time marching (see implicit scheme; semi-implicit scheme), 190, 212, 399
time step, 190, 211
time-space domain (see domain), 181
time-space grid (see grid), 199
Toeplitz matrix, 276
tree, 112
triangulation, 128, 259
Turing, 3
machine, 3, 136
type (see variable), 7
typedef (see "typedef"), 78
"typedef"
cubic triangle, 317
matrix, 87, 441
point, 78
quadratic triangle, 315
triangle, 290

uncomputable problem
practically, 3
uniqueness, 255
unsolvable problem
practically, 3
unstructured mesh (see mesh), 281
upwind scheme, 189

variable, 7, 15
concrete (see argument), 17
local, 12, 16, 58
in constructor, 54
in function, 40
in loop, 19
return, 59, 66, 350
returned, 70
pass by address, 57
variable (see object; "const"), 54
vector, 75, 111, 439
dynamic, 94, 111, 442
virtual address (see address), 121
void, 11, 12
"void", 11, 12
Von Neumann, 3

wave equation, 425
weak formulation, 250
discrete, 260
in elasticity, 404
weak sense, 220
well-posed, 251, 252, 259
in elasticity, 404, 405
while (see "while"), 17
"while", 17, 18
write (see "printf", "fprintf"), 13